Metabolomics, Proteomics and Gene Editing Approaches in Biofertilizer Industry

Sukhminderjit Kaur • Vagish Dwibedi •
Pramod Kumar Sahu
Editors

Metabolomics, Proteomics and Gene Editing Approaches in Biofertilizer Industry

Volume II

Springer

Editors
Sukhminderjit Kaur
Department of Biotechnology
Chandigarh University
Mohali, Chandigarh, India

Vagish Dwibedi
Department of Biotechnology
Chandigarh University
Mohali, Chandigarh, India

Pramod Kumar Sahu
Agricultural Microbiology
ICAR National Bureau of Agriculturally
Important Microorganisms
Mau Nath Bhanjan, Uttar Pradesh, India

ISBN 978-981-97-2909-8 ISBN 978-981-97-2910-4 (eBook)
https://doi.org/10.1007/978-981-97-2910-4

This Springer imprint is published by the registered company Springer Nature Singapore Pte Ltd.
The registered company address is: 152 Beach Road, #21-01/04 Gateway East, Singapore 189721, Singapore

Preface

Biofertilizers refers to the live or latent microbial inoculants capable of enhancing plant nutrition and growth through various direct and indirect mechanisms. The past few decades have seen biofertilizers gaining immense ground as crop productivity faces great challenges due to ever-increasing populations and environmental threats. Researchers are looking for alternatives to enhance crop production through sustainable approaches. This book is an attempt to consolidate recent advancements related to biofertilizers onto a single platform.

The venture of the book, entitled *Metabolomics, Proteomics and Gene Editing Approaches in Biofertilizer Industry*, second volume, is to present details of cutting-edge research in the field of biofertilizers and plant-microbe interactions that will help readers understand how microbes play a significant role as biofertilizers in boosting plant production in limited inorganic nutrients. It covers general strategies for harnessing integrated technologies: omics, proteomics, and metabolomics for the development of potential biofertilizers. New techniques for enhancing the efficacy and quality of biofertilizers have been elaborated in this book through different chapters contributed by experts in the field of agricultural microbiology.

The process of gene editing tools for engineering beneficial microorganisms in biofertilizers and the role of synthetic biology to improve plant growth and resistance to stress through the use of genetic engineering techniques are also elaborated in the second volume. This volume also gives insights into the role of bioinformatics strategies for analyzing metabolomics, proteomics, and gene editing data and the role of machine learning and intellectual property rights in the biofertilizer industry. The information in this book regarding the role of different microorganisms and the applications of nanoformulations used for nutrient management will be helpful for designing next-generation bio-formulations to assist plant growth.

We believe this book will help provide a substantial number of pieces of evidence that underline the genomic basis of nutrient management by microbes. Essential information will be provided regarding the genomic and metabolomic background of biofertilizers. This collective work is distinct because of our focus on diverse emerging technologies which are high-throughput, scalable, and applicable to different countries regardless of their socio-economic conditions. We consider this a sincere effort to highlight the underutilized potential in advanced technologies in the abatement of dynamic issues in sustainable biofertilizers. This book will improve the current state of knowledge and should inspire researchers and innovators to advance the current interdisciplinary knowledge into technologies that are readily available and effectively minimize hazards associated with chemical fertilizers.

Biofertilizers are the key to sustainable agriculture, and it has become the foremost priority of plant biologists to mine soil microbial diversity with respect to their spatiotemporal, biochemical, and physiological structure and their beneficial effects on plants through modern techniques like metagenomics, proteomics, metabolomics, and the like. Hence, the knowledge in these areas is vast and often does not reach every hand. An edited book like the present one provides updated knowledge to a researcher, and it is equally challenging to compile different studies in a book. First of all, we would like to acknowledge all the contributing authors of chapters in this book for bringing such an exhaustive compilation to a wider readership. All the editors gratefully acknowledge their parent organization for all the support and encouragement rendered. Thanks are also due to all those researchers whose original work has formed the basis of this compilation; without the endless efforts of researchers, science and society could not progress. We are also thankful to the team at Springer for bringing this compilation to the scientific fraternity.

Mohali, Chandigarh, India — Sukhminderjit Kaur
Mohali, Chandigarh, India — Vagish Dwibedi
Mau Nath Bhanjan, Uttar Pradesh, India — Pramod Kumar Sahu

Contents

Editors and Contributors

About the Editors

Sukhminderjit Kaur is working as Professor in the Department of Biotechnology, Chandigarh University, Mohali, India. She received her doctorate in 2008 from PGIMER, Chandigarh, India. She has experience of more than 20 years in academics and research and has published more than 58 research papers and various book chapters (16). Two patents have been granted to her credit for biofertilizer formulation. She has edited various books also with reputed publishers. She has guided four doctorate students. She is a member of numerous scientific/professional societies including Indian Science Congress, Association of Microbiologists of India, and Gastrointestinal Society of India. Her work is inclined toward exploring microorganisms in various applications such as biofertilizers development using different consortium of plant growth promoting bacteria, probiotics, and bioremediation.

Vagish Dwibedi obtained his Ph.D. from Thapar Institute of Engineering and Technology, Patiala, India. He has conducted research projects and consultancy work in plant-microbe interaction, bioassay-guided drug discovery and development, sustainable agriculture, and wastewater treatment. Presently, he is working as a visiting postdoctoral scientist at the Agriculture Research Organization, Ministry of Israel (ARO), Rishon Lezhion, Israel. Previously, he served for 2 years as an Assistant Professor in the Department of Biotechnology, Chandigarh University, Gharuan Mohali, Punjab. Formerly, he worked as Research Scientist at AGPHARM BIOINNOVATIONS LLP, incubated at Thapar Institute of Engineering and Technology, Patiala, India. Dr. Dwibedi works toward developing screening platforms for different biological activities such as antimicrobial, antioxidant, anticancer, or finding novel molecules that interfere in the mechanism of development of diseases such as Alzheimer's dementia, Parkinson's disease, obesity, anti-gout (arthritis), and type 2 diabetes.

Pramod Kumar Sahu is currently working as Scientist (Sr. Scale) at ICAR-National Bureau of Agriculturally Important Microorganisms, Uttar Pradesh, India. Dr. Sahu graduated from IGKVV Raipur and obtained master's and doctorate from the University of Agricultural Sciences, Bengaluru, India. His fields of interest are plant-endophyte interactions, soil microbiome, bioinoculants, biological control, salinity stress alleviation, and confocal scanning laser microscopy. He has 10 years of research experience and has published 62 research/review papers and 22 book chapters. He has developed methodologies for reversal of phenotypic conversion, fluid bed dried formulation, and screening of bacterial endophytes for bio-control activities. He was conferred with the National Academy of Agricultural Sciences (NAAS)—Young Scientist Award 2021, the Association of Microbiologists of India (AMI)—Young Scientist Award 2022, Dr. B. P. Pal's Prize Gold Medal, Dr. G. Rangaswamy Gold Medal, and Young Scientist Award in Agricultural Microbiology-2018 and many other awards to his repute.

Contributors

Aya A. M. Abdellatif Central Laboratory of Organic Agriculture (CLOA), Agricultural Research Center (ARC), Giza, Egypt

Mai N. Amer Department of Chemistry of Natural and Microbial Products, Pharmaceutical and Drug Industries Research Institute, National Research Centre, Giza, Egypt

Shinar Athwal Department of Biotechnology, Thapar Institute of Engineering and Technology, Patiala, Punjab, India

T. Chethan Department of Agricultural Entomology, University of Agricultural Sciences, GKVK, Bengaluru, India

Deepshikha Department of Biotechnology, Chandigarh University Gharuan, Mohali, India

Vagish Dwibedi Department of Biotechnology, Chandigarh University, Mohali, Chandigarh, India

Ibrahim M. K. Elmaghraby Central Laboratory of Organic Agriculture, Agricultural Research Center, Giza, Egypt

Maissara M. K. Elmaghraby Central Laboratory of Organic Agriculture (CLOA), Agricultural Research Center (ARC), Giza, Egypt

Seema Garcha Department of Microbiology, Punjab Agricultural University, Ludhiana, India

Kashish Garg Department of Biotechnology, UIBT, Chandigarh University, Mohali, Punjab, India

Doha Gebily Central Laboratory of Organic Agriculture, Agricultural Research Center, Giza, Egypt

V. Govindasamy Division of Microbiology, ICAR-Indian Agricultural Research Institute, New Delhi, India

Sunil Jadhav Division of Seed Science and Technology, ICAR-Indian Agricultural Research Institute, New Delhi, India

K. Jagadeesh Chandra Bose Department of Biotechnology, UIBT, Chandigarh University, Mohali, Punjab, India

Swapnil Kajale Institute of Soil, Water and Environmental Sciences, Volcani Research Center, Agricultural Research Organization, Rishon LeZion, Israel

Sowmiya Karuna Department of Agricultural Microbiology, University of Agricultural Sciences, Bengaluru, India

Gursharan Kaur Department of Biotechnology, UIBT, Chandigarh University, Mohali, Punjab, India

Hardeep Kaur Department of Biotechnology, UIBT, Chandigarh University, Mohali, Punjab, India

Harleen Kaur Department of Biotechnology, Thapar Institute of Engineering and Technology, Patiala, Punjab, India

Sukhminderjit Kaur Department of Biotechnology, Chandigarh University, Mohali, Chandigarh, India

T. Kavya Division of Microbiology, ICAR-Indian Agricultural Research Institute, New Delhi, India

S. Krishna Kumar Department of Agricultural Entomology, University of Agricultural Sciences, Bengaluru, India

Anil Kumar School of Chemical and Metallurgical Engineering, University of the Witwatersrand, Johannesburg, South Africa

Deepak Kumar Department of Biotechnology, Chandigarh University Gharuan, Mohali, India

Duppala Manoj Kumar Agricultural College, Bapatla, ANGRAU, Guntur, India

Hemant Kumar Department of Biotechnology, UIBT, Chandigarh University, Mohali, Punjab, India

Ritik Kumar Department of Biotechnology, Chandigarh University Gharuan, Mohali, India

Savani Ajit Kumar ICAR-NAARM, Hyderabad, India

Haralakal Keerthi Kumari Department of Agricultural Microbiology, University of Agricultural Sciences, GKVK, Bengaluru, India

Shanu Magotra Department of Biotechnology, UIBT, Chandigarh University, Mohali, Punjab, India

Sakshi Manhas Department of Forensic Science, Chandigarh University, Mohali, Punjab, India

Yalavarthi Nagaraju ICAR-National Bureau of Agriculturally Important Microorganisms, Mau, Uttar Pradesh, India

Neelam Prabha Negi Department of Biotechnology, UIBT, Chandigarh University, Mohali, Punjab, India

Nidhi Department of Biotechnology, UIBT, Chandigarh University, Mohali, Punjab, India

Praveen S. Patted ICAR-Indian Institute of Seed Science, Mau, Uttar Pradesh, India

K. Praveen Department of Agricultural Microbiology, University of Agricultural Sciences, Bengaluru, India

M. Priya Division of Microbiology, Indian Agricultural Research Institute, New Delhi, India

Palak Rana Department of Biotechnology, UIBT, Chandigarh University, Mohali, Punjab, India

Nitu Rani University Centre for Research and Development, Chandigarh University, Mohali, Punjab, India

Richa Department of Microbiology, Bundelkhand University, Jhansi, India

Narashans Alok Sagar University Centre for Research and Development, Chandigarh University, Mohali, Punjab, India

Department of Biotechnology, Chandigarh University, Mohali, Punjab, India

Pramod Kumar Sahu Agricultural Microbiology, ICAR National Bureau of Agriculturally Important Microorganisms, Mau Nath Bhanjan, Uttar Pradesh, India

Sakshi Department of Biotechnology, Chandigarh University Gharuan, Mohali, India

Samredhi Department of Microbiology, Punjab Agricultural University, Ludhiana, India

Jyoti Sarwan Department of Biotechnology, UIBT, Chandigarh University, Mohali, Punjab, India

Adarsh Sharma Department of Biotechnology, UIBT, Chandigarh University, Mohali, Punjab, India

Sonia Sharma Department of Biotechnology, UIBT, Chandigarh University, Mohali, Punjab, India

Gajendra B. Singh Department of Biotechnology, UIBT, Chandigarh University, Mohali, Punjab, India

Geetha Singh Division of Microbiology, ICAR-Indian Agricultural Research Institute, New Delhi, India

G. R. Sneha Division of Microbiology, ICAR-Indian Agricultural Research Institute, New Delhi, India

Gurleen Kaur Sodhi Department of Biotechnology, Thapar Institute of Engineering and Technology, Patiala, Punjab, India

J. Sushmitha Division of Biochemistry, ICAR-Indian Agricultural Research Institute, New Delhi, India

Babita Thakur Department of Biotechnology, Chandigarh University, Mohali, Punjab, India

Joythi Thapa Department of Biotechnology, UIBT, Chandigarh University, Mohali, Punjab, India

Nazim Uddin Department of Biotechnology, UIBT, Chandigarh University, Mohali, Punjab, India

Konderu Niteesh Varma Division of Microbiology, ICAR-Indian Agricultural Research Institute, New Delhi, India

Priyanka Verma Department of Forensic Science, Chandigarh University, Mohali, Punjab, India

D. Vijaysri Division of Microbiology, ICAR-Indian Agricultural Research Institute, New Delhi, India

Harleen Kaur Walia Department of Biotechnology, Thapar Institute of Engineering and Technology, Patiala, Punjab, India

Yashika Yadav Department of Biotechnology, Chandigarh University, Mohali, Punjab, India

Biofertilizers: An Imminent Approach for Sustainable Agriculture

1

Seema Garcha and Samredhi

Abstract

In the current scenario of ever-increasing population, the demand for a sustainable agriculture supply is evident to combat food crisis. Excessive use of synthetic fertilizers has taken a toil on human and environment health putting the ecosystem in a state of turmoil. To abate the situation, biofertilizers are an excellent choice for sustainable farming. The use of plant growth promoting microorganisms has its own benefits when it comes to soil and plant health and enhanced crop yield. But challenges to use biofertilizers include the lack of sufficient availability of the product and lack of awareness in regard to concentration, time, and method of application. Also, public investments in bio-inputs are substantially lower compared to the significant subsidy provided for the chemical inputs. Central government initiatives like NITI Aayog, Paramparagat Krishi Vikas Yojana (PKVY), Mission Organic Value Chain Development for North Eastern Region (MOVCDNER) and National Food Security Mission (NFSM) provide financial assistance and subsidies to predispose the interest of farmers towards organic farming. Under the Soil Health Card Program, the Government of India has also been pushing Integrated Nutrient Management (INM) for the prudent use of pesticides to limit excessive use of fertilizers and pesticides. Conventional biofertilizer formulation limit the shelf life and efficacy of biofertilizers whereas advanced techniques like pelletized formulation using carrier material like compost and biochar are useful in improving the shelf life and efficacy of the biofertilizer. Other than this, encapsulation is another way to further increase shelf life using alginate-based microcapsules of biofertilizers which assist in boosting the cell resistance against unfavorable temperatures and pH. Additives such as glyc-

S. Garcha (✉) · Samredhi
Department of Microbiology, Punjab Agricultural University, Ludhiana, India
e-mail: sgarcha@pau.edu

S. Kaur et al. (eds.), *Metabolomics, Proteomics and Gene Editing Approaches in Biofertilizer Industry*, https://doi.org/10.1007/978-981-97-2910-4_1

erol, chitin, peat, and skim milk are also providing promising results influencing the cell viability and enhanced growth for a better field performance.

Keywords
Biofertilizer · Central policies · Bioformulation · Encapsulation · Additives

1.1 Introduction

Escalating human population is putting a lot of pressure on agricultural lands to produce enormous amount of food. To combat this food crisis, farmers resort to chemical means to effectively generate agricultural productivity. Though the results are quick and effective but this long-term use of chemical fertilizers has resulted in soil degradation, environmental pollution, and deterioration of human health and putting the ecosystem in state of a turmoil. In conventional farming systems, synthetic NPK fertilizers are used extensively to meet plant nutritional requirements and increase agricultural output. But only a limited dose of fertilizer which is around 30–40% actually gets utilized by the crop leaving the rest in the soil and to contributing in polluting the environment. Furthermore, chemical fertilizers contain toxic metals and radionuclides which are not easy to degrade, making them persistent contaminants in the nature. Eutrophication of water sources is another major issue associated with the use of excessive chemical fertilizers (Kumar et al. 2022). Soil microbial community is an important aspect in ecosystem for soil nutrient cycling and nutrient uptake. Long-term chemical fertilization has affected the soil fertility and microbiota showcasing diminished presence of beneficial microbes like proteobacteria which were found to be more prevalent in naturally cultivated rhizospheric zone of walnut trees than the chemically cultivated ones (Bai et al. 2020).

Excessive applications of potassium fertilizers in the soil have disrupted the delicate balance of nutrients by plants leading to reduced productivity and crop losses. Another threat to the environment posed by chemical fertilizers includes the emission of greenhouse gases (GHG) altering the global warming potential (GWP). Synthetic fertilizers stimulate soil acidification, posing a risk to underground water, weakening of plants roots and making them more susceptible to phytopathogens (Randive et al. 2021). This indiscriminate use of synthetic fertilizers destroys the key fundamentals of agricultural production jeopardizing long-term food security. Ecology suffers when nutrient ratios are out of equilibrium due to the excessive use of highly concentrated mineral fertilizers. Currently, the primary issue in agriculture is to prevent further declines in crop yield and soil health (Bhatt et al. 2019).

1.2 Solution to the Problem—Biofertilizers

Plant rhizospheric zone contains crucial microbial communities such as bacteria, archea, fungi, oocytes, nematodes, and viruses. While some maybe pathogenic, a variety among them are plant growth promoting microbes responsible for protecting the plant against biotic and abiotic stress and providing improved crop yield and productivity. Using the beneficial microbial communities as biofertilizers in sustainable farming is a novel approach for agricultural production. These are live bacterial or fungal formulations used for enhancing plant and soil health (Suhag 2016). Biofertilizers aid in enhancing plant biomass, mineral content (potassium, phosphorus, iron), shoot and root length, seed germination, and atmospheric nitrogen fixation (Glick and Gamalero 2021). Utilizing biofertilizers lowers the high cost of synthetic purchases meeting the global demand for environment friendly crop production. Also, biofertilizers supply plenty micronutrients, organic matter, phytohormone secretion along with mitigating the detrimental effects of chemical fertilizers (Mahanty et al. 2017). Admist the blatant use of inorganic fertilizers, abiotic stress on crops is also amplified by climatic variability which further lowers agricultural output. Numerous abiotic and biotic stress factors, such as drought, soil salinity, temperature fluctuations, water retention capacity, heavy metal toxication, and various plant pathogens reduce the crop yield and quality. Biofertilizers provide nutrition through natural processes such as Zn, K, and P solubilization, siderophore production, nitrogen fixation, and hydrolytic enzymes to protect the plant against stress conditions (Chaudhary et al. 2022).

Plant diseases in agriculture are conventionally controlled by the use of chemical pesticides and fungicides. Haphazard use of these pesticides is toxic to environment. Beneficial microbes are being explored in this facet as biopesticides and certain microbial strains for use as biofertilizer formulations to protect the plant against phytopathogens. Effective microbes to be studied against pesticide toxicity include *Azotobacter*, *Enterobacter*, *Paenibacillus*, *Pseudomonas*, and *Bacillus* (Shahid et al. 2019). Microbial strains which are antagonistic against variety of phytopathogens responsible for plant diseases such as root rot, necrosis, spotting, wilting, canker, etc., are recognized for their valuable contribution as biological control agents. Moreover, formulations constituting microbial consortium where combined use of more than one strains exhibit multiple mechanisms for biocontrol activity to fight against complex plant diseases. Moreover, studies reveal that consortium boosts the transcriptional activation of various metabolic pathways, which then have a significant cascading effect on the host plant's signaling for defense against various plant pathogens and some arthropod pests (Dhir 2017).

Application of *P. fuorescens* strain in Cameroon's acidic soils boosted shoot length, grain yield, plant dry weight, and seed phosphorus content in maize (Nosheen et al. 2021). When *Bacillus thuringiensis* was applied as a biofertilizer on *Abelmoschus esculentus* plants in the field, it significantly improved shoot and root length, seed germination, fruit weight, leaf diameter, seed weight as well as the total dry and fresh weight in the treated plants as compared to the non-treated ones (Bandopadhyay 2020). Integrated biofertilizer application of *Bacillus amyloliquefacien*, *Nostocmucorum*, and *Saccharomyces cerevisiae* on Cowpea (*Vigna unguiculata*) was assessed for enzymatic activities, plant growth, and crop yield. Treated plants showed significant increase in dehydrogenase enzyme activity, chlorophyll content, plant dry weight, pod length, grain yield, and vascular bundle length (Omer et al. 2023). In field studies, cyanobacterial isolates positively influenced plant health and rice productivity. Within 4 months, substantial increase in plant length was observed. Plants treated with the bioinoculant were healthier and greener than the untreated control. Panicles gathered at the ripening stage were observed for the weight and number of grains per panicle and the results recorded a significant increase in both the parameters (Iniesta-Pallarés et al. 2021). Arbuscular mycorrhizal fungi (AMF) form symbiotic associations with plants thereby acting as bioameliorators which positively impact the rhizospheric characteristics of soil as well as protecting the plant during stress conditions like drought and heat. Results available show that AMF plays a very important role in enhancing the nutrient uptake and stimulating the production of phytohormones like gibberellins and auxins (Suhag 2016). In leguminous plants, *Rhizobium* in the root nodule fixes atmospheric nitrogen. The plant utilizes nitrogen to synthesize amino acids, vitamins, and nucleic acids. The use of rhizobia in nitrogen fixation reduces the dependency of leguminous plants on synthetic N fertilizers (Fasusi et al. 2021).

Biofertilizers are classified into several categories based on their purpose and mode of action. Nitrogen fixing, potassium solubilizing, phosphorus solubilizing, and plant growth promoting rhizobacteria (PGPR) are the most commonly used of them all. These include *Bacillus*, *Azotobacter*, *Pseudomonas*, *Rhizobium*, *Azospirillum*, cynobacteria, and mycorrhiza (Bhavya and Geetha 2021). For instance, nitrogen fixing biofertilizers classified as free living, symbiotic, and associative. These include *Azotobacter*, *Clostridium*, *Anabaena*, *Nostoc*, etc.

1.2.1 Classification of Fertilizers

Fertilizers are optimally classified into different categories based on their nature, source, and physical state. Upon application in the soil, fertilizers have the capacity to alter the soil pH. Therefore, based on their nature, they can be classified as acidic or basic fertilizers (Randive et al. 2021). Acidic residue is deposited in the soil due to acidic fertilizers and alkaline residue is deposited in the soil due to basic fertilizers. Fertilizers are further classified as natural, chemical, and microbial fertilizers. Natural biofertilizers are also known as traditional biofertilizers which are obtained from natural biogenic materials like cow dung or urine (Raj et al. 2014). These include fish-based biofertilizers, cold salt fertilizers, plant waste matter, composting matter, etc. which are rich in macronutrients such as potassium, nitrogen, and phosphorous (Randive et al. 2021). Chemical biofertilizers, on the other hand, are known as inorganic or synthetic fertilizers manufactured in the factories. Chemical synthesis is used for the production of these. Mixed fertilizers are made up of a number of components, either used alone or in conjunction with other ingredients. However, in reality, fertilizers rich in phosphate, potassium, and nitrogen are frequently utilized (Nangle et al. 2017). The most appropriate approach is biofertilizers which are also known as microbial inoculants or microbial fertilizers. These biofertilizers are made of microorganisms, which create an environment in the soil that is favorable to plant life and help in the synthesis of plant nutrients. These include nitrogen fixing biofertilizers such as *Rhizobium*, *Azolla*, *Azospirillum*; phosphorous solubilizers like

Bacillus, *Pseudomonas*; zinc solubilizers like *Thiobacillus*, *Thioxidans*, and organic matter decomposers like *Trichoderma* (Nosheen et al. 2021).

Biofertilizers based on the physical state includes carrier-based biofertilizers and liquid biofertilizers. In the production of solid carrier-based biofertilizers, a carrier is used as a medium to carry the required microorganism while maintaining their viability during the transportation. Peat, charcoal, and perlite are some of the commonly used carriers (Amenaghawon et al. 2021). A variety of carrier materials including sodium alginate, compost, peat, clay, wheat bran, wood ash, rice husk were tested. Sodium alginate was the most effective among all (Sakpirom et al. 2021). A suitable carrier must have high water holding capacity, be easy to sterilize and transport, have low production cost and should not be toxic to microorganisms. In addition to these, carriers must also possess excellent adhesion quality for seed inoculation (Deaker et al. 2004). An appropriate carrier material must be used to ensure adequate compatibility and quality. A study conducted to determine the best carrier material to support the microbial population of dry land inoculants (*Bacillus* sp., *Delftia tsuruhatensis* strain D9, and *Delftia sp.* strain MS2As2) was carried out. It was reported that the most efficient carrier material for preserving the quantity of live inoculant cells was a mixture of 3% molasses, 1% potassium sorbate, 3% glycerol along with 1% Tween-20 enhanced with 1% nutrient. This formulation met the primary requirements for biofertilizers and was able to maintain the inoculant population up to the 12th week, with a viable count of 21.60×10^8 CFU mL^{-1} (Harahap et al. 2023). Liquid biofertilizers are liquid bioformulations that have the required microorganism present in a dormant state. Liquid biofertilizers are known to be better than carrier-based biofertilizers because of longer shelf life and better tolerance to unfavorable conditions (Chaudhary et al. 2020). Moreover, the chances of contamination in the liquid biofertilizers are lower than that of carrier-based biofertilizers. The packaging material cost required under carried based is also saved here deeming it cost effective. Under the favorable conditions of soil, these dormant formulations become active and readily multiply in the rhizospheric zone of the plants. Seed treatment, root dipping, and direct soil application are the ways a liquid biofertilizer is used (Singh et al. 2021). The results of a study for the evaluation of shelf life showed that, in comparison to liquid-based biofertilizers, carried-based biofertilizers had higher contamination and low viable count. Liquid biofertilizer was able to maintain a sufficient viable count for up to 6 months, whereas carrier-based biofertilizers were able to maintain the viable count for only up to 3 months (Shravani 2019).

1.3 Advances in Biofertilizers

Biofertilizers include a range of microorganisms that colonizes the rhizospheric zone easing the uptake of nutrients and making them more accessible to the root hairs of the plants. Biofertilizers are widely recognized for their reliability and affordability. They are the most effective tools for agricultural production and a safe substitute for the toxic artificial fertilizers (Dasgupta et al. 2021). The solution for a sustainable agriculture demands for an advancement in the present day biofertilizer practices to enhance the biofertilizer technology.

1.3.1 Nanofertilizers

Nanofertilizers involve the use of nanoparticle technology or nanomaterials. They administer the required nutrients to a plant at a nano-scale. They enhance productivity by promoting plant growth and development (Chhipa 2017). These nanoparticles are less than 100 nm in size to serve the purpose of supplying vital nutrients. They address the issue of relatively low nutrient use efficiency (NUE) and hazardous environmental impact due to synthetic fertilizers which is a major hindrance in practicing sustainable agriculture (Dimkpa and Bindraban 2017). Appropriate application of nanofertilizers provides nutrients to the plant in a regulated manner while also reducing environmental impact, increasing NUE, minimizing leaching and volatilization. These fertilizers which are also known as "smart fertilizers" are now being recognized as a potential alternative against the conventional ones because their nanomaterials are highly reactive upon interaction with fertilizers enhancing their absorption and uptake of essential nutrients which are required for effective plant growth (Prasad et al. 2017). Additionally, when combined with microorganisms as "nanobiofertilizers," these nanofertilizers can be used to increase abiotic stress tolerance and offer the host with additional advantages (Zulfiqar et al. 2019). Combining nanofertilizers and biofertilizers has several benefits and creates new possibilities for sustainable agriculture. In addition to enhancing agricultural production, nanobiofertilizers are also known to reduce mineral loss and maximize yield. Moreover, by employing microbial integration through a bio-organic component, enhanced plant growth, rhizoremediation, and disease resistance is observed. Nanobiofertilizers also improve soil fertility and moisture levels along with providing the plants with required essential nutrients (Fazelian and Yousefzadi 2022). In a study, a nanobiofertilizer was developed using biosynthesis of onion silver nanoparticles (AgNP'S) using the onion extract as a reducing agent. When compared to control, plants fed with 15 ml/l of nanobiofertilizer had greater fresh weight and vigor. Tomato and brinjal plants benefit greatly from the application of the synthetic nanobiofertilizer made from onion extract. The overuse of chemical fertilizers, contamination of the environment and expenditures associated with farm management can all be decreased by using these kinds of nanobiofertilizers (Gosavi et al. 2020). A lightweight nanobiofertilizer was developed using immobilized bacterial endospores (*Panebacilluspolymyxa*) in activated carbon beads and Fe-carbon nanofibers (Fe-CNFs) coated with acylated homoserine lactone (AHL). It was tested on both leguminous (*Cicer arietinum*) and non-leguminous (*Triticum aestivum*) plants. The findings demonstrate a significant ($p < 0.05$) increase in biomass, root length, chlorophyll, and protein contents in plants grown for 30 days following nanocomposite biofertilizer (NCB) supplementation. Furthermore, after 21 days of growth with NCB, the plants showed a considerable ($p < 0.05$) ability to withstand root rot in wheat and chickpea caused by *Fusarium oxysporum* sp. *Cicero* and *Cochliobolus sativus*, respectively (Gahoi et al. 2021).

Nanofertilizers are made using the material obtained from synthetic fertilizers or extraction from plant reproductive or vegetative parts by means of chemical, mechanical, and physical processes (Qureshi et al. 2018). The physical process such as radiation, ultra-sonication, mechanical pressure, and thermal or electrical energy are used for melting and condensing for fabrication of nanoparticles, whereas the chemical process includes sol-gel method, micro-emulsion technique, hydrothermal synthesis polymer synthesis, plasma enhanced deposition method, and chemical vapor synthesis (Satyanarayana and Reddy 2018). The nanoparticles are synthesized via top-down approach (where bulk particles are broken into smaller bits using external force) or bottom-up approach (where gas and liquid molecules are combined and gathered together). Other than this, polymers are used for stabilizing and encapsulating these nanofertilizers. Nutrients are encapsulated by covering them in a thin layer of nanoparticle (Yaseen et al. 2020). Chemically synthesized nanoparticles were characterized and administered using randomized complete block design (RCBD) and were administered in plants using soil and foliar application. The results showed that ZnO as a nanofertilizer improved maize cultivar growth and extract production and has the potential to be used as a nanofertilizer for crops grown in Zn-deficient soils to increase crop quantitative and qualitative value (Azam et al. 2022).

Another approach for the production of nanofertilizers is the biosynthesis approach where biological process is used for the fabrication of nanofertilizers. The biological method is much more environment friendly dependable. Molecules such as pigments, proteins, enzymes, phenolic, and alkaloid compounds obtained from microbes and plants are used for nanoparticles production (Prasad et al. 2017). Larger surface area, increased reactivity, and improved contact are some of the advantages provided via this technique. Stabilizing agents such as proteins, carbohydrates, enzymes, and phytochemicals such as flavonoids, terpenoids, phenolics, and cofactors are also used. Additionally, it was found that a variety of microorganisms and plant extracts might be used to accomplish the biosynthesis of a nanofertilizer (Yaseen et al. 2020). An increasing amount of focus is being paid to pursue for newer techniques for the biosynthesis of nanofertilizers.

Nanotechnology has enormous potential to change many facets of human life but continuous improvement, ethical measures and safety must be applied before fabricating a nanoparticle. Risks such as nanoparticle toxicity associated with the implementation of this technology must be assessed before the application. As a result, a comprehensive research and series of experiments must be conducted before fully accepting this technology (Fatima et al. 2021).

1.3.2 Biofilm-Based Biofertilizers

Biofertilizers are biological products which can potentially reduce dependency on chemical fertilizers. However, the major drawback is the inadequate survival of viable cells in the soil. A feasible cue would be using the cells which are immobilized in biofilms as this would ensure protracted cell survival and better suitable environment for the bacteria and its metabolites to interact with the plant. Biofilm-based biofertilizers are another potential alternative to synthetic or conventional fertilizers. Biofilm is a structure developed by microbial assemblage encased in a matrix that the bacteria build on their own for the ease of communication. In terms of applications, they are already in use for purification of water by entrapment of pathogens in the polluted water and processes like bioremediation (Armbruster and Parsek 2018). Biofilm application is done to encourage plant growth and its development.

In terms of soil, fungus and bacteria live together and share micro-niches. The metabolic processes of an microorganism is influenced by other microorganisms. This co-existence positively impacts the plant growth and associated ecosystem. Comparing mono or mixed cultures without biofilm development to fungal–bacterial biofilms, which are bacterial biofilms connected to fungal surfaces, has demonstrated improved nutrient uptake, plant growth, and tolerance against environmental stress (Hassani et al. 2018). In the rhizospheric zone, biofilms aid in enhancing water retention, bacterial biomass, stimulating the exudation of roots for their response against stress (Kasim et al. 2016). Biofilms biodegrade heavy metals and organic contaminants in soil. Their application involves either direct spraying on the aerial regions of the plants as mixed inoculum or as beneficial bacterial biofilm inoculum. They can also be incorporated into the soil as inocula (Gui et al. 2017). As a result, these interactions not only promote plant development but also enhance soil health by increasing nutrient cycling. Multiple-specie biofilms need more attention because they have the potential to create polysaccharides and other bioactive substances that have a favorable influence on soil health and plant growth on a bigger scale (Pandit et al. 2020).

There is not enough naturally occurring biofilms in the soil to make a noticeable difference. Therefore, it is essential to generate biofilms in vitro and then use them as biofertilizers or biofilm-based fertilizer (BFBF) to increase agricultural output in an environmentally responsible way. In a study conducted with regard to rice production, application of biofilm-based fertilizer (BFBF) reduced the requirement of chemical fertilization to up to 50% while simultaneously boosting grain yield by up to 25%. These results contributed toward a healthy relation of BFBF's with microorganisms, plants, and soil (Premarathna et al. 2021).

Using glycerol as the only carbon source, the planktonic *Bacillus subtilis subsp. spizizenii* was able to create a thick biofilm in a study conducted on *Lactuca sativa* biofertilization. The immobilized cells portrayed positive PGPR results, increasing the root growth by 59% and aerial growth by 39% (Galelli et al. 2015).

Another aspect to biofilm-based biofertilizers is sequestration of organic carbon. Research is in progress for analyzing the increase in soil organic carbon while also enhancing the crop yield by using these biofilm biofertilizers. In four districts of Sri Lanka, the use of biofilm biofertilizer was compared with prevalent practice of applying chemical fertilization. Infrared spectroscopy was implied to study the stabilization of sequestered soil carbon. The outcome was that biofilm biofertilizer produced soils with stronger organomineral complexation. Soil respiration was also reduced down to about 40% in comparison to standard chemical fertilization. This could be because of enhanced mineral surface-reactive metabolites and low priming effect, respectively, in biofilm biofertilizer which contributed in mitigating the effects of climate change. In summary, farmers can effectively replace their use of chemical fertilizers with the biofilm biofertilizer practice upon adequate research (Premarathna et al. 2023).

1.3.3 Encapsulation Technology

To be employed as slow release fertilizers (SRF's), encapsulation is done in a compound which is made from natural or synthetic polymer. The newest technology in fertilizer development is polymers-coated nutrient fertilizers. Superabsorbent polymers have garnered significant since they utilize less water decreasing the load of frequent irrigation. These polymers are also effective in lowering the rate of plant mortality. Additionally, they increase plant growth rates and soil fertilizer retention (Ma and Wen 2020). This technology created using synthetic polymers portrays impressive outcomes in terms of continuous yet slow release but this ongoing utility is limited by their high cost and inability to decompose. To address these issues, biodegradable polymers are utilized as an environmentally friendly substitute for synthetic polymer-based ones. These polymers are derived from naturally occurring substances that are safe for living things to consume, such as corncob and cotton stalks (Iftime et al. 2019).

Polymer seed coating is another popular approach where a seed is coated in a polymer to protect it from biotic and abiotic stress conditions. Crop productivity is significantly influenced by the quality of the seeds. To improve seed quality, techniques such as treating seeds, improving the soil, modifying genes, etc. must be applied. In order to produce healthy crops, seed treatment serves as the first line of defense against soil-borne and seed-borne infections. The term "seed treatment" refers to the act of applying or covering seeds with biological or chemical agents. It is a seed coating in which the polymer, which contains various materials (pesticides, nutrients, etc.), sticks firmly to the seed resulting in the increase in nutrient use efficiency and reduced pollution (Krishnamoorthy and Rajiv 2018). Biopolymer like chitosan is extensively used in this method. In another example a mixture made of polyethylene glycol and chitosan that was used as a seed coating in the carrier for *Trichoderma* strains (Abd El-Aziz et al. 2022).

1.4 Constraints in Biofertilizer Technology

Despite being an organic source of almost all important crop nutrients, the current application rate of biofertilizers is still limited. Lack of information, insufficient extension services, global, national, regional, and local agricultural policies, institutional changes, etc., are key barriers to the implementation of sustainable agricultural methods. Therefore, it becomes important to discuss the constraints faced by various farmers for the use of biofertilizers and comprehend the reasons behind the same (Linares Quero et al. 2022).

One of the major limitations stems from the lack of knowledge and understanding among farmers about the use and importance of biofertilizers. Lack of knowledge regarding the concentration, duration, and manner of application leads to limited use of biofertilizers in modern agricultural fields. Other limitations arise from lack of suitable organic farming training to farmers and insufficient organic farming expertise making it a significant barrier to the implementation of organic farming (Jangid et al. 2012). Recent studies have concluded that majority of small-scale farmers have primary to graduate education, medium land holding, low scientific orientation, and lower mass media exposure relating to the use of biofertilizers. Studies concluded that a large number of small-scale farmers indicated a lack of technical expertise in using biofertilizers and thus it can be stated that biofertilizer consumption may be increased to great extent if they were trained about the proper practices surrounding its use (Diptesh and Chauhan 2016).

According to various studies, lack of awareness as well as familiarity about the efficacy and effectiveness of biofertilizers compared to the tried and tested inorganic fertilizers leads to further inability of farmers to incorporate biofertilizers to their crops. Limited capacity of government's ambiguous market support also leads to a reduced awareness of organic biofertilizer making them unable to compete with the well-established inorganic fertilizer industry and the massive marketing strategies applied by private fertilizer companies (Baconguis et al. 2012).

Institutional restrictions and a shortage of competent human capital in research institutes and private enterprises have always been key impediments to mass production and commercialization of many biotechnological advances. With the lack of technical advances, poor grade inoculants are produced without a fundamental grasp of microbiological processes also leading to shorter shelf lives. Other constraints in technology such as a scarcity of high-quality carrier material as well as a lack of trained and technical staff in production facilities affect the overall production and application of biofertilizers (Dhar et al. 2009).

To successfully commercialize biofertilizers, there is a need to broaden the existing resource base by locating and incorporating more viable strains, work on enhancing shelf life of the product, develop better production technologies and better-quality control systems (Rai et al. 2023). Many biofertilizers are required to be developed and constantly modified according to the environmental conditions. Variations in seasonal demands, simultaneous cropping activities, and differing time periods of sowing or planting in different locations lead to major environmental constraints on the use of biofertilizers (Kumar et al. 2017). Due to these considerable changes in agroclimatic conditions and different requirements for specific soils (alkaline, sodic, acidic, etc.) varying throughout the country, single biofertilizer cannot be produced to give equal productivity everywhere.

To address the obstacles, there is a need to improve dissemination of information, strengthening the government's role in fostering collaboration among various components of the bio-innovation system. Farmers currently have a significant need for information and they are eager to learning more about the practice of utilizing biofertilizers and how they relate to controlling plant pests and diseases, improving crop quality and fostering competitiveness. The benefits of employing biofertilizers should be aligned with the information that farmers require the most from pre-to post-harvest (Kassem et al. 2021). Participation of agricultural cooperatives to provide subsidies on organic fertilizers and enhance the existing awareness can have favorable effects in influencing farmers toward choosing biofertilizers over chemical fertilizers (Wang et al. 2018).

1.5 Conclusions and Future Prospects

The estimated value of the global biofertilizer market is 2.3 billion US dollars, with projections showing a 3.9 billion-dollar growth by 2025. Biofertilizer technology despite being having potential still faces significant obstacles preventing their widespread usage in agriculture (Mitter et al. 2021). Efforts must be given to investigate the success and failures in biofertilizer technologies. For this there is a need of high levels of innovation and active participation in cutting-edge scientific research and development, public demonstration programs, and encouraging private organizations and new policymakers to take an active role in the production and marketing of biofertilizer (Kumawat et al. 2021) Working on the limitations will pave the way for advanced biofertilizer technology giving the agriculture more sustainable approach.

References

Abd El-Aziz ME, Salama DM, Morsi SM, Youssef AM, El-Sakhawy M (2022) Development of polymer composites and encapsulation technology for slow-release fertilisers. Rev Chem Eng 38(5):603–616

Amenaghawon AN, Anyalewechi CL, Kusuma HS (2021) Fabrication approaches for biofertilizers. In: Inamuddin, Ahamed MI, Boddula R, Rezakazemi M (eds) Biofertilizers. https://doi.org/10.1002/9781119724995.ch16

Armbruster CR, Parsek MR (2018) New insight into the early stages of biofilm formation. Proc Natl Acad Sci 115(17):4317–4319

Azam M, Bhatti HN, Khan A, Zafar L, Iqbal M (2022) Zinc oxide nano-fertiliser application (foliar and soil) effect on the growth, photosynthetic pigments and antioxidant system of maize cultivar. Biocatal Agric Biotechnol 42:102343

Baconguis R, Peñalba L, Paunlagui M (2012) Mapping the innovation system of biofertilisers: constraints and prospects to enhance diffusion. Am Eurasian J Agric Environ Sci 12(9):1185–1195

Bai YC, Chang YY, Hussain M, Lu B, Zhang JP, Song XB, Lei XS, Pei D (2020) Soil chemical and microbiological properties are changed by long-term chemical fertilisers that limit ecosystem functioning. Microorganisms 8(5):694

Bandopadhyay S (2020) Application of plant growth promoting Bacillus thuringiensis as biofertiliser on Abelmoschus esculentus plants under field condition. J Pure Appl Microbiol 14(2):1287–1294

Bhatt MK, Labanya R, Joshi HC (2019) Influence of long-term chemical fertilisers and organic manures on soil fertility-a review. Univ J Agric Res 7(5):177–188

Bhavya K, Geetha A (2021) Plant growth promoting rhizobacteria. Adv Agric Sci 61:87
Chaudhary T, Dixit M, Gera R, Shukla AK, Prakash A, Gupta G, Shukla P (2020) Techniques for improving formulations of bioinoculants. 3 Biotech 10:1–9
Chaudhary P, Singh S, Chaudhary A, Sharma A, Kumar G (2022) Overview of biofertilisers in crop production and stress management for sustainable agriculture. Front Plant Sci 13:930340
Chhipa H (2017) Nanofertilizers and nanopesticides for agriculture. Environ Chem Lett 15:15–22. https://doi.org/10.1007/s10311-016-0600-4
Dasgupta D, Kumar K, Miglani R, Mishra R, Panda AK, Bisht SS (2021) Microbial biofertilizers: recent trends and future outlook. Recent Adv Microb Biotechnol:1–26
Deaker R, Roughley RJ, Kennedy IR (2004) Legume seed inoculation technology—a review. Soil Biol Biochem 36(8):1275–1288
Dhar DW, Saxena S, Singh NK (2009) BGA biofertilizer: production, constraints and future perspectives. In: Mallik CP, Wadhwani C, Kaur B (eds) Crop breeding and biotechnology. Pointer Publishers, Jaipur, pp 209–226
Dhir B (2017) Biofertilizers and biopesticides: eco-friendly biological agents. In: Kumar R, Sharma A, Ahluwalia S (eds) Advances in environmental biotechnology. Springer, Singapore. https://doi.org/10.1007/978-981-10-4041-2_10
Dimkpa CO, Bindraban PS (2017) Nanofertilisers: new products for the industry? J Agric Food Chem 66(26):6462–6473
Diptesh C, Chauhan NM (2016) Constraints faced by biofertiliser users. Gujarat J Ext Educ 49
Fasusi OA, Cruz C, Babalola OO (2021) Agricultural sustainability: microbial biofertilisers in rhizosphere management. Agriculture 11(2):163
Fatima F, Hashim A, Anees S (2021) Efficacy of nanoparticles as nanofertiliser production: a review. Environ Sci Pollut Res 28:1292–1303
Fazelian N, Yousefzadi M (2022) Nano-biofertilisers for enhanced nutrient use efficiency. In: Nano-enabled agrochemicals in agriculture. Academic Press, pp 145–158
Gahoi P, Omar RA, Verma N, Gupta GS (2021) Rhizobacteria and acylated homoserine lactone-based nanobiofertiliser to improve growth and pathogen defense in Cicer arietinum and Triticum aestivum plants. ACS Agric Sci Technol 1(3):240–252
Galelli ME, Sarti GC, Miyazaki SS (2015) Lactuca sativa biofertilization using biofilm from Bacillus with PGPR activity. J Appl Hortic 17:3
Glick BR, Gamalero E (2021) Recent developments in the study of plant microbiomes. Microorganisms 9(7):1533
Gosavi VC, Daspute AA, Patil A, Gangurde A, Wagh SG, Sherkhane A, AnandraoDeshmukh V (2020) Synthesis of green nanobiofertiliser using silver nanoparticles of Allium cepa extract short title: green nanofertiliser from Allium cepa. IJCS 8(4):1690–1694
Gui H, Purahong W, Hyde KD, Xu J, Mortimer PE (2017) The arbuscular mycorrhizal fungus Funneliformismosseae alters bacterial communities in subtropical forest soils during litter decomposition. Front Microbiol 8:1120
Harahap RT, Azizah IR, Herdiyantoro D, Setiawati MR, Simarmata T (2023) Shelf life of biofilm producing PGPR liquid inoculants in different enriched media composition. Agro Bali Agric J 6(2):278–285
Hassani MA, Durán P, Hacquard S (2018) Microbial interactions within the plant holobiont. Microbiome 6:1–17
Iftime MM, Ailiesei GL, Ungureanu E, Marin L (2019) Designing chitosan based eco-friendly multifunctional soil conditioner systems with urea controlled release and water retention. Carbohydr Polym 223:115040
Iniesta-Pallarés M, Álvarez C, Gordillo-Cantón FM, Ramírez-Moncayo C, Alves-Martínez P, Molina-Heredia FP, Mariscal V (2021) Sustaining rice production through biofertilization with N2-fixing cyanobacteria. Appl Sci 11(10):4628
Jangid MK, Khan IM, Singh S (2012) Constraints faced by the organic and conventional farmers in adoption of organic farming practices. Indian Res J Ext Educ 2:28–32

Kasim WA, Gaafar RM, Abou-Ali RM, Omar MN, Hewait HM (2016) Effect of biofilm forming plant growth promoting rhizobacteria on salinity tolerance in barley. Ann Agric Sci 61(2):217–227

Kassem HS, Alotaibi BA, Aldosri FO, Muddassir M (2021) Exploring the relationship between information-seeking behavior and adoption of biofertilisers among onion farmers. Agronomy 11(6):1258

Krishnamoorthy V, Rajiv S (2018) Tailoring electrospun polymer blend carriers for nutrient delivery in seed coating for sustainable agriculture. J Clean Prod 177:69–78

Kumar R, Kumawat N, Sahu YK (2017) Role of biofertilisers in agriculture. Popular kheti 5(4):63–66

Kumar S, Sindhu SS, Kumar R (2022) Biofertilisers: an ecofriendly technology for nutrient recycling and environmental sustainability. Curr Res Microb Sci 3:100094

Kumawat KC, Nagpal S, Sharma P (2021) Present scenario of bio-fertilizer production and marketing around the globe. In: Biofertilizers. Woodhead Publishing, pp 389–413

Linares Quero A, IraguiYoldi U, Gava O, Schwarz G, Povellato A, Astrain C (2022) Assessment of the common agricultural policy 2014–2020 in supporting agroecological transitions: a comparative study of 15 cases across Europe. Sustain For 14(15):9261

Ma X, Wen G (2020) Development history and synthesis of super-absorbent polymers: a review. J Polym Res 27:1–12

Mahanty T, Bhattacharjee S, Goswami M, Bhattacharyya P, Das B, Ghosh A, Tribedi P (2017) Biofertilisers: a potential approach for sustainable agriculture development. Environ Sci Pollut Res 24:3315–3335

Mitter EK, Tosi M, Obregón D, Dunfield KE, Germida JJ (2021) Rethinking crop nutrition in times of modern microbiology: innovative biofertilizer technologies. Front Sustain Food Syst 5:606815

Nangle SN, Sakimoto KK, Silver PA, Nocera DG (2017) Biological-inorganic hybrid systems as a generalized platform for chemical production. Curr Opin Chem Biol 41:107–113

Nosheen S, Ajmal I, Song Y (2021) Microbes as biofertilizers, a potential approach for sustainable crop production. Sustainability 13(4):1868. https://doi.org/10.3390/su13041868

Omer RM, Hewait HM, Mady E, Yousif SK, Gashash EA, Randhir R, Ashmawi AE, El-Taher AM, Al-Harbi NA, Randhir TO (2023) Chemical, anatomical, and productivity responses of cowpea (Vigna unguiculata L.) to integrated biofertiliser applications with PGPR, cyanobacteria, and yeast. Sustainability 15(9):7599

Pandit A, Adholeya A, Cahill D, Brau L, Kochar M (2020) Microbial biofilms in nature: unlocking their potential for agricultural applications. J Appl Microbiol 129(2):199–211

Prasad R, Bhattacharyya A, Nguyen QD (2017) Nanotechnology in sustainable agriculture: recent developments, challenges, and perspectives. Front Microbiol 8:1014

Premarathna M, Seneviratne G, Ketipearachchi KG, Pathirana A, Karunaratne RKC, Balasooriya WK, Fonseka K (2021) Biofilm biofertiliser can reinstate network interactions for improved rice production. Ceylon J Sci 50(3):235–242

Premarathna M, Jayasekara APDA, Rathnathilaka ATD, Ekanayake SNB, Warnakulasooriya WMKDS, Abeysinghe DC, Seneviratne G (2023) Biofilm biofertiliser stabilizes sequestered paddy soil carbon while cutting down chemical fertilisers: answers for climate and fertiliser issues. Eurasian Soil Sci:1–9

Qureshi A, Singh DK, Dwivedi S (2018) Nano-fertilisers: a novel way for enhancing nutrient use efficiency and crop productivity. Int J Curr Microbiol App Sci 7(2):3325–3335

Rai PK, Rai A, Sharma NK, Singh T, Kumar Y (2023) Limitations of biofertilisers and their revitalization through nanotechnology. J Clean Prod:138194

Raj A, Jhariya MK, Toppo P (2014) Cow dung for eco-friendly and sustainable productive farming. Environ Sci 3(10):201–202

Randive K, Raut T, Jawadand S (2021) An overview of the global fertiliser trends and India's position in 2020. Miner Econ:1–14

Sakpirom J, Nunkaew T, Khan E, Kantachote D (2021) Optimization of carriers and packaging for effective biofertilisers to enhance Oryza sativa L. growth in paddy soil. Rhizosphere 19:100383

Satyanarayana T, Reddy SS (2018) A review on chemical and physical synthesis methods of nanomaterials. Int J Res Appl Sci Eng Technol 6(1):2885–2889

Shahid M, Zaidi A, Ehtram A, Khan MS (2019) In vitro investigation to explore the toxicity of different groups of pesticides for an agronomically important rhizosphere isolate Azotobactervinelandii. Pestic Biochem Physiol 157:33–44

Shravani K (2019) Evaluation of shelf life and quality of carrier and liquid based biofertilisers. Int J Microbiol Res, ISSN:0975–5276

Singh D, Thapa S, Geat N, Mehriya ML, Rajawat MVS (2021) Biofertilisers: mechanisms and application. In: Biofertilisers. Woodhead Publishing, pp 151–166

Suhag M (2016) Potential of biofertilisers to replace chemical fertilisers. Int Adv Res J Sci Eng Technol 3(5):163–167

Wang Y, Zhu Y, Zhang S, Wang Y (2018) What could promote farmers to replace chemical fertilisers with organic fertilisers? J Clean Prod 199:882–890

Yaseen R, Ahmed AIS, Omer AM, Agha MKM, Emam TM (2020) Nano-fertilisers: bio-fabrication, application and biosafety. Novel Res Microbiol J 4(4):884–900

Zulfiqar F, Navarro M, Ashraf M, Akram NA, Munné-Bosch S (2019) Nanofertiliser use for sustainable agriculture: advantages and limitations. Plant Sci 289:110270

Futuristic Approaches in Biofertilizer Industry: Challenges, Opportunities, and Future Directions

2

Hardeep Kaur, Shinar Athwal, and Kashish Garg

Abstract

Over the years conventional fertilizers have been replaced by biofertilizers owing to their negative impact on humans and the environment. Biofertilizers are an integral part of agricultural industries as they are responsible for maintaining the soil ecosystem and increasing the crop yield. Also, bio-based fertilizers have environment-friendly aspects that promote sustainable agriculture. The biofertilizers industry is at its boom and is experiencing an addition of new approaches with time. The future of organic fertilizers lies in the growth of microbial, nano-sized, and liquid biofertilizers as they have the potential to bring a green change in the crop industry. The most used microbial biofertilizers are nitrogenous (rhizobacteria, Azospirillum), fungal (mycorrhizae, penicillium), and endophytic (Pseudomonas, Rahnella). Nanobiofertilizers are composed of nanoparticles (gold, silver, iron-oxide, zinc) or nanocomposites (nanoforms of organic compounds). Concerning liquid biofertilizers the commonly used solutions (solubilizers) are nitrogen, potassium, and phosphate, which are the essential components of the fertilizers. This review gives a deep insight into the emerging forms of biofertilizers (microbial, nano, and liquid) along with their impact on the soil system and crop production. All the mentioned forms exhibit properties that increase and/or maintain the microbial, and nutrient balance in the soil consequently improving the crop yield. Not only this but also these biofertilizers help to preserve the essential components of nature and thus promote green agricul-

H. Kaur (✉) · K. Garg
Department of Biotechnology, UIBT, Chandigarh University, Mohali, Punjab, India
e-mail: hardeep.uibt@cumail.in

S. Athwal
Department of Biotechnology, Thapar Institute of Engineering and Technology, Patiala, Punjab, India
e-mail: sathwal_msc23@thapar.edu

S. Kaur et al. (eds.), *Metabolomics, Proteomics and Gene Editing Approaches in Biofertilizer Industry*, https://doi.org/10.1007/978-981-97-2910-4_2

ture. Although the upcoming biofertilizers have certain limitations which will be discussed in the articles along with this the futuristic direction will also be reflected upon to deal with the challenges faced while employing biofertilizers.

Keywords
Biofertilizers · Nano fertilizers · Microbial fertilizers · Liquids fertilizers · Sustainable agriculture · Green approach

2.1 Introduction

Biofertilizers play a pivotal role in the progress of the sustainable agriculture approach. It is a technique that uses fertilizers but in the form of biological matters to increase the overall positive impact it has on the environment and humans (Mahanty et al. 2017; Malusà et al. 2016). Chemical fertilizers are hazardous and responsible for causing natural damage and negatively impacting human health so biofertilizers present an effective way to deal with this problem. Fertilizers are important sources of microbial communities and plant growth so it is important to achieve better development of crops (Mącik et al. 2020; Mahapatra et al. 2022). Over the years various research studies have been performed to understand the applicability of biofertilizers and raw materials which can be used to develop green fertilizers (Nosheen et al. 2021). The fertilizers industry is the most important one in the agricultural sector as it can promote the shoot length, root length, leaf size, quality, and yield of the crop (Kaur and Purewal 2019; Reddy et al. 2020; Pirttilä et al. 2021; Mahmud et al. 2021; Singh et al. 2021) (Fig. 2.1). Since biofertilizers have so many advantageous points so considerable amount of research has been conducted on the development of efficient biofertilizers (Anli et al. 2020; Dal Cortivo et al. 2020; Du et al. 2022).

Innovations in the field of biofertilizers are related to the advancement in the material properties of the fertilizer. Improvised fertilizers used in today's scenario are based on microbes, nanotechnology, and liquid sprays (Seenivasagan and Babalola 2021; Puglia et al. 2021). Microbial fertilizers such as nitrogen-fixing, fungus, and endophytic organisms are used in biofertilizers to fix atmospheric nitrogen, solubilize phosphate, and increase the availability of minerals in the soil environment (Riaz et al. 2021; Dasgupta et al. 2021; Zainuddin et al. 2022; Negi et al. 2021). Additionally, the integration of nanotechnology into fertilization materials brings about effective results. Gold, zinc, iron, manganese nanoparticles, and other nanocomposites can be utilized to improve the overall development of the host (Bhattacharyya et al. 2020; Dong et al. 2019; Sambangi et al. 2022). Liquid fertilizers have also gained considerable attention as they are handy and easy to use and commercially available as potassium, phosphorus, and nitrogen solubilizers (Ma et al. 2020; Barzee et al. 2019; Sharma et al. 2023a). This article includes the recent advancements in biofertilizers, namely, microbial, nano, and liquid along with this the limitations of the same have been discussed in detail. Biofertilizer is a way of promoting green agriculture to follow the path of sustainable development.

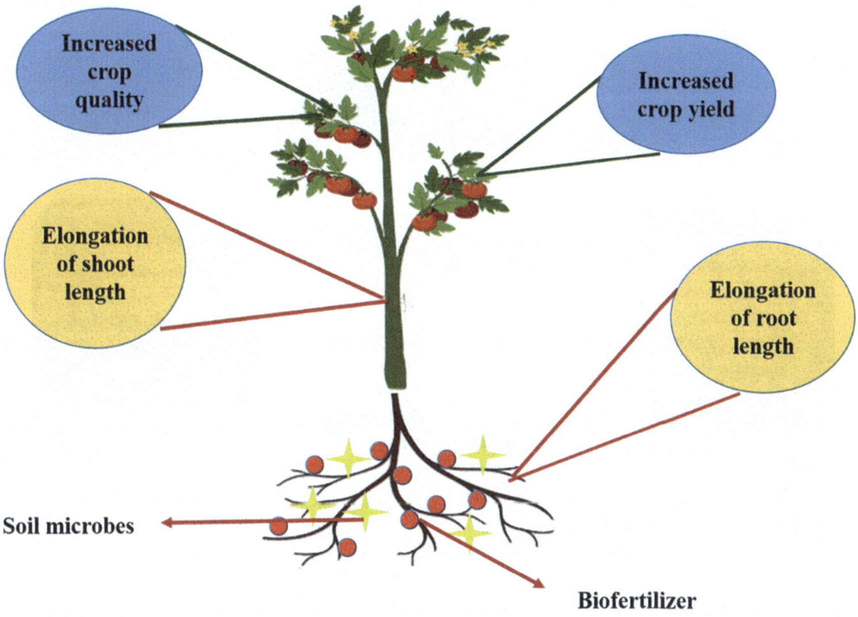

Fig. 2.1 Biofertilizers can promote the shoot length, root length, leaf size, quality, and yield of the crop

2.2 An Emerging Trend of Microbes in the Biofertilizer Industry

The microbes that are widely used to form biofertilizers are nitrogen-fixing, fungus, and endophytic organisms. Nitrogen-fixing bacteria includes *Azotobacter, rhizobium, Azospirillum, and Klebsiella*, fungal species consists of arbuscular mycorrhiza fungi, *Trichoderma*, and saprophytic species while endophytic species includes *Bacillus* sp. and *Piriformospora indica* as shown in (Fig. 2.2).

2.2.1 Types of Microbial Fertilizers

2.2.1.1 Nitrogen-Fixing

Different microbial species can be used to develop enhanced biofertilizers such as *Azotobacter, rhizobium, Azospirillum, and Klebsiella*. These fertilizers help to fix atmospheric nitrogen with the associated plants and are capable of uptake fixed nitrogen along with decreased losses by leaching, volatilization, and denitrification. Several nitrogen-fixing bacteria have been used to develop biofertilizers (Thomas and Singh 2019). *Azotobacter* species uses classic molybdenum-containing enzymes for biological nitrogen fixation (BNF). It also can produce more than one alternative nitrogenase in the absence of molybdenum. Some studies depicted that *Azotobacter* increases nitrogen accumulation as it produces phytohormone-like compounds that

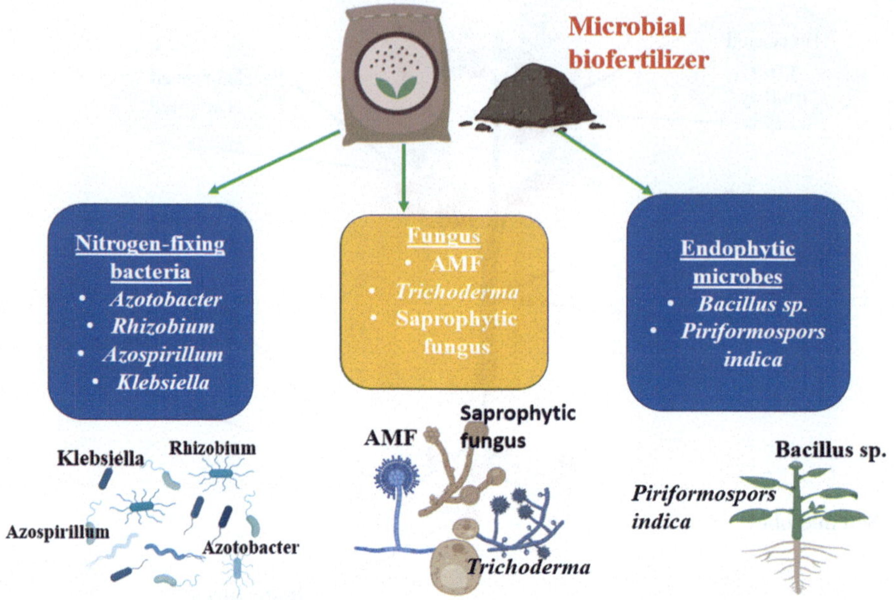

Fig. 2.2 The microbes which are widely used to form biofertilizers are nitrogen-fixing, fungus, and endophytic organisms. Nitrogen-fixing bacteria include *Azotobacter, rhizobium, Azospirillum, and Klebsiella*, fungal species consists of arbuscular mycorrhiza fungi, *Trichoderma*, and saprophytic species, while endophytic species includes *Bacillus* sp. and *Piriformospora indica*

alter morphology and plant growth (Aasfar et al. 2021). The availability of rhizodeposition and soil nitrogen modifies the microbial environment of nitrogen-fixing bacteria. Moreover, moderate application of nitrogen enhances the symbiotic fixation of nitrogen in *Astragalus sinicus L.* (rhizosphere). In legume-rhizobia platforms, BNF is optimized by the interchange of organic N and C nutrients between bacteria and plants (host) (Li et al. 2019). The incorporation of nitrogenous bacteria with composites can develop an efficient delivery system. For example, Herrera et al. immobilized hydrotalcite/alginate composite on *Streptomyces* sp. To produce a biological composite that functioned as a bacterial reservoir and controlled discharge matter to be utilized as biofertilizers (Velázquez-Herrera et al. 2022b). BNF is a conventional nitrogen-fixing technique that with further improvisations can be beneficial for the agricultural industry.

2.2.1.2 Fungal

The most common fungal microbe used in biofertilizers is mycorrhiza. Mycorrhiza is a phosphate adsorbing or mobilizing fertilizer. In nature, it is present in a mutualistic symbiotic state with the plant (host) where both the associated bodies are benefited from each other. Other fungi such as *Azospirillum, Trichoderma*, and saprophytic species are also utilized to develop biological fertilizers (Thomas and Singh 2019). To study the effect of arbuscular mycorrhiza fungi (AMF), Elmostapha et al. inoculated Argan seedlings with different species of AFM. The experiment revealed that AFM had a considerable effect on plant height (33–70%), root dry weight, aerial dry weight, chloroplast content, and mineral proportions (Elmostapha et al. 2022). Various saprophytic fungi are capable of mobilizing phosphorus from insoluble form via different mechanisms. Species such as *Alternaria alternata, Rhizopus stolonifer var. stolonifer*, and *Aspergillus Niger* were the best-performing fungal strains in the soil to solubilize phosphate residues. Saprophytic fungi store novel mineral form of phosphorus which accumulates in mycelium and further assimilate phosphorus in organic form (Ceci et al. 2018). Moreover, fungus-based fertilizers can be used as herbicides, for example, Harzianum (A and B) produced from *Trichoderma brevicompactum* can control vital mono- and dicotyledon weeds even at low concentrations (Yin et al. 2020). Thus fungal microbes can be used to enrich the fertilizer in phosphate solubilizing entities.

2.2.1.3 Endophytic

Endophytes are those microbial species that reside within the tissues of plants. These microbes are not harmful rather they protect from pathogenic organisms and nourish the plants. The endophytic community consists of fungi, algae, actinomycetes, bacteria, and transgenic microbial strains like *Piriformospora indica* and *Bacillus species* (Martínez-Hidalgo et al. 2021). The attributes of endophytes can be used to develop biofertilizers. The beneficial aspects of endophytes can be used in the case of non-leguminous plants in which they can conduct multi-dimensional mechanisms BNF process. Bacillus sp. is one promising endophytic strain that has various positive attributes that can produce efficacious biofertilizers. It can fix a considerable amount of atmospheric nitrogen and can improve crop yield and quality along with the production of phytohormone (auxins) (Hossain et al. 2022). Treatment of plants with rhizobium fertilizers can increase the presence of bacterial endophytes as in the case of *Oryza sativa* where the microbial availability increased after 55 days of incubating the rice plant in the biofertilizers (Jha et al. 2020). Thus endophytic microbes are a reliable source of nourishment for plants by embedding it into a fertilizers form.

2.2.2 Impact of Microbial Biofertilizers on Crop Production

Microbial biofertilizers have a significant impact on the yield and maintenance of agricultural products so they are greatly employed in improving crop production. A brief insight into various microbial fertilizers has been discussed in Table 2.1. Various studies have been conducted to realize the potential of microbes as biofertilizers. Masuda et al. speculated that the addition of ferric-iron oxide (electron acceptor) in iron-reducing bacteria for respiration can increase the nitrogen-fixing ability of *Geobacter* and *Anaeromyxobacter* (Masuda et al. 2021; Odoh et al. 2020). Similarly, Hala applied *Azospirillum brasilense* and *Klebsiella pneumonia* along with urea (fertilizer) toward the upland growth of rice plants. Nitrogen-fixing bacteria when inoculated into upland rice resulted in increased tiller formation, and

Table 2.1 A brief insight into the world of microbial fertilizers with their characteristics, impact, and certain immobilizations

Microorganism	Substrate/ immobilized composites	Characteristics features	Impact	Reference
Geobacter and Anaeromyxobacter	Ferric-iron oxide (electron acceptor)	Powdered iron increases soil nitrogen-fixing ability	More rice yield, low burden of environmental nitrogen, decreased input of fertilizer	Masuda et al. (2021)
Azospirillum brasilense and Klebsiella pneumonia	Urea	Nitrogen production increased from 0.13 to 0.22 g	Increased tiller formation, production of nitrogen, root, and shoot dry weight	Hala (2020)
Azotobacter	*Aspergillus niger*	Nitrogen fixation as per gram of carbon was 35.08 mg of nitrogen was formed.	Enhanced leaf length/width, fruit size, fruit yields, and plant height in comparison to the untreated plant	Din et al. (2019)
Streptomyces sp.	Hydrotalcite/ alginate composite material	Increased plant length in just 14 days	Controlled bacterial delivery	Velázquez-Herrera et al. (2022b)
Rhizobium mayense	Ammonium sulfate	pH 7, 1% glucose at 30 °C	Increased chloroplast content	Shameem et al. (2023)
Trichoderma guizhouense	Inoculum of *Trichoderma guizhouense* NJAU4742	Interaction between inoculated species and indigenous soil microbes	Increased biomass of cabbage plant	Wang et al. (2023)
AMF	Rhizosphere rich soil	It decreased the availability of pathogenic fungi present in soil like *F. solani, Fusarium oxysporum, and Candidatus Solibacter* a bacterial species	Increased continuous-cropping of American ginseng by enhancing inoculation rate of AMF	Liu et al. (2020)
AMF	Monospecific isolates of AMF	Plant length ranged between 33 and 70%	Increased chloroplast and mineral content	Elmostapha et al. (2022)

(continued)

Table 2.1 (continued)

Microorganism	Substrate/ immobilized composites	Characteristics features	Impact	Reference
Saprophytic fungi	–	Saprophytic fungi stores novel mineral form of phosphorus which accumulates in mycelium	Solubilize phosphate residues	Ceci et al. (2018)
Bacillus sp.	–	Produce phytohormone-like auxins and convert N2 to ammonia for plant	Increase crop yield and quality	Hossain et al. (2022)
Endophytic bacteria	Canola plant	Require greenhouse conditions, sporulated rods	Well-formulated biofertilizers due to sporulating rod structure	Martínez-Hidalgo et al. (2021)

production of nitrogen, root, and shoot dry weight. Regression analysis revealed the relationship between root dry-weight with plant total nitrogen production (Hala 2020). Din et al. reported *Azotobacter* sp. (SR-4) a nitrogen-fixing bacteria via the Kjeldahl method and found that the bacteria increased the efficiency of nitrogen fixation as per gram of carbon was 35.08 mg of nitrogen was formed. The bacteria-based fertilizer considerably enhanced leaf length/width, fruit size, fruit yields, and plant height in comparison to the untreated plant. Integration of *Azotobacter* with *Azospirillum* can enhance the overall performance of microbial fertilizer (Din et al. 2019). Fungal microbes have remarkable phosphate solubilizing properties which can enrich the biofertilizers. Wang et al. tracked the growth of the cabbage plant and fungal community in the soil to understand the effect of *Trichoderma guizhouense* on the production of plants. The study suggested that Trichoderma enhanced the biomass of cabbage plants, and this increase was dependent on the changes in the composition of fungal communities consisting of relatively high and abundant indigenous fungal groups in the soil system (Wang et al. 2023). Likewise, Liu et al. improved the growth of American ginseng by AMF biofertilizers. The authors indicated that AMF fertilizer increased the continuous cropping of American ginseng by enhancing the inoculation rate of AMF by employing rhizosphere-beneficial fungi and bacteria. It decreased the availability of pathogenic fungi present in soil like *F. solani, Fusarium oxysporum, and Candidatus Solibacter* a bacterial species (Liu et al. 2020). Faridian et al. worked on different phosphate mobilizing fungi and determined that granular phosphate fertilizer included all forms of phosphate decomposing fungi which acted best at temperatures ranging between 28 and 40 °C thereby revealing optimization techniques to construct efficient phosphate biofertilizers (Faridian et al. 2023). To study endophytic microbes Singh et al. conducted research to determine the availability of endophytes in the Saccharum species to sustainably grow the crops of sugarcane. A maximum number of these microbes are present in the root and a minimum in the leaf. Different communities resided in different parts of the sugarcane (Singh et al. 2022). Fungal endophytes like *Porostereum spadiceum* AGH786 can produce gibberellins (bioactive GA3) to sustain the growth of soybeans in a salt-affected region. Endophytic fungus minimizes the salinity effect by regulating endogenous phytohormone of the seed along with this it enhances endogenous levels of genistein and daidzein in the seed of soybean to boost growth and NaCl-stress tolerance in soybean (Hamayun et al. 2017). Therefore, microbial biofertilizers are an effective source of promoting the production and quality rate of crops for commercial purposes.

2.2.3 Challenges Faced while Using Microbial Fertilizers

Microbial biofertilizers are quite potent as fertilizers but despite having so many advantages these fertilizers have certain drawbacks which need attention. Microbial fertilizers are prone to contamination by other microbial bodies during production, packaging, delivery, or strong processes. To ensure contamination-free fertilizer, improvements in technology and instrumentation are of utmost need. Also, more vial microbial species are needed to be used commercially to increase its large-scale production (Rai et al. 2023). Alternative or improved species must be studied and discovered to improve the applicability of microbial fertilizers. For example, Hilário et al. suggested endophytic Diaporthe as a better alternative to bacterial and fungal endophytes (Hilário and Gonçalves 2022). Undoubtedly microbial fertilizers have the potential to overcome the faults of chemical fertilizers but more advancement in certain areas of microbial fertilizer can develop much better biofertilizers.

2.3 Integration of Nanotechnology with Biofertilizers

2.3.1 Nanoparticles Employed to Develop Nanobiofertilizers

2.3.1.1 Gold Nanoparticles

Gold nanoparticles have several applications in biosensing, adsorption, and various biomedical and industrial purposes. One such application of gold nanoparticles is in developing biofertilizers. Panichikkal et al. synthesize gold nanoparticles (AuNPs) from *Bacillus subtilis* SJ15. The nanoparticles were triangular, hexanol, and spherical with sizes ranging between 12 and 32 nm. 50 μg/mL of AuNPs significantly enhanced the production of indole-3-acetic acid (phytohormone) by P. monteilii and showed increased growth at this value (Panichikkal et al. 2019). On the same line, Panichikkal et al. augmented chitosan with gold nanoparticles for the generation of indole-3-acetic acid by *Pseudomonas aeruginosa.* 100 μg/mL of AuNPs and 5 mg/mL of chitosan nanoparticles enhanced the production of phytohormones in plants, and increased shoot length and fresh weight (Panichikkal and Krishnankutty 2022). Thus AuNPs are preferentially reliable nanostructures with significant ability to promote plant growth.

2.3.1.2 Iron-Oxide Nanoparticles

Iron-oxide nanoparticles (FeNPs) are another nanoforms that can positively impact plant development when used in fertilizers. Gahoi et al. formulated a composite consisting of iron, carbon, and endospore of *Panebacillus polymyxa* augmented over activated carbon resins. The nanocomposite remarkably enhanced the root length, chlorophyll, protein content, and biomass of the host plant. The iron composite resisted root rot of wheat and Fusarium wilt of chickpea caused by *Cochliobolus sativus* and *Fusarium oxysporum f.* sp. *Cicero*, respectively (Gahoi et al. 2021). The combination of iron oxide with Rhizobium species greatly improved the germination rate of the seed, with enhanced dry weight and length of the host's organs in comparison to stand-alone entities. The chlorophyll content increased to 39.2 mg/g of fresh weight proline proportions decreased by 53% concerning control (Saleem and Khan 2023). Iron nanoparticles have a prominent impact on the chlorophyll content of the plant as well it enhance the length of the plant.

2.3.1.3 Zinc Nanoparticles

Zinc is one of the commonly used nanoparticles used in the formation of biofertilizers either stand-alone or in combination with another platform. Yasmin et al. integrated zinc oxide nanoparticles (ZnO/NPs) with the biological fertilizer to increase the tolerance of salt in safflower. The ZnO/NPs enhanced the salinity tolerance by increasing the antioxidant function of enzymes and minimizing the concentrations of malondialdehyde and proline (Yasmin et al. 2021). Integrated units of zinc oxide and iron oxide (ZnO-FeO) reduced electrical conductivity to 17%, increased peroxidase activity to 27%, and increased polyphenol oxidase to 66% under scarcity of water. ZnO-FeO in combination with Azotobacter increased the growth yield by 88% (Seyed Sharifi et al. 2020). Thus zinc nanoparticles can be embedded in nanoforms to fabricate the ultimate nano-unit which is tolerant to extreme conditions of salt and water.

2.3.1.4 Nanocomposites

Nanocomposites are the nanoplatform comprising two or more units combined to develop a functional unit. Various nanocomposites have been employed to produce efficient biofertilizers. Masouleh et al. prepared magnetic nanocomposites made from chitosan polymer. The magnetic unit was prepared through chemical and physical processes to reduce the generation rate cyclamen. The use of biofertilizers did not cause any abnormalities, by accelerated flowering despite genotypic changes. 30 mg L^{-1} nanochitosan was the best concentration to enhance the yield of flowering (Shafiei-Masouleh 2022). Immobilization of bacterial species on Layered double hydroxide (LDH)/SBS – 15 can act as biofertilizers to enhance the growth and development of lettuce. The formed nanocomposites exhibit spongy agglomerates with sizes ranging between 0.5 and 2 μm and had a specific surface area value of 375 m^2/g. It was an efficient nanocomposite as it was able to increase the length of the plant in just 14 days (Velázquez-Herrera et al. 2022a). The acylated homoserine lactones (AHL) present in bacteria can encapsulate a nanocomposite within itself for sustained and precise delivery of the nanoparticles at the target site. The zeta potential on nanofiber (negative) is opposite to that on AHL thereby facilitating the binding of two components (Gupta et al. 2019). Thus, nanocomposites are better-performing systems with the capability to be used at large scale.

2.3.2 Effect of Nanobiofertilizers on the Soil System and Agriculture

Nanotechnology is the new emerging trend in the field of science be it any part, biomedical, industrial, technology, or agricultural. In the area of agriculture, one such desirable application of nanoparticles is in the formulation of biofertilizers (Table 2.2). As many nanoparticles can promote the growth and development of plants; however, applicability can be specific to one host or at times can be broad-spectrum utility (Mary Isabella Sonali et al. 2022). Researchers around the world have been conducting different experiments to estimate the effect of nanoparticles on plant growth and yield. For instance, Jafararid et al. studied the impact of bio-based manganese nanoparticles on growth, physiological response, and phytochemical classification of a plant *Hypericum perforatum* L. (shoot culture). The outcome of the study showed that manganese nanocomposite enhanced hypericin and pseudo-hypericin while reducing terpenoids (Jafarirad et al. 2021). Stand-alone availability of selenium (Se) nanoparticles in the soil can have progressive effects in comparison to conventional biofertilizers as the addition of nanoparticles brings about change in the microbial communities. Se nanoparticles deposited 8.8% nitrogen fixation, 44.9% nitrite reducers, and 20.4% streptomycin producers thereby increasing the growth of the host plant (Sonali et al. 2023). Additionally, algal nanoparticles are a good source for developing excellent biofertilizers as they exhibit the property to retain phosphorus and nitrogen in the soil, preventing the run-off of the essential nutrients (Joshi et al. 2019). As a result, nanotechnology is an effective technique to commercially use biofertilizers without disturbing the immediate environmental resources.

2.3.3 Limitations of Nanobiofertilizers

Nanofertilizers have various advantages as it had high bioavailability, positive impact on microbial communities, and high delivery efficiency. Despite these attributes, nanofertilizers have certain drawbacks. The cost of production for nanofertilizers is quite high as nanotechnology in itself is an expensive technology. Also, aggregation of nanoparticles is a major issue that can inhibit the application of the nano unit (Sharma et al. 2023b). Nanofertilizers can be toxic and the toxicity levels must be ensured before the application. It can be highly reactive owing to its small size or phototoxic for certain species (Zulfiqar et al. 2019). So pre-laboratory evaluations are important for safe utilization of nanofertilizers.

Table 2.2 Brief discussion about the nanofertilizers

Nanostructure	Synthesis	Characteristics of nanoparticles	Impact on plant	Reference
Gold nanoparticles	*Bacillus subtilis* SJ15	Nanoparticles were triangular, hexanol, and spherical with sizes ranging between 12 and 32 nm	50 µg/ML of AuNPs significantly enhanced the production of indole-3-acetic acid (IAA)	Panichikkal et al. (2019)
Gold nanoparticles and chitosan nanoparticles	Biological	–	100 µg/ML of AuNPs and 5 mg/mL of chitosan nanoparticles produced IAA	Panichikkal and Krishnankutty (2022)
Iron-carbon-endospore	Suspension polymerization	Porous carbon = 10–50 µm size The nanoparticles appeared as bright spots	Enhanced the root length, chlorophyll, protein content, and biomass of the host plant	Gahoi et al. (2021)
Iron-oxide	*Syzigium cumini*	Crystalline structure, surface plasmon resonance = 300–700 nm	The chlorophyll content increased to 39.2 mg/g of fresh weight proline proportions decreased by 53%	Saleem and Khan (2023)
Zinc-oxide	*Carica papaya*	17 mg/mL concentration of ZnO/NPs to	Increased antioxidant function of enzymes and minimize concentrations of malondialdehyde and proline	Yasmin et al. (2021)
Zn-Fe oxide	–	Size = 10–30 nm, specific surface area = 30 m^2/grain	Reduced electrical conductivity = 17%, increased peroxidase activity = 27% and increased polyphenol oxidase = 66% under scarcity of water	Seyed Sharifi et al. (2020)
Magnetic nanoparticles	Chitosan	Size ≤ 20 nm	30 mg L^{-1} nanochitosan was the best concentration to enhance the yield of flowering	Shafiei-Masouleh (2022)

(continued)

Table 2.2 (continued)

Nanostructure	Synthesis	Characteristics of nanoparticles	Impact on plant	Reference
LDH/SBA-15 nanocomposite	*Streptomyces* sp.	Size = 0.5–2 μm Specific surface area = 375 m^2/g	Increased the length of the plant in just 14 days	Velázquez-Herrera et al. (2022a)
MnO_2	*Hypericum perforatum*	Size = 20–50 nm	Enhanced hypericin and pseudo-hypericin while reducing terpenoids.	Jafarirad et al. (2021)
Selenium nanoparticles	*Pseudomonas libanesis*	Size = 220–330 nm	8.8% of nitrogen fixation, 44.9% nitrite reducers	Sonali et al. (2023)
Cesium oxide	*Pseudomonas azotoformans and Pseudomonas gessardi*	Size = 25.24 nm.	75 ppm of nanocomposite promoted shoot and root length	Mary Isabella Sonali et al. (2022)
Manganese and iron	Bacterial supernatant	Agglomerated spherical structures Size = 250–300 nm	Enhanced germination rate and root growth	de França Bettencourt et al. (2020)

2.4 Liquid Biofertilizers and its Role in Plant Growth

Liquid fertilizers are widely utilized as UAN, i.e., mixture of urea and ammonium nitrate and can be in any form (phosphorus, nitrogen, and potassium) (Fig. 2.3). It is composed of anhydrous ammonia (82mass% N), superphosphoric acid (0-70-0 grade), and ammonia-reacted superphosphoric acid to form liquid fertilizer (11-37-0 grade or 10-34-0 grade) (Falls and Siegel 2005). Several scientists have worked with liquid fertilizers to understand their application in crop development and production. Ansari et al. used liquid fertilizers involving phosphate solubilizer or Azotobacter species on *Cicer aeritinum* (chickpea) and found that though the yield of the crop increased by the addition of biofertilizer the quality of the product was not good as the required protein content in chickpea was not available (Ansari et al. 2015). Microbial alga like *Spirulina platensis* and *Chlorella vulgaris* when incorporated with liquid fertilizer increased the growth rate of root length, root nodules, plant leaves, plant height, leaf zone index, and weight as the microalgal species exhibited maximum elemental contents and chemical compositions (Dineshkumar et al. 2022). Kumar et al. added diazotrophic bacteria in the formulation of liquid biofertilizer and proposed that compost tea a liquid carrier can be used in the production of aqueous biofertilizer with excellent properties to generate ammonia and IAA (Kumar et al. 2019). Further fish scales can be an organic source of liquid biofertilizer. Bhuimbar et al. produced liquid biofertilizer from fish scales which revealed higher concentrations of flavonoid, chlorophyll, and free proteins (Bhuimbar and Dandge 2023). Another sustainable source of liquid fertilizer is

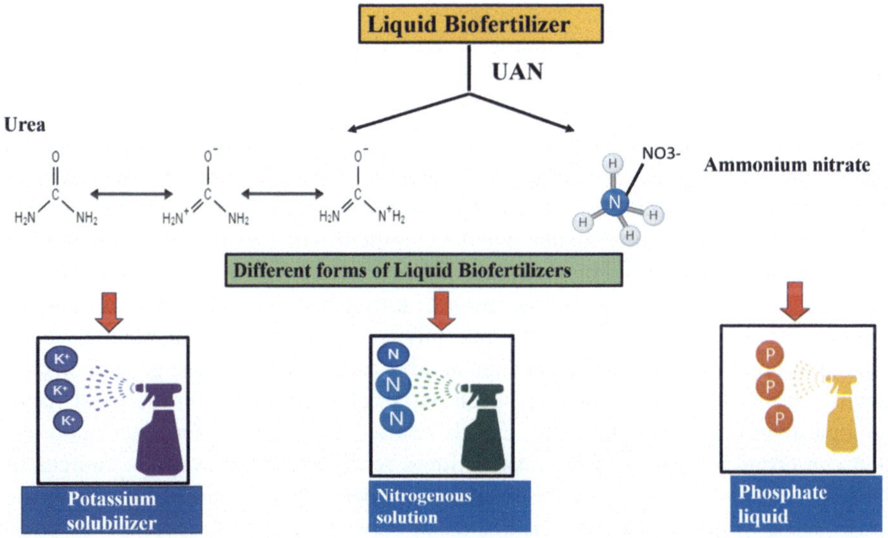

Fig. 2.3 Liquid biofertilizers are widely utilized as UAN, i.e., mixture of urea and ammonium nitrate. The various used liquid nitrogenous are potassium solubilizers, nitrogenous solutions, and phosphate liquid

dairy wastewater which has a relatively high content of phosphate (36.44%), reduced nitrate (70.77%), and phosphate (36.44%) (Gogoi et al. 2022). Thus, liquid fertilizers are a very good source of biofertilizer provided the fact that the initial source of it and immobilization of other composites increases its overall applicability.

2.5 A Forthcoming Scenario of the Biofertilizer Industry

Biofertilizer is a blooming industry along with being an integral part of promoting bioeconomy. The upcoming prospects for biofertilizers are strong and can revolutionize the agricultural sector (Amoo et al. 2021; Barman et al. 2019; Yong et al. 2021). Considering the microbial biofertilizers new bioavailable strains can be further discovered which can be used for the large-scale production of biological fertilizers. The presently available microbial strains can also be explored to increase the efficiency of these strains to fix nitrogen or other growth hormones. Nanotechnology is one of the most recent technologies which can bring ground-breaking improvement in sustainable agriculture. Most of the nanoparticles either stand-alone or in combination with another composite can result in the growth of roots, shoots, leaves, and overall development of a plant. Focus can be made on developing more stable nanoparticles which might retain their functionality for a longer time. Liquid fertilizers too are convenient biofertilizers however act more effectively in the presence of certain substrates and can be explored more to find more compatible binding composites for better outcomes. Therefore, more solution-focused research in the field of biofertilizers and bring about desired changes for commercial purposes.

2.6 Conclusion

In a nutshell, it can be said that biofertilizers are a good source of nutrition and growth factors to achieve the complete development of crops. Not only the biofertilizers helps to meet the nutrition need of the people but also maintains the ecological balance thereby promoting sustainable development. The innovations in the field of microbial, nano, and liquid biofertilizers have brought excellent changes in the area of fertilizers. Microbial fertilizers consist of nitrogen-fixing bacteria (symbiotic or non-symbiotic), fungi (AMF), and endophytic organisms (within plant tissues). Each microbial species is different from another and exhibits unique growth mechanisms out of which nitrogen-fixing bacteria are efficient ones. On the other hand, nanoparticles and nanocomposites are also quite efficacious biofertilizer materials that can promote plant growth. Lastly, liquid fertilizers are convenient sources of biofertilizers though they need substrates for better performance. However, the drawback demonstrated by biofertilizers can be worked upon to increase the commercial, large-scale applicability of biofertilizers which can be offered by even a common person.

References

Aasfar A et al (2021) Nitrogen fixing azotobacter species as potential soil biological enhancers for crop nutrition and yield stability. Front Microbiol 12(February):1–19. https://doi.org/10.3389/fmicb.2021.628379

Amoo AE et al (2021) In: Babalola OO (ed) Biofertilizer: an eco-friendly approach for sustainable crop production BT—food security and safety : African perspectives. Springer International Publishing, Cham, pp 647–669. https://doi.org/10.1007/978-3-030-50672-8_32

Anli M et al (2020) Biofertilizers as strategies to improve photosynthetic apparatus, growth, and drought stress tolerance in the date palm. Front Plant Sci 11. https://doi.org/10.3389/fpls.2020.516818

Ansari MF, Tipre DR, Dave SR (2015) Efficiency evaluation of commercial liquid biofertilizers for growth of Cicer aeritinum (chickpea) in pot and field study. Biocatal Agric Biotechnol 4(1):17–24. https://doi.org/10.1016/j.bcab.2014.09.010

Barman S, Das S, Bhattacharya SS (2019) In: Singh Jay Shankar, Singh DPBT-N, F. D., M. B. and B (eds) Chapter 17—The prospects of bio-fertilizer technology for productive and sustainable agricultural growth. Elsevier, pp 233–253.

Barzee TJ et al (2019) Digestate biofertilizers support similar or higher tomato yields and quality than mineral fertilizer in a subsurface drip fertigation system. Front Sustain Food Syst 3. https://doi.org/10.3389/fsufs.2019.00058

Bhattacharyya C et al (2020) Chapter 11—Biofertilizers as substitute to commercial agrochemicals. In: Prasad MNV (ed) Agrochemicals detection, treatment and remediation. Butterworth-Heinemann, pp 263–290. https://doi.org/10.1016/B978-0-08-103017-2.00011-8

Bhuimbar MV, Dandge PB (2023) Production of organic liquid biofertilizer from fish waste and study of its plant growth promoting effect. Proc Natl Acad Sci India Section B Biol Sci 93(1):235–243. https://doi.org/10.1007/s40011-022-01413-8

Ceci A et al (2018) Saprotrophic soil fungi to improve phosphorus solubilisation and release: in vitro abilities of several species. Ambio 47(s1):30–40. https://doi.org/10.1007/s13280-017-0972-0

Dal Cortivo C et al (2020) Effects of seed-applied biofertilizers on rhizosphere biodiversity and growth of common wheat (Triticum aestivum L.) in the field. Front Plant Sci 11. https://doi.org/10.3389/fpls.2020.00072

Dasgupta D et al (2021) In: Surajit De Mandal, Ajit Kumar Passari, M. B (eds) Chapter 1—Microbial biofertilizers: recent trends and future outlook. Academic Press, pp 1–26. https://doi.org/10.1016/B978-0-12-822098-6.00001-X

de França Bettencourt GM et al (2020) Green biosynthesis of single and bimetallic nanoparticles of iron and manganese using bacterial auxin complex to act as plant bio-fertilizer. Biocatal Agric Biotechnol 30:101822. https://doi.org/10.1016/j.bcab.2020.101822

Din M et al (2019) Production of nitrogen fixing Azotobacter (SR-4) and phosphorus solubilizing Aspergillus niger and their evaluation on Lagenaria siceraria and Abelmoschus esculentus. Biotechnol Rep 22:e00323. https://doi.org/10.1016/j.btre.2019.e00323

Dineshkumar R et al (2022) Microalgal liquid biofertilizer and biostimulant effect on green gram (Vigna radiata L) an experimental cultivation. Biomass Convers Biorefinery 12(8):3007–3027. https://doi.org/10.1007/s13399-020-00857-0

Dong L et al (2019) Biofertilizers regulate the soil microbial community and enhance Panax ginseng yields. Chin Med 14(1):20. https://doi.org/10.1186/s13020-019-0241-1

Du T-Y et al (2022) Positive effects of organic fertilizers and biofertilizers on soil microbial community composition and walnut yield. Appl Soil Ecol 175:104457. https://doi.org/10.1016/j.apsoil.2022.104457

Elmostapha O et al (2022) Application of arbuscular mycorrhizal fungi isolates from semi-arid Mediterranean ecosystems as biofertilizers in Argan tree development. J Soil Sci Plant Nutr 22(1):944–955. https://doi.org/10.1007/s42729-021-00704-9

Falls JH, Siegel SA (2005) Fertilizers. In: Worsfold P, Townshend A, Poole CBT-E, of A. S. (Second E.) (eds) Encyclopedia of analytical science. Elsevier, Oxford, pp 1–8. https://doi.org/10.1016/B0-12-369397-7/00150-3

Faridian L et al (2023) An exploratory research on the adoption of different phosphate-solubilizing fungi for production of phosphate biofertilizers. Geomicrobiol J 40(5):493–500. https://doi.org/10.1080/01490451.2023.2196288

Gahoi P et al (2021) Rhizobacteria and acylated homoserine lactone-based nanobiofertilizer to improve growth and pathogen defense in Cicer arietinum and Triticum aestivum plants. ACS Agric Sci Technol 1(3):240–252. https://doi.org/10.1021/acsagscitech.1c00039

Gogoi M et al (2022) Microbial bioconversion of dairy wastewater in packed bed biofilm reactor into liquid biofertilizer. Geomicrobiol J 39(3–5):249–258. https://doi.org/10.1080/01490451.2021.1980921

Gupta GS, Kumar A, Verma N (2019) Bacterial homoserine lactones as a nanocomposite fertilizer and defense regulator for chickpeas. Environ Sci Nano 6(4):1246–1258. https://doi.org/10.1039/C9EN00199A

Hala Y (2020) The effect of nitrogen-fixing bacteria towards upland rice plant growth and nitrogen content. IOP Conf Ser Earth Environ Sci 484:1. https://doi.org/10.1088/1755-1315/484/1/012086

Hamayun M et al (2017) Gibberellins producing endophytic fungus Porostereum spadiceum AGH786 rescues growth of salt affected soybean. Front Microbiol 8(APR):1–13. https://doi.org/10.3389/fmicb.2017.00686

Hilário S, Gonçalves MFM (2022) Endophytic diaporthe as promising leads for the development of biopesticides and biofertilizers for a sustainable agriculture. Microorganisms 10(12):1–21. https://doi.org/10.3390/microorganisms10122453

Hossain GMA et al (2022) In: Islam MT, Rahman M, Pandey P (eds) Enhanced nutrient accumulation in non-leguminous crop plants by the application of endophytic bacteria bacillus species BT—Bacilli in agrobiotechnology: plant stress tolerance, bioremediation, and bioprospecting. Springer International Publishing, Cham, pp 349–364. https://doi.org/10.1007/978-3-030-85465-2_16

Jafarirad S et al (2021) Impacts of manganese bio-based nanocomposites on phytochemical classification, growth and physiological responses of Hypericum perforatum L. shoot cultures. Ecotoxicol Environ Saf 209:111841. https://doi.org/10.1016/j.ecoenv.2020.111841

Jha PN et al (2020) Alterations in the endophyte-enriched root-associated microbiome of rice receiving growth-promoting treatments of urea fertilizer and rhizobium biofertilizer. Microb Ecol 79(2):367–382. https://doi.org/10.1007/s00248-019-01406-7

Joshi M et al (2019) In: Pudake RN, Chauhan N, Kole C (eds) Applications of algal nanoparticles in agriculture BT—nanoscience for sustainable agriculture. Springer International Publishing, Cham, pp 265–280. https://doi.org/10.1007/978-3-319-97852-9_12

Kaur P, Purewal SS (2019) In: Giri B et al (eds) Biofertilizers and their role in sustainable agriculture BT—biofertilizers for sustainable agriculture and environment. Springer International Publishing, Cham, pp 285–300. https://doi.org/10.1007/978-3-030-18933-4_12

Kumar A et al (2019) Stenotrophomonas: a versatile diazotrophic bacteria from the rhizospheric soils of Western Himalayas and development of its liquid biofertilizer formulation. Vegetos 32(1):103–109. https://doi.org/10.1007/s42535-019-00013-8

Li Y, Pan F, Yao H (2019) Response of symbiotic and asymbiotic nitrogen-fixing microorganisms to nitrogen fertilizer application. J Soils Sediments 19(4):1948–1958. https://doi.org/10.1007/s11368-018-2192-z

Liu N et al (2020) Arbuscular mycorrhizal fungi biofertilizer improves American ginseng (Panax quinquefolius L.) growth under the continuous cropping regime. Geoderma 363(December 2019):114155. https://doi.org/10.1016/j.geoderma.2019.114155

Ma Y, Shen Y, Liu Y (2020) Food waste to biofertilizer: a potential game changer of global circular agricultural economy. J Agric Food Chem 68(18):5021–5023. https://doi.org/10.1021/acs.jafc.0c02210

Mącik M, Gryta A, Frąc M (2020) In: Sparks, D. L. B. T.-A. in A (ed) Chapter two—biofertilizers in agriculture: an overview on concepts, strategies and effects on soil microorganisms. Advances in agronomy, 162: pp 31–87. https://doi.org/10.1016/bs.agron.2020.02.001
Mahanty T et al (2017) Biofertilizers: a potential approach for sustainable agriculture development. Environ Sci Pollut Res 24(4):3315–3335. https://doi.org/10.1007/s11356-016-8104-0
Mahapatra DM, Satapathy KC, Panda B (2022) Biofertilizers and nanofertilizers for sustainable agriculture: phycoprospects and challenges. Sci Total Environ 803:149990. https://doi.org/10.1016/j.scitotenv.2021.149990
Mahmud AA et al (2021) Biofertilizers: a Nexus between soil fertility and crop productivity under abiotic stress. Curr Res Environ Sustain 3:100063. https://doi.org/10.1016/j.crsust.2021.100063
Malusà E, Pinzari F, Canfora L (2016) In: Singh DP, Singh HB, Prabha R (eds) Efficacy of biofertilizers: challenges to improve crop production BT—microbial inoculants in sustainable agricultural productivity: Vol. 2: functional applications. Springer India, New Delhi, pp 17–40. https://doi.org/10.1007/978-81-322-2644-4_2
Martínez-Hidalgo P et al (2021) Identification of canola roots endophytic bacteria and analysis of their potential as biofertilizers for canola crops with special emphasis on sporulating bacteria. Agronomy 11:9. https://doi.org/10.3390/agronomy11091796
Mary Isabella Sonali J et al (2022) Application of a novel nanocomposite containing micronutrient solubilizing bacterial strains and CeO2 nanocomposite as bio-fertilizer. Chemosphere 286:131800. https://doi.org/10.1016/j.chemosphere.2021.131800
Masuda Y et al (2021) Enhancement of the nitrogen-fixing activity of paddy soils owing to iron application. Soil Sci Plant Nutr 67(3):243–247. https://doi.org/10.1080/00380768.2021.188862 9
Negi YK et al (2021) Enhancement in yield and nutritive qualities of strawberry fruits by the application of organic manures and biofertilizers. Sci Hortic 283:110038. https://doi.org/10.1016/j.scienta.2021.110038
Nosheen S, Ajmal I, Song Y (2021) Microbes as biofertilizers, a potential approach for sustainable crop production. Sustain For 13:4. https://doi.org/10.3390/su13041868
Odoh CK et al (2020) In: Yadav AN et al (eds) Fungal biofertilizers for sustainable agricultural productivity BT—agriculturally important fungi for sustainable agriculture: volume 1: perspective for diversity and crop productivity. Springer International Publishing, Cham, pp 199–225. https://doi.org/10.1007/978-3-030-45971-0_9
Panichikkal J, Krishnankutty RE (2022) Chitosan and gold nanoparticles supplementation for augmentation of Indole-3-acetic acid production by Rhizospheric Pseudomonas aeruginosa and plant growth enhancement. Curr Microbiol 79(6):185. https://doi.org/10.1007/s00284-022-02850-4
Panichikkal J et al (2019) Biogenic gold nanoparticle supplementation to plant beneficial pseudomonas monteilii was found to enhance its plant probiotic effect. Curr Microbiol 76(4):503–509. https://doi.org/10.1007/s00284-019-01649-0
Pirttilä AM et al (2021) Biofertilizers and biocontrol agents for agriculture: how to identify and develop new potent microbial strains and traits. Microorganisms 9:4. https://doi.org/10.3390/microorganisms9040817
Puglia D et al (2021) The opportunity of valorizing agricultural waste, through its conversion into biostimulants, biofertilizers, and biopolymers. Sustain For 13:5. https://doi.org/10.3390/su13052710
Rai PK et al (2023) Limitations of biofertilizers and their revitalization through nanotechnology. J Clean Prod 418:138194. https://doi.org/10.1016/j.jclepro.2023.138194
Reddy GC et al (2020) In: Varma A, Tripathi S, Prasad R (eds) Biofertilizers toward sustainable agricultural development BT - plant microbe symbiosis. Springer International Publishing, Cham, pp 115–128. https://doi.org/10.1007/978-3-030-36248-5_7
Riaz U et al (2021) In: Hakeem KR et al (eds) Plant growth-promoting rhizobacteria (PGPR) as biofertilizers and biopesticides BT—microbiota and biofertilizers: a sustainable continuum for plant and soil health. Springer International Publishing, Cham, pp 181–196. https://doi.org/10.1007/978-3-030-48771-3_11

Saleem S, Khan MS (2023) Phyto-interactive impact of green synthesized iron oxide nanoparticles and rhizobium pusense on morpho-physiological and yield components of greengram. Plant Physiol Biochem 194:146–160. https://doi.org/10.1016/j.plaphy.2022.11.013

Sambangi P et al (2022) Nano-biofertilizers on soil health, chemistry, and microbial community: benefits and risks. Proc Indian Natl Sci Acad 88(3):357–368. https://doi.org/10.1007/s43538-022-00094-1

Seenivasagan R, Babalola OO (2021) Utilization of microbial consortia as biofertilizers and biopesticides for the production of feasible agricultural product. Biology 10:11. https://doi.org/10.3390/biology10111111

Seyed Sharifi R et al (2020) Effects of biofertilizers and nano zinc-iron oxide on yield and physicochemical properties of wheat under water deficit conditions. Commun Soil Sci Plant Anal 51(19):2511–2524. https://doi.org/10.1080/00103624.2020.1845350

Shafiei-Masouleh S-S (2022) Use of magnetic Nano-chitosan as bio-fertilizer to reduce production period in three cyclamen cultivars. J Soil Sci Plant Nutr 22(1):281–293. https://doi.org/10.1007/s42729-021-00647-1

Shameem MR et al (2023) Rhizobium mayense sp. Nov., an efficient plant growth-promoting nitrogen-fixing bacteria isolated from rhizosphere soil. Environ Res 220:115200. https://doi.org/10.1016/j.envres.2022.115200

Sharma A et al (2023a) Consolidated bioprocessing of biodegradable municipal solid waste for transformation into biofertilizer formulations. Biomass Convers Biorefinery. https://doi.org/10.1007/s13399-023-04110-2

Sharma B et al (2023b) Nano-biofertilizers as bio-emerging strategies for sustainable agriculture development: potentiality and their limitations. Sci Total Environ 860:160476. https://doi.org/10.1016/j.scitotenv.2022.160476

Singh D et al (2021) In: Rakshit A et al (eds) Chapter 12—biofertilizers: mechanisms and application. Woodhead Publishing, pp 151–166. https://doi.org/10.1016/B978-0-12-821667-5.00024-5

Singh RK et al (2022) Unraveling nitrogen fixing potential of endophytic diazotrophs of different Saccharum species for sustainable sugarcane growth. Int J Mol Sci 23(11). https://doi.org/10.3390/ijms23116242

Sonali JMI et al (2023) Effect of biogenic selenium nanocomposite on okra and metagenomic analysis of bacterial community associated with rhizosphere and phyllosphere. Waste Biomass Valorization. https://doi.org/10.1007/s12649-023-02215-3

Thomas L, Singh I (2019) Microbial biofertilizers: types and applications:1–19. https://doi.org/10.1007/978-3-030-18933-4_1

Velázquez-Herrera FD, Lobo-Sánchez M, Fetter G (2022a) LDH/SBA-15 nanocomposite containing nitrogen-fixing bacteria as an efficient biofertilizer. Mater Today Commun 31:103832. https://doi.org/10.1016/j.mtcomm.2022.103832

Velázquez-Herrera FD et al (2022b) Novel bio-fertilizer based on nitrogen-fixing bacterium immobilized in a hydrotalcite/alginate composite material. Environ Sci Pollut Res 29(21):32220–32226. https://doi.org/10.1007/s11356-021-17943-z

Wang Y et al (2023) Biodiversity of the beneficial soil-borne fungi steered by Trichoderma-amended biofertilizers stimulates plant production. npj Biofilms Microbiomes 9(1):1–9. https://doi.org/10.1038/s41522-023-00416-1

Yasmin H et al (2021) Combined application of zinc oxide nanoparticles and biofertilizer to induce salt resistance in safflower by regulating ion homeostasis and antioxidant defence responses. Ecotoxicol Environ Saf 218:112262. https://doi.org/10.1016/j.ecoenv.2021.112262

Yin M et al (2020) Herbicidal efficacy of harzianums produced by the biofertilizer fungus, Trichoderma brevicompactum. AMB Express 10(1):4–11. https://doi.org/10.1186/s13568-020-01055-x

Yong JJJY et al (2021) Prospects and development of algal-bacterial biotechnology in environmental management and protection. Biotechnol Adv 47:107684. https://doi.org/10.1016/j.biotechadv.2020.107684

Zainuddin N et al (2022) Effect of integrated biofertilizers with chemical fertilizers on the oil palm growth and soil microbial diversity. Biocatal Agric Biotechnol 39:102237. https://doi.org/10.1016/j.bcab.2021.102237

Zulfiqar F et al (2019) Nanofertilizer use for sustainable agriculture: advantages and limitations. Plant Sci 289. https://doi.org/10.1016/j.plantsci.2019.110270

Metabolomics: A Paradigm Shift in Understanding Biofertilizers Dynamics

3

Nitu Rani and Narashans Alok Sagar

Abstract

This chapter delves into the crucial role of biofertilizers in sustainable agriculture and highlights the transformative potential of metabolomics in unraveling intricate biochemical processes. Biofertilizers, comprised of beneficial microorganisms, address global concerns surrounding food security and environmental sustainability. These formulations actively contribute to soil fertility, plant growth, and resilience through symbiotic interactions with host plants. Various biofertilizer formulations and associated challenges are discussed, emphasizing the emergence of metabolomics as a potent tool for decoding plant–microbe interactions. Metabolomics allows the identification of unique metabolic signatures, elucidation of plant response mechanisms, detection of biochemical signaling, assessment of nutrient uptake, and evaluation of stress responses. The chapter explores techniques and methodologies in metabolomics, providing a roadmap for its application in unraveling the dynamic interactions between plants and plant growth-promoting rhizobacteria (PGPR). Additionally, the significance of rhizosphere metabolomics in understanding biofertilizer-PGPR dynamics is discussed. Looking ahead, future prospects in metabolomics research involve advancements in analytical methods, integration with omics technologies, the utilization of bioinformatics tools, and an expanded focus on ecological and environmental impacts. This evolving field of research is poised to optimize

N. Rani (✉)
University Institute of Biotechnology, Chandigarh University, Mohali, Punjab, India
e-mail: nitu.agri@cumail.in

N. A. Sagar
University Institute of Biotechnology, Chandigarh University, Mohali, Punjab, India

University Centre for Research and Development, Chandigarh University, Mohali, Punjab, India

S. Kaur et al. (eds.), *Metabolomics, Proteomics and Gene Editing Approaches in Biofertilizer Industry*, https://doi.org/10.1007/978-981-97-2910-4_3

agricultural practices, contribute to the preparation of tailored biofertilizers, and support the establishment of resilient and sustainable food production systems globally.

Keywords

Metabolomics · PGPR · Biofertilizer · Bioformulations

3.1 Introduction

In contemporary agriculture, there is a notable movement toward sustainable practices to tackle the issues of global food security and environmental responsibility. Biofertilizers, heralded as eco-friendly alternatives to conventional fertilizers, play a prime role in this transformative journey (Singh et al. 2016; Rani et al. 2023). Comprising useful microbes or their metabolic by-products, biofertilizers stand out as promising solutions. These bio-based formulations incorporate either live or latent cells of microorganisms, offering a multifaceted approach to improving soil fertility. Beyond this, they actively contribute to enhancing plant growth and resilience through processes such as potassium and phosphate solubilization, nitrogen fixation, zinc mobilization, and the production of hydrogen cyanide (HCN) and siderophores (Etesami and Adl 2020; Rani et al. 2022).

Functioning by colonizing the rhizospheric roots of host plants, biofertilizers establish a symbiotic relationship that goes beyond mere nutrient enrichment (Uzoh and Babalola 2018; Sati et al. 2020). This intricate interaction not only nurtures the soil ecosystem but also fosters a harmonious coexistence between agriculture and the environment. As we tackle the challenges of today's agriculture, the judicious adoption of biofertilizers emerges as a beacon toward a sustainable and resilient future for our global food systems. Functioning by colonizing the rhizospheric roots of host plants, several plant growth-promoting rhizobacteria, from genera such as *Rhizobium*, *Azospirillum*, *Bacillus*, and *Pseudomonas*, have emerged as possible and potential biofertilizers (Verma et al. 2019; Nitu et al., 2020).

Biofertilizers are prepared using a variety of formulation techniques, such as liquid formulations, metabolite-based formulations, solid carrier-based formulations (both inorganic and organic), and synthetic formulations (Kalita et al. 2015; Rani et al. 2021). The formulation process often involves the application of the fed-batch cultivation technique, utilizing high-density cultures of microbial species, such as *Pseudomonas* spp. (Mutturi et al. 2017). Microbial inoculants, including *Frankia*, *Azospirillum*, *Pseudomonas*, and *Enterobacter*, are cultivated through scale-up formulations on liquid or solid nutrient mediums (Saif et al. 2021; Rani et al., 2020). To aid in colonization, these microbes produce various inorganic and organic chemicals, including citric, tartaric, lactic, gluconic, oxalic, acetic, and 2-keto-gluconic acids (Sharma et al. 2020a, b; Rani et al. 2023). Recent findings highlight the impact of volatile compounds released by microbial communities, influencing plant growth (Fincheira and Quiroz 2018). Advanced methods such as genetic engineering, gene editing, etc. are being increasingly employed to understand microbial interactions. Additionally, in silico methods are being utilized to analyze such complex interactions (Kumar et al. 2016; Choudhary et al. 2018).

Metabolomics, an innovative analytical approach, has emerged as a transformative tool in unraveling the complexities of cellular processes. The standardization of extraction and analytical techniques has propelled this advancement, with crucial roles of gas chromatography (GC), electronic ionization, and mass spectrometry (MS), alongside compound annotation in databases. In the realm of non-targeted metabolomics, common approaches for pattern recognition include the utilization of methodologies like independent components analysis and principal components analysis (Eshawu and Ghalsasi 2023). It is worth noting that metabolites provide a more accurate representation of the phenotype when compared to mRNA transcripts or proteins. This reflects the amalgamation of gene expression, protein interaction, and other regulatory processes (Teh et al. 2017). Recent advancements in the field underscore the growing significance of metabolic engineering and computer modeling in predicting genetic manipulations within metabolic networks. Additionally, contemporary biotechnological tools, such as OptKnock, constraint-based modeling (CBM), and biological network modeling have become indispensable for forecasting the capabilities of bioinoculants (Benckiser et al. 2019).

Interestingly, the integration of metabolomics in the study of biofertilizers allows for a holistic understanding of the biochemical changes occurring in both plant hosts and the microbial agents within the formulations (Chaudhary and Shukla 2019). Recent advancements in metabolomics present an opportunity to optimize biofertilizer formulations by deciphering the intricate signaling pathways and metabolic crosstalk between plants and beneficial microbes (Chaudhary and Shukla 2019; Sharma et al. 2020a, b).

This chapter explores biofertilizers' vital role in sustainable agriculture. Integrating metabolomics unravels complex biochemical interactions, offering insights into optimizing formulations and decoding signaling pathways. This transformative approach, emphasizing biofertilizers, holds promise for shaping a resilient and sustainable global food system.

3.2 Biofertilizers and Challenges in Understanding its Dynamics

The interpretation of the term "biofertilizer" is diverse, encompassing a range of meanings from seaweed extracts to microbial mixtures, as highlighted by El-Ghamry et al. (2018). Scientific definitions also vary, covering a spectrum from green manures to plant extracts (Vessey 2003). The concept has evolved over time, reflecting an increased understanding of soil microorganism-plant associations. Initially, it referred to living microorganisms colonizing the rhizosphere or plant interior, contributing to growth by improving nutrient supply. Fuentes-Ramírez and Caballero-Mellado (2006) expanded the definition to include living microorganisms that exert beneficial effects on the growth of the plant via various mechanisms, including the suppression of phytopathogens.

Scientifically, a biofertilizer is a single microbe that encourages the development of plants. In agronomic contexts, it refers to a product composed of beneficial strains in a carrier with features for storage and effective application to soil or plants, potentially enabling the addition of substances enhancing microorganism activity (Mącik et al. 2020). Biofertilizer application has a pivotal role in promoting the development of the plant without environmental harm (Mishra and Arora 2016; Umesha et al. 2018). Various biofertilizer types are compromising of PGP microbes either alone or in consortia promoting the overall growth of the crop through different direct (production of phytohormones, siderophores) and indirect mechanisms (secretion of enzymes, ammonia and HCN) as demonstrated in Fig. 3.1. Previous studies have indicated increased crop yields with biofertilizer inoculation. These formulations contribute to soil fertility, improve structure, and offer protection against drought, diseases, and pests (Mishra and Arora 2016; Umesha et al. 2018).

While biofertilizers hold promise in sustainable agriculture, several challenges persist, including the selection of appropriate microbes, limited shelf life, susceptibility to high temperatures, and complexities in storage and transportation. Continuous advancements in technology are underway to tackle these challenges, aiming to establish biofertilizers as essential components within sustainable agricultural ecosystems.

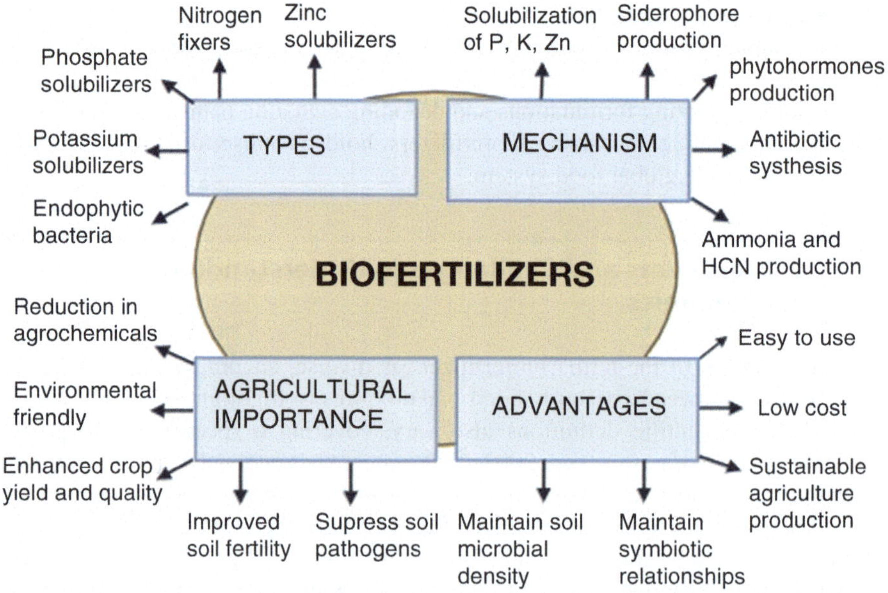

Fig. 3.1 Comprehensive overview of biofertilizers

3.2.1 Navigating Biofertilizer Challenges Together

Understanding the dynamics of biofertilizers presents a set of challenges that necessitate careful consideration and exploration. One significant hurdle lies in the selection and complexity of microbial communities (PGPR) within biofertilizers. These communities comprise diverse species with intricate interactions, making it challenging to decipher the specific roles and contributions of individual microorganisms (Ibáñez et al. 2023). Moreover, the variability in environmental conditions, such as soil types and climates, adds another layer of complexity, impacting the performance of biofertilizers. The lack of standardized testing methodologies and assessment criteria further complicates the evaluation of biofertilizer efficacy (Salam et al. 2023). Additionally, limited knowledge about the survival and persistence of beneficial microorganisms in different soil environments poses a challenge, hindering the establishment of optimal application strategies.

The other challenges include difficulties in selecting suitable carrier mediums for biofertilizers to marketing-related hurdles such as inadequate transportation and limited farmer awareness, hindering widespread acceptance. Field application constraints manifest as a slow establishment of inoculum in the root region, resulting in delayed and sometimes ineffective crop responses. Ensuring the quality control and biosafety of biofertilizers adds another layer of complexity, given the regulatory intricacies and the need for standardized procedures (Kaur et al. 2024).

Addressing these challenges requires a multidisciplinary approach, combining microbiology, agronomy, and environmental science, to unravel the intricacies of biofertilizer dynamics and pave the way for their more effective and widespread use in sustainable agriculture. Indeed metabolomics serves as a powerful tool in the selection of PGPR for biofertilizer development by offering insights into the metabolic activities, interactions, and signaling mechanisms of these bacteria (Chaudhary and Shukla 2019). This information aids researchers in identifying and optimizing PGPR strains to improve nutrient absorption, which can establish agricultural sustainability via increasing nutrient uptake, improving stress tolerance, and promoting overall plant health.

3.3 Unraveling Metabolic Dynamics: A Deep Dive into Metabolomics

Metabolomics has an important role in the identification of plant growth-promoting rhizobacteria (PGPR) for incorporation into biofertilizers. It provides valuable insights into the metabolic profiles of these microorganisms and their interactions with plants (Chaudhary and Shukla 2019). Metabolomics entails the comprehensive analysis of small-molecule metabolites within a biological system, revealing intricate details about diverse biochemical pathways and the functional responses of organisms. In the context of selecting PGPR, metabolomic approaches empower researchers to evaluate the metabolic footprint of these bacteria, enhancing the understanding of their interactions with plant roots and the surrounding soil environment (Salam et al. 2023).

Through metabolomic analyses, researchers can identify the specific metabolites produced by PGPR during interactions with plants. This information is crucial for determining the potential benefits of PGPR on the development of the plant as well stress tolerance. Metabolomics also aids in identifying the unique metabolic signatures associated with PGPR strains that exhibit superior plant growth-promoting activities. By comparing the metabolomic profiles of different PGPR strains, researchers can pinpoint those with the most favorable characteristics for enhanced plant-–microbe interactions.

Furthermore, metabolomics facilitates the identification of key metabolites involved in plant–microbe signaling and communication. PGPR often establishes symbiotic relationships with plants through the secretion of various metabolites that influence plant physiology. Metabolomic studies have a crucial contribution in unraveling the complex signaling pathways and biochemical alterations in plants triggered by plant growth-promoting rhizobacteria (PGPR). This information is crucial in the selection of strains that can effectively enhance plant growth and health (Salam et al. 2023).

Metabolomics enables us to understand and explore plant nutritional status and fertilizer research. It achieves this by providing a comprehensive analysis of the biochemical pathways and metabolites involved in nutrient uptake, assimilation, and utilization (Salam et al. 2023). This analytical approach enables the identification and quantification of various metabolites, allowing researchers to evaluate the availability of essential nutrients in plants. Through deciphering metabolic profiles, scientists can gain insights into the intricate interactions between plants and fertilizers, leading to the optimization of fertilizer utilization rates. Metabolomics facilitates the assessment of nutrient deficiencies and excesses, guiding the development of targeted fertilization strategies to enhance crop productivity while minimizing environmental impact.

3.3.1 Techniques in Metabolomics

Metabolomics represents a swiftly advancing technology that employs metabolite profiling and fingerprinting methodologies. This exploration transcends disciplinary boundaries, intertwining with genomics, transcriptomics, and proteomics, facilitating the identification of phenotypic traits. Metabolomics methodology involves scrutinizing concentrations of various bicompounds, encompassing lipids, proteins, carbohydrates, organic acids, and medicinally relevant compounds within distinct samples. Employing advanced technologies such as solid-state NMR, liquid chromatography-mass spectrometry (LC-MS), capillary electrophoresis-mass spectrometry (CE-MS), gas chromatography-mass spectrometry (GC-MS), and mass spectrometry (MS), metabolomics operates at the microscopic level, providing intricate details about small molecules. Each technique varies in its capacity to reveal types and numbers of metabolites, necessitating a customized approach for studying specific metabolic landscapes.

In metabolomics, separation methods are tailored to diverse metabolites. For instance, ionic and secondary metabolites are separated by capillary electrophoresis and liquid chromatography, respectively and gas chromatography is used for volatile components. Ultra-performance liquid chromatography (UPLC) stands out as a dominant technique in metabolomics due to its high sensitivity and efficiency compared to high-performance liquid chromatography (HPLC). Mass analyzer techniques like time-of-flight mass spectrometry (TOF-MS), Fourier transform ion cyclotron resonance-mass spectrometry (FTICR-MS), and orbitrap mass spectrometry (Orbitrap-MS) are commonly employed for precise metabolite profiling due to their rapid scan times (Zhalnina et al. 2018).

GC–MS, widely utilized in electron impact and metabolomics applications, plays a significant role, particularly in metabolite fingerprinting for identifying metabolic signatures during stress conditions. This analytical technique excels in detecting biological samples with complex matrices, offering exceptional separation and resolution through its use of the electron impact (EI) ionization source (Zhalnina et al. 2018).

Instead, LC-MS is utilized to investigate, thermolabile, polar, and high molecular weighted metabolites. It eliminates the need for derivatization by employing soft ionization sources like atmospheric pressure chemical ionization, often known as electrospray ionization (ESI). Plant metabolomics studies make considerable use of the coupling of MS with LC because of its sensitivity for the elucidation of biological data that is important to metabolic study of microbes (Jorge et al. 2016).

Investigating plant samples through metabolomics is increasingly acknowledged as a complement to protein profiling techniques. The integration of MS-based methods allows for the thorough quantification of metabolites, establishing metabolomics as a robust technology that can rival or even surpass traditional methods for genomic or proteomic studies. The potential of metabolomic analyses has been demonstrated in plant metabolomics studies involving various food crops. However, it is crucial to assess their robustness in plant-metabolomic investigations across a broader spectrum of crops beyond rice, maize, soybean, and wheat (Hu et al. 2014; Rao et al. 2014; Kusano et al. 2015; Francki et al. 2016).

Metabolite fingerprinting can be subjected to the analysis of variance, principal component analysis, discriminant function analysis, self-organizing maps, and hierarchical clustering as statistical and pattern recognition techniques. A diverse set of analytical techniques, including MS, FT-IR, and NMR spectroscopy, is employed for metabolite fingerprinting. While MS offers higher sensitivity than NMR, posing a challenge in detecting low quantities of cellular metabolites, NMR provides less resolving power than MS. Despite generating complex spectra, MS proves advantageous in identifying a broad spectrum of secondary metabolites. Different approaches to metabolite fingerprinting have been reported, including the utilization of IR spectroscopy for capturing low molecular weight metabolomes.

Metabolic compounds also play an important role in the rapid fermentation of forage proteins. In a specific study, PCA was utilized to assess deviations in cultivars and changes in metabolite fingerprinting between second growth and spring growth. Seasonal variations resulted in changes in metabolic compounds such as glucose, choline, and malic acid. Metabolomics aids in distinguishing stimulated metabolites based on data analysis, metabolic pathways, and models, contributing to the comprehension of metabolic engineering in microbes. The metabolome reveals cellular processes controlling the phenotypic nature of cells, tissues, or organisms. Notably, the biochemical status of microorganisms is monitored for gene function through metabolomic analysis.

The metabolomic analysis follows a series of standardized procedures, including experiment design, specialized techniques for sample collection, metabolism quenching, optimized metabolite extraction, optional chemical derivatization, and the utilization of MS with chromatography or NMR. The subsequent stages of data analysis involve various techniques, such as data alignment, filtering, imputation, statistical analysis, annotation, and pathway analysis. Figure 3.2 illustrates the flow of biological data at the genome level through metabolomics.

Moreover, metabolomic studies play a crucial role in advancing the development of biological fertilizers by providing insights into the impact of microbial

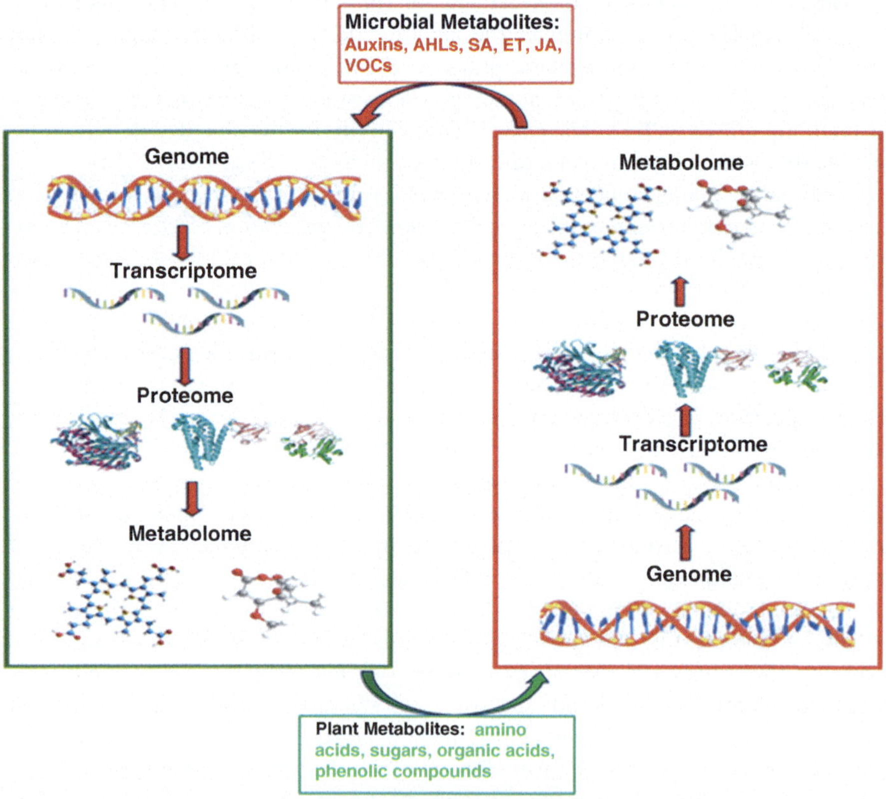

Fig. 3.2 Flow of biological data at genome level via metabolome

communities on plant metabolism and nutrient acquisition. Essentially, metabolomics serves as a powerful tool in unraveling the complexities of plant–fertilizer interactions, thereby fostering precision agriculture practices for sustainable and efficient nutrient management in crop production.

3.3.2 Metabolomics: Unraveling the Intricacies of Plant–Microbe Interactions

Metabolomics offers valuable insights into the biochemical changes occurring during these symbiotic relationships between plants and PGPR. This method encompasses a thorough examination of metabolites in a biological system, providing researchers with the capability to identify and measure the small molecules participating in plant–PGPR interactions. Some of the highlights are given as follows:

(a) **Identification of Metabolic Signatures:** Metabolomic studies help identify unique metabolic signatures associated with the presence of PGPR. By comparing the metabolite profiles of plants with and without PGPR inoculation, researchers can pinpoint specific compounds or pathways influenced by the beneficial bacteria. This information aids in understanding the biochemical changes triggered by PGPR, including alterations in primary and secondary metabolites.
(b) **Elucidation of Plant Response Mechanisms:** Metabolomics provides insights into how plants respond to PGPR inoculation at the biochemical level. Changes in metabolite concentrations, such as protein molecules, sugars, and organic acids can be indicative of the plant's response to the presence of PGPR. Understanding these response mechanisms is crucial for elucidating the pathways through which PGPR enhances plant development, uptake of nutrients, and resilience to stress.
(c) **Detection of Biochemical Signaling:** Metabolomic approaches contribute to the identification of biochemical signaling molecules exchanged between plants and PGPR. Some PGPR produce signaling compounds, such as phytohormones and volatiles that can modulate plant growth and development. Metabolomics helps uncover these signaling molecules and their impact on the plant's metabolic network.
(d) **Assessment of Nutrient Uptake and Utilization:** Metabolomics aids in assessing the impact of PGPR on nutrient uptake and utilization by plants. Changes in the levels of metabolites related to nutrient metabolism, such as nitrogen-containing compounds, can provide insights into the efficiency of nutrient acquisition facilitated by PGPR.
(e) **Evaluation of Stress Responses:** Metabolomics contributes to our understanding of how PGPR as enhance stress resilience in plants. Stress-related metabolites, including osmoprotectants like proline and specific sugars, are often identified through metabolomic analyses (Llanes et al. 2018; Anzano et al. 2021). These metabolites act as key players in mitigating the adverse effects of

stressors like drought, salinity, or pathogen attacks. Metabolomics provides a detailed picture of how PGPR-induced changes in metabolite levels fortify plants against environmental stressors, promoting resilience in challenging conditions (Table 3.1).

Metabolomics extends its reach into the exploration of allelopathy among PGPR and mycorrhiza, as exemplified by studies on *Setaria italica*. Utilizing NMR spectroscopy, researchers delve into the metabolic intricacies, revealing the impact of endomycorrhizal consortiums and *Pseudomonas* TLC6-6-5.4 on plant biomass, height, and chlorophyll content. The intricate dance of metabolic variances and upregulated pathways, such as glyoxylate and dicarboxylate metabolism, showcases the transformative potential of these microbial interactions on maize seedlings raised in unique environments (Dhawi et al. 2016, 2018).

3.3.3 Rhizosphere Metabolomics: Unveiling Biofertilizer-PGPR Dynamics

An emerging area of plant metabolomics, i.e., rhizosphere metabolomics entails a thorough investigation of the rhizosphere's whole metabolome as well as its microbes. The objective is to comprehend the dynamics of living together and the resulting physiological impacts on both plants and microbes. This study carefully investigates the symbiotic or pathogenic interactions between rhizosphere organisms. The complex and dynamic nature of the ecosystem presents challenges for metabolomics studies, considering the wide variety of metabolites, the constant discharge of rhizodeposits, microbe secretions, and the competition among rhizomicrobes. The introduction of biofertilizers further complicates the comprehensive metabolic analyses of the region. Metabolomics, gaining prominence alongside traditional transcriptomic and proteomic methods, contributes to a broader understanding of plant-pathogen studies from a metabolomics perspective. Understanding the underlying dynamics of the rhizosphere is challenging due to its chemical variability, despite advancements in genomic research that define microbial populations in the rhizosphere through high-throughput sequencing.

Recent studies in rhizosphere metabolomics exemplify diverse methodologies. For instance, Smercina et al. (2020) investigated switchgrass rhizosphere chemistry using hydroponics with nitrogen and diazotroph bacteria. NMR spectroscopy revealed increased metabolite concentrations in nitrogen-supplemented plants, suggesting a minimal impact of diazotrophs on overall rhizosphere chemistry. Another study by Zhalnina et al. (2018) utilized UHPLC-MS to assess the role of plant exudates in rhizosphere composition under hydroponics. While successful exudate extraction occurred, the study lacked actual microbes and predicted microbial behavior was analyzed separately.

Addressing knowledge gaps, Pétriacq et al. (2017) revealed that polar and nonpolar rhizosphere metabolites can be extracted from soils containing *Arabidopsis thaliana* and *Zea mays* using the untargeted metabolite profiling approach in

Table 3.1 Impact of PGPR on plants—role of metabolic studies

S. no	PGPR	Metabolomic studies	Plant	Reference
1	*Bacillus subtilis* and *Paenibacillus alvei* (T22)	Metabolic reprogramming, impacting amino acids, organic acids, and phenolic compounds, thereby making plant adaptations to stripe rust infection	*Triticum aestivum* L.	Mashabela et al. (2023)
2	*Rhizoglomus irregulare BEG72 and Funneliformis mosseae BEG234, Pseudomonas strain So_08*	Metabolomics revealed the beneficial effects of PGPR and AMF in mitigating Cd and Zn stress in tomato plants, influencing physiological parameters and nutrient balance	*Lycopersicon esculentum*	Zhang et al. (2023)
3	*Enterobacter bugendensis* WRS7	GC-MS metabolite profiling emphasized the role of metabolites, indicating potential for metabolic engineering in developing drought-resistant bioinoculants	–	Arora et al. (2023)
4	*E. mori strain C3.1 Paenibacillus (T22) and Bacillus subtilis* strains	Diverse compound accumulation (organic acids, phenylpropanoids, lipids, and benzenoids) with roles in plant–microbe interactions and growth promotion	*Triticum aestivum* L.	Mashabela et al. (2022a, b)
	Bacillus subtilis MBI600	The stimulation of defense-related genes (peroxidase, endo-1,3(4)-beta-glucanase, PR-4, thaumatin-like) in biocontrol-related pathways	*Cucumis sativus* L.	Samaras et al. (2022)
5	*Pseudomonas oryzihabitans* AXSa06	Metabolomics identified salt tolerance signatures: amino acids (aspartate, threonine, serine, glutamate), crucial genes (ethylene, abscisic acid), and ion antiporters	*Lycopersicon esculentum*	Mellidou et al. (2021)
6.	*Pseudomonas fluorescens* N04, *P. koreensis* N19, *Paenibacillus alvei* T19, and *Lysinibacillus sphaericus* T22	Alterations in hydroxycinnamates, benzoates, flavonoids, and glycoalkaloids enhancing defense responses	*Lycopersicon esculentum*	Mhlongo et al. (2020)
7	*E. mori strain C3.1*	Metabolomics revealed crucial symbiotic metabolites (myo-inositol, monopalmitin)	*Lycopersicon esculentum*	Kalozoumis et al. (2021)

(continued)

Table 3.1 (continued)

S. no	PGPR	Metabolomic studies	Plant	Reference
8.	*Bacillus subtilis*, *Bacillus thuringiensis*, and *Bacillus megaterium*	Accumulation of specific compounds, such as proline and riboflavin, illustrates the metabolic adjustments crucial for chickpea growth and resilience	*Cicer arietinum* L.	Khan et al. (2019)
9	*Pseudomonas aeruginosa* (M1) and *Burkholderia gladioli* (M2)	Metabolomic changes, such as modified antioxidants, improved photosynthesis, and elevated protective compounds, revealed key role in plant–biocontrol interaction	*Lycopersicon esculentum* Mill.	Khanna et al. (2019)
10	*Pseudomonas* and arbuscular mycorrhizal fungi	Effects on root nutrient levels and defense compound suppression. Estimating diverse secondary metabolite production for biocontrol agent potential	*Triticum durum* Desf	Saia et al. (2015)

non-sterile rhizosphere soils. This method minimized damage to root and its rhizomicrobes, providing a closer representation of the rhizosphere's environmental setting. Thus, metabolomics, especially in conjunction with biofertilizer applications, offers a potent approach to study rhizosphere dynamics, fostering a deeper understanding of the intricate chemistry driving plant–microbe interactions in non-sterile soils. Ongoing efforts aim to refine methods, providing a more holistic view of the total rhizosphere ecosystem and facilitating the exploration of plant–microbe interactions in diverse environments (Mashabela et al. 2022a, b).

3.3.4 Metabolomics Unveiling Plant Metabolite Changes Induced by Biofertilizer-PGPR Interactions

The significance of specific metabolites in primary metabolism (sugars, amino acids, and organic acids) and secondary metabolites (lipids, phytohormones, terpenes, flavonoids, and alkaloids) in plant–microbe interactions has long been recognized. Recently, plant pathologists have embraced global metabolite analysis through metabolomics, revealing the intricate nature of these interactions and plant metabolism. Plant metabolism, a delicate and highly responsive process to external stimuli, undergoes rapid fluctuations during plant–microbe interactions, activating response mechanisms. The integration of biofertilizers adds complexity, forming tripartite interactions that necessitate the examination and characterization of involved metabolites (Mashabela et al. 2022a, b).

Metabolite content variations in plant roots, root exudates, and shoots following biofertilizer-plant growth-promoting rhizobacteria (PGPR) inoculations have been analyzed in many studies using metabolomics approaches. For example, metabolic reprogramming of 49 metabolites, including amino acids and chemicals linked to lipid metabolism, was found in Sorghum bicolor treated with *Paenibacillus alvei*, suggesting a multicomponent defense response in plants primed with biofertilizer. In comparison to their non-primed counterparts, the research demonstrated improved control of major and secondary metabolic pathways (Carlson et al. 2019). Akram et al. (2019) investigated metabolome changes in tomato plants that react to the halotolerant biofertilizer strain *B. megaterium* A12 under salt stress. Significant changes in phytohormone concentrations were observed, indicating stress tolerance and increased concentrations of plant growth-promoting phytohormones suggested the biofertilizer's contribution to plant growth. The study demonstrated that *B. megaterium* A12 could potentially ameliorate salinity stress in tomatoes.

A recent study by Mhlongo et al. (2020) inoculated tomato plants with four distinct PGPR strains to investigate their priming effects. Utilizing *P. fluorescens* N04, *P. koreensis* N19, *P. alvei* T22, and *Lysinibacillus sphaericus* T19, metabolite extraction from leaf, root, and stem tissues in 80% methanol was followed by UHPLC-MS analysis, unveiling defense-related metabolic reprogramming. Chemometric analysis highlighted differential adaptations in PGPR-treated plants secondary metabolites and aromatic amino acids in comparison to controls identifying potential biomarkers for PGPR-induced plant priming. The study suggests that the differential modulation of these metabolites results from the plant's response to external stimuli from PGPR, leading to a primed state and enhanced defense against future diseases.

Other studies explored metabolic perturbations in maize plants treated with *Azospirillum lipoferum*, CRT1, and sugar cane plants interacting with the fungal pathogen *Sporisorium scitamineum*. These investigations emphasized the impact of biofertilizers on defense-related hydroxycinnamic acids and primary metabolism pathways (Rozier et al. 2016). Thus, the integration of metabolomics in biofertilizer-PGPR studies provides valuable insights into the complex molecular interactions occurring in plants under various stresses, offering potential solutions for enhancing plant growth and stress tolerance.

3.4 Future Prospects and Emerging Trends in Metabolomics Research

There are a lot of exciting opportunities to transform sustainable agriculture with metabolomics research in the context of biofertilizer dynamics. New developments in this area are moving us closer to a thorough comprehension of the complex metabolic processes taking place in the plant–soil–microbe nexus. It is expected that sophisticated analytical methods like mass spectrometry and nuclear magnetic resonance spectroscopy will improve the accuracy and range of metabolite detection, allowing scientists to learn more about the metabolic changes by biofertilizers. A comprehensive systems biology approach will be further enhanced by the integration of omics technologies, including proteomics and genomics, which will help to untangle the intricate network of biochemical reactions set off by the use of biofertilizers. In order to handle and interpret the enormous information created, bioinformatics tools will be crucial in helping to identify the metabolites unique to each biofertilizer. Furthermore, metabolomics research is probably going to go beyond plants in the future to investigate the ecological and environmental effects of using biofertilizers. As this sector develops, the knowledge acquired will have the potential to optimize agricultural practices, promote the creation of customized biofertilizers, and support resilient and sustainable food production systems.

3.5 Conclusion

In conclusion, the integration of metabolomics into the study of biofertilizers and plant–microbe interactions offers profound insights into the intricate biochemical processes crucial for sustainable agriculture. Biofertilizers, driven by beneficial microorganisms, provide multifaceted solutions to soil fertility, plant growth, and resilience. Metabolomics emerges as a versatile and powerful tool, enabling the identification of key metabolites and unraveling dynamic responses in plants subjected to biofertilizer treatments. The future of metabolomics research holds great promise, with ongoing advancements expected to refine analytical methods, foster multidisciplinary approaches, and explore broader ecological implications. As we navigate challenges and opportunities in the biofertilizer landscape, metabolomics stands as a beacon guiding the optimization of formulations, the decoding of signaling pathways, and the enhancement of our understanding of plant–microbe interactions. This trajectory positions metabolomics at the forefront of shaping resilient and sustainable agricultural practices, thereby contributing significantly to global food systems and environmental well-being.

References

Akram W, Aslam H, Ahmad SR, Anjum T, Yasin NA, Khan WU, Ahmad A, Guo J, Wu T, Luo W et al (2019) Bacillus megaterium strain A12 ameliorates salinity stress in tomato plants through multiple mechanisms. J Plant Interact 14:506–518

Anzano A, Bonanomi G, Mazzoleni S, Lanzotti V (2021) Plant metabolomics in biotic and abiotic stress: a critical overview. Phytochem Rev:1–22

Arora S, Babele PK, Jha PN (2023) Biochemical and metabolic signatures are fundamental to drought adaptation in PGPR Enterobacter bugandensis WRS7. Mol Omics 19(8):640–652

Benckiser G, Kumar K, Hartmann A, Honermeier B (eds) (2019) Plant-microbe-insect interactions in ecosystem management and agricultural praxis. Frontiers Media SA

Carlson R, Tugizimana F, Steenkamp PA, Dubery IA, Labuschagne N (2019) Differential metabolic reprogramming in Paenibacillus alvei-primed Sorghum bicolor seedlings in response to fusarium pseudograminearum infection. Meta 9:150

Chaudhary T, Shukla P (2019) Bioinoculant capability enhancement through metabolomics and systems biology approaches. Brief Funct Genomics 18(3):159–168

Choudhary DK, Kumar M, Prasad R, Kumar V (eds) (2018) In silico approach for sustainable agriculture. Springer, Singapore, pp 161–167

Dhawi F, Datta R, Ramakrishna W (2016) Mycorrhiza and heavy metal resistant bacteria enhance growth, nutrient uptake and alter metabolic profile of sorghum grown in marginal soil. Chemosphere 157:33–41

Dhawi F, Datta R, Ramakrishna W (2018) Metabolomics, biomass and lignocellulosic total sugars analysis in foxtail millet (Setaria italica) inoculated with different combinations of plant growth promoting bacteria and mycorrhiza. Plant Sci 8(8):10–26814

El-Ghamry AM, Mosa AA, Alshaal TA et al (2018) Nanofertilizers vs. biofertilizers: new insights. Environ Biodivers Soil Secur 2:51–72

Eshawu AB, Ghalsasi VV (2023) Metabolomics of natural samples: a tutorial review on latest technologies. J Sep Sci 2300588

Etesami H, Adl SM (2020) Plant growth-promoting rhizobacteria (PGPR) and their action mechanisms in availability of nutrients to plants. In: Phyto-Microbiome in stress regulation, Springer, Singapore, pp 147–203

Fincheira P, Quiroz A (2018) Microbial volatiles as plant growth inducers. Microbiol Res 208:63–75

Francki MG, Hayton S, Gummer J, Rawlinson C, Trengove RD (2016) Metabolomic profiling and genomic analysis of wheat aneuploid lines to identify genes controlling biochemical pathways in mature grain. Plant Biotechnol J 14(2):649–660

Fuentes-Ramírez LE, Caballero-Mellado J (2006) Bacterial biofertilizers. In: Siddiqui ZA (ed) *PGPR:* biocontrol and biofertilization. Springer, Netherlands, pp 143–172

Hu C, Shi J, Quan S, Cui B, Kleessen S, Nikoloski Z, Wang J (2014) Metabolic variation between japonica and indica rice cultivars as revealed by non-targeted metabolomics. Sci Rep 4:5067

Ibáñez A, Garrido-Chamorro S, Vasco-Cárdenas MF, Barreiro C (2023) From lab to field: biofertilizers in the 21st century. Horticulturae 9(12):1306

Jorge F, Rodrigues J, Caldana C, Schmidt R, van Dongen J, Thomas-Oates J, António C (2016) Mass spectrometry-based plant metabolomics: metabolite responses to abiotic stress. Mass Spectrom Rev 35:620–649

Kalita M, Bharadwaz M, Dey T, Gogoi K, Dowarah P, Unni BG, Saikia I (2015) Developing novel bacterial based bioformulation having PGPR properties for enhanced production of agricultural crops. Indian J Exp Biol 53(1):56–60

Kalozoumis P, Savvas D, Aliferis K, Ntatsi G, Marakis G, Simou E et al (2021) Impact of plant growth-promoting rhizobacteria inoculation and grafting on tolerance of tomato to combined water and nutrient stress assessed via metabolomics analysis. Front Plant Sci 12:670236

Kaur S, Dwibedi V, Sahu PK, Kocher GS (eds) (2024) Metabolomics, proteomes and gene editing approaches in biofertilizer industry, Springer Nature

Khan N, Bano A, Rahman MA, Guo J, Kang Z, Babar MA (2019) Comparative physiological and metabolic analysis reveals a complex mechanism involved in drought tolerance in chickpea (Cicer arietinum L.) induced by PGPR and PGRs. Sci Rep 9(1):2097

Khanna K, Sharma A, Ohri P, Bhardwaj R, Abd-Allah EF, Hashem A, Ahmad P (2019) Impact of plant growth promoting rhizobacteria in the orchestration of Lycopersicon esculentum mill. Resistance to plant parasitic nematodes: a metabolomic approach to evaluate defense responses under field conditions. Biomol Ther 9(11):676

Kumar V, Baweja M, Singh PK, Shukla P (2016) Recent developments in systems biology and metabolic engineering of plant–microbe interactions. Front Plant Sci 7:1421

Kusano M, Yang Z, Okazaki Y, Nakabayashi R, Fukushima A, Saito K (2015) Using metabolomic approaches to explore chemical diversity in rice. Mol Plant 8(1):58–67

Llanes A, Andrade A, Alemano S, Luna V (2018) Metabolomic approach to understand plant adaptations to water and salt stress. In: Plant metabolites and regulation under environmental stress. Academic Press, pp 133–144

Mącik M, Gryta A, Frąc M (2020) Biofertilizers in agriculture: an overview on concepts, strategies and effects on soil microorganisms. Adv Agron 162:31–87

Mashabela MD, Piater LA, Dubery IA, Tugizimana F, Mhlongo MI (2022a) Rhizosphere tripartite interactions and PGPR-mediated metabolic reprogramming towards ISR and plant priming: a metabolomics review. Biology 11(3):346

Mashabela MD, Tugizimana F, Steenkamp PA, Piater LA, Dubery IA, Mhlongo MI (2022b) Untargeted metabolite profiling to elucidate rhizosphere and leaf metabolome changes of wheat cultivars (Triticum aestivum L.) treated with the plant growth-promoting rhizobacteria Paenibacillus alvei (T22) and Bacillus subtilis. Front Microbiol 13:971836

Mashabela MD, Tugizimana F, Steenkamp PA, Piater LA, Dubery IA, Terefe T, Mhlongo MI (2023) Metabolomic evaluation of PGPR defence priming in wheat (Triticum aestivum L.) cultivars infected with Puccinia striiformis f. sp. tritici (stripe rust). Front Plant Sci 14:1103413

Mellidou I, Ainalidou A, Papadopoulou A, Leontidou K, Genitsaris S, Karagiannis E et al (2021) Comparative transcriptomics and metabolomics reveal an intricate priming mechanism involved in PGPR-mediated salt tolerance in tomato. Front Plant Sci 12:713984

Mhlongo MI, Piater LA, Steenkamp PA, Labuschagne N, Dubery IA (2020) Metabolic profiling of PGPR-treated tomato plants reveal priming-related adaptations of secondary metabolites and aromatic amino acids. Meta 10:210

Mishra J, Arora NK (2016) Bioformulations for plant growth promotion and combating phytopathogens: a sustainable approach. In: Bioformulations: For sustainable agriculture, Springer, New Delhi, pp 3–33

Mutturi S, Sahai V, Bisaria VS (2017) Fed-batch cultivation for high density culture of Pseudomonas spp. for bioinoculant preparation. Modern tools and techniques to understand microbes, 381–400

Nitu R, Rajinder K, Sukhminderjit K (2020) Zinc solubilizing bacteria to augment soil fertility—a comprehensive review. Int J Agric Sci Vet Med 8:38–44

Pétriacq P, Williams A, Cotton A, McFarlane AE, Rolfe SA, Ton J (2017) Metabolite profiling of non-sterile rhizosphere soil. Plant J 92:147–162. [CrossRef] [PubMed]

Rani N, Kaur R, Kaur S (2021) Present scenario of plant growth promoting microbes and technologies used for biofertilizer development. In: Plant-microbe dynamics: recent advances for sustainable agriculture. CRC Press, Boca Raton, FL, pp 121–134

Rani N, Kaur G, Kaur S, Mutreja V, Upadhyay SK, Tripathi M (2022) Comparison of diversity and zinc solubilizing efficiency of rhizobacteria obtained from solanaceous crops under polyhouse and open field conditions. Biotechnol Genet Eng Rev:1–22

Rani N, Chauhan A, Kaur S, Solanki MK, Tripathi M, Jain D et al (2023) Molecular mechanistic of Zn-solubilizing bacteria for agronomic Eminence: recent updates and futuristic development. J Plant Growth Regul:1–15

Rao A, Ahmad SD, Sabir SM, Awan SI, Shah AH, Abbas SR, Shafique S, Khan F, Chaudhary A (2014) Potential antioxidant activities improve salt tolerance in ten varieties of wheat (Triticum aestivum L.). Am J Plant Sci 4:69–76

Rozier C, Hamzaoui J, Lemoine GD, Czarnes S, Legendre L (2016) Field-based assessment mechanisms of maize yield enhancement by Azospirillum lipoferum CRT1. Sci Rep 7:7416

Saia S, Ruisi P, Fileccia V, Di Miceli G, Amato G, Martinelli F (2015) Metabolomics suggests that soil inoculation with arbuscular mycorrhizal fungi decreased free amino acid content in roots of durum wheat grown under N-limited, P-rich field conditions. PLoS One 10(6):e0129591

Saif S, Abid Z, Ashiq MF, Altaf M, Ashraf RS (2021) Biofertilizer formulations. Biofertilizers: Study and Impact, 211–256

Salam U, Ullah S, Tang ZH, Elateeq AA, Khan Y, Khan J et al (2023) Plant metabolomics: an overview of the role of primary and secondary metabolites against different environmental stress factors. Life 13(3):706

Samaras A, Kamou N, Tzelepis G, Karamanoli K, Menkissoglu-Spiroudi U, Karaoglanidis GS (2022) Root transcriptional and metabolic dynamics induced by the plant growth promoting rhizobacterium (PGPR) Bacillus subtilis Mbi600 on cucumber plants. Plan Theory 11(9):1218

Sati D, Pandey SC, Pande V, Upreti S, Gouri V, Joshi T et al (2020) Toward an enhanced understanding of plant growth promoting microbes for sustainable agriculture. In: Recent advancements in microbial diversity. Academic Press, pp 87–112

Sharma D, Gahtyari NC, Chhabra R, Kumar D (2020a) Role of microbes in improving plant growth and soil health for sustainable agriculture. In: Advances in plant microbiome and sustainable agriculture: diversity and biotechnological applications, Springer, Singapore, pp 207–256

Sharma M, Sudheer S, Usmani Z, Rani R, Gupta P (2020b) Deciphering the omics of plant-microbe interaction: perspectives and new insights. Curr Genomics 21(5):343–362

Singh M, Dotaniya ML, Mishra A, Dotaniya CK, Regar KL, Lata M (2016) Role of biofertilizers in conservation agriculture. In: Conservation agriculture: an approach to combat climate change in Indian Himalaya, Springer, Singapore, pp 113–134

Smercina DN, Evans SE, Friesen ML, Tiemann LK (2020) Impacts of nitrogen addition on switchgrass root-associated diazotrophic community structure and function. FEMS Microbiol Ecol 96:fiaa208

Teh HF, Neoh BK, Ithnin N, Daim LDJ, Ooi TEK, Appleton DR (2017) Omics and strategic yield improvement in oil crops. J Am Oil Chem Soc 94:1225–1244

Umesha S, Singh PK, Singh RP (2018) Microbial biotechnology and sustainable agriculture. In: Biotechnology for sustainable agriculture. Woodhead Publishing, pp 185–205

Uzoh IM, Babalola OO (2018) Rhizosphere biodiversity as a premise for application in bioeconomy. Agric Ecosyst Environ 265:524–534

Verma M, Mishra J, Arora NK (2019) Plant growth-promoting rhizobacteria: diversity and applications. In: Environmental biotechnology: for sustainable future, Springer, Singapore, pp 129–173

Vessey JK (2003) Plant growth promoting rhizobacteria as biofertilizers. Plant Soil 255:571–586

Zhalnina K, Louie KB, Hao ZL, Monsoori N, Da Rocha UN, Shi S, Cho H, Karaoz U, Loque D, Bowen PB et al (2018) Dynamic root exudate chemistry and microbial substrate preferences drive patterns in rhizosphere microbial community assembly. Nat Microbiol 3:470–480

Zhang L, Zuluaga MYA, Pii Y, Barone A, Amaducci S, Miras-Moreno B et al (2023) A pseudomonas plant growth promoting rhizobacterium and arbuscular mycorrhiza differentially modulate the growth, photosynthetic performance, nutrients allocation, and stress response mechanisms triggered by a mild zinc and cadmium stress in tomato. Plant Sci 337:111873

Metabolite-based Bioformulation: Next Generation of Biofertilizers

4

Richa

Abstract

"Next-generation bioformulations" signify to cutting-edge techniques to formulate biological products or molecules in order to improve their transmission methods, stability, safety, and efficacy. Maintaining effectiveness, increasing shelf-life, and ensuring consistent efficacy all depend on the stability of the product. At each stage of product life, next-generation bioformulations minimize their environmental impact, use eco-friendly components, and prioritize sustainability. For the development of non-toxic, renewable, bio-based, and biodegradable production, practices of next-generation bioformulations are required. Metabolite-based formulations are a significant component of next-generation bioformulations. They frequently have beneficial qualities, biological benefits, which are obligatory for biochemical processes. Microbes and plants are among the fundamental natural sources of various metabolites. The development of metabolite-based formulations that have been designed to treat particular diseases or pathogens is aided by the identification of novel biological substance, which is only possible by the complete monitoring of metabolites within biological systems and advances in metabolomics. Metabolite based bioformulation protects and nourishes plants against various diseases and pathogens through various mechanisms like hormone regulation, nutritional enhancement, nitrogen fixation, formation of plant innate resistance.

Keywords

Biofertilizer · Bioformulation · Metabolite · Next-generation formulation · Plant growth

Richa (✉)
Department of Microbiology, Bundelkhand University, Jhansi, India

S. Kaur et al. (eds.), *Metabolomics, Proteomics and Gene Editing Approaches in Biofertilizer Industry*, https://doi.org/10.1007/978-981-97-2910-4_4

4.1 Introduction

Implementation of chemical-based pesticides and fertilizers has been a general approach to increase agricultural output to satisfy the growing universal demand for food. However, their continual application has had negative effects on society including human and animal, as a whole and the environment (Baweja et al. 2020). The only thing that chemical-based agriculture has produced are dry, unfertile soils with the loss of advantageous microbes, especially in soil now well acknowledged (Seneviratne and Kulasooriya 2013).

Broad-spectrum microorganisms are thought of as bioagents that improve the health of crops and soil. They are mostly used to create bioformulations, which have the potential to directly or indirectly affect plant growth in a number of ways (Lugtenberg and Kamilova 2009). Microbes that are both beneficial to the environment and highly successful at improving soil beneath challenging circumstances are known as bioinoculants. Bioformulation applied for various purposes, such as sustainable agriculture production and environmental management. Usually bioformulations contain living microorganisms, mainly bacteria, fungi, or archaea, which boost nutrient availability and reception by plants. These bioformulations can be functional to seeds, soil, or plant surfaces directly in the form of biofertilizers, which are eco-friendly and sustainable replacements to chemical fertilizers that harness the potential of advantageous microorganisms to improve soil fertility and promote plant development. Biofertilizers provide a universal approach to nutrient management, promoting overall plant health and productivity.

Biofertilizers function by encouraging symbiotic connections between plants and particular microbes, in lieu of standard fertilizers, which typically deliver critical nutrients directly to plants. The native microbial population has the capacity to alter the compound makeup and physical characteristics of the soil through formulation strategies. A favourable and reciprocal impact is produced by formulation, which promotes growth of roots as well as shoots they safeguard the ecosystem from unfavourable circumstances and aid in the preservation of soil fertility. Microbial associations are impacted by the methods employed to increase soil microbial communities. While bioinoculant protects the soil from adverse circumstances and benefits farmers and consumers financially, its usage is sustainable (Kumar et al. 2016; Singh et al. 2011). While formulating commercial biofertilizers, it is preferable to select and employ the most efficient microbial strains, considering the fact that there are still a lot of unanswered essential aspects about the effective application of biofertilizers (Kaari et al. 2023).

4.2 Understanding and Addressing Essential Factors for Next-generation Bioformulations

Several factors such as strain type, soil conditions, climate, crop type, etc. can influence the effectiveness and success of bioformulations. For the successful development and application of bioformulations, understanding and addressing essential factors is crucial aspect. Interrelated application of biofertilizers with other sustainable agricultural practices can contribute to enhanced soil fertility and reduced reliance on synthetic fertilizers. Apart from the application of certain agricultural methods that negatively impact soils in both a financial and environmental sense, it is crucial to enhance the inoculation efficiency in numerous manners (Bajsa et al. 2013). Use of precision microbial consortia, where specific combinations of beneficial microorganisms are carefully selected to enhance plant growth under particular conditions is possible with implementing next-generation formulation.

These microbial consortia may include various strains of bacteria, fungi, and other microorganisms that work synergistically to promote plant health and nutrient uptake. Addressing some crucial elements can help in generating novel biological products or develop next-generation bioformulations which are as follows:

4.2.1 Microbial Strain Selection

Selection of microbial strains is one of the crucial sections. Different strains display variations in their effectiveness under specific conditions. Factors such as microbial viability, stability, and compatibility with the target application need to be considered.

4.2.2 Formulation Components

The ingredients used in the formulation play a substantial role, which includes carriers, stabilizers, nutrients, and other additives. The formulation should provide an appropriate environment for the growth and endurance of the microorganisms.

4.2.3 Environmental Conditions

Temperature, humidity, pH, and other environmental aspects directly influence the performance of bioformulations. Microorganisms may have specific requirements for growth and activity, and formulations must consider these conditions.

4.2.4 Storage Stability

The stability of bioformulations during storage is crucial for their effectiveness. Formulations need to maintain the viability and activity of microorganisms over time. Proper storage conditions, such as temperature and humidity, are essential aspects.

4.2.5 Compatibility with Chemical Inputs

In agricultural bioformulations, compatibility with other inputs such as fertilizers and pesticides is important. Some microbial strains may be sensitive to certain chemicals, so formulation design should consider potential interactions.

4.2.6 Application Methods

The method of application can influence the success of bioformulations. Whether applied as seed treatments, soil amendments, foliar sprays, or other methods, the formulation should facilitate the effective delivery of microorganisms to the target site.

4.2.7 Target Organism and Host Specificity

Bioformulations are often designed for specific target organisms, such as pests or pathogens. The formulation should be tailored according to the nature of the target and should not harm non-target organisms.

4.2.8 Regulatory Compliance

Acquiescence with regulatory requirements is essential, especially in agriculture. Bioformulations must meet safety and efficacy standards set by regulatory authorities.

4.2.9 Economic Feasibility

The cost-effectiveness of bioformulations is a critical factor. The production, formulation, and application costs should be balanced against the potential benefits.

4.3 Constituents Exploration

Cell-based formulation is a traditional method of formulation, which have several draw backs like limited shelf life, short spell efficiency, limited awareness of constituents (Richa 2023). Continuous exploration of the involved ingredients is necessary for developing effective bioformulations. Exploration practice involves developing novel supply chains, refining formulations, and exploring novel microbial strains. Due to certain limitations, unfortunately ecologically sound bioformulations have yet to triumph over the agro-market. In order to earn the vigorous productive agricultural as well as trust of the final consumers, there are some issues that need to be addressed, specifically long-term retention of microbial populations and vitality, unpredictable field performance, insufficient soil organic carbon, skewed effectiveness, and short shelf life (Arora 2015). Previous studies of Tewari and Arora (2016) addressed the flaws of traditional cell-based formulation by the development of metabolite-based bioformulation (Tewari and Arora 2016).

Figuring out solutions requires learning from the field's observations as well as anticipating the next generation of investigation to produce more effective bioinoculants by comprehending the route that technology is now headed. Cell-based formulations hold a significant market share among ecologically friendly goods. Because of some hindrances, bioformulations consequently fall significantly behind chemicals in the market, accounting for fewer than 5% of all products used as insecticides or crop fertilizers. Plant growth-promoting (PGP) microorganisms are frequently incorporated in compositions that are cell-based (Arora 2015). Technological exploration shows how additives and metabolites work with organisms to improve the product's efficacy and dependability. Development and application of products containing microorganisms or their metabolic by-products/substances have the potential to be employed to counter phytopathogens, supply nutrients, or act as bioregulators (Arora et al. 2010).

4.4 Microbial Metabolites-based Bioformulation

Microbes stimulate plant growth either directly or indirectly, through intermediate products termed as "'metabolites". These metabolites are produced during metabolic process within microbial cells. These metabolic process catalyzed by different enzymes that take place naturally within cells. Metabolites are chemical components with a low molecular weight (MW) of less than 1k Da that are essential for microbial metabolism (Pinu et al. 2017). An individual microorganism's whole set of metabolites is recognized as its 'metabolome' (Aldridge and Rhee 2014) and stands for three different bases: outside the cellular media, the cell itself, and the growing environment. Since the metabolome and phenotype are precisely connected, they serve as good indicators of an organism's biological function (Macintyre et al. 2014). Because of unlimited life spans, metabolites can be created, absorbed, broken down, or expelled (Villas-Boas et al. 2007).

Though metabolites are primarily formed during metabolic processes, but they are also transformed into other materials. The difference in metabolite synthesis and transformation rates determines the total amount of metabolites available in a microbial cell (Pinu et al. 2017). Metabolites serve a variety of functions in living cells, such as energy production, structural support, signalling, catalysis, defence, and interacting with other living organisms.

Functions of metabolic substances usually execute hormone regulation, nutrition uptake, and water equilibrium, i.e., the dissolved form of inorganic and organic phosphates, nitrogen fixation, establishment of plant innate resistance as a defence against pathogens; and possibly the microbial synthesis of antibiotic substances or enzymes (Calvo et al. 2014; Elnahal et al. 2022). Activities of bacteria, majorly from soil, are thought to be extremely promising because of their adaptability to different metabolic conditions, resilience to drought, ability to provide nutrients (like phosphate) that improve growth, and capability to produce an extensive series of bioactive metabolites, including vitamins, antibiotics, plant growth factors, and enzymes. Many fungi and bacterial genera, containing *Trichoderma*, *Azospirillum, Klebsiella, Azotobacter, Bacillus, Enterobacter, Pseudomonas, Kocuria, Streptomyces, Brady rhizobium*, *Mesorhizobium*, *Rhizobium*, and mycorrhiza appear to have the capacity to stimulate plant growth (Afridi et al. 2021; Arora and Mishra 2016; Compant et al. 2005; Doolotkeldieva et al. 2015).

Metabolite-based bioformulation reflected as next-generation bioformulation, which have more effective mechanism. Metabolite-based bioformulations represent the next generation of bioformulations, utilizing microorganisms to produce a broad range of bioactive metabolites through more effective processes (Faddetta et al. 2023). When combined with PGP microorganisms (Fig. 4.1), the modification of metabolites can be far more efficient than conventional cell-based formulations (Arora and Mishra 2016).

Metabolite-based formulation involves the usage of metabolites, which are minor molecules or substances generally produced by microorganisms, plants, or other living organisms, as a basis for formulating products with specific applications in agriculture, medicine, or industrial processes. The utilization of metabolites in bioformulations is an innovative approach that leverages the bioactive compounds produced by living organisms for various beneficial purposes. Metabolites play an

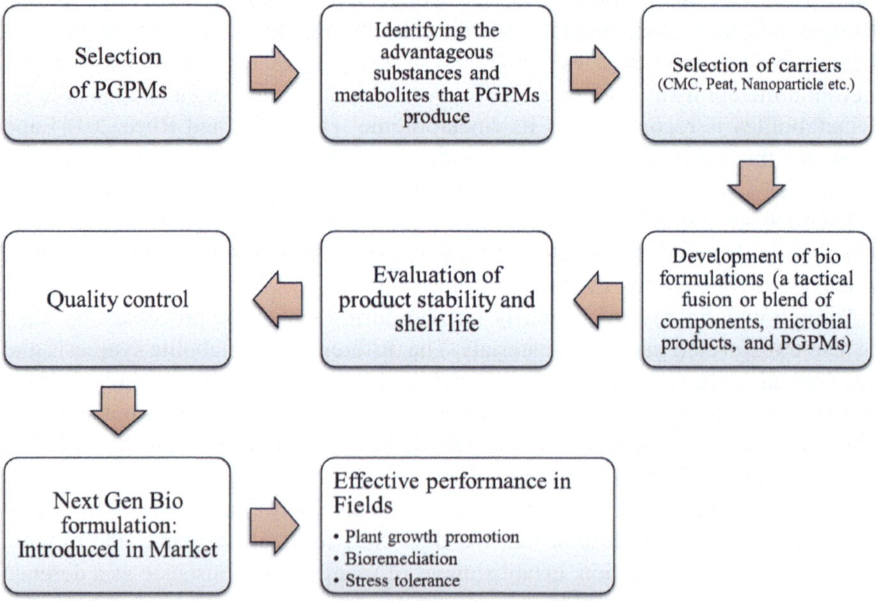

Fig. 4.1 Process flow-diagram to develop next-generation bio formulations

important role in biofertilizers formulation as they are the products of the metabolic activities of microorganisms within the biofertilizer (Tewari and Arora 2016).

As a whole, metabolite-based formulations are a potential approach for next-generation bioformulations, providing a range of possibilities, customized management plans, and creative solutions to severe ailments. Metabolite-based formulation covers microbial metabolites that support in providing bioregulators, necessary nutrients, and challenge phytopathogens. Flavanoids are one of the earliest biomolecules utilized in assortment with rhizobial inoculants to increase nodulation. According to previous research, adding flavanoids when mixed with rhizobial inoculation improves nodulation, nitrogen fixation, and resistance to abiotic threats (Oldroyd 2013). Mostly *Mesorhizobium*, *Rhizobium*, *Pseudomonas*, *Trichoderma* as well as mycorrhizal fungal strains are used for metabolite-based formulation (Tewari and Arora 2016). Streptomyces species is the most prevalent supplier of bioactive compounds, which not only accelerate plant growth and development but also protest to both biotic and abiotic stress (Compant et al. 2005; Doolotkeldieva et al. 2015; Kaari et al. 2023).

In particular, Kocuria genus is receiving a huge attention as a PGP bacterium owing to capacity to adapt to many metabolic pathways, potential to withstand extreme temperatures, and are capable to produce indole-3-acetic acid (IAA) (Faddetta et al. 2023). It has also been shown in the past that under extreme circumstances, the rhizobial strain's flavonoid biomolecules promote nitrogen fixation and nodulation. Lipo-chitooligosaccharide biomolecules are secreted by rhizobial microbes that form relationships to leguminous host plants. According to reports, lipo-chitooligosaccharide promotes symbiosis in fields lacking in rhizobia. Novozymes has considerably enhanced agricultural yield for each leguminous and non-leguminous crops by using flavonoids and lipo-chitooligosaccharide promoter. Similar to nod factors, mycorrhizal fungi also release myc factors, which promote symbiosis and open the signal transduction channel (Maillet et al. 2011).

Recent research has revealed that, in disastrous and stressful environments, dispersal of exopolysaccharide (EPS) by bacteria (which promote plant growth) especially Rhizobia and Pseudomonas, increases root colonization, biofilm formation, and nodulation (Wang et al. 2019). A crucial component of microbial resistance are the exopolysaccharide provided by Rhizobium. Strains are prevented from osmotic shock, high pH, radiation, desiccation, predators, and harmful compounds by using EPS as bioformulation essentials. Research reports have shown that adding initial molecules, like tryptophan, improved the synthesis of IAA and increased plant matter, grain production, and the development of root hairs. Growth was enhanced by PGPR supplemented with l-methionine, an essential metabolite and progenitor of ethylene. Nutrients including molasses, amino acids, and starch effluent have reportedly been employed as beneficial additives. According to Timmusk et al. (2014), these additives can increase the favourable strain's adaptability and durability in the hostile soil environment (Timmusk et al. 2014).

As per Arthurs and Dara (2019), there are 356 functional biological pesticides products legally authorized in the USA were employed against parasites such as nematodes, insects, and mites. Out of the 356 components, 57 are derived from microorganisms. This emphasizes the role that microorganisms play in biological protection, which was accomplished by directly inoculating the soil with the live hostile strain (Arthurs and Dara 2019). According to Berini et al. (2018), an amalgamation of microbial enzymes and metabolites from biocontrol cultures can be framed, created, and improved this strategy will get around some restrictions on the convention, care, and storage of living things (Berini et al. 2018).

In previous studies, phosphate-solubilizing bacteria, or PSBs, have been shown to emit biosurfactants with antibacterial, emulsifying, hydrating substance, antiviral, and anti-insecticidal properties (Arora and Mishra 2016). Liquid bioinoculants with biosurfactants are primarily sprayed onto the plant's above-ground sections. Liquid bioinoculants release phagostimulatory and stimuli substances, such as pheromones, glutamate, sucrose, and molasses, which attract and destroy phytopathogens. Antimicrobial microorganisms that produce diacetylphloroglucinol, pyrrolnitrin, fanzines, and omission A include fluorescent Pseudomonas. Analogous studies have been conducted on Bacillus species concerning antibiotic production. Bacillus species have also been investigated for their potential to produce antibiotics. Furthermore, a number of antiphytopathogenic characteristics demonstrated by secondary metabolites of Bacillus and Pseudomonas (Perez-Garcia et al. 2011).

4.5 Metabolites: Biologically Active Substances

According to Villas-Boas et al. (2007), biological material like microbial cells typically contain three different types of metabolites and can be discovered both within the cell and beyond the cell are water soluble (polar), water insoluble (nonpolar), and volatiles. In cells primary metabolism and secondary metabolism have independent activities, but they are linked closely because primary metabolism produces the energy and pioneer metabolites, needed for secondary metabolism (Villas-Boas et al. 2007). According to Barkal et al. (2016), microbial secondary metabolism is an abundant supplier of biologically active substances (Barkal et al. 2016). The fundamental biochemical mechanisms of secondary metabolism include those that produce PKs, isoprenes, peptides, and aromatic compounds (Ruiz et al. 2010).

Metabolic substances are extremely diverse, despite the variety is limited to atoms rather than molecules (Aldridge and Rhee 2014). They can also be categorized according to their action (primary or secondary) and origin (intrinsic or extrinsic). External metabolites come from sources outside of microbial cells, whereas intrinsic metabolites derived inside a cell (Vaidyanathan 2005). Manufacturing of primary or secondary metabolites, subjected to the growth phase of the microbial cells. The synthesis and analysis of metabolites (anabolic and catabolic activity) are linked to the metabolic paths of primary metabolites, whereas poor development ratio and a stress reaction are linked to the metabolic pathways of secondary metabolites (Pinu et al. 2017). Since the former metabolite categorizations are widely employed in the previous studies, there is much room for interpretation because the lines dividing them are not clearly defined (Vaidyanathan 2005).

4.5.1 Primary Metabolites

Primary metabolites are chemicals formed throughout the growth and development processes and involved in the primary metabolic processes. These metabolites are the intermediate products of anabolic metabolism. Some of the primary metabolites manufactured industrially, which involved amino acids, vitamins, organic acids, out of which alcohol is the foremost primary metabolite manufactured on a huge scale, industrially.

Primary metabolites are constantly being produced by bacteria while bacterial cell is proliferating. Important cellular functions including growth, development, and reproduction depend on them in addition to being necessary for survival of microbial. According to Pande and Kost (2017), auxotrophy occurs while these necessary metabolites cannot be produced by bacteria, and the organism will perish unless the metabolite is provided externally (Pande and Kost 2017).

Primary metabolites remain intact throughout phyla and kingdoms (Karlovsky 2010), most microbial species create primary metabolites that are almost identical to one another (Vaidyanathan 2005). Primary metabolites tend to accumulate in small numbers and at low concentrations, because they transform within cells more quickly than secondary metabolites; but this doesn't happen all the time because microbes might produce an excessive amount of intermediate metabolites from primary metabolic processes, which they could then release into the surrounding fluid (Pinu et al. 2017).

4.5.2 Secondary Metabolites

These are the organic substances produced by microbes that do not directly contribute to the organism's growth, development, or reproduction—all of which are necessary for the survival of the leading microbes. Secondary metabolites often serve specific ecological functions. Rather, they are seen as necessary because they offer some form of endurance and advantage to the manufacturing organisms, such as enhancing the nutrients accessible intake or defence against threats from the environment (Breitling et al. 2013; Sansinenea and Ortiz 2011). Nutritional restriction typically drives secondary metabolism. Secondary metabolites are generated throughout the static stage of bacterial development when resources become scarce and trash gathers, it follows that metabolites generated during this development stage should have more biological activities (Ruiz et al. 2010).

Certain secondary metabolites possess antimicrobial characteristics and have the ability to impede the proliferation of harmful microbes. This helps to prevent infections that are spread through the soil. Adding microbes to biofertilizers results in the construction of secondary metabolites that are antibacterial or antifungal, which can help in the management of plant diseases and improve general plant health (Pinu et al. 2017).

Systemic resistance against infections can be induced in plants by certain secondary metabolites that activate the plant's defence mechanisms. Generally phenolic compounds flavonoids (mainly genistein, hesperitin, and naringenin), lipo-chitooligosaccharides (LCOs), terpenoids, alkaloids, phytohormones such as auxins, gibberellins, cytokinins, ethylene, etc. are well identified secondary metabolites, which play a variety of functions, such as signaling, antibacterial activities, and interaction with the plant immune system. Novel secondary metabolites such as enzyme inhibitors, pheromones, medications, and so forth are derived from streptomycetes and related actinomycetes (Morel et al. 2016).

Certain secondary metabolites can stimulate nitrogen-fixing activity in symbiotic bacteria, enhancing the process of biological nitrogen fixation. Biofertilizers with phosphate-solubilizing microorganisms producing secondary metabolites can improve phosphorus availability to plants. Some secondary metabolites help microorganisms adapt to environmental stressors, such as extreme temperatures or changes in soil conditions. Biofertilizers with microorganisms producing stress-tolerant secondary metabolites may remain viable and effective under diverse environmental conditions Biofertilizers containing microorganisms that produce ISR-inducing secondary metabolites contribute to the plant's capacity to protect itself against diseases (Choudhary et al. 2007) .

Secondary metabolites are involved in microbial communication through quorum sensing, influencing the behaviour of microbial populations. Microorganisms especially bacteria in biofertilizers that utilize quorum sensing and secondary metabolite production can influence the dynamics of the microbial community in the rhizosphere (Barnard et al. 2007). Many bacterial activities, including those responsible for in the construction of secondary metabolites, a disease's and a harmonious relationship, can be brought underneath the quorum sensing regulon, irrespective of the specific strain of the bacteria. Understanding the roles of secondary metabolites in microbial interactions is essential for designing effective bioformulations-based biofertilizers that contribute to sustainable and environmental friendly agricultural practices.

4.5.3 Soluble and Volatile Organic Compounds (VOCs)

Some microbes release volatile organic compounds (VOCs) that can affect plant development and act as signalling molecules in plant–microbe relations. VOCs are organic compounds that have great vapour pressure and can easily evaporate into the air.

In the context of bioformulation, VOCs created by microorganisms, particularly bacteria and fungi, can play various roles, induce systemic resistance in plants, have antimicrobial properties, inhibit the growth of pathogens in the neighbouring environment as well as act as plant growth regulators and promoters. They may influence various aspects of plant growth, including seed germination, root development, improved nutrient cycling, and overall biomass production, leading to increased crop yields. Microorganisms use VOCs for intercellular communication and signalling, which show a major role in establishing advantageous interactions between microorganisms and plants, establish a healthy rhizosphere environment by facilitating communication and cooperation among microbes (He et al. 2022). Stimulate the activity of other beneficial microorganisms in the soil to leading to and soil health. Soil bacteria produces two basic forms of secondary metabolites that are known as soluble metabolites and volatile organic compounds (VOCs), which were recently explored (Tyc et al. 2017). Nearly 120 volatile compounds, involving alcohols, ketones, esters, and PGPs, have been identified from Streptomyces species (Schöller et al. 2002). Multiple applications of these VOCs in agriculture have been documented, including the management of phytopathogens and PGP consequences (Olanrewaju and Babalola 2019).

Polarity and the metabolites' capacities for evaporation and their diffusion serve as the root for this taxonomy. Soluble metabolites are unique in their bioactivities and are water soluble owing to their extreme polarity, however, volatile organic compounds (VOCs) can permeate and evaporate via openings that are occupied with air or water. Soluble secondary metabolites include bacteriocins, lipopeptides, and PKs and volatile organic compounds (VOCs) consist of pyrazines, indole, and several sulphur-containing substance (Tyc et al. 2017).

Some VOCs have been stated to repel nematodes, providing a natural defence against these plant-parasitic worms, protect plant roots from damage. Application of VOCs in bioformulations are areas of ongoing research, and the specific effects of VOCs can vary depending on the microbial species involved.

It is worthwhile to remember that, even though microorganisms in biofertilizers might produce antimicrobial chemicals that are useful, their consumption should be carefully controlled to prevent possible harm to natural environments and non-target microbes. Sustainable agriculture practices often involve a holistic approach that considers both the positive and potential drawbacks of biofertilizers with antimicrobial properties.

4.6 Major Aspects of Metabolite-Based Bioformulation

Metabolite-based biofertilizers offer innovative solutions in sustainable agriculture by fastening the power of microbial metabolites to improve soil fertility, nutrient uptake, and plant growth (Lugtenberg and Kamilova 2009). Some key applications of metabolite-based biofertilizers (Fig. 4.2) are as follows:

4.6.1 Plant Growth-Promoting Metabolites

Metabolites produced by several microbes can improve response to stress, nutrients endorsement, and vegetative growth. Formulating biofertilizers or biostimulants based on these plant growth-promoting metabolites can improve crop yields and overall agricultural productivity.

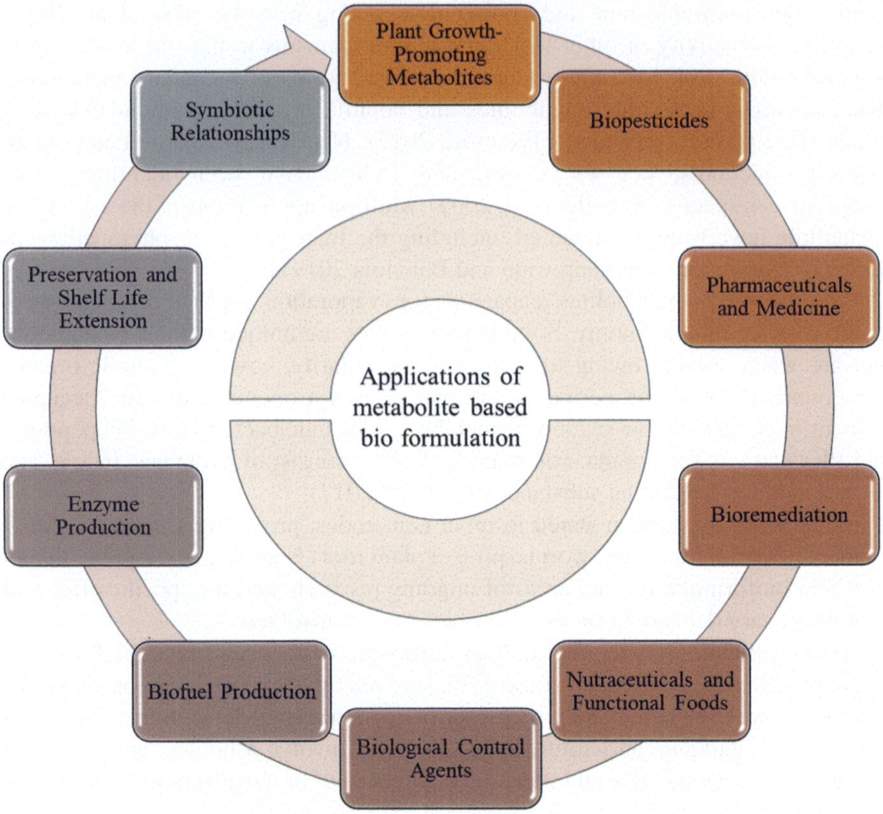

Fig. 4.2 Schematic diagram showing different application of metabolite-based bioformulations

4.6.2 Biopesticides

Metabolites with pesticidal properties can be used in the preparation of biopesticides. These bioformulations can provide an ecological substitute to chemical pesticides, helping to manage pests and diseases in crops.

4.6.3 Pharmaceutical and Industrial Applications

Metabolite-based bioformulations can be designed for pharmaceutical and industrial applications. Some microorganisms produce bioactive metabolites with antimicrobial properties. These metabolites can be used in effective treatments and development of fruitful formulations.

4.6.4 Bioremediation

Metabolites produced by certain microorganisms have the ability to degrade or detoxify pollutants. Bioformulations based on these metabolites can be employed in bioremediation processes to clean up contaminated environments.

4.6.5 Nutraceuticals and Functional Foods

Metabolites with health-promoting properties can be incorporated into formulations for the production of nutraceuticals or functional foods. These products can provide additional health benefits beyond basic nutrition.

4.6.6 Biological Control Agents

It is possible to develop metabolite bioformulations for the biological management of plant diseases. Certain metabolites can act as antagonists to pathogens, contributing to the management of soil-borne and foliar diseases.

4.6.7 Biofuel Production

Metabolites from microorganisms involved in fermentation processes can be harnessed for biofuel production. Bioformulations can be optimized to enhance the efficiency of biofuel conversion processes.

4.6.8 Enzyme Production

Some microorganisms produce metabolites that stimulate the production of enzymes with industrial applications. Bioformulations can be designed to expand the yield and stability of these enzymes for numerous industrial methods.

4.6.9 Preservation and Shelf-life Extension

Metabolites with antimicrobial properties can be used in bioformulations for food preservation. These formulations can help extend the shelf life of perishable goods by constraining the growth of spoilage microorganisms.

4.6.10 Symbiotic Relationships

Metabolite-based bioformulations can exploit symbiotic relationships between microorganisms and plants. Formulations can be designed to enhance the exchange of beneficial metabolites between plant roots and microbial partners, fostering mutualistic interactions.

Development of metabolite-based bioformulations (Table 4.1) involves understanding the specific metabolites produced by microorganisms and optimizing their delivery and stability for field applications. This approach holds potential for sustainable and eco-friendly solutions.

4.7 Type of Metabolic Substances

Various plant growth-promoting substances or metabolites play key role in refining soil condition, fertility, and crop yield. The specific types of metabolites (Table 4.2) produced by beneficial microorganisms in biofertilizers include the following substances:

4.7.1 Exopolysaccharides (EPS)

Exopolysaccharides (EPS) are complex, high-molecular-weight polysaccharides produced by microorganisms, including bacteria and fungi. These polymers are secreted into the extracellular environment and can form a slimy or gel-like substance. Exopolysaccharides (EPS) produced by certain bacteria which helps to improve soil structure by enhancing aggregation. This contributes to water retention, aeration, and nutrient availability (Sandhya and Ali 2015).

In the context of biofertilizer formulation, EPS produced by certain microorganisms can have several important roles. EPS are additional significant metabolites released by a variety of PGPR, such as pseudomonads and rhizobia, that support nodulation, root the invasion, the formation of biofilms, the defence of established cells subject to strain, the removal of toxic substances, and the function of a carbon source (Tewari and Arora 2014).

Large-scale, inexpensive, and simple production of EPS is possible. Rhizobia, are known to create large amounts of these biological materials, are known to be beneficial from the defence that EPS provides for nitrogenase. EPS-modified bioformulations shield the cells and also facilitate the bioinoculant's rhizosphere occupation. EPS can be very helpful in defending plant roots from abiotic threats (Sandhya and Ali 2015). Therefore, EPS can be added to bioformulations as modifications or even employed to carry slow-release bioprotectant microorganisms.

According to Seneviratne et al. (2011), microorganisms protected by EPS holders can withstand radiation, drastic pH changes, osmotic shock, dehydration dangerous chemicals in soil, and even predators. Exopolysaccharides can enhance soil accumulation, leading to the formation of constant soil structures or masses. This

Table 4.1 Applications of metabolites in biofertilizer formulations

Type of metabolites (based on application)	Function	Description
Phosphate-solubilizing metabolites	Nutrient mobilization	Plants can access more bound phosphate in the soil by solubilizing it with organic acids and enzymes produced by certain microorganisms. This helps improve phosphorus uptake by plants
Nitrogen-fixing metabolites	Nitrogen fixation and assimilation	Nitrogen-fixing bacteria, like rhizobia and cyanobacteria, produce metabolites like ammonia and amino acids during nitrogen fixation. These compounds contribute to increased soil nitrogen levels and provide plants with a readily available nitrogen source
Plant growth-promoting substances	Indirect plant growth promotion	Metabolites like phytohormones (auxins, cytokinins, gibberellins) and siderophores produced by certain beneficial microbes act as plant growth regulators. They promote root development, improve nutrient acceptance, and increase overall plant health
Stress-inducible metabolites	Stress tolerance	Stress-Inducible Metabolites: Some microorganisms produce metabolites that help plants tolerate environmental stress, such as drought or salinity. These metabolites can include osmoprotectants and antioxidants that protect plants from stress-induced damage
Antimicrobial metabolites	Biocontrol and disease suppression	Certain microbes produce metabolites with antimicrobial properties, facilitating to conquer soil-borne pathogens and shelter plants from diseases. These metabolites can include antibiotics, lytic enzymes, and volatile organic compounds
Exopolysaccharides (EPS)	Improved soil structure	Some bacteria produce EPS, which can improve soil structure by enhancing aggregation. This contributes to better water retention, aeration, and nutrient availability in the rhizosphere
Organic acid production	Enhanced nutrient absorption	Organic acids such as citric acid, malic acid, and gluconic acid manufactured by certain microbes. Who can upsurge nutrient solubility in the soil, facilitating better nutrient absorption by plant roots
Buffering metabolites	pH regulation	Microbial metabolites can act as buffers, helping to regulate soil pH. This is important for optimizing nutrient availability to plants, as the solubility of many nutrients is pH-dependent
Secondary metabolites	Biostimulation of indigenous microbes	Some microorganisms produce secondary metabolites that act as signalling molecules, influencing the behaviour of other microorganisms in the rhizosphere. This can lead to a more beneficial microbial community for plant growth

Table 4.2 Metabolic substances produced by beneficial microorganisms and their applications

Metabolite/biomolecule	Secreted by	Application	References
Flavanoids (Novozyme (Denmark)	–	Enhancement of nodulation, ability to withstand abiotic stress	Morel et al. (2016), Oldroyd (2013)
Lipo-chitooligosaccharides (LCOs)/nod factors	Root modulating bacteria	LCOs are the signal substances which are necessary for rhizobia and legume roots to form a symbiotic relationship	Morel et al. (2016), Oldroyd (2013)
Myc factors	Mycorrhizae	Myc factors are able to induce mycorrhizal associations with plants by activating the signal transduction pathway or common symbiosis (SYM) pathway in the host species	Maillet et al. (2011)
Exopolysaccharides (EPS)	PGPR including rhizobia and pseudomonads	EPS help with nodulation, biofilm formation or root the invasion, protecting newly developed cells under stress, removing toxins, serving as a carbon source, etc.	Sandhya and Ali (2015), Seneviratne et al. (2011), Tewari and Arora (2014)
Tryptophan (Phytohormone, indole acetic acid, IAA)	–	Significant rise in IAA synthesis, as well as improvements in grain yield, plant matter, root hair development, and even disease resistance	Zahir (2010)
L-methionine (precursor of ethylene)	–	Enhanced growth	Arora and Mishra (2016)
Phytohormones (gibberellic acid and cytokinins)	–	Good yield in the field	Arora and Mishra (2016), Morel et al. (2016)
Phosphates, potash rocks, sulphur or insoluble zinc	–	Good health	Arora and Mishra (2016)
Crude rock phosphate and potassium	–	Create sulphuric acid, which provides micronutrients and promotes phosphates and sulphates to dissolve	Stamford et al. (2008)
Chitin	–	Increased chitinase activity, which in turn increased PGP and biocontrol activity by giving carbon to the opponents	Arora et al. (2007)

(continued)

Table 4.2 (continued)

Metabolite/biomolecule	Secreted by	Application	References
Nutrients/amino acids/ cellulose/starch wastewater/molasses	–	Useful supplements in the bioinoculants	Arora (2015), Brar et al. (2006)
Biosurfactants	–	Moistening/spreading substances, lubricants, antiviral, antibacterial, and insecticidal properties	Farrar and Ridgway (1995), Thavasi et al. (2014)
Antibiotics	–	Develop a range of antifungal and antibacterial substances	Haas and Défago (2005), Raaijmakers et al. (2002)

improves soil porosity, aeration, water infiltration and retention and contributes to the development of a well-structured soil, creating an environment conducive to root growth and nutrient availability. This is particularly beneficial during dry periods, as it helps plants withstand drought stress (Seneviratne et al. 2011).

EPS shows a key part in the formation of biofilms, which are complex microbial communities surrounded in a medium of EPS. Biofilms can adhere to root surfaces, promoting beneficial interactions between microorganisms and plant roots, establish a beneficial microbial community around plant roots, promoting nutrient cycling. EPS can chelate essential nutrients, which includes the chelation of cations such as calcium, magnesium, and iron, making them more available to plants.

4.7.2 Antimicrobial Compounds

4.7.2.1 Antibiotics

Beneficial microbes produce antibiotics that are capable to conquer the growth of harmful soil-borne pathogens. Biofertilizer-producing microorganisms may release antibiotics (phenazines, diacetylphloroglucinol, pyoluteorin, pyrrolnitrin, and oomycin A, zwittermicin, and kanosamine, etc.) that have specific inhibitory effects on pathogenic bacteria or fungi in the soil. Antimicrobial substances that include pyrrolnitrin, fanzines, omission A, and diacetylphloroglucinol are synthesized by microorganisms analogous to fluorescent Pseudomonas. Bacillus species have also been investigated for their potential to produce antibiotics (Perez-Garcia et al. 2011).

4.7.2.2 Lytic Enzymes

Some microbes manufacture enzymes that can break down the cell walls of competing microorganisms, leading to their lysis and death. Enzymes that destroy the cell walls of pathogenic organisms are known as lytic enzymes.

4.7.3 Siderophores

Iron is a micronutrient that is crucial for several biochemical practices in plants. Siderophores are produced by certain microbes to chelate iron, making it more available to plants. By facilitating the uptake of iron, siderophores contribute to the mobilization of this essential nutrient. This is mainly significant in soils where iron availability is limited. Siderophores are minor, high affinity iron-chelating combinations formed by microorganisms, including bacteria and fungi. These compounds are secreted into the surrounding environment to scavenge iron, an essential nutrient, from the soil. In the context of biofertilizer preparation, the fabrication of siderophores by certain microorganisms plays a crucial role (Saha et al. 2013).

In soil which is iron-stressed, siderophore-producing microorganisms (SPM) generate a variety of chemicals that chelate iron and reduce plant stress. Microbes that create siderophores engage in biofertilizer and biocontrol activities for plants; as a result, SPM is an environmentally favourable marker for sustainable agriculture and is suitable for high yield and cultivation on damaged land. Siderophores have a high affinity for iron ions, which are often available in the soil in insoluble forms that are not easily accessible to plants. They chelate iron, forming soluble complexes that can be taken up by microorganisms (Singh et al. 2022).

Siderophore-producing microorganisms in biofertilizers improve the availability of iron in the rhizosphere, making it more accessible for plant uptake, improve nutrient availability to plants, leading to enhanced growth and development. Iron is involved in chlorophyll synthesis and other metabolic processes critical for plant health. Adequate iron availability helps prevent iron deficiency and promotes overall plant vigour. Siderophore-producing microbes in biofertilizers contribute to improved plant condition by ensuring an adequate supply of iron, particularly in iron-deficient soils. Siderophores can be categorized according to a variety of factors, including their source, their cyclic and linear structures, and the makeup of their efficient collections. Siderophores are categorized as hydroxamate siderophore, catecholate type siderophore, carboxylate type siderophore, and miscellaneous ligand siderophore based on their functional groups (Singh et al. 2022).

Some siderophores exhibit antagonistic properties against certain pathogenic microbes. They can compete for iron, limiting the availability of this essential nutrient to harmful pathogens. The production of siderophores is a valuable trait in biofertilizer microorganisms, as it enhances their capability to support plant development, improve nutrient availability, as well as contribute to overall soil health. Including biofertilizers with siderophore-producing strains in agricultural practices can be part of a sustainable approach to nutrient management and plant nutrition. This formulation's expensive price and massive amounts manufacture are its main disadvantages. The next-generation formulations may additionally incorporate the strain that promotes plant development with chemicals and vehicles that boost microbial metabolism. In this composition, attractants, additives, and accelerators comprise primary and secondary metabolites as well (Chaudhary et al. 2020).

4.7.4 Osmoprotectants

To combat osmotic strains, a variety of bacteria, seaweeds, and plants might gather a variety of substances known as compatible liquids or osmoprotectants. Bacteria employ a variety of survival tactics, such as the synthesis and build-up of polyhydroxyalkanoates (PHA) or osmolytes. Bacteria that are osmo-adapted add osmolytes, like trehalose or glycine betaine, and are more resistant to desiccation than non-osmo-adapted cells, which significantly increases the rate of plant growth (Bonaterra et al. 2005). Because PHA gives cells the capacity to withstand a range of damaging physical and chemical stimuli, cells with greater PHA contents can live longer than those with lesser quantities (Kadouri et al. 2005; Morel et al. 2016). Examples of these compounds include sugar alcohols (inositol, mannitol), carbohydrates (trehalose), amino acids (proline, glutamate, etc.), and tertiary sulphonium compounds (glycine betaine, for example) (Ejaz et al. 2020; Slama et al. 2015).

These are the compounds that support organisms, particularly microorganisms, cope with osmotic stress, environmental pressure, such as drought or salinity. Osmotic stress occurs when there is a major difference in solute concentration between a cell and its external environment, leading to water movement across the cell membrane. Osmoprotectants help maintain cell turgor pressure, prevent water loss, and protect cellular structures under conditions of high osmolarity.

In the context of biofertilizer production, osmoprotectants play a role in refining the stress tolerance of microorganisms. Proline, glycine, betaine, trehalose are three different osmoprotectants. Proline is a common amino acid that serves as a major osmoprotectant in many microorganisms and plants. Glycine betaine is a quaternary ammonium compound that helps maintain cell osmolarity. Trehalose is another disaccharide sugar that helps to maintain the structural integrity of cells under stress. Proline analogues and derivatives are synthetic compounds designed to mimic the osmoprotective effects of natural osmolytes. Furthermore, an extensive variety of osmoprotectant substances found in algae, including ions such as Ca2+, amino acids, betaines, carbohydrates, flavonoids, phenolic acids, phytohormones, polyols, and tertiary sulphonium, might boost resistance to stress (Carillo et al. 2020). Proline and glycine betaine help plants and microorganisms maintain water balance and survive periods of water scarcity. Biofertilizers with osmoprotectant-producing strains can be beneficial in arid or drought-prone regions by enhancing plant and microbial resilience to water stress. Osmoprotectants play a key role in microbial and plant adaptation to saline environments by counteracting the osmotic effects of high salt concentrations. In biofertilizers, osmoprotectant-producing microorganisms can contribute to improved nutrient accessibility and plant growth in saline soils. Osmoprotectants help microorganisms and plants survive extreme temperatures by stabilizing cellular structures and preventing damage from freezing or heat. Biofertilizers with osmoprotectant-producing strains are more likely to remain viable and active under temperature extremes, ensuring their effectiveness in the field.

Osmoprotectant-producing strains in biofertilizers support the establishment of a favourable microbial community around plant roots, contributing to nutrient cycling and improved plant health. Osmoprotectants released by microorganisms can stimulate the growth and activity of other advantageous microbes in the soil. Incorporation of osmoprotectant-producing microorganisms into biofertilizer formulations enhances the stress tolerance of these microorganisms and their aptitude to encourage plant growth beneath challenging environmental conditions. This contributes to sustainable agriculture by improving nutrient availability and promoting plant health in diverse and often stressful agroecosystems.

4.7.5 Vitamins and Growth Factors

Rhizobacteria and plant pathogens rely on the rhizo-deposits, which contain vital carbon sources such as sugars, organic acids, amino acids, fatty acids, phenolics, nucleotides, sterols, and vitamins (Sasse et al. 2018). Vitamins and growth factors are essential organic compounds that play critical part in the growth, development, and metabolism of microorganisms.

In biofertilizer production, incorporation of vitamins and growth factors can be beneficial for enhancing the sustainability, activity, and effectiveness of the microorganisms used in the formulation. Addition of vitamins and growth factors to biofertilizer formulations ensures that the microorganisms have the necessary nutrients for their metabolic processes, promoting cell growth, and maintaining high viability.

Vitamins and certain growth factors contribute to the stress tolerance of microorganisms. Including these compounds in biofertilizers helps the microbes withstand environmental stresses such as temperature fluctuations, drought, or nutrient limitations. Some microorganisms in biofertilizers produce beneficial metabolites, such as enzymes and organic acids, that aid in nutrient solubilization and availability. Vitamins and growth factors support the biosynthesis of these metabolites, enhancing nutrient mobilization and positively influence the nutrient acquisition by plants. Microorganisms in biofertilizers contribute to the production of these compounds, potentially enhancing nutrient uptake by plant roots. Microbes in biofertilizers form plant growth-promoting (PGP) substances such as auxins, cytokinins, and gibberellins. Vitamins and growth factors support the biosynthesis of these compounds, which are important to improved plant growth and development. Some microorganisms in biofertilizers produce secondary metabolites with beneficial properties, such as antimicrobial or biocontrol activities. Vitamins and growth factors are necessary for the biosynthesis of these secondary metabolites and contribute to the efficiency of biocontrol agents in suppressing soil-borne diseases. Biofertilizers enriched with these compounds can enhance the antagonistic activities of beneficial microorganisms against plant pathogens. They support the colonization of plant roots by beneficial microbes which is essential for establishing symbiotic or associative relationships that contribute to plant health and nutrient endorsement (Kumar et al. 2022).

Adding vitamins and growth factors can improve the stability and longevity of biofertilizer formulations, extending their shelf life and maintaining the viability of microbial populations during storage. Vitamins and growth factors contribute to the rhizosphere competence of biofertilizer microorganisms, enhancing their ability to establish and maintain beneficial interactions with plant roots. Incorporating vitamins and growth factors into biofertilizer formulations is a next-generation approach to regulate the performance of microorganisms and maximize their positive impression on plant growth and soil health. These compounds are essential for the overall functionality of the microorganisms and their ability to interact synergistically with plants in the agroecosystem.

The specific types and amounts of metabolites production depend on the microbial species present in the biofertilizer formulation. Researchers and manufacturers aim to optimize the production of these metabolites to enhance the effectiveness of biofertilizers in promoting plant growth and improving soil condition. Additionally, advancements in genetic engineering and microbial strain selection aim to tailor microbial strains for enhanced metabolite production in biofertilizers.

In biofertilizer formulations, understanding and manipulating the production of these metabolites are critical for designing effective products that enhance soil fertility, promote plant growth, and contribute to sustainable agriculture. Research in this field continues to explore ways to optimize the production of specific metabolites for improved biofertilizer performance.

4.8 Instances of Metabolites

Some examples of metabolites are as follows.

4.8.1 Bacteriocins

Bacteriocins serve as only the precursors of a restricted number of metabolic processes, they show greater deposition quantities than primary metabolites (Covington et al. 2017). They may gather inside microbial cells or are discharged into the extracellular media (Pinu et al. 2017). Genes that control the enzymes needed for the manufacture of secondary metabolites are typically chromosomal. These metabolites can be synthesized either ribosomally (like bacteriocins) or nonribosomally like PKs (Ruiz et al. 2010). In an investigation conducted by Ayed et al. (2015) antibacterial properties of bacteriocins produced by *B. amyloliquefaciens* An6 has been found most effective at the beginning of the static stage (Ayed et al. 2015).

4.8.2 Dimethyl Disulphide

Dimethyl disulphide is a very interesting volatile substances produced by bacteria. With the trade name "Palandin", it is applied as a fumigant beside nematodes and soil-borne infections and has been licenced by the US Environmental Protection Agency (Tyc et al. 2017). Hydrolytic enzymes and secondary metabolites are two significant bioactive substances that Bacillus species create. They have been found to be bactericidal, antifungal, and antibacterial in numerous investigations. In addition it also shows nematicidal activities, however, their nematocides' exact actions are yet unknown. While some research suggests the nematicidal activity seen in Bacillus species is related to the generation of secondary metabolites and/or hydrolytic enzymes, not all studies have thoroughly examined these metabolites or enzymes, so their precise perspectives and categorization it is still unclear.

4.8.3 Cyclic Lipopeptides

Many secondary metabolite gene groups, including nonribosomal peptide synthetases and polyketide synthetases, are found in species belonging to the Bacillus genus. These enzymes are responsible for the synthesis of various lipopeptides and PKs (Aleti et al. 2015). Such microbes are the primary manufacturers of cyclic lipopeptides, which belong to four families: kurstakin, fengycin, surfactin, and iturin (Frikha-Gargouri et al. 2017; Hathout et al. 2000; Rangarajan and Clarke 2016).

4.8.4 Polyketides

These are a broad part of secondary metabolites that are created nonribosomally (Frikha-Gargouri et al. 2017) through enzymes called PK synthetases (Aleti et al. 2015). This kind of metabolites includes polyketides (Mondol et al. 2013; Palazzini et al. 2016). According to Mondol et al. (2013), they have pharmaceutical qualities and a variety of structures. Difficidin, bacillaene, and macrolactins are among the PKs made by Bacillus species (Aleti et al. 2015; Palazzini et al. 2016; Sobhy et al. 2021).

4.9 Constraint of Metabolites-Based Bioformulations

This formulation's high price and large-scale production are its main disadvantages. Future metabolite formulations also combine the strain that promotes plant development with substances and carriers that boost microbial activity. Additionally, this mixture contains stimuli, additives, and stimulants that are primary and secondary metabolites (Chaudhary et al. 2020).

Indeed, previous research are recommending the application of formulations devoid of cells (based on metabolites), especially when dealing with common or persistent human diseases like *Pseudomonas aeruginosa*, Bacillus sp., or fungi like *Purpureocillium lilacinum* (Cavello et al. 2015; Khare and Arora 2010).

Although *P. aeruginosa* is a renowned PGPR that produces a variety of PGP metabolites, its direct application (in biological inoculations) is severely limited due to its status as a human disease. Utilizing formulations free of cells derived from these microorganisms may offer the answer. Cell-free extracts can have a wide range of metabolites; therefore, it will be challenging to list all of the constituents in these kinds of formulations. Nevertheless, there are concerns over the ingredients of these formulations. Issues have also been raised about detoxification issues and the necessary proportions of the most active metabolites (Zacky and Ting 2015). Prior to the large-scale production and implementation of cell-free formulations, these concerns must be resolved.

4.10 Future Prospective

Studies and experiments conclusively show that an evolution in thinking is needed in order to use metabolites and additives in bioformulations. Due to their viability or low cost, several of these biomolecules are already used in bioformulations; other additions of this type offer good value despite requiring inputs at extremely low quantities. The metabolites should be incorporated into the manufacturing processes of the next generation, as this will not only increase their performance and shelf life but also expand their role and level of acceptance worldwide. Because these metabolite-based formulations work in a variety of geographical and climatic conditions, they are more widely applicable and favoured by consumers and the agro-market. The modified formulations with the addition of bacterial cells and metabolites will be a dependable and sustainable technology that improves the action's quality on multiple levels. These future bioformulations can be defined as "amalgamation of the plant growth promoting microbe(s), a carrier and additives which aid the microbe in one way or the other. These additives can be the metabolite(s) or nutrients (for microbes or plants). Metabolites can be in the form of protectants, adjuvants, attractants, stimulants, anti-microbials or precursors of biological origin". Nowadays, microbial metabolites are employed in many different sectors across the globe. We remain concerned about their application in bioformulations and agriculture, nevertheless. It is time to release the reins and employ the metabolites to boost the inserted PGP microorganisms' survivability, activity, and overall performance.

The development of DNA sequencing methods will highlight the importance of soil microbial communities' role in the soil ecosystem and their capacity for metabolic processes (Kumar et al. 2017). The primary emphasis is on the methods for isolating, formulating, and understanding the interactions between the domestic growth-promoting bacteria and unfavourable stress conditions so that they can be utilized as bioinoculants (Kaushal and Wani 2016).

Whenever employed as inoculating organism, microbes increase their vulnerability to biotic and abiotic stressors as well as diseases, in addition to promoting growth (Dangi et al. 2018; Imam et al. 2016). Although there remain plenty of opportunity for improvement in terms of leveraging cutting-edge technologies to enhance the performance of bioinoculants, doing so will require a more cohesive application of concepts for sustainable agriculture (Chaudhary et al. 2020). Metabolites produced by certain microbes exhibit the ability to degrade various pollutants and contaminants present in soil, water, and air. Bioformulations containing these metabolites can be used for bioremediation purposes, facilitating the detoxification and removal of environmental pollutants, including hydrocarbons, heavy metals, and chemical-based pesticides. Metabolites derived from microbial sources have shown promising therapeutic potential for treating various human diseases and medical conditions. Bioformulations containing these metabolites can be developed into pharmaceutical products for managing infections, inflammation, metabolic disorders, and other health-related issues. Metabolite-based bioformulations can be formulated using microbial consortia consisting of multiple beneficial microorganisms that synergistically produce a different variety of metabolites with complementary functions. These kind of formulations can offer enhanced efficacy and versatility compared to single-strain bioformulations, providing comprehensive solutions for agriculture, environmental remediation, and healthcare applications. Overall, metabolite-based bioformulations represent a promising avenue for harnessing the metabolic capabilities of microorganisms to address various challenges in agriculture, healthcare, and environmental management. Global improvements in food security, public health, and environmental sustainability could be made possible by further study and development in this area, which has plenty of possibilities.

4.11 Conclusion

Metabolite-based formulations are a potential direction for next-generation bioformulations, providing a range of options, tailored management plans, and creative ways to confrontation challenging issues or disease in agriculture. Metabolite-based bioformulations have gained significant interest in various fields including agriculture, medicine, and environmental science due to their potential in enhancing crop production, refining human health, and mitigating environmental pollution. Currently metabolite-based bioformulations are increasingly pursued. Metabolites produced by beneficial microbes such as bacteria and fungi often exhibit biocontrol properties against plant pathogens and pests. Bioformulations containing these metabolites can serve as eco-friendly alternatives to chemical pesticides, promoting sustainable agriculture practices while minimizing environmental risks. Certain metabolites synthesized by beneficial microbes can stimulate plant growth and development by improving nutrient uptake, improving stress tolerance, and modulating plant hormone levels. Bioformulations enriched with such metabolites can contribute by amplified crop yields, improved soil fertility, and reduced dependency on synthetic fertilizers.

References

Afridi MS, Van Hamme JD, Bundschuh J, Sumaira, Khan MN, Salam A, Waqar M, Munis MFH, Chaudhary HJ (2021) Biotechnological approaches in agriculture and environmental management - bacterium Kocuria rhizophila 14ASP as heavy metal and salt- tolerant plant growth-promoting strain. Biologia 76(10):3091–3105. https://doi.org/10.1007/s11756-021-00826-6

Aldridge BB, Rhee KY (2014) Microbial metabolomics: innovation, application, insight. Curr Opin Microbiol 19:90–96. https://doi.org/10.1016/j.mib.2014.06.009

Aleti G, Sessitsch A, Brader G (2015) Genome mining: prediction of lipopeptides and polyketides from Bacillus and related Firmicutes. Comput Struct Biotechnol J 13:192–203. https://doi.org/10.1016/j.csbj.2015.03.003

Arora NK (2015) Plant microbes Symbiosis: applied facets. Springer, India

Arora NK, Mishra J (2016) Prospecting the roles of metabolites and additives in future bioformulations for sustainable agriculture. In: Applied soil ecology, vol 107. Elsevier B.V, pp 405–407. https://doi.org/10.1016/j.apsoil.2016.05.020

Arora NK, Kim MJ, Kang SC, Maheshwari DK (2007) Role of chitinase and β-1,3-glucanase activities produced by a fluorescent pseudomonad and in vitro inhibition of Phytophthora capsici and Rhizoctonia solani. Can J Microbiol 53(2):207–212. https://doi.org/10.1139/w06-119

Arora NK, Khare E, Maheshwari DK (2010) Plant growth promoting Rhizobacteria: constraints in bioformulation, commercialization, and future strategies. In: Maheshwari DK (ed) Bacteria and plant health. Springer, pp 97–116. https://doi.org/10.1007/978-3-642-13612-2_5

Arthurs S, Dara SK (2019) Microbial biopesticides for invertebrate pests and their markets in the United States. J Invertebr Pathol 165:13–21. https://doi.org/10.1016/j.jip.2018.01.008

Ayed HB, Maalej H, Hmidet N, Nasri M (2015) Isolation and biochemical characterisation of a bacteriocin-like substance produced by bacillus amyloliquefaciens An6. J Glob Antimicrob Resist 3(4):255–261. https://doi.org/10.1016/j.jgar.2015.07.001

Bajsa N, Morel MA, Brana V, Castro-Sowinski S (2013) The effect of agricultural practices on resident soil microbial communities: focus on biocontrol and biofertilization. In: de Bruijn FJ (ed) Molecular microbial ecology of the rhizosphere, vol 2. Wiley-Blackwell, Hoboken, pp 687–700

Barkal LJ, Theberge AB, Guo C-J, Spraker J, Rappert L, Berthier J, Brakke KA, Wang CCC, Beebe DJ, Keller NP, Berthier E (2016) Microbial metabolomics in open microscale platforms. Nat Commun 7(1):10610. https://doi.org/10.1038/ncomms10610

Barnard AM, Bowden SD, Burr T, Coulthurst SJ, Monson RE, Salmond GP (2007) Quorum sensing, virulence and secondary metabolite production in plant soft-rotting bacteria. Philos Trans R Soc Biol Sci 362(1483):1165–1183. https://doi.org/10.1098/rstb.2007.2042

Baweja P, Kumar S, Kumar G (2020) Fertilizers and pesticides: their impact on soil health and environment:265–285. https://doi.org/10.1007/978-3-030-44364-1_15

Berini F, Katz C, Gruzdev N, Casartelli M, Tettamanti G, Marinelli F (2018) Microbial and viral chitinases: attractive biopesticides for integrated pest management. Biotechnol Adv 36(3):818–838. https://doi.org/10.1016/j.biotechadv.2018.01.002

Bonaterra A, Camps J, Montesinos E (2005) Osmotically induced trehalose and glycine betaine accumulation improves tolerance to desiccation, survival and effi- cacy of the postharvest biocontrol agent Pantoea agglomerans EPS125. FEMS Microbiol Lett 250:1–8

Brar SK, Verma M, Tyagi RD, Valero JR, Al E (2006) Recent advances in downstream processing and formulations of Bacillus thuringiensis based biopesticides. Process Biochem 41:323–342

Breitling R, Ceniceros A, Jankevics A, Takano E (2013) Metabolomics for secondary metabolite research. Metabolites 3(4):1076–1083. https://doi.org/10.3390/metabo3041076

Calvo P, Nelson L, Kloepper JW (2014) Agricultural uses of plant biostimulants. Plant Soil 383(1–2):3–41. https://doi.org/10.1007/s11104-014-2131-8

Carillo P, Ciarmiello LF, Woodrow P, Corrado G, Chiaiese P, Rouphael Y (2020) Enhancing sustainability by improving plant salt tolerance through macro- and micro-algal biostimulants. Biology 9(9):253. https://doi.org/10.3390/biology9090253

Cavello IA, Crespo JM, García SS, Zapiola JM, Luna MF, Cavalitto SF (2015) Plant growth promotion activity of keratinolytic fungi growing on a recalcitrant waste known as "hair waste.". Biotechnol Res Int 2015:1–10. https://doi.org/10.1155/2015/952921

Chaudhary T, Dixit M, Gera R, Shukla AK, Prakash A, Gupta G, Shukla P (2020) Techniques for improving formulations of bioinoculants. 3 Biotech 10(5):1–9. https://doi.org/10.1007/s13205-020-02182-9

Choudhary DK, Prakash A, Johri BN (2007) Induced systemic resistance (ISR) in plants: mechanism of action. Indian J Microbiol 47(4):289–297. https://doi.org/10.1007/s12088-007-0054-2

Compant S, Duffy B, Nowak J, Clément C, Barka EA (2005) Use of plant growth-promoting bacteria for biocontrol of plant diseases: principles, mechanisms of action, and future prospects. Appl Environ Microbiol 71(9):4951–4959. https://doi.org/10.1128/AEM.71.9.4951-4959.2005

Covington BC, McLean JA, Bachmann BO (2017) Comparative mass spectrometry-based metabolomics strategies for the investigation of microbial secondary metabolites. Nat Prod Rep 34(1):6–24. https://doi.org/10.1039/C6NP00048G

Dangi AK, Sharma B, Khangwal I, Shukla P (2018) Combinatorial interactions of biotic and abiotic stresses in plants and their molecular mechanisms: systems biology approach. Mol Biotechnol 60(8):636–650. https://doi.org/10.1007/s12033-018-0100-9

Doolotkeldieva T, Bobusheva S, Konurbaeva M (2015) Effects of streptomyces biofertilizer to soil fertility and Rhizosphere's functional biodiversity of agricultural plants. Adv Microbiol 05(07):555–571. https://doi.org/10.4236/aim.2015.57058

Ejaz S, Fahad S, Anjum MA, Nawaz A, Naz S, Hussain S, Ahmad S (2020) Role of Osmolytes in the mechanisms of antioxidant defense of plants. Sustain Agric Rev 39:95–117. https://doi.org/10.1007/978-3-030-38881-2_4

Elnahal ASM, El-Saadony MT, Saad AM, Desoky E-SM, El-Tahan AM, Rady MM, AbuQamar SF, El-Tarabily KA (2022) The use of microbial inoculants for biological control, plant growth promotion, and sustainable agriculture: A review. Eur J Plant Pathol 162(4):759–792. https://doi.org/10.1007/s10658-021-02393-7

Faddetta T, Polito G, Abbate L, Alibrandi P, Zerbo M, Caldiero C, Reina C, Puccio G, Vaccaro E, Abenavoli MR, Cavalieri V, Mercati F, Palumbo Piccionello A, Gallo G (2023) Bioactive metabolite survey of actinobacteria showing plant growth promoting traits to develop novel biofertilizers. Meta 13:3. https://doi.org/10.3390/metabo13030374

Farrar RR, Ridgway RL (1995) Enhancement of activity of bacillus thuringiensis Berliner against four Lepidopterous insect pests by nutrient-based Phagostimulants1. J Entomol Sci 30(1):29–42. https://doi.org/10.18474/0749-8004-30.1.29

Frikha-Gargouri O, Ben Abdallah D, Ghorbel I, Charfeddine I, Jlaiel L, Triki MA, Tounsi S (2017) Lipopeptides from a novel Bacillus methylotrophicus 39b strain suppress Agrobacterium crown gall tumours on tomato plants. Pest Manag Sci 73(3):568–574. https://doi.org/10.1002/ps.4331

Haas D, Défago G (2005) Biological control of soil-borne pathogens by fluorescent pseudomonads. Nat Rev Microbiol 3(4):307–319. https://doi.org/10.1038/nrmicro1129

Hathout Y, Ho Y-P, Ryzhov V, Demirev P, Fenselau C (2000) Kurstakins: a new class of lipopeptides isolated from Bacillus t huringiensis. J Nat Prod 63(11):1492–1496. https://doi.org/10.1021/np000169q

He Y, Guo W, Peng J, Guo J, Ma J, Wang X, Zhang C, Jia N, Wang E, Hu D, Wang Z (2022) Volatile organic compounds of streptomyces sp. TOR3209 stimulated tobacco growth by upregulating the expression of genes related to plant growth and development. Front Microbiol 13. https://doi.org/10.3389/fmicb.2022.891245

Imam J, Singh PK, Shukla P (2016) Plant microbe interactions in post genomic era: perspectives and applications. Front Microbiol 7. https://doi.org/10.3389/fmicb.2016.01488

Kaari M, Manikkam R, Annamalai KK, Joseph J (2023) Actinobacteria as a source of biofertilizer/biocontrol agents for bio-organic agriculture. J Appl Microbiol 134:2. https://doi.org/10.1093/jambio/lxac047

Karlovsky P (2010) Secondary metabolites in soil ecology. Springer-Verlag, Heidelberg, Germany

Kaushal M, Wani SP (2016) Plant-growth-promoting rhizobacteria: drought stress alleviators to ameliorate crop production in drylands. Ann Microbiol 66(1):35–42. https://doi.org/10.1007/s13213-015-1112-3

Khare E, Arora NK (2010) Effect of Indole-3-acetic acid (IAA) produced by Pseudomonas aeruginosa in suppression of charcoal rot disease of chickpea. Curr Microbiol 61(1):64–68. https://doi.org/10.1007/s00284-009-9577-6

Kadouri D, Jurkevitch E, Okon Y, Castro-Sowinski S (2005) Ecological and agricultural significance of bacterial polyhydroxyalkanoates. Crit Rev Microbiol 31:55–67

Kumar V, Baweja M, Singh PK, Shukla P (2016) Recent developments in systems biology and metabolic engineering of plant–microbe interactions. Front Plant Sci 7. https://doi.org/10.3389/fpls.2016.01421

Kumar BK, Ismail S, Patil VD (2017) Role of microbial solubilisers on major nutrient uptake—a review. J Pharmacogn Phytochem 6(5):641–644

Kumar S, Diksha, Sindhu SS, Kumar R (2022) Biofertilizers: an ecofriendly technology for nutrient recycling and environmental sustainability. Curr Res Microb Sci 3:100094. https://doi.org/10.1016/j.crmicr.2021.100094

Lugtenberg B, Kamilova F (2009) Plant-growth-promoting Rhizobacteria. Ann Rev Microbiol 63(1):541–556. https://doi.org/10.1146/annurev.micro.62.081307.162918

Macintyre L, Zhang T, Viegelmann C, Martinez I, Cheng C, Dowdells C, Abdelmohsen U, Gernert C, Hentschel U, Edrada-Ebel R (2014) Metabolomic tools for secondary metabolite discovery from marine microbial symbionts. Mar Drugs 12(6):3416–3448. https://doi.org/10.3390/md12063416

Maillet F, Poinsot V, André O, Puech-Pagès V, Haouy A, Gueunier M, Cromer L, Giraudet D, Formey D, Niebel A, Martinez EA, Driguez H, Bécard G, Dénarié J (2011) Fungal lipochitooligosaccharide symbiotic signals in arbuscular mycorrhiza. Nature 469(7328):58–63. https://doi.org/10.1038/nature09622

Mondol M, Shin H, Islam M (2013) Diversity of secondary metabolites from marine bacillus species: chemistry and biological activity. Mar Drugs 11(8):2846–2872. https://doi.org/10.3390/md11082846

Morel MA, Cagide C, Castro-Sowinski S (2016) The contribution of secondary metabolites in the success of bioformulations. In: Bioformulations: for sustainable agriculture. Springer India, pp 235–250. https://doi.org/10.1007/978-81-322-2779-3_13

Olanrewaju OS, Babalola OO (2019) Streptomyces: implications and interactions in plant growth promotion. Appl Microbiol Biotechnol 103(3):1179–1188. https://doi.org/10.1007/s00253-018-09577-y

Oldroyd GED (2013) Speak, friend, and enter: signalling systems that promote beneficial symbiotic associations in plants. Nat Rev Microbiol 11(4):252–263. https://doi.org/10.1038/nrmicro2990

Palazzini JM, Dunlap CA, Bowman MJ, Chulze SN (2016) Bacillus velezensis RC 218 as a biocontrol agent to reduce Fusarium head blight and deoxynivalenol accumulation: genome sequencing and secondary metabolite cluster profiles. Microbiol Res 192:30–36. https://doi.org/10.1016/j.micres.2016.06.002

Pande S, Kost C (2017) Bacterial unculturability and the formation of intercellular metabolic networks. Trends Microbiol 25(5):349–361. https://doi.org/10.1016/j.tim.2017.02.015

Perez-Garcia A, Romero D, De Vicente A (2011) Plant protection and growth stimulation by microorganisms: biotechnological applications of bacilli in agriculture. Curr Opin Biotechnol 22(2):187–193

Pinu F, Villas-Boas S, Aggio R (2017) Analysis of intracellular metabolites from microorganisms: quenching and extraction protocols. Meta 7(4):53. https://doi.org/10.3390/metabo7040053

Raaijmakers JM, Vlami M, de Souza JT (2002) Antibiotic production by bacterial biocontrol agents. Antonie Van Leeuwenhoek 81:537–547. https://doi.org/10.1023/A:1020501420831

Rangarajan V, Clarke KG (2016) Towards bacterial lipopeptide products for specific applications—a review of appropriate downstream processing schemes. Process Biochem 51(12):2176–2185. https://doi.org/10.1016/j.procbio.2016.08.026

Richa (2023) Liquid bio-fertilizers: prospects and challenges. In: Metabolomics, proteomes and gene editing approaches in biofertilizer industry. Springer Nature, Singapore, pp 77–99. https://doi.org/10.1007/978-981-99-3561-1_5

Ruiz B, Chávez A, Forero A, García-Huante Y, Romero A, Sánchez M, Rocha D, Sánchez B, Rodríguez-Sanoja R, Sánchez S, Langley E (2010) Production of microbial secondary metabolites: regulation by the carbon source. Crit Rev Microbiol 36(2):146–167. https://doi.org/10.3109/10408410903489576

Saha R, Saha N, Donofrio RS, Bestervelt LL (2013) Microbial siderophores: a mini review. J Basic Microbiol 53(4):303–317. https://doi.org/10.1002/jobm.201100552

Sandhya V, Ali SZ (2015) The production of exopolysaccharide by Pseudomonas putida GAP-P45 under various abiotic stress conditions and its role in soil aggregation. Microbiology 84(4):512–519. https://doi.org/10.1134/S0026261715040153

Sansinenea E, Ortiz A (2011) Secondary metabolites of soil Bacillus spp. Biotechnol Lett 33(8):1523–1538. https://doi.org/10.1007/s10529-011-0617-5

Sasse J, Martinoia E, Northen T (2018) Feed your friends: do plant exudates shape the root microbiome? Trends in plant science 23(1):25–41

Schöller CEG, Gürtler H, Pedersen R, Molin S, Wilkins K (2002) Volatile metabolites from actinomycetes. J Agric Food Chem 50(9):2615–2621. https://doi.org/10.1021/jf0116754

Seneviratne G, Kulasooriya SA (2013) Reinstating soil microbial diversity in agroecosystems: the need of the hour for sustainability and health. Agric Ecosyst Environ 164:181–182. https://doi.org/10.1016/j.agee.2012.10.002

Seneviratne G, Jayasekara APDA, De Silva MSDL, Abeysekera UP (2011) Developed microbial biofilms can restore deteriorated conventional agricultural soils. Soil Biol Biochem 43(5):1059–1062. https://doi.org/10.1016/j.soilbio.2011.01.026

Singh JS, Pandey VC, Singh DP (2011) Efficient soil microorganisms: A new dimension for sustainable agriculture and environmental development. Agric Ecosyst Environ 140(3–4):339–353. https://doi.org/10.1016/j.agee.2011.01.017

Singh P, Chauhan PK, Upadhyay SK, Singh RK, Dwivedi P, Wang J, Jain D, Jiang M (2022) Mechanistic insights and potential use of siderophores producing microbes in rhizosphere for mitigation of stress in plants grown in degraded land. Front Microbiol 13. https://doi.org/10.3389/fmicb.2022.898979

Slama I, Abdelly C, Bouchereau A, Flowers T, Savouré A (2015) Diversity, distribution and roles of osmoprotective compounds accumulated in halophytes under abiotic stress. Ann Bot 115:433–447

Sobhy MA, Tehseen M, Takahashi M, Bralić A, De Biasio A, Hamdan SM (2021) Implementing fluorescence enhancement, quenching, and FRET for investigating flap endonuclease 1 enzymatic reaction at the single-molecule level. Comput Struct Biotechnol J 19:4456–4471. https://doi.org/10.1016/j.csbj.2021.07.029

Stamford NP, Lima RA, Lira MA, Santos CRS (2008) Effectiveness of phosphate and potash rocks with Acidithiobacillus on sugarcane yield and their effects on soil chemical attributes. World J Microbiol Biotechnol 24(10):2061–2066. https://doi.org/10.1007/s11274-008-9710-x

Tewari S, Arora NK (2014) Multifunctional exopolysaccharides from Pseudomonas aeruginosa PF23 involved in plant growth stimulation, biocontrol and stress amelioration in sunflower under saline conditions. Curr Microbiol 69(4):484–494. https://doi.org/10.1007/s00284-014-0612-x

Tewari S, Arora NK (2016) Exopolysaccharides based bioformulation from Pseudomonas aeruginosa combating saline stress. Recent Trends PGPR Res Sustain Crop Product:93

Thavasi R, Marchant R, Banat IM (2014) 15 Biosurfactant applications in agriculture. Biosurfactants: Production and Utilization—Processes Technologies, and Economics, 159, 313

Timmusk S, Abd El-Daim IA, Copolovici L, Tanilas T, Kännaste A, Behers L, Nevo E, Seisenbaeva G, Stenström E, Niinemets Ü (2014) Drought-tolerance of wheat improved by rhizosphere bacteria from harsh environments: enhanced biomass production and reduced emissions of stress volatiles. PLoS One 9(5):e96086. https://doi.org/10.1371/journal.pone.0096086

Tyc O, Song C, Dickschat JS, Vos M, Garbeva P (2017) The ecological role of volatile and soluble secondary metabolites produced by soil bacteria. Trends Microbiol 25(4):280–292. https://doi.org/10.1016/j.tim.2016.12.002

Vaidyanathan S (2005) Profiling microbial metabolomes: what do we stand to gain? Metabolomics 1(1):17–28. https://doi.org/10.1007/s11306-005-1104-6

Villas-Boas SG, Roessner U, Hansen MAE, Smedsgaard J, Nielsen J (2007) In: Hoboken NJ (ed) Metabolome analysis: an introduction. John Wiley & Sons, New York

Wang Y, Ren W, Li Y, Xu Y, Teng Y, Christie P, Luo Y (2019) Nontargeted metabolomic analysis to unravel the impact of di (2-ethylhexyl) phthalate stress on root exudates of alfalfa (Medicago sativa). Sci Total Environ 646:212–219. https://doi.org/10.1016/j.scitotenv.2018.07.247

Zacky FA, Ting ASY (2015) Biocontrol of Fusarium oxysporum f.sp. cubense tropical race 4 by formulated cells and cell-free extracts of Streptomyces griseus in sterile soil environment. Biocontrol Sci Tech 25(6):685–696. https://doi.org/10.1080/09583157.2015.1007921

Zahir ZA (2010) Substrate-dependent auxin production by rhizobium phaseoli improves the growth and yield of Vigna radiata L. under salt stress conditions. J Microbiol Biotechnol 20(9):1288–1294. https://doi.org/10.4014/jmb.1002.02010

Gene Editing Tools for Engineering Beneficial Microorganism in Biofertilizer

5

T. Kavya, Haralakal Keerthi Kumari, Geetha Singh, V. Govindasamy, D. Vijaysri, J. Sushmitha, Konderu Niteesh Varma, G. R. Sneha, T. Chethan, and Sunil Jadhav

Abstract

The exponentially growing global population has demand for sustainable agriculture practices to increase food production. To achieve higher production many approaches have been employed, but they are time consuming and are not eco-friendly. Biofertilizers which are produced from naturally occurring algae, fungi, bacteria are receiving increased interest over chemical fertilizer due to their holistic and sustainable approach to agriculture, promoting soil health, environmental sustainability, and cost-effectiveness. Their benefits make them a valuable tool in modern farming practices, aiming for a more sustainable and responsible use of natural resources. However, their effectiveness can be limited by natural constraints and environmental factors. Advances in gene editing technologies have helped in efficiently deploying the microbiome to increase crops nutrient acquisition and resistance to abiotic and biotic stress. This chapter provides a comprehensive overview of the gene editing tools, particularly focusing on CRISPR-Cas9, TALENs, and zinc finger nucleases, their applications in mod-

T. Kavya (✉) · G. Singh · V. Govindasamy · D. Vijaysri · K. N. Varma · G. R. Sneha
Division of Microbiology, ICAR-Indian Agricultural Research Institute, New Delhi, India

H. K. Kumari
Department of Agricultural Microbiology, University of Agricultural Sciences, GKVK, Bengaluru, India

J. Sushmitha
Division of Biochemistry, ICAR-Indian Agricultural Research Institute, New Delhi, India

T. Chethan
Department of Agricultural Entomology, University of Agricultural Sciences, GKVK, Bengaluru, India

S. Jadhav
Division of Seed Science and Technology, ICAR-Indian Agricultural Research Institute, New Delhi, India

S. Kaur et al. (eds.), *Metabolomics, Proteomics and Gene Editing Approaches in Biofertilizer Industry*, https://doi.org/10.1007/978-981-97-2910-4_5

ifying microbial genomes. We discuss various case studies showcasing successful applications of gene editing technologies to enhance nitrogen fixation, phosphate solubilization, and nutrient uptake capacities of these microorganisms. These examples showcase how engineered microorganisms have been designed to thrive in diverse soil conditions, resist environmental stressors, and interact harmoniously with crops, ultimately promoting sustainable agriculture practices by providing nutrients. This chapter also discusses about the regulatory aspects and ethical considerations surrounding the use of gene-edited tools in shaping a sustainable and beneficial microbial world.

Keywords

Gene editing tools · Meganucleases · Zinc finger nucleases · TALENs · CRISPR-Cas9

5.1 Introduction

Persistent global insecurity is a critical concern that likely to worsen with earth's shifting climate, burgeoning population growth, and the diminishing availability of arable land. The current concerning forecasts indicating that sustainable increase in the global population impose significant challenges on food security. To the escalating demand of food requirement for increasing global population, a focus on increasing food production often resulting in the improper application of chemical fertilizers potentially causing the harmful results on soil quality (Nath et al. 2023). Another approach to elevate food production is through the development of high yielding varieties via genetic improvement. However, the major drawback is that this process is time consuming. The application of microbial inoculants, functioning as biofertilizers, biopesticides, and plant growth promoters, is considered the most promising approach to meeting agricultural productivity needs without compromising the environment and soil health. This approach has been extensively utilized to enhance crop growth, particularly in developing and emerging nations, as highlighted by Bashan et al. (2014). The use of beneficial microbes for improving plant growth, productivity, and health will be an auspicious eco-friendly sustainable approach for the forthcoming green upheaval. These microbial inoculants also offer a substitute to chemical-based fertilizers, which are widely criticized for their detrimental effects on human health and the environment. The beneficial microbes present in the biofertilizers such as nitrogen-fixing bacteria, mycorrhizal fungi, phosphorus, potassium, and sulfur solubilizers are mostly found in the rhizosphere will supply nutrients to plants at the time of nutrient deficiency and also, they influence the plant growth through the synthesis of phytohormones. These microbes by their antifungal and antibacterial activities also offer protection to plants from harmful pathogens by producing extracellular enzymes, siderophores, HCN, and also these beneficial microbes compete for nutrients making the environment deprived of nutrients, which in turn hinders the establishment of pathogens. An example includes biocontrol strains of *Trichoderma* that attack plant-pathogenic fungi (Sood et al. 2020). Agricultural stress is rising because of climate change in addition to other natural and man-made variables, which lowers agricultural productivity. Under these circumstances, microorganisms have been shown to favor plants and may aid in the body's defense against stress through a variety of direct and indirect ways. These microbes are directly applied to soil, on foliar parts of plants or as seed coating. At present, bioinoculants are predominantly found in the form of individual entities. However, there is a growing trend toward formulating them as consortia comprising multiple bacteria and fungi, showcasing synergistic plant growth-promoting (PGP) traits, as highlighted by Bashan et al. (2020). Nonetheless, a significant information gap persists regarding the effectiveness of plant growth-promoting (PGP) microbes when comparing laboratory studies to their actual performance in the field. The practical utility of many bioinoculants identified in laboratory settings remains uncertain, and questions arise about whether they may have unintended effects on nontarget organisms, including humans, once implemented in field conditions. The processes for creating and using microbial inoculation are always being developed and improved upon. Nevertheless, the efficacy of beneficial microbes alone may not sufficient for marketable usage, necessitating the improvement of their performance. Achieving this goal demands the utilization of high-quality strains.

Molecular biology tools such as genomics, metagenomics, metabolomics, transcriptomics, and proteomics are frequently used in the field of biofertilizer industry particular to incorporate specific beneficial traits or genes in the wild type microbial strains to get better performing microbial strain in a particular product of interest (Kumar et al. 2023). New genetic engineering tools are continually being developed for genetic transformation and strain improvement of beneficial microorganisms that are designed to address current environmental challenges by reducing the use of synthetic fertilizers. An alternative and promising approach known as transgenic technology accelerates outcomes by introducing foreign genetic material. In this technique, engineered endonucleases induce a double-strand break (DSB) in the DNA, prompting repair through inherent mechanisms. Accordingly, these tools can be used to enhance the plant growth-promoting traits like increasing nitrogen fixation, phosphate solubilization, potassium solubilization, zinc solubilization, siderophore production, hydrogen cyanide production, NH_3 production, and indole acetic acid production of the beneficial microorganism used as a biofertilizers. Genome manipulation can also be best used to improve the biocontrol ability of microbes by inserting a particular gene into microbes lacking such genes (Ali et al. 2020). However, its widespread utilization is restricted by regulatory consideration. In this context, gene editing is getting much attention through which we can edit genomes of particular microbe in a specific manner without introducing gene of foreign origin. The most widely used gene editing tools for engineering beneficial microorganism in biofertilizer are CRISPR-Cas9, TALENs, and zinc finger nucleases (Fig. 5.1). Most commonly, these tools are employed to modify the expression of specific genes, either enhancing or reducing their function. This editing of particular genes and enzymes in microorganisms is done to facilitate the production of a diverse

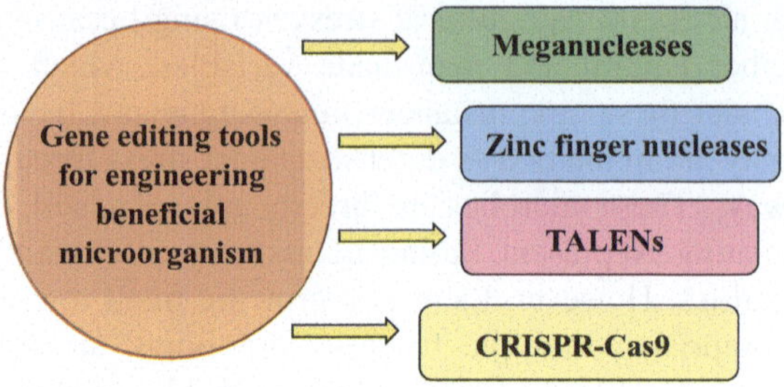

Fig. 5.1 Different gene editing tools used for engineering beneficial microorganism

range of nutrients, hormones, enzymes, molecules, and compounds, as outlined by Sudheer et al. (2020).

5.2 History of Genome Editing in Microorganisms

Every living thing in the cosmos has genetic information that determines how it will grow, develop, operate, and reproduce. This information is present in a segment of DNA whose functional unit is called as gene. The deoxyribonucleic acid (DNA), is a polymeric molecule composed of four types of nucleotides. The nucleotides arrangements on genes will carry the instruction for synthesis of ribonucleic acids (RNA) and proteins necessary for normal functioning of living organisms. However, a change in nucleotide sequences can enhance or reduce the expression of genes which can be achieved by the process called mutation. Historically, the study of genetics has relied on identifying and evaluating natural mutations. Before the discovery of DNA, in the late 1920s and early 1930s, Lewis John Stadler pioneered the introduction of random mutations in plants by subjecting them to radiation bombardment. In the mid-20th century Mendel, Avery, and Morgan relied on these traditional methods for inducing changes in the DNA. Muller and Auerbach increased this rate of mutations using chemical treatment or radiation. Later, transposon-based insertions were used for some organisms to induce mutation. However, all these technologies induced genomic changes randomly, this became the major drawback in their wide spread utilization. After several years, the encounter of the genetic material deoxyribonucleic acid by Watson and Crick in 1953 made several scientists to realize the significance of genetic material in biological characteristics and phenotypes, and this groundbreaking finding also sparked the notion of synthetic DNA modification. During the twilight seventies, groundbreaking gene targeting (GT) trials conducted in the yeast *Saccharomyces cerevisiae* opened the new possibilities for genome engineering, as documented by Hinnen et al. 1978. Subsequently, the identification of restriction enzymes and the advancement of transformation techniques enabled Rudolf Jaenisch to produce the first genetically modified mouse in 1974, marking a revolutionary moment in the field of genomic engineering (Jaenisch and Mintz 1974). The majority of the restriction enzymes recognize short palindromic DNA sequences to cleave the target DNA. But its application is restricted to use in mammalian cells because these enzymes distinguish short sequences, slicing many sites in the target DNA. In 1976 and 1978, the first instances of creating genetically modified bacteria took place. Specifically, genetically modified *Escherichia coli* bacteria were engineered to produce the hormones somatostatin and insulin respectively (Goeddel et al. 1979; Itakura et al. 1977). Initially, many researchers used homologous recombination to modify the specific genes within the cell genomes (Smithies et al. 1984; Musunuru 2017). However, the occurrence of getting homologous recombination with a custom-made template is too low with only $\sim 10^{-6}$ to 10^{-9} in eukaryotes. Consequently, there arose a necessity for precise genetic engineering techniques to augment the frequency of homologous recombination at targeted sites. This has challenged many scientists to find more precise and effective techniques to produce finely tuned genetic circuits in the living organisms. It quickly became apparent that inducing a double-strand break (DSB) in the interior, the region of homology significantly boosted the rate of homologous recombination (HR) within the gene (Fernández et al. 2017; Smithies et al. 1984). Hence, the advantages in integration efficiency seemed to be associated with the promotion of DSBs in the donor DNA template.

Recent era of genome editing utilizes the tools that employ nucleases to generate double-strand DNA breaks at anticipated positions in the genome. Soon after the double-strand break, cellular machinery activates DNA repair mechanism which can significantly enhance the efficiency of genome alteration by orders of magnitude. Alterations to the gene can often occur at frequencies exceeding 1-within-10, contrary to constituting 1-in-a-million catastrophes. Following this, a series of nuclease-based genome editing were exploited to boost the efficiency of gene erasure in beneficial microorganism which in turn helps regulation of biosynthetic gene clusters and helps in improving metabolic output and phenotypes. The repairing of double-strand breaks (DSBs) induced by nuclease-based gene editing technique at a specific site in the genome was made using foreign DNA that produces the desired edit. The cell will repair DSBs through three major ways like nonhomologous end joining (NHEJ), microhomology-mediated end joining (MMEJ), and homology-directed repair (HDR) (Gao 2021). Nonhomologous end joining (NHEJ) is a default patch-up lane operating universally across diverse cell types and continuously. This repair process is less perfect which will lead to semirandom addition or deletion of DNA base pairs, commonly referred to as "indels." When indels occur within the encrypting sequence of a gene, they can induce either frameshift mutations or in-frame insertions or deletions. In the prior mutation addition and deletion of nucleotide base in the DNA sequence will disturb the corresponding portion of the amino acid sequence, resulting in the premature termination of the gene's protein product. But later involves the addition or removal of amino acids from the protein, potentially affecting its function or causing a complete loss of functionality (Bibikova et al. 2002). Both will completely change the normal function of protein in the cell. In contrast to nonhomologous end joining (NHEJ), homology-directed repair (HDR) is limited to actively dividing cells, specifically in S phase or G2 phase of the cell cycle. DNA repair by HDR fallouts in the unchanging alteration into the genome because it uses mending template exhibiting similarity to the DNA sequences surrounding the double-strand break. Typically, the repair template is a replicated chromatid with an identical sequence or a matching chromosome. Unlike NHEJ, HDR mediated DSB repair mechanism by cell is more precise. Among the protein-based tools available for genome editing, meganucleases flagged the way as the initial genome editing tools, well before other well-known nucleases used in genome editing nucleases like zinc finger nucleases (ZFN); transcription activator-like effector nucleases (TALEN); clustered regularly interspaced short palindromic repeats (CRISPR) and their associated proteins (Cas) tools (Musunuru 2017). Despite the subsequent advancements, all these tools share a common principle—they induce double-strand breaks (DSBs) at specific genomic sequences, subsequently repaired by endogenous DNA repair mechanisms. In topical years, CRISPR-Cas based gene editing has arisen as a groundbreaking technology, revolutionizing precise and controlled genetic modifications in various microbes such as *Escherichia coli*, *Saccharomyces cerevisiae*, *Lactobacillus reuteri*, *Clostridium beijerinckii*, *Streptococcus pneumonia*, and *Staphylococcus aureus* (Doudna and Charpentier 2014).

5.3 Different Types of Gene Editing Tools

5.3.1 Gene Editing Using Meganucleases

Several gene targeting techniques tend to rely on site-specific recombination tools made possible by transposases or recombinases. Recombinase-or transposase-facilitated inclusion is generally cell-type agnostic in terms of effectiveness as contrasted with homologous recombination. The latter is especially helpful because homologous gene targeting techniques are often ineffective in inactive or post-mitotic cells. Improving the efficiency of homologous DNA targeting has been the focus of many research projects. This can be accomplished in a few ways: The first strategy uses "array optimization," which is the process of altering the focus vector design in order to greatly increase efficacy. The second method relies on site-specific recombination. Finally, third method employs sequence-specific endonucleases like meganucleases to stimulate homologous gene targeting. In particular, this strategy offers a wide range of applications because it may be used for directed gene addition or silencing in addition to gene repair (Pâques and Duchateau 2007). Techniques such employing RNA-DNA oligonucleotides (RDOs), small fragment homologous replacement (SFHR), and triplex forming oligonucleotides (TFOs) have all been included in matrix refinement (Gruenert et al. 2003), and another type of optimization involving the use of Adeno-Associated Virus (AAV) as vectors for the repair matrix. While initially promising, these approaches were eventually deemed inefficient in many experiments. These approaches were thought to be promising at first, but numerous tests later showed that they were ineffective.

Meganuclease (MegNs) also called as homing endonucleases which were discovered in the late 1980s, were found in all forms of microbial life and also it is a part of self-splicing rDNA introns found *S. cerevisiae* mitochondria, which can generate DNA modifications. MegNs facilitate lateral mobility of genetic elements within an organism. This process is referred to as "homing" and gives the name homing endonuclease to these enzymes. They can recognize and cut large DNA sequences from 12 to 40 base pairs. These meganucleases and are composed of five different families based on their sequences and motifs: LAGLIDADG, GIY-YIG, His-Cys box, H–N–H, and PD-(D/E) XK (Danilo et al. 2022). The story of meganuclease started with the identification of HO nuclease that mediates mating-type switching by inducing cleavage and recombination at nuclear MAT locus in certain species of *Saccharomyces* yeasts (Klar et al. 1984). One of the most studied meganuclease is I-SceI, encoded by a mitochondrial intron. It plays a crucial role in the homing process, facilitating the spread of this intron to intronless copies of the same genes (Chevalier and Stoddard 2001; Jacquier and Dujon 1985). The function of this meganuclease was first discovered in the yeast *Sacharomyces cerevisiae*, marking the onset of a new era in gene targeting. Another meganuclease I-CreI, which functions as a self-splicing chloroplast rDNA intron in the unicellular algae *Chlamydomonas reinhardtii*. It has been employed to induce intrachromosomal recombination of inverted repeats in *Arabidopsis* (Chiurazzi et al. 1996) and of direct repeats in *Drosophila* (Rong et al. 2002), respectively. On the other hand, the primary disadvantage of spontaneously generated meganucleases is the requirement to introduce a known breakage site into the area concerned. Although there are certain methods like ΦC31 integrase which exploits pseudo-sites within a genome, but such approaches are not always feasible in therapeutic cases where targeting demands a higher precision level. The inherent challenge of precision targeting has led to the exploration of engineered meganucleases, capable of addressing this issue and should allow for a broader use of homologous recombination-based technology. The evident limitation of target sequences associated with naturally occurring meganucleases would ultimately be overcome by producing engineered meganucleases with customized recognition sites. A prime instance of one of these designed meganucleases, which came from I-CreI, has been applied to high-frequency target recombination at the SCID gene in the cells of humans (Grizot et al. 2009). In the meantime, in rat and mice embryos, an additional manufactured I-CreI that was intended to disrupt a 22-bp stretch of the Rag1 gene showed successful targeting events (Ménoret et al. 2013). Consequently, to create a unique MegN, per new genetic engineering target needs to go through a preliminary protein design step. Therefore, even with the large number of MegNs that are currently accessible, the likelihood of discovering an enzyme specifically addresses a particular locus is quite low, and the process of producing tailored MegNs is still extremely difficult and inefficient. Consequently, MegNs have few routine applications in genome editing and their use with these endonucleases has proven to be technically difficult (Khalil 2020).

5.3.2 Gene Editing Using Zinc Finger Nucleases

The ability of a zinc finger protein motif paired alongside the FokI endonuclease domain to function as a site-dependent nuclease that cuts DNA at precisely specified places in vitro was first shown in 1996. Although each zinc finger domain of this chimeric protein (zinc finger nuclease, ZFN) identifies a single nucleotide triplet, the protein has a modular framework. This technique served as the foundation for cell culture editing. By creating a break in both strands at the location selected for alteration, ZFNs start these processes. Sequence changes encoded in this donor may be integrated into the genome by homology-directed repair (HDR) if an external repair framework is also provided (Liang et al. 1998).

Individual ZFNs in the early investigations bound a 9-bp target with three fingers, allowing ZFN dimers (the active species) to specify 18 bp of DNA per cleavage site. In order to specify longer and rarer cleavage targets, extra fingers-up to six per ZFN-have been added in more recent research. ZFNs have been largely successful because of the FokI domain, which has multiple properties that facilitate targeted cleavage inside complicated genomes (Vanamee et al. 2001). Due to the weak nature of this interaction, two consecutive and independent binding events are necessary for the cleavage by FokI as part of a ZFN. These events must also occur in the proper orientation and spacing to allow for the synthesis of dimers (Table 5.1).

5.3.3 Gene Editing Using Transcription Activator-like Effector Nucleases (TALENs)

The *Xanthomonas* genus of bacteria is studied in relation to the development history of this system. The extensive investigation on this bacteria was motivated by the fact that these bacteria causes plant diseases that inflict serious economic harm to agriculture, specifically to rice, pepper and tomato crops. Effector proteins, also known as transcription activator-like effectors, or TALEs, are secreted by the bacteria into the cytoplasm of plant cells. These proteins alter internal processes and heighten the plant cell's vulnerability to the pathogen. After more research into the functioning mechanisms of executioner proteins, it was discovered that they could potentially imitate eukaryotic transcription factors to bind DNA and activate the encoding of their target genes (Boch et al. 2009). The nuclear localization signal is the domain that initiates the transcription of the target gene in addition to the central domain where DNA binds, make up the TALE proteins. Two research groups cracked the code allowing TALE proteins to recognize the target DNA. Shortly, the ability of these proteins to bind to DNA was first reported in 2007. It was shown that the DNA-binding domain is composed of monomers, each of which binds a single nucleotide in the target nucleotide sequence. These monomers are 34 amino acid residues that are tandem repeated; two of these residues are very variable and are found at positions 12 and 13. They allow a particular nucleotide to be recognized. Since various repeat variable di-residues (RVDs), may link to several nucleotides with varying efficiency, this code is degenerate (Brian et al. 2013). The identical nucleotide, thymidine, is always present in the intended DNA molecule prior to the 5' end of a sequence that is attached by a TALE monomer, and this affects the binding effectiveness. It is known as a half-repeat because the final tandem recur, which

Table 5.1 Comparison of gene editing tools

ZNF	TALEN	CRISPR-Cas9
Fist tested in *Xenopus laevis* (African clawed frog)	Rat	*Escherichia coli*
Endonuclease domain of ZNF is acquired the from the bacterial restriction enzyme FokI	Endonuclease domain of TALEN is acquired the from the bacterial restriction enzyme FokI	CRISPR systems utilize their native endonuclease, typically a Cas protein, such as Cas9
DNA-binding domain is ZF protein	TALE protein	Guide RNA
It recognizes the sequence contain G base as follows; 5′-GNNGNNGNN-3′	It recognizes the sequence starting from 5′-T and ending with A-3′	It recognizes the sequence immediately followed by PAM sequences such as 5′-NGG-3′, 5′-NGA-3′, and 5′-YTN-3′
Zinc finger proteins has the specificity toward triple base in the target sequence	TALE has specificity toward a single-base in the target sequence	Guide RNA will decide the specific target region
ZFN depends on protein-DNA interactions to identify the specified target genomic sequences, through distinct amino acid/ nucleotide codes	TALEN also depends on protein-DNA interactions to identify the specified target genomic sequences, through distinct amino acid/ nucleotide codes	CRISPR system depends on its own natural partner, a small RNA molecule, to match the target DNA sequence using the standard nucleotide pairing code as given by Watson-and-Crick
Multiple gene modification is not possible at one time because a new ZFN protein need to designed for each DNA target site	Multiple gene modification is not possible at one time because a new TALEN protein need to designed for each DNA target site	Multiple gene modifications can be introduced at a single time because this system involves a guide RNA nuclease that can be designed for different target sites by simply changing the guide RNA sequence
ZFNs is difficult to design and high cost as compared to TALEN	High specificity and it is easy to design TALE-DNA binding arrays	High specificity and it is easy to design guide RNA

attaches a nucleotide at the 3'-end of the detection site, only consists of 20 nitrogenous residues.

A straightforward "protein-DNA code" that associates individual bases in a target-binding site with modular DNA-binding TALE repeat domains has made it possible for researchers to quickly and efficiently create TALENs, which has sparked a lot of interest and excitement. A very conserved repeat that is taken from naturally occurring TALEs encoded by the proteobacteria *Xanthomonas* serves as the basic building block for engineering the DNA-binding domain of TALENs (Boch and Bonas 2010). By binding to genomic DNA and changing transcription in host plant cells, these TALEs are injected into the cells using a Type III secretion system, which facilitates the colonization of harmful bacteria. Till now exploitation of this system is restricted only to pathogens associated with plant diseases but it also has wide scope in exploiting beneficial microorganisms like biofertilizers (Table 5.1).

5.3.3.1 Unique Characteristics Associated with TALENS:

- Specificity: Scientists and engineers can use this basic technique to design and build TALEs for genome change, as TALEs offer an established DNA base-pair option. Each DNA base pairs are recognized by a protein repeat tandem, which consists of two alpha helices connected by a solenoid-shaped loop of three RVD residues. There are six conventional RVDs namely, NG, HD, NI, NK, NH, and NN which are frequently used in construction of protein repeat tandem. Each RVDs have association with different nitrogenous base pare along with different strength.
- Sensitivity against Methyl: It is possible to efficiently reengineer TALE proteins so that they are sensitive to certain chemical changes in DNA. Since methylated cytosine and thymine have a considerable structural similarity, NG can bind to methylated cytosine even though the standard RVD HD is not effective at doing so (Deng et al. 2012).
- Incoherence nature of TALEN'S: Due to the degeneracy of the TALEN coding, specific RVDs exhibit a wide range of binding efficiencies when binding to different nucleotides. The TALENs' capacity to bind the repeat variable di-residues of NN (for A and G) and NS (for A, C, and G) allows them to express degeneracy for the intended DNA.

5.3.4 Gene Editing Using Clustered Regularly Interspaced Short Palindromic Repeats/CRISPR Associated Proteins (CRISPR/CAS)

CRISPR-Cas9 was developed based on a model of the innate genome modification system used by bacteria as an immune reaction. To create what are known as CRISPR arrays, virus-infected bacteria take little pieces of the virus's DNA and rearrange it inside their own DNA in specific ways. The bacteria "remember" the viruses (or similarly related ones) thanks to the CRISPR arrays. To create RNA segments that can attach to DNA sequences on the viruses in the event that they launch another attack, the bacteria use CRISPR arrays. When the bacteria use Cas9 or a similar enzyme to divide the DNA, the virus becomes inoperable. Researchers have modified DNA using this immune protection system. They synthesize small RNA segments with a brief "guide" motif that attaches (binds) to a particular target sequence in a cell's DNA, much to the RNA fragments that bacteria make from the CRISPR array (Nemudryi et al. 2014). The Cas9 enzyme is also attached to this guide RNA. Similar to how it works in bacteria, the Cas9 enzyme copies DNA at the precise spot in cells once the guide RNA identifies the desired DNA sequence. While the most frequently used enzyme is Cas9, other enzymes, such as Cpf1, can also be employed. After the DNA is cut, scientists can alter or add genetic material by using the cell's built-in DNA repair machinery (Horvath and Barrangou 2010).

5.3.4.1 Basic Components of CRISPR/Cas and Their Function

A single-guide RNA (sgRNA) and an RNA-guided Cas9 endonuclease are the two primary parts of the CRISPR-Cas9 system. Each of the two nuclease domains that the Cas9 protein comprises, HNH and RuvC, cleaves one strand of the target double-stranded DNA. The combination of tracrRNA and crRNA is simplified into single-guide RNAs (sgRNAs). Together, the sgRNA and Cas9 nuclease generate the Cas9 ribonucleoprotein (RNP), which have the ability to attach and break the targeted DNA. Moreover, the binding of the Cas9 protein to the target DNA requires the presence of a protospacer adjacent motif (PAM) sequence (Ran et al. 2013).

5.3.4.2 A General Scheme of the Structure of Cas9 Nuclease and its Lobes and Domains

Two separate lobes make up the structures of Cas9 in the apo state: the nuclease (NUC) lobe, which contains the more variable C-terminal domain (CTD) and the conserved HNH and split RuvC nuclease domains. The lobe of alpha-helical recognition (REC). The arginine-rich bridging helix and a disordered linker generate the first of two linking segments that further join the two lobes. The three alpha-helical domains that make up the REC lobe (Hel-I, Hel-II, and Hel-III) are not structurally comparable to any other known protein. In addition, the extended CTD has a fold unique to Cas9 and PAM-interacting sites that are necessary for PAM interrogation. Nevertheless, the apo-Cas9 structure's PAM-recognition region is mainly disorganized, suggesting that the enzyme is maintained in an inactive state and is unable to identify target DNA before attaching to a guide RNA (Jiang and Doudna 2017).

The Cas9 RuvC nuclease domain indicates that RuvC is most likely to cleave the nontarget DNA strand using a two-metal-ion catalytic mechanism since it has structural similarities with members of the retroviral integrase superfamily that are distinguished by an RNase H fold. On the other hand, the HNH nuclease domain most likely uses a one-metal-ion mechanism for target-strand DNA cleavage and adopts the hallmark ββα-metal fold common by other HNH endonucleases. An active DNA surveillance complex made up of Cas9 and guide RNA (a natural crRNA–tracrRNA or a sgRNA) is necessary to achieve site-specific DNA recognition and cleavage. DNA target specificity is conferred by the 20-nt spacer sequence of crRNA, and Cas9 recruitment is significantly aided by tracrRNA (Doudna and Charpentier 2013)

5.3.4.3 Characterization of the Immune Response Using CRISPR-Cas

In accordance with the effector proteins that are encrypted, two classes, 1 and 2, comprise the most recent classification of CRISPR-Cas systems, which is represented in Fig. 5.2. Class 1 CRISPR-Cas systems function with multi subunit effector assemblies made up of four to seven Cas proteins that are distributed unevenly in stoichiometry. This framework, which makes up around 90% of all known CRISPR-cas loci, is widely distributed in bacteria and archaea, including all hyperthermophiles. The other ten percent are classified as class 2, which is almost solely prevalent in bacteria.

There are at present three types in every class: types II, V, and VI in class 2 and types I, III, and IV in class 1. A distinctive signature protein called Cas3 is present

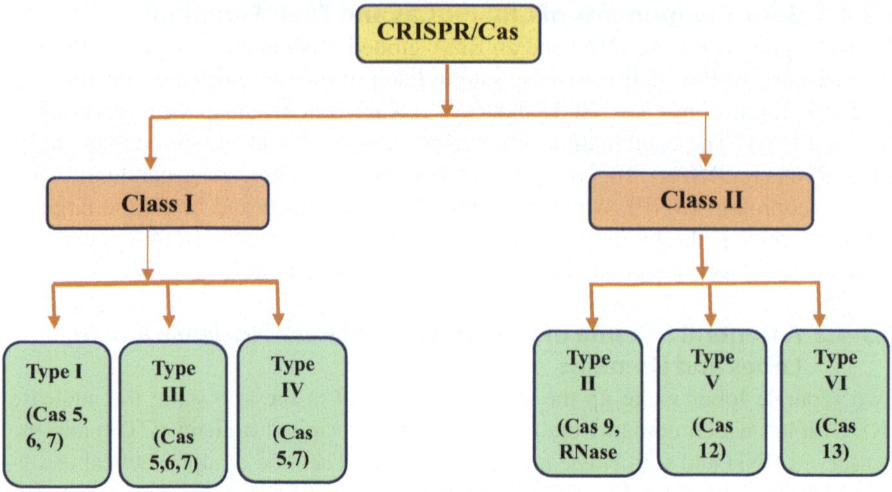

Fig. 5.2 Grouping the CRISPR/Cas system

in type I, Cas9 is present in type II, and Cas10 is present in type III, making it easy to differentiate between the three types. Each type is then divided into several subtypes according to additional signature genes and distinctive gene combinations (Shmakov et al. 2017).

5.3.4.4 Different Cas Enzymes Used in Gene Editing

The present-day CRISPR-Cas categorization hierarchy comprises three main categories (I, II, and III) as well as the less frequent but distinctly different type IV at the top. The characteristic arrangement of the corresponding loci and the corresponding signature genes serve as the basis for differentiating the various CRISPR-Cas types. Typically, type I platforms are encoded by a single operon that includes the cas1 and cas2 genes, the genes encoding the effector complex or tumble subunits (large, small, and often fused to the large subunit), the genes encoding cas5 and cas7, and the cas6 gene, which is specifically involved in the processing of pre-crRNA transcripts.

The signature gene of type II CRISPR-Cas systems, cas9, encodes a multidomain protein required for the development of the crRNA that combines all the functions of effector complexes with target DNA cleavage (Table 5.2). Type II systems are also known as streptococcus-like, Nmeni subtype, and "HNH" systems. Supplementary to the cas9 gene, the pervasive cas1 and cas2 genes are present in each CRISPR-Cas site of this subtype.

Table 5.2 Differences in CRISPR/Cas enzymes

Characters	Cas9	Cas12	Cas13
Nucleic acid	DNA	DNA	RNA
Restriction site	Protospacer adjacent motif (Pam)	Protospacer adjacent motif (Pam)	Protospacer Flanking Site (Pfs)
Type of nick	Double-Stranded break with blunt end	Double-Stranded break with sticky end	Degraded RNA
Length of spacer	17-21 nucleotides	16-24 nucleotides	25-35 nucleotides
Special features	Collateral cleavage is not observed	Occurrence of collateral cleavage	Occurrence of collateral cleavage
Applications	Gene editing	Gene editing	RNA knockdown

5.3.5 Application of Gene Editing Technologies in Beneficial Microorganism

Nitrogen is the most important nutrient to agricultural crops for its growth and development. It is most frequently applied to agricultural fields as a synthetic fertilizer in the forms of urea, nitrate, or ammonia. Recently a diazotrophic microbes which can able to fix nitrogen from the atmosphere through the activity of enzyme nitrogenase are viewed as one of the potential solutions for lowering our dependence on industrial fertilizers and developing toward sustainable production to feed the increasing population. However, due to the lack of availability of model microbial strains which can efficiently fix the atmospheric nitrogen their commercial use is restricted. Many researchers primarily concentrated on *Azotobacter vinelandii* as a potential biofertilizer strain because, unlike other diazotrophs it can able to fix nitrogen aerobically, it can also be co cultured with other algal strains. One drawback to *A. vinelandii* is that it is a free-living soil diazotroph that does not generally form tight associations with higher land plants (Das 2019). As an alternative to this, endophytic diazotrophic bacteria usually found within the tissue of the sugarcane plants and also in other crops such as Arabidopsis, wheat, sorghum, and coffee can be used a potential biofertilizer strain (Saravanan et al. 2008). Dietz et al. (2024) engineered the endophyte *G. diazotrophicus* to produce high extracellular nitrogen in the form of ammonium approximately 18mM ammonium compared compared to wild type *G. diazotrophicus* which is below 50 μM. The significant increase in the external nitrogen is obtained by dual amtB deletion and combined Q-linker with the dual amtB gene deletion from wild type strain. However, that the amount of ammonium that can be obtained from *G. diazotrophicus* using only the dual amtB deletions is far higher than what has been reported for other strains. Engineering nitrogen reductase (nifH) of *Rhizobium etli* as a nifHDK operon under the nifHc promoter that had only a truncated version of the operon resulted in enhanced nitrogenase activity (58% on average), increased plant weight (32% on average), increased nitrogen content and seed yield (36%) in common bean plants (Peralta et al. 2004). Sharma et al. (2016) has successfully transferred phytase gene appA from *Escherichia coli* K12 into the *Ensifer meliloti* 1020. Furthermore, transgenic *Ensifer meliloti* showed an increase in phosphate mineralization properties that ultimately improved maize plant growth. In another study, the deletion of *ropAe* gene predicted to encode an outer membrane protein (OMP) with a β–barrel channel structure that may be involved in Cu transport from the *Rhizobium etli*CFN42 plasmid pRet42e (505 kb) has enhanced their survival in the field supplied with copper-based fungicides.

In another study, Cu tolerance by *R. etli* was enhanced upon the deletion of a plasmid carrying ropAe gene encoding an outer membrane protein with a β-barrel channel structure that likely facilitates Cu transport (González-Sánchez et al. 2018).

References

Ali R, Zulaykha KD, Sajjad N (2020) Genetically modified microbes as biofertilizers. Bioremediation and biotechnology, Vol 4: Techniques for noxious substances remediation, 275–293.

Bashan Y, de-Bashan LE, Prabhu SR, Hernandez JP (2014) Advances in plant growth-promoting bacterial inoculant technology: formulations and practical perspectives (1998–2013). Plant Soil *378*:1–33

Bashan Y, Prabhu SR, de-Bashan LE, Kloepper JW (2020) Disclosure of exact protocols of fermentation, identity of microorganisms within consortia, formation of advanced consortia with microbe-based products. Biol Fertil Soils *56*:443–445

Bibikova M, Golic M, Golic KG, Carroll D (2002) Targeted chromosomal cleavage and mutagenesis in Drosophila using zinc-finger nucleases. Genetics *161*(3):1169–1175

Boch J, Bonas U (2010) Xanthomonas AvrBs3 family-type III effectors: discovery and function. Annu Rev Phytopathol **48**:419–436

Boch J, Scholze H, Schornack S, Landgraf A, Hahn S, Kay S, Bonas U (2009) Breaking the code of DNA binding specificity of TAL-type III effectors. Science 326(5959):1509–1512

Brian ML, Andrew CM, Carlos FB (2013) Directed evolution of the TALE N-terminal domain for recognition of all 5′ bases. Nucleic Acids Res 41(21):9779–9785

Chevalier BS, Stoddard BL (2001) Homing endonucleases: structural and functional insight into the catalysts of intron/intein mobility. Nucleic Acids Res *29*(18):3757–3774

Chiurazzi M, Ray A, Viret JF, Perera R, Wang XH, Lloyd AM, Signer ER (1996) Enhancement of somatic intrachromosomal homologous recombination in Arabidopsis by the HO endonuclease. Plant Cell *8*(11):2057–2066

Danilo B, Montes É, Archambeau H, Lodé M, Rousseau-Gueutin M, Chèvre AM, Mazier M (2022) I-SceI and customized meganucleases-mediated genome editing in tomato and oilseed rape. Transgenic Res:1–19

Das HK (2019) Azotobacters as biofertilizer. Adv Appl Microbiol 108:1–43

Deng D, Yin P, Yan C, Pan X, Gong X, Qi S, Xie T, Mahfouz M, Zhu JK, Yan N, Shi Y (2012) Recognition of methylated DNA by TAL effectors. Cell Res 22(10):1502–1504

Dietz BR, Olszewski NE, Barney BM (2024) Enhanced extracellular ammonium release in the plant endophyte *Gluconacetobacter diazotrophicus* through genome editing. Microbiol Spectr *12*(1):e02478–e02e23

Doudna JA, Charpentier E (2013) Genome editing. The new frontier of genome engineering with CRISPR-Cas9. Science 346:1258096. https://doi.org/10.1126/science.1258096

Doudna JA, Charpentier E (2014) The new frontier of genome engineering with CRISPR-Cas9. Science 346(6213):1258096

Fernández A, Josa S, Montoliu L (2017) A history of genome editing in mammals. Mamm Genome *28*(7-8):237–246

Gao C (2021) Genome engineering for crop improvement and future agriculture. Cell *184*(6):1621–1635

Goeddel DV, Kleid DG, Bolivar F, Heyneker HL, Yansura DG, Crea R et al (1979) Expression in Escherichia coli of chemically synthesized genes for human insulin. Proc Natl Acad Sci *76*(1):106–110

González-Sánchez A, Cubillas CA, Miranda F, Dávalos A, García-de los Santos A (2018) The ropAe gene encodes a porin-like protein involved in copper transit in *Rhizobium etli* CFN42. MicrobiologyOpen 7(3):e00573

Grizot S, Smith J, Daboussi F, Prieto J, Redondo P, Merino N et al (2009) Efficient targeting of a SCID gene by an engineered single-chain homing endonuclease. Nucleic Acids Res *37*(16):5405–5419

Gruenert DC, Bruscia E, Novelli G, Colosimo A, Dallapiccola B, Sangiuolo F, Goncz KK (2003) Sequence-specific modification of genomic DNA by small DNA fragments. J Clin Invest *112*(5):637–641

Hinnen A, Hicks JB, Fink GR (1978) Transformation of yeast. Proc Natl Acad Sci *75*(4):1929–1933

Horvath P, Barrangou R (2010) CRISPR/Cas, the immune system of bacteria and archaea. Science **8**:167–170

Itakura K, Hirose T, Crea R, Riggs AD, Heyneker HL, Bolivar F, Boyer HW (1977) Expression in Escherichia coli of a chemically synthesized gene for the hormone somatostatin. Science 198(4321):1056–1063

Jacquier A, Dujon B (1985) An intron-encoded protein is active in a gene conversion process that spreads an intron into a mitochondrial gene. Cell 41(2):383–394

Jaenisch R, Mintz B (1974) Simian virus 40 DNA sequences in DNA of healthy adult mice derived from preimplantation blastocysts injected with viral DNA. Proc Natl Acad Sci *71*(4):1250–1254

Jiang F, Doudna AJ (2017) CRISPR–Cas9 structures and mechanisms. Annu Rev Biophys **46**:505–529

Khalil AM (2020) The genome editing revolution. J Genet Eng Biotechnol 18(1):1–16

Klar AJS, Strathern JN, Abraham JA (1984) Involvement of double-strand chromosomal breaks for mating-type switching in Saccharomyces cerevisiae. Cold Spring Harb Symp Quant Biol 49:77–88. Cold Spring Harbor Laboratory Press

Kumar A, Baldia A, Rajput D, Kateriya S, Babu V, Dubey KK (2023) Multiomics and optobiotechnological approaches for the development of microalgal strain for production of aviation biofuel and biorefinery. Bioresour Technol 369:128457

Liang F, Han M, Romanienko PJ, Jasin M (1998) Homology-directed repair is a major double-strand break repair pathway in mammalian cells. Proc Natl Acad Sci 95:5172–5177

Ménoret S, Fontanière S, Jantz D, Tesson L, Thinard R, Rémy S et al (2013) Generation of Rag1-knockout immunodeficient rats and mice using engineered meganucleases. FASEB J 27(2):703–711

Musunuru K (2017) The hope and hype of CRISPR-Cas9 genome editing: a review. JAMA Cardiol *2*(8):914–919

Nath A, Bhuyan P, Gogoi N, Deka P (2023) Pesticides and chemical fertilizers: role in soil degradation, groundwater contamination, and human health. In: Xenobiotics in urban ecosystems: sources, distribution and health impacts. Springer International Publishing, Cham, pp 131–160

Nemudryi AA, Valetdinova KR, Medvedev SP, Zakian SM (2014) TALEN and CRISPR/Cas genome editing systems: tools of discovery. Acta Naturae 6(3):19–40

Pâques F, Duchateau P (2007) Meganucleases and DNA double-strand break-induced recombination: perspectives for gene therapy. Curr Gene Ther *7*(1):49–66

Peralta H, Mora Y, Salazar E, Encarnación S, Palacios R, Mora J (2004) Engineering the nifH promoter region and abolishing poly-β-hydroxybutyrate accumulation in *Rhizobium etli* enhance nitrogen fixation in symbiosis with Phaseolus vulgaris. Appl Environ Microbiol 70(6):3272–3281

Ran FA, Hsu PD, Wright J, Agarwala V, Scott DA (2013) Genome engineering using the CRISPR-Cas9 system. Nat Protoc 8:2281–2308

Rong YS, Titen SW, Xie HB, Golic MM, Bastiani M, Bandyopadhyay P, Olivera BM, Brodsky M, Rubin GM, Golic KG (2002) Targeted mutagenesis by homologous recombination in D. melanogaster. Genes Dev 16(12):1568–1581

Saravanan VS, Madhaiyan M, Osborne J, Thangaraju M, Sa TM (2008) Ecological occurrence of *Gluconacetobacter diazotrophicus* and nitrogen-fixing Acetobacteraceae members: their possible role in plant growth promotion. Microbial Ecol 55:130–140

Sharma V, Kumar A, Archana G, Kumar GN (2016) *Ensifer meliloti* overexpressing *Escherichia coli* phytase gene (AppA) improves phosphorus (P) acquisition in maize plants. Sci Nat 103:1–10

Shmakov S, Smargon A, Scott D, Cox D, Pyzocha N, Yan W, Abudayyeh OO, Gootenberg JS, Makarova KS, Wolf YI, Severinov K, Zhang F, Koonin EV (2017) Diversity and evolution of class 2 CRISPR-Cas systems. Nat Rev Microbiol **15**:169–182

Smithies O, Koralewski MA, Song KY, Kucherlapati RS (1984) Homologous recombination with DNA introduced into mammalian cells. Cold Spring Harb Symp Quant Biol 49:161–170

Sood M, Kapoor D, Kumar V, Sheteiwy MS, Ramakrishnan M, Landi M et al (2020) Trichoderma: the "secrets" of a multitalented biocontrol agent. *Plants 9*(6):762

Sudheer S, Bai RG, Usmani Z, Sharma M (2020) Insights on engineered microbes in sustainable agriculture: biotechnological developments and future prospects. Curr Genomics 21(5):321–333

Vanamee ES, Santagata S, Aggarwal AK (2001) FokI requires two specific DNA sites for cleavage. J Mol Biol 309:69–78

Methods from the Field of Synthetic Biology that Aim to Improve Plant Growth and Resistance to Stress through the Use of Genetic Engineering

6

K. Jagadeesh Chandra Bose, Sukhminderjit Kaur, Sonia Sharma, Jyoti Sarwan, and Nazim Uddin

Abstract

The expanding worldwide issues of food security and environmental sustainability can now be addressed with the help of synthetic biology. The modification of plant genomes to promote growth and enhance stress responses has become a major area of research in agriculture. By fine-tuning the stress response through the engineering of novel regulatory targets, comprehending posttranslational changes, and preserving hormone homeostasis, researchers have successfully increased plant tolerance to diverse abiotic stresses, such as salinity and drought. Additionally, crops with improved growth, productivity, and stress tolerance have been created using synthetic biology techniques that are salt- and drought-tolerant. The creation of stress-tolerant crops is one of the primary areas where synthetic biology is having a substantial influence. Crop productivity is seriously threatened by climate change and different environmental stressors such as drought, salt, and severe temperatures. Synthetic biology enables the creation of plants with enhanced stress tolerance by selectively altering genes and pathways involved in the stress response. The higher photosynthesis, decreased susceptibility to disease, and enhanced water usage efficiency of these modified plants ultimately increase crop resilience. Additionally, synthetic biology is essential in overcoming the difficulties associated with sustainable agriculture. Synthetic biology assists in minimizing the negative environmental effects of contemporary agriculture by designing plants to fix nitrogen more effectively, require fewer chemical fertilizers, and be more resistant to pests and infections. This increases crop yields while reducing the damaging consequences of agriculture

K. Jagadeesh Chandra Bose (✉) · S. Sharma · J. Sarwan · N. Uddin
Department of Biotechnology, UIBT, Chandigarh University, Mohali, Punjab, India

S. Kaur
Department of Biotechnology, Chandigarh University, Mohali, Chandigarh, India

S. Kaur et al. (eds.), *Metabolomics, Proteomics and Gene Editing Approaches in Biofertilizer Industry*, https://doi.org/10.1007/978-981-97-2910-4_6

on ecosystems and biodiversity. The study of the function of polyamines (PAs) in plant growth and stress responses has also benefited from synthetic biology. It has been demonstrated that PAs, whether used exogenously or generated internally through genetic engineering, have a favourable impact on plant growth, productivity, and stress tolerance. The study of plant growth and stress response has also greatly benefited from the work of metabolomics, a subfield of plant science. Metabolomics has aided in the development of practical applications for improving crop quality and modifying the metabolic processes of plants by examining the genetic and biochemical pathways behind plant growth, development, and stress responses. Additionally, the processes underlying plant performance under hypoxia-related environmental challenges have been revealed by synthetic biology. To better understand the role of oxygen biology in plants and make suggestions for increasing plants' resistance to hypoxia-related stressors, researchers have used a synthetic biology perspective. Additionally, the use of synthetic biology to increase stress tolerance in microbial biotransformation systems provides a potent method for enhancing the capabilities of already-existing systems. The potential of synthetic biology to transform plant breeding and lessen the difficulties faced by climate change and environmental stresses in the agriculture sector will be highlighted in this chapter.

Keywords
Synthetic biology · CRISPR · PGPR · Biostimulants · Genome editing

6.1 Introduction

Improvements in crop production are a significant concern that needs to be addressed in the context of the current scenario, which involves a rapidly increasing population and fluctuating climate. The agricultural sector in the twenty-first century is plagued by persistent difficulties brought on by a variety of both biotic and abiotic pressures, with insects and pests being one of the most significant of these forces. The stagnation of output can be recognized to the fact that insects are responsible for the loss in potential yields. Although operating control mechanisms are in place, many commercially significant crops have a wide range of yield losses. Farmers, on the other hand, have used chemical based pesticides to control insect pests. The extensive application of these pesticides can have negative consequences for both humankind and the environment. Due to these issues, researchers all around the world have focused their efforts on developing innovative and environmentally friendly methods of pest management. The primary objective of modern agriculture has been to attain increased yields while making use of the land and resources that are already available to increase agricultural sustainability and ensure international food security. As a result of the development of genetic engineering, rDNA technology, and plant biotechnology, the administration of pest control has expanded to new heights. The genetic and genome modification of a variety of crop plants with numerous insect resistance genes (Bacillus thuringiensis Bt-ICPs) and the ongoing application of bt gene-introduced crops have demonstrated a reliable impact on the technology's ability to increase production and ensure its long-term viability. There has been a significant amount of concern regarding the development of resistance in insect pests to the cry toxins. The presence of mutations in the genes that code for receptor molecules causes the interaction between the toxin and the insect to be disrupted, which ultimately leads to the development of a population of insect pests that are resistant to the toxin. The utilization of combinatorial ICPs and chimeric toxins, which are generated by the process of domain swapping of toxins, are also being utilized to achieve the goals of delivering extended effectiveness against insect-based pests and stopping resistance. As an alternative, there are various techniques and methodologies, such as RNA interference (RNAi), which are focused on silencing specific genes that are implicated.

6.2 What Kind of Biology Is Synthetic?

As discussed in Fig. 6.1, synthetic biology is interdisciplinary field that associates the ideas from biology, engineering, chemistry, and computer science to create and construct new biological entities, systems, or functions that do not occur in nature or to redesign existing biological systems for specific goals. Synthetic biology is

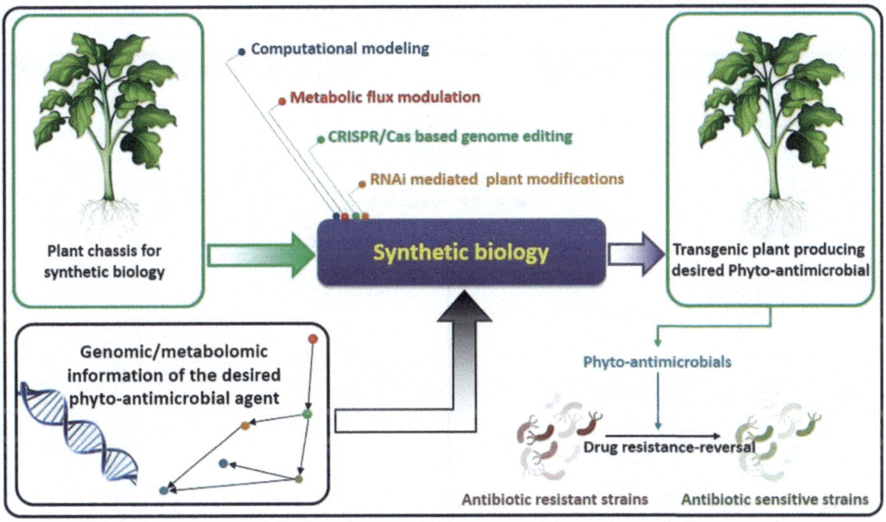

Fig. 6.1 Antimicrobial resistance (AMR) is a global communal health concern caused by improper antimicrobial medication use in a variety of industries. Potential treatments have been found as phyto-antimicrobials, which target significant microbial drug-resistance determinants. However, their full potential is hampered by limited distribution, low intracellular concentrations, and environmental factors such as environmental change and over-exploitation of plant sources. Synthetic biology technologies such as metabolic engineering, genome editing, RNA interference, and systems biology could be used to create phyto-antimicrobials. This viewpoint review emphasizes the importance of phyto-antimicrobials in the post-antibiotic age (Tiwari et al. 2021)

also responsible for the development of new biological systems. The arena of Synthetic Biology encompasses not only the creation and development of new biological components, technologies, and systems but also the redesign of natural biological systems that are already in existence to make them more useful. Through the application of engineering principles to biology, the purpose of synthetic biology is to make it possible to create artificial biological systems that can perform the activities that are needed. There are many different applications for it, and it is dependent on several different technologies. Synthesis and editing of DNA synthesis and editing of DNA is an essential part of synthetic biology. Synthesis and editing of DNA are synonymous terms. The precision and effectiveness of DNA manipulation have been significantly improved by the introduction of technologies such as CRISPR-Cas9 (Fig. 6.2).

The use of computational biology tools is an essential component in the process of designing, modelling, and analysing synthetic biological systems. Synthetic biology makes use of computation. Researchers can more accurately predict the behaviour of modified biological components with the assistance of bioinformatics. To achieve standardization, it is necessary to create a collection of biological components that are interchangeable and may be utilized in the construction of synthetic structures in a modular approach. The predictability and dependability of synthetic biology are both improved by this method. The process of designing and constructing genomic circuits and metabolic pathways is frequently performed in the field of

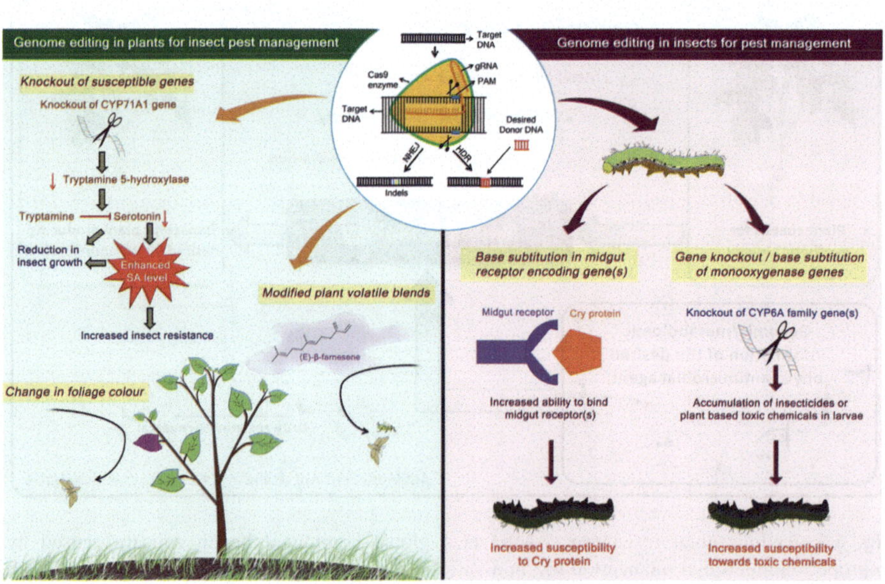

Fig. 6.2 CRISPR-based gene editing strategies in plants and insects for pest control. By knocking down genes that are prone to insect pests, adjusting plant volatile mixtures, and altering the colour of the leaf, it has been established that plant genome editing may be used to create insect pest resistance. Genetic engineering can be used to change the Cry protein binding receptors and turn off detoxifying enzymes in insects so that they are more susceptible to plants. (Tyagi et al. 2020)

computational and synthetic biology. Specific tasks, such as sensing environmental stimuli and responding with a desired output, can be designed into these circuits so that they can perform specific functions through engineering. Applications of synthetic biology can be found in a wide variety of sectors, including medicine, biotechnology, environmental science, materials science, and energy, to name a few. Examples of this include the construction of synthetic tissues for medical purposes, the development of modified microorganisms for the generation of biofuels, and the design of biological sensors for the monitoring of the environment. Synthetic biology is a discipline that poses ethical problems regarding the production of artificial living forms, as well as the possible threats to the environment and safety that are linked with manufactured organisms. For the purpose of establishing guidelines for safe synthetic biology procedures, researchers in the area are actively addressing these challenges and attempting to produce recommendations. Synthetic biology has the potential to revolutionize many different parts of science and industry by providing new tools for building and directing biological systems to accomplish certain goals. The subject is dynamic, with ongoing research contributing to its expansion and the creation of fresh applications. Furthermore, the discipline is constantly evolving. Synthetic biological systems are being developed for use in diagnostics, drug delivery, and personalized treatment. Biofuel generation, industrial enzyme optimization, and sustainable agriculture are all areas that could benefit from enhanced microbial strains. Organisms that have been engineered for cleaning up pollutants and treating waste. Manufacturing of innovative materials, including bio-based polymers and materials that are compatible with living organisms.

The application of computational and synthetic biology allows for the creation of organisms that can produce biofuels and improving the efficiency of the production of renewable energy. More recent developments in synthetic biology include the development of vaccinations. The quick development of mRNA-based vaccines during the COVID-19 pandemic exemplifies the promise of synthetic biology in the design of vaccines. Development has been made in the development of synthetic biology technologies for cell treatments, including the creation of modified immune cells for the treatment of chronic cancer. The development of biological circuits for information processing and computation has made significant progress. While this is going on, there are a few problems and ethical considerations that are related to safety and biosecurity. In addition, there are worries over the unintended consequences of altered organisms, and there is a requirement for effective safety measures. There have been very few discussions and ethical issues about the ethical implications of the creation of synthetic life forms and the responsible application of synthetic biology. The domain, synthetic biology continues to make strides towards greater precision and predictability. The incorporation of synthetic biology into other domains, such as nanotechnology and artificial intelligence, among others.

6.3 Improving the Growth of Plants and the Response of Plants to Stress

Improving plant growth and its ability to respond to stress is essential for several reasons, particularly in the context of global food security, environmental sustainability, and climate and environmental change. Enhancing plant development and resistance to stress is necessary for a variety of reasons, and there are many of them. The number of people living on this planet is rapidly growing, and it is anticipated that by the year 2050, it will have surpassed 9 billion. It is necessary to enhance agricultural output and output to meet the demands of a growing population and ensure global food security with elevated population growth levels. Climate change is causing environmental pressures to occur more frequently and with greater severity. These stresses include droughts, floods, heatwaves, and changes in the patterns of precipitation. To be able to produce crops constantly, plants need to be able to withstand various pressures. In addition to the stressors that are caused by climate, plants are also subject to abiotic stresses such as salinity, an imbalance in the nutrients in the soil, and heavy metal poisoning. By improving their stress response mechanisms, plants are better able to thrive in difficult conditions. Agriculture is a significant user of freshwater resources around the world. Especially in areas that are experiencing a lack of water, it is needed to develop crops that are either more water-efficient or can grow in settings where there is a limited amount of water available. Increasing the efficiency with which plants use nutrients can lessen the amount of fertilizers that are required and lower the amount of pollution that is released into the environment. Increasing a plant's resistance to stress and its capacity for development can result in more sustainable agricultural methods, which in turn reduces the negative impact that farming has on the environment. Increasing agricultural yields and resilience has a direct influence on the livelihoods of farmers, particularly those who live in vulnerable locations where agriculture is their primary source of income. It is possible to reduce the amount of pressure that is placed on natural ecosystems to be converted into farmland by enhancing the efficiency of agriculture and thus contributing to the preservation of biodiversity. To improve plant growth, it may be necessary to enhance the nutritional content of crops, as well as address concerns connected to malnutrition and deficiencies in nutrients. It is possible to genetically modify certain plants so that they may take in and store toxins from the soil, thereby contributing to the improvement of environmental conditions in places that are contaminated. Since new plant pathogens are constantly being discovered, it is becoming increasingly important to cultivate crops that have improved resistance to diseases in order to keep agricultural production consistent. Researchers are currently engaged in the process of developing a variety of approaches to accomplish these objectives. These approaches include conventional breeding techniques, genetic engineering, and precision agricultural techniques. To contribute to a global agriculture system that is more sustainable and resilient, the goal is to grow crops that can grow in a variable environmental circumstance, are resistant to pests and diseases, and provide higher yields while using fewer resources.

6.4 Increasing the Growth of Plants

The process of enhancing plant growth involves several different factors, including the utilization of substances that promote plant growth, the selection of arbuscular mycorrhizal fungal species that are effective, and the comprehension of the metabolic capabilities of the rhizosphere microbiome (Babalola et al. 2021; Barea and Azcón-Aguilar 1982; Fasusi et al. 2021). Furthermore, the significance of the plant growth-promoting agents, rhizobacteria (PGPR) in improving plant development through the production of secondary metabolites has been well-documented (Raghav et al. 2019). This is because PGPR encourages the creation of secondary metabolites. Furthermore, it has been demonstrated that the application of plant growth regulators and hormones, such as auxins, cytokinin, and gibberellins, can promote the growth of plants, particularly in polluted soil (Chouychai et al. 2015; Hassan et al. 2023; Soliman 2019). It has also been discovered that the utilization of phytohormones and calcium peroxide in the process of phytoremediation can enhance the efficiency of plant rehabilitation (Hassan et al. 2023). Furthermore, it has been established that the production of growth-regulating chemicals by soil microorganisms, such as vesicular-arbuscular mycorrhizal fungus, has been associated with the enhancement of plant development (Barea and Azcón-Aguilar 1982). In addition, it has been demonstrated that the application of plasma and UV light irradiation to seeds can stimulate the growth of plants (Hayashi et al. 2015). According to Siswanti and Khairunnisa (2021) and Parakhia et al. (2014), the utilization of biofertilizers and the comprehension of the genetic composition of endophytic bacteria have been linked to enhanced plant growth and the ability to withstand abiotic challenges. Salinity, drought periods, and other negative elements can have a strong influence on plant growth, which highlights the necessity of addressing these problems to enhance plant growth (Mauer et al. 2018; Yahmi et al. 2021). It is vital to remember that adverse circumstances can have a considerable impact on plant growth. Furthermore, the fact that plant hormones have a role in encouraging growth in lichen-forming fungi further highlights the complex interaction that exists between hormones and the growth of microorganisms, which can have ramifications for the overall health and development of plants (Wang et al. 2010). In conclusion, the process of improving plant development is a multidimensional process that requires the administration of substances that promote plant growth, the selection of microbial species that are efficient, the study of the metabolic capabilities of the rhizosphere microbiome, and the utilization of plant growth regulators and hormones. Additionally, it is essential to manage environmental stressors and unfavourable influences to promote optimal plant growth.

6.5 There Are Several Benefits to Promoting Plant Growth

Enhancing the development of plants has been shown to provide a multitude of benefits and different methods have been investigated to accomplish this goal. It has been demonstrated that humic compounds that are formed from organic waste can significantly improve the growth of the roots, the leaves, and the shoots, as well as stimulate the germination of a wide variety of crop species. Additionally, the application of calcium and potassium is useful in boosting the activity of antioxidant enzymes and solutes, which ultimately results in increased plant growth even when the plant is subjected to difficult conditions. Additionally, the utilization of brassinosteroids has been associated with the enhancement of the activation of Rubisco and the expression of photosynthetic relevant genes in plants, which in turn leads to the promotion of photosynthesis and plants' growth. It has been discovered that the administration of proline and trehalose from an exogenous source can enhance the growth of plants as well as their physiological responses during times of stress and recuperation. In addition, it has been demonstrated that interactions between mycorrhizal fungi and bacteria can enhance plant growth in the presence of both biotic and abiotic stress situations. Furthermore, it has been demonstrated that the absorption of torrefied biomass into soil yields improvements in the ground's physical, chemical, and biological qualities, which ultimately results in increased plant development.

As an additional point of interest, it has been discovered that carbon generated from amino acids, such as proline, can boost plant development by enhancing the conversion and accumulation of biomass. It has also been observed that the application of proline from an exogenous source can improve the growth of plants and their physiological parameters when they are subjected to stress. In addition, it has been demonstrated that the buildup of free proline is a necessary component for the growth of the plants that are subjected to drought stress. In addition, the utilization of bacteria that promote plant growth has been brought to light as a significant factor in the enhancement of plant growth. The administration of nitrogen fertilizer has been discovered to boost both the development of the plant and development of the plant's vegetative tissue, which ultimately increases the plant's output. In addition, the utilization of biostimulants, such as sodium metasilicate, has proven to be beneficial in enhancing the development and productivity of plants under soil conditions that are difficult to cultivate.

In conclusion, enhancing plant growth through a variety of methods, such as the application of growth-promoting substances, interactions between bacteria and mycorrhizal fungi, along with the utilization of biostimulants, offers a multitude of benefits, including improved physiological responses, enhanced photosynthesis, and increased accumulation of biomass. When it comes to fostering plant development and productivity, these tactics are quite important, particularly when the plant is subjected to stressful conditions.

The process of enhancing plant growth involves several different factors, including the utilization of substances that promote plant growth, the selection of arbuscular mycorrhizal fungal species that are effective, and the comprehension of the metabolic capabilities of the rhizosphere microbiome (Babalola et al. 2021; Barea and Azcón-Aguilar 1982; Fasusi et al. 2021). Furthermore, the significance of plant growth-promoting rhizobacteria (PGPR) in improving plant development through the production of secondary metabolites has been well-documented (Raghav et al. 2019). This is because PGPR encourage the creation of secondary metabolites. Furthermore, it has been demonstrated that the application of plant growth regulators and hormones, such as auxins, gibberellins, and cytokinins, can stimulate the growth of plants, particularly in polluted soil (Chouychai et al. 2015; Hassan et al. 2023; Soliman 2019). It has also been discovered that the utilization of phytohormones and calcium peroxide in the process of phytoremediation can enhance the efficiency of plant rehabilitation (Yeh 2015). Furthermore, it has been established that the production of growth-regulating chemicals by soil microorganisms, such as vesicular-arbuscular mycorrhizal fungus, has been associated with the enhancement of plant development (Barea and Azcón-Aguilar 1982). In addition, it has been demonstrated that the application of plasma and UV light irradiation to seeds can stimulate the growth of plants (Hayashi et al. 2015). According to Siswanti and Khairunnisa (2021) and Parakhia et al. (2014), the utilization of biofertilizers and the comprehension of the genetic composition of endophytic bacteria have been linked to enhanced plant growth and the ability to withstand abiotic challenges.

Salinity, drought periods, and other negative elements can have a substantial influence on plant growth, which highlights the necessity of addressing these problems to enhance plant growth (Mauer et al. 2018; Yahmi et al. 2021). It is crucial to remember that adverse circumstances can have a considerable impact on plant growth. Furthermore, the fact that plant hormones have a role in encouraging growth in lichen-forming fungi further highlights the complex interaction that exists between hormones and the growth of microorganisms, which can have ramifications for the overall health and development of the plants (Wang et al. 2010).

In conclusion, the process of improving plant development is a multidimensional process that requires the administration of substances that promote plant growth, the selection of microbial species that are efficient, the study of the metabolic capabilities of the rhizosphere microbiome, and the utilization of plant growth regulators and hormones. Additionally, it is essential to manage environmental stressors and unfavourable influences to promote optimal plant growth.

6.6 CGP's Stress Response in Plants and their Needs

Increasing the susceptibility of plants to stress is essential for ensuring the productivity and sustainability of agricultural production. There have been many different approaches investigated to improve the responses of plants against stress. Some of these approaches include the utilization of transcription factors, rhizosphere microorganisms, and genetic engineering.

It has been revealed that some of the transcription factors, for instance NAC (NAM, ATAF, and CUC), can improve the levels of both the drought resistance and salt tolerance in rice. This highlights the potential of these transcription factors for the improvement of plant stress tolerance, as stated by Hu et al. (2006). Additionally, rhizosphere microbes, including both plant growth-promoting bacteria (PGPB) and plant growth-promoting fungi (PGPF), have been identified as essential partners for enhancing plant stress tolerance, demonstrating the importance of microbial interactions in promoting plant growth and conferring their resistance towards both biotic and abiotic stresses (De Zelicourt et al. 2013). Approaches that include genetic engineering, such as the overexpression of genes, have also confirmed their potential in the process of increasing the stress tolerance of vegetables. An example of this would be the overexpression of the transcriptional coactivator Multiprotein Bridging Factor 1c, which has been shown to increase plant tolerance to environmental stress conditions. This provides a potent method for improving stress responses in transgenic plants (De Zelicourt et al. 2013). In a similar vein, the deletion of gene families has been associated with increased drought and salinity tolerance in plants. This finding offers insights into prospective targets for genetic modification to improve stress tolerance (Chen et al. 2021b). Furthermore, the utilization of chemicals, such as melatonin and raffinose, has been explored to determine whether or not they have the capability of boosting the stress tolerance of plants. According to research (Huang et al. 2019), it has been demonstrated that exogenous melatonin can reduce the effects of oxidative damage and protect photosystem II in maize seedlings that are experiencing drought stress. This finding highlights the role that melatonin plays as a regulator of stress responses in plants. In addition, the overexpression of raffinose synthase has been linked to increased drought tolerance in maize and Arabidopsis plants. This was accomplished by either the synthesis of raffinose or the hydrolysis of galactinol. This finding demonstrates the potential of particular metabolic pathways in the improvement of stress tolerance (Li et al. 2020).

In general, the investigation of transcription factors, interactions between microorganisms, genetic engineering, and certain substances provides useful insights into the process of increasing the levels of stress tolerance traits in plants. These methods lay groundwork for the development of robust crop types that can endure a wide range of environmental stress factors, which will ultimately contribute for the growth of sustainable agriculture and ensure food security.

6.7 The Benefits of the Stress Response in Plants—The Requirement for CGP

It is possible to expand plant ability and productivity by enhancing the plant's stress response, which offers several benefits and contributes to improved plant performance in demanding environmental conditions. Adaptation to environmental stress is accomplished by plants through the integration of several hormone-response pathways, as proven by research conducted by Ha et al. (2014). Additionally, it has been revealed that a combination of biotic and abiotic stress has a beneficial impact on plant performance by lowering the plant's vulnerability to biotic stress (Ben Rejeb et al. 2014). This highlights the possible advantages that plants may derive from their reaction to stress.

According to Suzuki et al. (2005), transgenic plants that express certain genes have demonstrated increased tolerance to osmotic stress and heat stress, even in the conditions, when both pressures were combined. This finding suggests that genetic engineering may have the potential to improve plants' ability to withstand stress. Additionally, overexpression of certain genes, such as cytosolic copper/zinc superoxide dismutase, has been associated with higher tolerance to drought stress, suggesting the possibility of genetic manipulation in boosting the stress tolerance (Prashanth et al. 2008).

Furthermore, it was revealed that beneficial bacteria can reverse the negative consequences of plant stress responses by producing the effector proteins and increasing antioxidant levels in plants. This finding highlights the significance that microbial interactions play in developing stress tolerance in the plants, as proven by numerous research experiments (Bastías et al. 2022). Additionally, chemical priming has been recognized as a potentially effective method for enhancing the salt stress tolerance of plants (Sako et al. 2021).

According to Yamano et al. (2018), the benefits of stress-tolerant crops become apparent when the crop is subjected to the stresses that it is tolerant against. This highlights the practical value of incorporating stress response enhancement into plant cultivation. Furthermore, the development of heat stress tolerance has been connected to increased pollen viability in transgenic plants, highlighting the potential influence that stress response enhancement could have on the reproductive performance of plants (Deng et al. 2020).

In a nutshell, the enhancement of the stress responses in plants by the use of genetic engineering, interactions between microbes, and chemical priming offers a multitude of benefits, such as greater plant performance, tolerance to different stresses, and increased reproductive success. Based on these findings, valuable insights have been gained into the potential of stress response augmentation to contribute to sustainable agriculture practices and food safety and security.

6.8 What Specific Ways Will the Application of Synthetic Biology Contribute to the Improvement of Plant Growth Potential?

Synthetic biology presents a great opportunity to improve plant development by utilizing advanced genetic engineering and biotechnological methods. This might be accomplished through the application of synthetic biology. Researchers can create plants to demonstrate improved growth traits, enhanced stress tolerance, and increased productivity by utilizing synthetic biology. Other benefits include increased output. The manipulation of metabolic pathways is one of the areas in which synthetic biology could make a substantial contribution to the enhancement of plant development. In synthetic biology, the manufacture of useful molecules, such as terpenoids, is made possible through the introduction or modification of metabolic pathways in plants. These compounds have the potential to positively influence both the growth and development of plants, as (Chen et al. 2021a) have demonstrated. In addition, the engineering of full metabolic pathways in transgenic plants has the potential to improve both the production of huge amounts of biomass and the overall performance of the plant (Krichevsky et al. 2012).

Moreover, synthetic biology offers a platform for the generation of genetically modified plants that have greater drought avoidance or tolerance, which contributes to improved plant growth in settings where there is a limited amount of water available (Yang et al. 2020). Under this strategy, engineering principles are utilized to design and implement genetic alterations that bestow specific characteristics related to stress tolerance and growth augmentation. These modifications are then introduced into the population. Synthetic biology, in addition to metabolic engineering, makes it possible for plant-based pharmaceutical compounds to be expressed in microbial systems. This provides an alternative method for the creation of useful bioactive molecules that have the potential to positively influence the growth and the health of plants (Jain et al. 2012). Additionally, the utilization of synthetic biology technologies for the creation of specialized metabolites in plants gives prospects to improve the effective applications of synthetic biology to exploit metabolic pathways that contribute to the growth of plants and their responses to stress (Muhich et al. 2022).

Another way in which synthetic biology can assist to the enhancement of plant growth is through the development of sophisticated imaging tools that make it possible to observe fluorescent proteins in the canopies of plants. This capability makes it possible to monitor and manipulate the physiological processes of plants, which in turn provides insights that may be utilized to maximize the growth and the productivity of plants (Rigoulot et al. 2021). Furthermore, the application of synthetic biology has the promising potential to engineer the plants to produce compounds, such as resveratrol, in microbial systems. This allows for the expansion of the possibilities for the manufacture of bioactive molecules that can have a positive impact on plant growth in a sustainable and scalable manner (Song et al. 2021). Furthermore, synthetic biology offers a powerful toolkit for increasing plant growth through the engineering of metabolic pathways, the development of different stress-tolerant crops and varieties, the manufacture of valuable bioactive compounds, and the optimization of plant physiological processes. All these activities are accomplished through the utilization of synthetic biology. Using the principles of computational and synthetic biology, researchers can propel innovation in plant biotechnology and contribute to the creation of agricultural techniques that are more environmentally friendly.

6.9 In What Ways Will Genetic Engineering Contribute to the Enhancement of Plant Growth?

As shown in Fig. 6.3, through the manipulation of plant genomes, genetic engineering provides a potent method for boosting plant development. This is accomplished by introducing certain features that contribute to enhanced performance, stress tolerance, and production. Through the utilization of genetic engineering techniques, researchers can specifically target specific genes and metabolic pathways to build plants with desirable features, which eventually results in improved growth as well as the development abilities. The change of lignin content, which is a substantial component of plant cell walls, is one of the significant applications of genetic engineering in the improvement of plant development. It has been shown that fodder

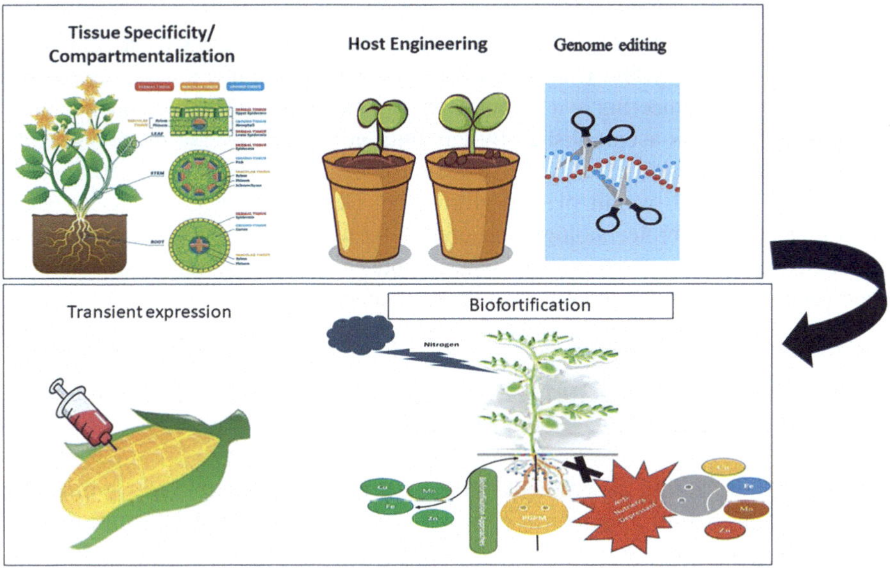

Fig. 6.3 Synthetic biology can be built on the foundation of plants. It will be very helpful to make progress in tissue-specific promoters, compartmentalization methods, plant host engineering, and genome editing so that plants can be used as a platform for making health-promoting small molecules and adding more nutrients to crops. (Barnum et al. 2021)

digestibility and pulping efficiency can be enhanced by the efforts of genetic engineering techniques that aim to alter the amount of lignin present. This, in turn, can lead to an increase in the amount of biomass produced and the development of plants (Li et al. 2008). Similarly, the genetic alteration of cell walls to improve biomass saccharification and lodging resistance has been observed as a potential technique for minimizing biomass recalcitrance and sustaining normal plant growth (Li et al. 2008). This strategy aims to boost the ability of plants to support their growth naturally. In addition, genetic engineering protocols makes it possible to manipulate metabolic pathways to enhance plant growth and resilience to stress. For example, it has been demonstrated that the expression of a fungal sterol desaturase in tomato plants led to improvements in drought tolerance, resistance to pathogens, and nutritional quality. This finding highlights the potential of genetic engineering to improve plant performance under stressful conditions (Kamthan et al. 2012). Additionally, the genetic modification of hemicellulose arabinose levels has been linked to improvements in both plant lodging resistance and biomass enzymatic digestibility, demonstrating the potential of genetic engineering to enhance plant mechanical strength and biomass production (Ouyang et al. 2017).

In addition, genetic engineering experiments makes it possible to introduce particular genes into plants that boost their production and tolerance to stress, which in turn contributes to an improvement in crop output and resilience (Wille et al. 2017).

Genetic engineering methods allow for the combination of features to boost resistance against insects and herbicides, which ultimately increases plant growth and productivity (Verma et al. 2022). This is accomplished by stacking multiple transgenes. In conclusion, genetic engineering is a strong technique that may be used to enhance plant development. This is accomplished by targeting genes and diverse metabolic pathways to introduce desirable features that are correlated with stress tolerance, increased biomass output, and overall plant performance. Genetic engineering can promote innovation in plant biotechnology and contribute to the production of durable and high-performing crop varieties. This is accomplished by the precise modification of plant genomes, which is accomplished through genetic engineering.

6.10 The Use of Synthetic Biology Has Resulted in Enhanced Plant Growth in Several Crops

Through the utilization of synthetic biology, it has been possible to significantly improve the growth of a variety of plant species (Fig. 6.4). As an illustration, synthetic biology has proven of great assistance in the development of metabolic engineering of bioenergy crops like sugarcane, to increase the crops' resistance to stress and rise their overall output (Shih et al. 2016). In addition, synthetic biology techniques have been utilized to expand the nutritionary quality of crops, which has resulted in an improvement in human health as a result of the consumption of biofortified crops (Pérez-Massot et al. 2013). As an additional point of interest,

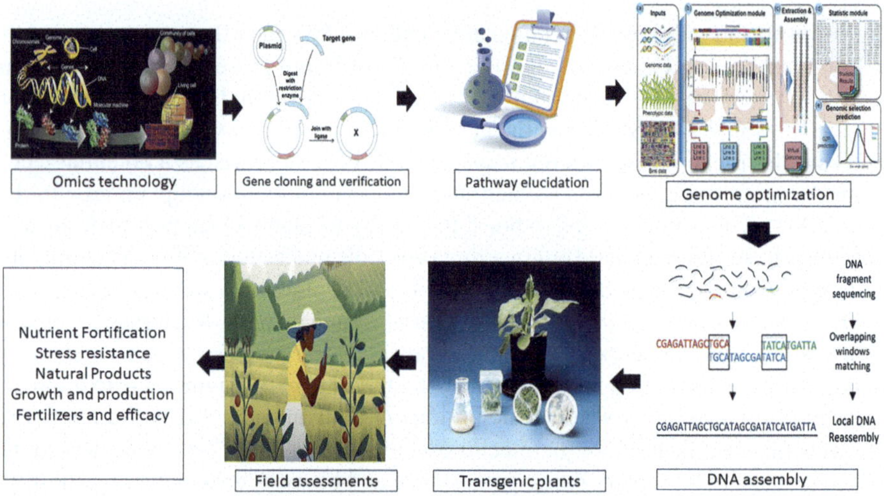

Fig. 6.4 Here, we'll go over the prospective and existing uses of plant synthetic biology techniques, mostly about crops. These include elements such as agricultural productivity, the production and quality of nutrients, the use of less fertilizer, plant-derived goods, and a few other uses in the pharmaceutical industry. Synthetic biology applications in agriculture show that genetic circuits, metabolic pathways, and plant structures can be altered to boost crop yields (Bi et al. 2022)

synthetic biology has been applied to alter Rubisco and produce photorespiratory bypasses, which has resulted in the augmentation of C3 photosynthesis in contemporary crop improvement efforts (Jin et al. 2023). According to Llorente et al. (2018), the synthetic biology has been connected to the production of multi-biofortified crops that have better nutritional qualities. These crops have the potential to give benefits to human nutrition and health.

In the context of abiotic stress and its tolerance, synthetic biology has been utilized to develop genetically engineered plants that have improved drought avoidance or tolerance, which has contributed to an increase in agricultural resilience in the face of severe environmental conditions. Furthermore, synthetic biology has been concerned to improve the stress tolerance and yield of a variety of plants. This is done in response to the growing requirements for crops that have progressed in terms of stress tolerance and productivity (Wágner 2012). In addition, synthetic biology has been utilized to improve the abiotic stress tolerance of several crop plants, which has contributed to the advancement of sustainable farming from the point of view of functional genomics (Gill et al. 2015). In a nutshell, synthetic biology has been essential in boosting plant development in a wide range of crops, which has led to improvements in stress tolerance, nutritional quality, and overall output. With these breakthroughs, the potential of synthetic biology to address agricultural difficulties and improve crop performance to justify the demands of a raising population has been demonstrated. I have included some instances of crops that have benefited from improved plant development because of the application of genetic engineering. The following are some examples of crops that have benefited from increased plant growth because of genetic engineering:

- Genetic engineering has been applied to tomato plants to enhance their resistance to drought, resistance to pathogens, and nutritional quality, as stated by Kamthan et al. (2012).
- In the case of rice, genetic engineering has been utilized to strengthen the plant's resistance to drought, which has resulted in an increased capacity to withstand harsh climatic conditions (Cui et al. 2016).
- Maize: According to Abidoye and Mabaya (2014), genetically modified maize has been designed to expand yields while simultaneously decreasing the amount of pesticides that are used. This has contributed to an increase in both productivity and sustainability.
- It has been demonstrated that the application and exploitation of genetic engineering in sugarcane farming has the potential to reduce the effects of biotic as well as abiotic stresses, hence contributing to increased stress tolerance and productivity (Verma et al. 2022).
- Carrot: Genetic engineering has been utilized to induce the formation of shikimic acid and phenolic antioxidants in damaged carrot material, which has contributed to increased nutritional quality and stress response (Becerra-Moreno et al. 2012). This has led to improved nutritional quality and nutritional response.

- Citrus sinensis, also known as the sweet orange, has been subjected to targeted genome editing with Cas9/sgRNA to improve characteristics. This demonstrates the promise of genetic engineering in citrus crops (Jia and Wang 2014).
- Transgenic turnip plants that express the BrOAT1 gene have demonstrated improved resistance to the salt stress. This finding highlights the potential of genetic engineering to enhance stress tolerance in Brassica crops (Jung et al. 2013). *Brassica rapa* is the scientific name for turnip. These examples highlight the numerous applications of genetic engineering in the enhancement of plant growth and stress tolerance across a wide range of crop species.

Through the utilization of synthetic biology, it has been possible to successfully obtain increased stress tolerance in a variety of crops. Enhanced stress tolerance was obtained in several crops by the application of synthetic biology, including the following:

- Approaches from the field of synthetic biology have been employed in the field of tomato research to promote plant stress tolerance and yield output in tomato plants. This has contributed to increased resilience in the face of harsh environmental conditions.
- Rice: Synthetic biology has been used to enhance the stress tolerance of rice, which has contributed to increased resilience and productivity in the face of adverse environmental conditions because of the increased resilience.
- The use of genetic modification and synthetic biology has been implemented to improve the stress tolerance of maize. This is done to reach the growing demand for crops that have increased resilience and production.
- Sugarcane: Synthetic biology has been utilized to reduce the effects of biotic and abiotic stresses in sugarcane farming, which has resulted in increased stress tolerance and productivity.
- Citrus sinensis, also known as the sweet orange, has been subjected to targeted genome editing with the use of synthetic biology techniques to improve characteristics. This serves as an example of the possible advantages of synthetic biology in citrus crops.
- The Brassica rapa plant, also known as the turnip, has been subjected to synthetic biology techniques to improve its stress tolerance, which has resulted in increased resilience in the face of challenging environmental conditions.

The examples highlight the numerous uses of synthetic biology in the enhancement of stress tolerance across a wide range of crop species, which ultimately contributes to increased resilience and production measures. Include a few examples of crops that have been genetically modified to have increased resistance to stress when grown.

Several crops, including those listed below, have been genetically modified to exhibit increased resistance to stress.

- Wheat: The application of genetic engineering has been done to improve drought tolerance in wheat, which has led to increased biomass output and decreased emissions of stress volatiles, according to Timmusk et al. (2014).
- Rice: The use and application of genetic engineering has been implemented to improve the stress tolerance of rice, which has resulted in increased resilience and productivity in the face of adverse environmental conditions (Cui et al. 2016).
- Transgenic soybeans that are drought-resistant have been developed through the use of genetic engineering to combat the negative effects that drought stress has on crop output (Bi et al. 2020).
- Genetic engineering and rDNA technologies have been used to improve maize's stress tolerance, which has contributed to increased resilience and productivity (Swain et al. 2023). Maize is commonly used in agricultural production.
- According to Kamthan et al. (2012), genetic engineering and rDNA technologies have been applied to improve plant stress tolerance and yield output in tomato plants. This has contributed to increased resilience in the face of severe environmental conditions.

The examples presented here illustrate the several ways in which genetic engineering can be applied to improve the stress tolerance of a varied range of agricultural species, hence contributing to increased resilience and productivity. How genetic engineering can be used to increase plant growth and the stress response through the application of synthetic biology approaches. Using genetic engineering, synthetic biology technologies have greatly contributed to the improvement of plant growth and the plant's reaction to stress. The utilization of cutting-edge biotechnological technologies has enabled researchers to make great headway in the augmentation of stress tolerance and productivity in a wide range of crop species. Some of the most important discoveries and methods in this area include the following:

- The introduction of stress-tolerant genes has been made possible by synthetic biology, which has made it possible to include stress-tolerant genes from a wide variety of plant species into current crops. This has resulted in an improvement of the stress-tolerant characteristics of these crops without the need for genetic alteration.
- The identification and introduction of genes that deliberate drought tolerance have been made possible by genetic engineering. This has resulted in improved plant growth and survival in transgenic wheat and rice (Cui et al. 2016). Genetic modification for drought tolerance helps plants become more drought resistant.
- Utilization of Rhizosphere Bacteria Synthetic biology has made it possible to use rhizosphere bacteria that have developed along with the plant roots in severe conditions. This has been done to improve drought tolerance and increase biomass output in wheat (Timmusk et al. 2014).
- Transcription Factor-Based Genetic Engineering: Synthetic biology approaches have mostly focused on the utilization of transcription factor-based modifications with genetic engineering to enhance agricultural tolerance to drought and other abiotic stresses.

- To improve plant abiotic stress tolerance, metabolic engineering has been used to promote the manufacture of stress-responsive metabolites, such as glycine betaine and melatonin (Pu et al. 2021). This has been accomplished through the application of synthetic biology in metabolic engineering.
- Genetic engineering has been used to engineer cereal crops for increased abiotic stress tolerance, which has contributed to greater resilience and productivity (Lata & Shivhare, 2021). This has been accomplished by genetic engineering.
- The development of stress-tolerant crop varieties by using synthetic biology technologies have sped up the process of developing stress-tolerant crop varieties. This is in response to the demand for types of crops that are more resilient and productive (Bakhsh et al. 2015).

The outcomes of this study highlight the considerable impact that synthetic biology and genetic engineering have had in increasing plant growth and stress response, giving viable options for sustainable food production and food security. The application of synthetic biology techniques has made a substantial contribution to the improvement of plant growth and the response to stress by genetic engineering. Researchers have been able to build plants with improved characteristics linked to stress tolerance, production, and resilience by utilizing technologies that are associated with synthetic biology. In several studies, it has been established that synthetic biology may be successfully applied to improve plant growth and stress response using genetic engineering. For example, synthetic biology techniques have been applied to optimize photorespiration to increase crop yield. This exemplifies the potential of genetic engineering to further plant growth and tolerance to stress, as demonstrated by South et al. (2018). In addition, the GoldenBraid 2.0 framework has made it possible to apply engineering principles to the process of plant genetic design (Sarrion-Perdigones et al. 2013). This background has offered a thorough DNA assembly framework for plant-based synthetic biology. Moreover, synthetic biology of plant natural products has provided insights into designed biosynthesis in plant cells, showing the potential of synthetic biology in boosting plant growth and stress response through metabolic engineering (Zhu et al. 2021). This highlights the promising applications of synthetic biology to improve plant growth and stress response. In addition, synthetic biology has played a significant role in the production of stress-tolerant plants through the use of genetic engineering, which has contributed to increased resilience and productivity in the face of harsh environmental conditions (Qu et al. 2021). When synthetic biology methods are combined with existing genetic engineering methods, it makes it possible to rationally develop crops that can handle abiotic stress. These opportunities are tempting and have the potential to transform the agricultural industry. This demonstrates the potential of synthetic biology in increasing the stress tolerance of plants (Lohani et al. 2022). In addition, synthetic biology has been utilized to build plants that are resistant to stress by the utilization of genetic feedback control and regulatory network rewiring. This demonstrates the potential of synthetic biology to enhance the stress response in plants (Foo et al. 2018). In conclusion, synthetic biology approaches have been significant in the enhancement of plant growth and stress response by genetic engineering. Researchers have been able to create plants with enhanced stress tolerance, productivity, and resilience by utilizing technologies from synthetic biology. This has contributed to the development of crop varieties that are both stress-resistant and high-performing.

6.11 Conclusion

Genome editing is steadily and constantly becoming the technique of choice for laboratories all over the world for the purpose of elucidating the function of genes as well as the translational utility of these genes. They are being utilized by a variety of agricultural development projects to reduce the effects of biotic and abiotic pressures on crops. Because of their great specificity, efficiency, cost-effectiveness, and time-effectiveness, editing procedures have developed into useful tools during their evolution. Even though they have been successful in combating diseases in commercially significant crops, their application in insect management has not been utilized to its full potential. Despite the cognitive effort that has been put into developing techniques for insect pest resistance in both insects and plants, the level of success that has been achieved has been quite low. Editing insects is an exciting alternative; nevertheless, it demands caution in the selection of traits that need to be environmentally benign in order to ensure that the food chain is not disrupted. The lack of availability of target genes has been the most significant obstacle, in contrast to the other pressures that have been encountered. Consequently, it is imperative that scientists all around the world place a strong emphasis on the identification of sources of resistance that can serve as a foundation for pest control methods. Because of their potential to lessen the impact of off-target impacts on the insects that are the focus of the efforts, several novel targets are now being considered. The extensive phenotyping and enrichment of the resistance gene pool is a significant nugget of information that needs to be addressed in a way that is innovative. For this purpose, it is necessary to conduct an analysis of the available germplasm, which should include wild relatives of certain crops, to determine how insects react to stress and to identify genes that are responsive to stress using numerous omics techniques. The nitty-gritty of insect control by genome editing is the multiplex editing of such discovered resistance genes with the application of high-throughput transformation techniques. Targeted mutations that lead to the transformation of sensitive plants into those that can control their respective pests are not far off in either space or time, given the fact that such studies are already in vogue. Together with regulatory considerations for gene-edited crops, these features have the potential to lead to the success of the technology, which would not only lead to advancements in scientific research but also to the acceptance of the technology by society.

References

Abidoye BO, Mabaya E (2014) Adoption of genetically modified crops in South Africa: effects on wholesale maize prices. Agrekon 53(1):104–123

Babalola OO, Molefe RR, Amoo AE (2021) Metagenome assembly and metagenome-assembled genome sequences from the rhizosphere of maize plants in Mafikeng, South Africa. Microbiol Resour Announc 10:8. https://doi.org/10.1128/mra.00954-00920

Bakhsh A, Ashfaq M, Ali A, Hussain M, Rasool G, Haider Z, Faraz RH (2015) Economic evaluation of different irrigation systems for wheat production in Rechna Doab, Pakistan

Barea JM, Azcón-Aguilar C (1982) Production of plant growth-regulating substances by the vesicular-arbuscular mycorrhizal fungus glomus mosseae. Appl Environ Microbiol 43(4):810–813

Barnum CR, Endelman BJ, Shih PM (2021) Utilizing plant synthetic biology to improve human health and wellness. Front Plant Sci 12:691462

Bastías DA, Balestrini R, Pollmann S, Gundel PE (2022) Environmental interference of plant–microbe interactions. Plant Cell Environ 45(12):3387–3398

Becerra-Moreno A, Benavides J, Cisneros-Zevallos L, Jacobo-Velazquez DA (2012) Plants as biofactories: glyphosate-induced production of shikimic acid and phenolic antioxidants in wounded carrot tissue. J Agric Food Chem 60(45):11378–11386

Ben Rejeb I, Pastor V, Mauch-Mani B (2014) Plant responses to simultaneous biotic and abiotic stress: molecular mechanisms. Plan Theory 3(4):458–475

Bi H, Zhao Y, Li H, Liu W (2020) Wheat heat shock factor TaHsfA6f increases ABA levels and enhances tolerance to multiple abiotic stresses in transgenic plants. Int J Mol Sci 21(9):3121

Bi B, Li W, Jiang Y, Du H (2022) Present and future prospects of crop synthetic biology. Crop Design 100017

Chen Y, Hu B, Xing J, Li C (2021a) Endophytes: the novel sources for plant terpenoid biosynthesis. Appl Microbiol Biotechnol 105(11):4501–4513

Chen S, Zhang N, Zhou G, Hussain S, Ahmed S, Tian H, Wang S (2021b) Knockout of the entire family of AITR genes in Arabidopsis leads to enhanced drought and salinity tolerance without fitness costs. BMC Plant Biol 21:1–15

Chouychai W, Kruatrachue M, Lee H (2015) Effect of plant growth regulators on phytoremediation of hexachlorocyclohexane-contaminated soil. Int J Phytoremediation 17(11):1053–1059

Cui Y, Wang M, Zhou H, Li M, Huang L, Yin X, Zhao G, Lin F, Xia X, Xu G (2016) OsSGL, a novel DUF1645 domain-containing protein, confers enhanced drought tolerance in transgenic rice and Arabidopsis. Front Plant Sci 7:2001

De Zelicourt A, Al-Yousif M, Hirt H (2013) Rhizosphere microbes as essential partners for plant stress tolerance. Mol Plant 6(2):242–245

Deng M, Wang Y, Kuzma M, Chalifoux M, Tremblay L, Yang S, Ying J, Sample A, Wang HM, Griffiths R (2020) Activation tagging identifies Arabidopsis transcription factor AtMYB68 for heat and drought tolerance at yield determining reproductive stages. Plant J 104(6):1535–1550

Fasusi OA, Amoo AE, Babalola OO (2021) Propagation and characterization of viable arbuscular mycorrhizal fungal spores within maize plant (Zea mays L.). J Sci Food Agric 101(14):5834–5841

Foo M, Gherman I, Zhang P, Bates DG, Denby KJ (2018) A framework for engineering stress resilient plants using genetic feedback control and regulatory network rewiring. ACS Synth Biol 7(6):1553–1564

Gill SS, Anjum NA, Gill R, Yadav S, Hasanuzzaman M, Fujita M, Tuteja N (2015) Superoxide dismutase—mentor of abiotic stress tolerance in crop plants. Environmental science and pollution research, 22, 10375–10394

Ha CV, Leyva-González MA, Osakabe Y, Tran UT, Nishiyama R, Watanabe Y, Tanaka M, Seki M, Yamaguchi S, Dong NV (2014) Positive regulatory role of strigolactone in plant responses to drought and salt stress. Proc Natl Acad Sci 111(2):851–856

Hassan ZM, Al-Sudani ZA, Atshan KA (2023) Effect of plant extract and plant hormonal accessories on growth and yield tomato. IOP Conf Ser Earth Environ Sci 1225:012081

Hayashi N, Ono R, Uchida S (2015) Growth enhancement of plant by plasma and UV light irradiation to seeds. J Photopolym Sci Technol 28(3):445–448

Hu H, Dai M, Yao J, Xiao B, Li X, Zhang Q, Xiong L (2006) Overexpressing a NAM, ATAF, and CUC (NAC) transcription factor enhances drought resistance and salt tolerance in rice. Proc Natl Acad Sci 103(35):12987–12992

Huang B, Chen Y-E, Zhao Y-Q, Ding C-B, Liao J-Q, Hu C, Zhou L-J, Zhang Z-W, Yuan S, Yuan M (2019) Exogenous melatonin alleviates oxidative damages and protects photosystem II in maize seedlings under drought stress. Front Plant Sci 10:677

Jain A, Bhatia P, Chugh A (2012) Microbial synthetic biology for human therapeutics. Syst Synth Biol 6:9–22

Jia H, Wang N (2014) Targeted genome editing of sweet orange using Cas9/sgRNA. PLoS One 9(4):e93806

Jin K, Chen G, Yang Y, Zhang Z, Lu T (2023) Strategies for manipulating rubisco and creating photorespiratory bypass to boost C3 photosynthesis: prospects on modern crop improvement. Plant Cell Environ 46(2):363–378

Jung YJ, Nou IS, Kang KK (2013) Transgenic tomato plants expressing BrOAT1 gene from Brassica rapa var. SUN-3061 Show enhanced tolerance to salt stress. Plant Breed Biotech 1(1):70–79

Kamthan A, Kamthan M, Azam M, Chakraborty N, Chakraborty S, Datta A (2012) Expression of a fungal sterol desaturase improves tomato drought tolerance, pathogen resistance and nutritional quality. Sci Rep 2(1):951

Krichevsky A, Zaltsman A, King L, Citovsky V (2012) Expression of complete metabolic pathways in transgenic plants. Biotechnol Genet Eng Rev 28(1):1–14

Lata C, Shivhare R (2021) Engineering cereal crops for enhanced abiotic stress tolerance. Proceedings of the Indian National Science Academy, 87(1):63–83

Li X, Weng JK, Chapple C (2008) Improvement of biomass through lignin modification. Plant J 54(4):569–581

Li T, Zhang Y, Liu Y, Li X, Hao G, Han Q, Dirk LM, Downie AB, Ruan Y-L, Wang J (2020) Raffinose synthase enhances drought tolerance through raffinose synthesis or galactinol hydrolysis in maize and Arabidopsis plants. J Biol Chem 295(23):8064–8077

Llorente B, Williams TC, Goold HD (2018) The multiplanetary future of plant synthetic biology. Genes 9(7):348

Lohani N, Singh MB, Bhalla PL (2022) Biological parts for engineering abiotic stress tolerance in plants. BioDes Res 2022:9819314

Mauer O, Rozmánek M, Houšková K (2018) Drought spells and their impact on the growth of young plantations established with the containerized planting stock. Acta Univ Agric Silvic Mendelianae Brun 66:89

Muhich AJ, Agosto-Ramos A, Kliebenstein DJ (2022) The ease and complexity of identifying and using specialized metabolites for crop engineering. Emerg Top Life Sci 6(2):153–162

Ouyang B, Gu X, Holford P (2017) Plant genetic engineering and biotechnology: a sustainable solution for future food security and industry. Plant Growth Regul 83:171–173

Parakhia MV, Tomar RS, Malaviya BJ, Dhingani RM, Rathod VM, Thakkar JR, Golakiya B (2014) Draft genome sequence of the endophytic bacterium Enterobacter spp. MR1, isolated from drought tolerant plant (Butea monosperma). Indian J Microbiol 54:118–119

Pérez-Massot E, Banakar R, Gómez-Galera S, Zorrilla-López U, Sanahuja G, Arjó G, Miralpeix B, Vamvaka E, Farré G, Rivera SM (2013) The contribution of transgenic plants to better health through improved nutrition: opportunities and constraints. Genes Nutr 8:29–41

Prashanth S, Sadhasivam V, Parida A (2008) Over expression of cytosolic copper/zinc superoxide dismutase from a mangrove plant Avicennia marina in indica rice var Pusa Basmati-1 confers abiotic stress tolerance. Transgenic Res 17:281–291

Pu YJ, Cisse EHM, Zhang LJ, Miao LF, Nawaz M, Yang F (2021) Coupling exogenous melatonin with Ca2+ alleviated chilling stress in Dalbergia odorifera T. Chen. Trees 35(5):1541–1554

Qu D, Show P-L, Miao X (2021) Transcription factor ChbZIP1 from alkaliphilic microalgae chlorella sp. BLD enhancing alkaline tolerance in transgenic Arabidopsis thaliana. Int J Mol Sci 22(5):2387

Raghav RS, Khichi DS, Singh Y (2019) In vitro bioassay of secondary metabolites of soybean (Glycine max) plant roots and their effect on growth of bacteria, hormones and plants. Indian J Agric Sci 89(3):531–533

Rigoulot SB, Schimel TM, Lee JH, Sears RG, Brabazon H, Layton JS, Li L, Meier KA, Poindexter MR, Schmid MJ (2021) Imaging of multiple fluorescent proteins in canopies enables synthetic biology in plants. Plant Biotechnol J 19(4):830–843

Swain R, Sahoo S, Behera M. Rout GR (2023) Instigating prevalent abiotic stress resilience in crop by exogenous application of phytohormones and nutrient. Frontiers in Plant Science, 14, 1104874

Sako K, Van Ha C, Matsui A, Tanaka M, Sato A, Seki M (2021) Transcriptome analysis of Arabidopsis thaliana plants treated with a new compound Natolen128, enhancing salt stress tolerance. Plan Theory 10(5):978

Sarrion-Perdigones A, Vazquez-Vilar M, Palací J, Castelijns B, Forment J, Ziarsolo P, Blanca J, Granell A, Orzaez D (2013) GoldenBraid 2.0: a comprehensive DNA assembly framework for plant synthetic biology. Plant Physiol 162(3):1618–1631

Shih PM, Liang Y, Loque D (2016) Biotechnology and synthetic biology approaches for metabolic engineering of bioenergy crops. Plant J 87(1):103–117

Siswanti DU, Khairunnisa NA (2021) The effect of biofertilizer and salinity stress on Amaranthus tricolor L. growth and total leaf chlorophyll content. BIO Web of Conferences,

Soliman AS (2019) Plant growth hormones. Cell Growth 1

Song J-T, Woo D-U, Lee Y, Choi S-H, Kang Y-J (2021) The semi-supervised strategy of machine learning on the gene family diversity to unravel resveratrol synthesis. Plan Theory 10(10):2058

South PF, Cavanagh AP, Lopez-Calcagno PE, Raines CA, Ort DR (2018) Optimizing photorespiration for improved crop productivity. J Integr Plant Biol 60(12):1217–1230

Suzuki N, Rizhsky L, Liang H, Shuman J, Shulaev V, Mittler R (2005) Enhanced tolerance to environmental stress in transgenic plants expressing the transcriptional coactivator multiprotein bridging factor 1c. Plant Physiol 139(3):1313–1322

Timmusk S, Abd El-Daim IA, Copolovici L, Tanilas T, Kännaste A, Behers L, Nevo E, Seisenbaeva G, Stenström E, Niinemets Ü (2014) Drought-tolerance of wheat improved by rhizosphere bacteria from harsh environments: enhanced biomass production and reduced emissions of stress volatiles. PLoS One 9(5):e96086

Tiwari P, Khare T, Shriram V, Bae H, Kumar V (2021) Plant synthetic biology for producing potent phyto-antimicrobials to combat antimicrobial resistance. Biotechnol Adv 48:107729

Tyagi S, Kesiraju K, Saakre M, Rathinam M, Raman V, Pattanayak D, Sreevathsa R (2020) Genome editing for resistance to insect pests: an emerging tool for crop improvement. ACS Omega 5(33):20674–20683

Verma KK, Song X-P, Budeguer F, Nikpay A, Enrique R, Singh M, Zhang B-Q, Wu J-M, Li Y-R (2022) Genetic engineering: an efficient approach to mitigating biotic and abiotic stresses in sugarcane cultivation. Plant Signal Behav 17(1):2108253

Wágner D, Fable E, Pitzschke A, Sauter O, Weisen H (2012) Understanding the core density profile in TCV H-mode plasmas. Plasma Physics and Controlled Fusion 54(8),085018.

Wang XY, Wei XL, Luo H, Kim JA, Jeon HS, Koh YJ, Hur J-S (2010) Plant hormones promote growth in lichen-forming fungi. Mycobiology 38(3):176–179. [Record #9 is using a reference type undefined in this output style]

Wille W, Pipper CB, Rosenqvist E, Andersen SB, Weiner J (2017) Reducing shade avoidance responses in a cereal crop. AoB Plants, 9(5), plx039

Yahmi O, Belouchrani AS, Drouiche N, Lounici H (2021) Effects of seaweed (Ulva lactuca) and freshwater (spirogyra sp.) algae extracts on the salt tolerance of the bean (Vicia faba L.). J Food Process Preserv 45(9):e15738

Yamano T, Malabayabas ML, Habib MA, Das SK (2018) Neighbors follow early adopters under stress: panel data analysis of submergence-tolerant rice in northern Bangladesh. Agric Econ 49(3):313–323. [Record #10 is using a reference type undefined in this output style]

Yang K, Gong XL, Li GC, Huang LQ, Ning C, Wang CZ (2020) A gustatory receptor tuned to the steroid plant hormone brassinolide in Plutella xylostella (Lepidoptera: Plutellidae). Elife, 9, e64114.

Yeh T (2015) The improvement of phytomediation on the treatment effectiveness of heavy metals with energy sunflower plants with calcium peroxide and phytohormones. J Environ Prot 6(10):1158

Zhu X, Liu X, Liu T, Wang Y, Ahmed N, Li Z, Jiang H (2021) Synthetic biology of plant natural products: from pathway elucidation to engineered biosynthesis in plant cells. Plant Commun 2(5):100229

From Data to Discoveries: Bioinformatics Strategies for Analyzing Metabolomics, Proteomics, and Gene Editing Data

7

Palak Rana, Gursharan Kaur, Harleen Kaur Walia, Gurleen Kaur Sodhi, Swapnil Kajale, and Vagish Dwibedi

Abstract

The integration of computational and experimental methodologies has propelled the field of bioinformatics toward a comprehensive understanding of biological systems. This review explores the pivotal role of bioinformatics in analyzing omics data, including genomics, proteomics, and metabolomics, to uncover novel insights into cellular functions. The combination of diverse omics layers enables researchers to simulate system behavior and predict responses to external stimuli, guiding experimental design and hypothesis generation. Challenges such as data variability and reproducibility issues are discussed, emphasizing the importance of standardized protocols and rigorous study designs. Advanced bioinformatics tools and resources, along with emerging technologies like next-generation sequencing and gene editing, provide unprecedented opportunities to unravel the complexities of biological systems. Ultimately, the integration of multi-omics data offers a powerful approach to elucidate cellular signaling mechanisms and facilitate the development of targeted therapeutic interventions.

P. Rana · G. Kaur
Department of Biotechnology, UIBT, Chandigarh University, Mohali, Punjab, India

H. K. Walia · G. K. Sodhi
Department of Biotechnology, Thapar Institute of Engineering and Technology, Patiala, Punjab, India

S. Kajale
Institute of Soil, Water and Environmental Sciences, Volcani Research Center, Agricultural Research Organization, Rishon LeZion, Israel

V. Dwibedi (✉)
Department of Biotechnology, Chandigarh University, Mohali, Chandigarh, India
e-mail: vagish@volcani.agri.gov.il

S. Kaur et al. (eds.), *Metabolomics, Proteomics and Gene Editing Approaches in Biofertilizer Industry*, https://doi.org/10.1007/978-981-97-2910-4_7

Keywords
Bioinformatics · Omics data · Genomics · Proteomics · Metabolomics

7.1 Introduction

The combination of computational and experimental developments has enabled us to simultaneously profile various regulatory factors such as genome, transcriptome, epigenome, chromatin conformation, and metabolome. All of them together are acknowledged as "omics" layers. The specially designed mathematical, statistical, and biostatistical software is crucial for managing, analyzing, and processing the data. Modern high-throughput omics measurement platforms have made it possible for biomedical studies to adopt a holistic, integrated approach to extract and utilize data on a comprehensive scale for understanding biological systems. This approach integrates data from various omics sources, such as genetics, proteomics, and metabolomics, to tackle complex systems biology challenges using machine learning-based predictive algorithms. These innovative methods enable the integration and analysis of omics data to discover novel markers, also known as biological markers (Reel et al. 2021).

Many branches of omics depict separate studies, such as genomic studies, which refer to DNA studies. In contrast, transcriptomic study involves mRNAs, proteomic study concerns proteins, and, as the name suggests, metabolomics involves the study of metabolites. In brief, genomics is the exploration of the genetic scheme of a cell with direct reference to DNA to spot specific genes of an individual (Fatumo et al. 2022). However, transcriptomics looks at the transcribed genetic materials, giving insight into actively expressed genes and providing information about the cellular-level processes (Mo and Jiao 2022). On the other hand, proteomics is essential in describing information flows within cells and organisms employing protein pathways and networks (Yang et al. 2020a, b). Using these different kinds of data like genomic, proteomic, and metabolomics, one can assemble a highly accurate mathematical model that simulates the functioning of the whole system under influence or interruption. They consider such things as the interaction of molecules, rate of chemical reaction and regulatory feedback loop involved. Through running simulations, the researchers will know their systems' reaction patterns against changes in surrounding areas. It exposes more ideas that also assist in guiding experiment design (Heiner 2023).

The successful outcome of a whole genome system requires that information obtained from different biomolecules that drive cellular activities must be consolidated. There are technical and physiological obstacles in bringing about convergence because it involves integrating and interpreting data from different omic fields. While genomics and transcriptional profiling are comprehensive, protein and metabolic profiles only include a small amount of the information generated by the omics technology. Complications arise, especially in the proteomics and metabolomics field, where irreproducibility becomes a problem. To illustrate, when measured by mass spectroscopy, a comparison between two equal samples of one proteome could have a protein overlap at approximately 60–70%. The variability and inherent uncertainty in every mass spectrometry determination make many models with global proteome and metabolome measurements unspecific (Veenstra 2021). Omics experiments usually utilize samples that could be non-random or more broadly defined, such as clinical practice/natural history studies. This is a critical point that should be considered while evaluating the significance of the evidence derived from the bioinformatics analysis. Also, there should be a well-developed study plan to ensure that all SOPs-controlled specimens and data collection follow best practices (Reimand et al. 2019). The plan should specify what clinical question is being addressed, the primary and secondary endpoints, and potential confounders that might require adjustments later for downstream bioinformatics and statistical analysis purposes. These boundaries also help estimate sample sizes to mitigate type I and II errors and define specific statistical tests (Shreffler and Huecker 2020).

Data and sample repositories (databases) are critical in this regard. Such databases may already have been developed alongside the previous collections or could even form part of forthcoming collections. This is where easy access to a data and sample database saves time in selecting cases and controls for Omics analysis (or verification of features). However, if there is no database for such documents, it is imperative to create a separate database. Among other things, this procedure entails looking beyond the usual considerations of events rate and follow-up time and ensuring proper technical design (Reimand et al. 2019).

Nonetheless, the developments achieved in extracting omics information during the last few years stand out notably. Although there are many chances to expand the horizon, the existing level of information remains adequate regarding the advancement of technologies that will integrate omic data from different domains into an overall picture of how cells or organisms are working. The first attempt at integration of the omics data utilizes transcriptomics and proteomics. They integrated this using microarray and Global Proteomics data (Mehmood et al. 2021).

If the relative fold change for a particular species is known, then comparing these datasets is relatively easy. Although we may expect a strong relationship between mRNA and protein levels based on simple biology, a direct comparison of transcripts and protein amounts shows a poor association. Nevertheless, some research has shown a high correlation between the amount of transcript and protein when grouped into appropriate functional categories (Mehmood et al. 2021). Coverage inconsistencies are among the most common issues in genomic, transcriptomic, proteomic, and metabolomic datasets. The data is usually gathered as if a pharmacogenomic study is done, whereby different genes, transcripts, and enzymes are checked out on how the body responds to foreign objects. As for the future, systems biology might have to build models incorporating empirical findings from all omics levels. These all-encompassing datasets will feed knowledge that will significantly create whole system models incorporating empiric and non-empiric data. This modeling capacity aligns with a critical objective of systems biology: building up predictive models for predicting what happens when an organism is affected either exogenously or endogenously. Since some aspects of a biological system will remain a mystery, non-empirical modeling will likely become a constituent part of current systems biology. Another cutting-edge analysis method is predictive modeling in systems biology, which brings insight into how biological systems function together. The move from simulated models to confirming their effectiveness through the proteome is critical toward verifying the models' integrity. Scientists will link the "dots" between the theoretical simulations and the real-life observations by using the proteomic data they incorporate in the modeling process, thereby validating the model (Heiner 2023).

This chapter discusses an overview of the dry lab analyzing techniques for studying proteome, transcriptome, genome, and metabolome data that come under the advanced bioinformatics techniques.

7.2 Basics of Metabolomics, Proteomics, and Gene Editing Data

In the past, scientists mainly focused on studying parts of living organisms without considering the complex web of interactions that shape the entire system's behavior. However, in systems biology, a different approach is taken. It involves looking at the system and using models to understand its functions. These models combine genomics, proteomics, and metabolomics to create representations. With these models, researchers can simulate how the system would respond to changes or disturbances. These simulations consider factors such as interactions, reaction rates, and regulatory feedback loops. Using these simulations, scientists can predict how the system would react to changes in its surroundings. This helps generate ideas for research and guides designing experiments (Heiner 2023).

Technology and development systems of biology are closely linked. The emergence of such technologies enabled understanding of system operations and led to a massive collection of data concurrently. Systems biology and technological developments also encouraged each other because they were only sometimes available. It is, however, essential to point out that system biology has also changed science's way of planning and implementing their studies (Veenstra 2021). In omics, three primary approaches are routinely used: genomics, proteomics, and metabolomics. However, these three methods, particularly within the system biology, more commonly termed pathways, networks or integrated biology, have seen significant paradigm shifts in studies. It introduces a novel way of developing theories that is quite beneficial because it enhances the usual theory generation system. However, network analysis and systems biology approaches could potentially integrate disparate data layers, relating molecular pathways to cell, tissue, and functional organization levels. Ideally, this approach can give us a clear picture of how integrative physiology works and why it is relevant when the whole organism becomes the focus issue in science (Mayer et al. 2011).

7.2.1 Metabolomics Data Analysis

Metabolomics involves a detailed computational study of metabolites. It analyzes thousands of biological entities' metabolite products and gives a global picture of the organism's metabolic and internal chemistry (Wishart 2019). Metabolomics data is stored in various databases with diverse information, such as NMR and MS spectra and metabolic pathways. In recent years, information found in metabolite databases has continually expanded as metabolomics studies have increased (Chen et al. 2022). Similarly, since metabolomics is an infant field, new methods of approach are still emerging, while a metabolomics database is also an emerging one that continues developing rapidly without end. The databases include data for concentration, localization, and related pathologies. The databases include MassBank, HMDB, Madison Metabolomics Consortium Database, LIPID MAPS, KEGG, and METLIN (Ren et al. 2015). Another computational MetaboAnalyst also provides several statistical, functional, and graphical tools. Some of these include methods such as PLS-DA(partial least squares discriminant analysis), for cluster analysis and visualization, MetPA(metabolic pathway analysis), MSEA (metabolite set enrichment analysis) and PCA (principal component analysis). Different approaches are applied to examine each subgroup of metabolites (compound detection) in the first step of metabolomics. There are primarily two primary tools used for metabolomics analysis: MS (mass spectroscopy) and NMR (nuclear magnetic resonance) spectroscopy techniques (Chong et al. 2019). In MS-based metabolomics, conducting a separation process for the analysis is customary. This step helps simplify the sample, making it possible to examine sets of molecules at different time points using MS. MS captures information in the form of a mass-to-charge ratio (m/z) and the relative intensity of the compound. Liquid chromatography (LC) and gas chromatography (GC) columns are the used separation techniques. Using MS separation methods, metabolites can be reliably identified with resolving capability (Alseekh et al. 2021).

Similarly, nuclear magnetic resonance (NMR) is a technique that utilizes the behavior of nuclei as they absorb and release energy in response to alterations in an external magnetic field. The spectral information obtained from NMR can be used to quantify the concentrations of metabolites and gain insights into their makeup (Nagana Gowda and Raftery 2019). After the step of compound detection, the process starts by converting signals into a format suitable for statistical analysis using spectra, chromatograms, or NMR data (data pre-processing). After that, data normalization is applied to minimize any biases caused by the system or technical factors. For this, metabolites are identified by comparing the data with a designated database and (quality control) samples employed to assess the variability of metabolite characteristics. In cases where a feature exhibits excessive variance, it will be excluded during data processing (Mattoli et al. 2022). Then, univariate and multivariate methods are performed for the statistical analyses to identify the metabolites that are significantly expressed. Finally, essential metabolites are associated with their context through enrichment, pathway analyses, and functional analyses. The metabolomics data can be combined with omics data like transcriptomics, proteomics or microbiome to gain insights into the mechanisms underlying pathophysiological processes (Chong et al. 2019).

7.2.2 Proteomics Data Analysis

The proteome includes a cell or organism's entire set of proteins synthesized through 20 amino acids. It is important to note that UniProt lists over 200 modified amino acids for eukaryotic organisms, underscoring that protein diversity is considerably more intricate than previously believed. However, it must be noted that it is one of the most challenging things to quantify. Therefore, it is essential to acknowledge that measuring protein stability and turnover remains challenging to calculate (Bartolec et al. 2023).

In particular, relative quantifications and post-translational modifications of proteins are widely applied in proteomics. Additionally, the post-translational modifications (PTMs) in proteins are detectable by MS as well. However, searching a database for all possible combinations of position/modifications may take much longer since these combinations are numerous. Various ways can increase the accuracy of the results; examples include PTMselect, ModifiComb, and G-PTM-D (Perchey et al. 2019). Similarly, another recent method of in-depth analysis of peptide mixtures comprising data-independent acquisition (DIA) MS has been developed. The unique aspect of this method is that unlike the conventional data-dependent acquisition (DDA) approach, where precursor ions are selected for fragmentation based on selection and dissociation, DIA captures all the fragmented ions at each point in time during chromatography within the specified m/z ranges (Yang et al. 2023). In addition, unlike in other mass spectrometry data collections, there is no bias toward any specific precursor ions since DIA becomes a more efficient source of raw spectrometric data for further systematic analyses. Different methods for analyzing the data have been created during the evolution of the DIA approach.

MSPLIT-DIA evaluates the spectrum matches via a normalized dot product when DIA spectra are compared and matched with a library database. Nevertheless, `it is worth noting that library-dependent approaches have limited capabilities since they rely on the prior understanding stored within the library. However, this shortcoming has been addressed with various library-free approaches (Yang et al. 2020a, b). These types of software usually regenerate false spectra by unraveling the extended spectra or using a group of correlated precursor ions. For example, although DIA-Umpire can separate multiplexed spectra, it is worth noting that the software FT-ARM (Zhang et al. 2020) is much more straightforward, while PECan (Ting et al. 2017) provides better confidence estimation for each query peptide. For example, a study suggested a quick, tailored approach of creating libraries in DIA-MS for atypical organisms just recently. The construction of libraries containing all possible peptides in silico represents a proteome, which is then refined using information extracted from public protein sequences (Searle et al. 2020).

All the strategies explained above are employed for peptide identification, after which protein interference will be analyzed. This stage sees long peptides offering more excellent informative value as their nature is precise. On the other hand, creating a dependable protein list using shorter peptides is often challenging because some of these peptides might be common to two or more proteins. These "degenerate peptides" typically present multiple equally valid options for assigning proteins. Most models of peptide synthesis postulate a minimum number of proteins required to account for the observed peptides (Santa-Coloma 2022). In peptide-based mass spectrometry quantification, researchers often face the challenge of estimating protein abundance using several peptides, even just one. To ensure reliable results from this pool of peptides, it is customary to establish minimum peptide count thresholds (Chen et al., 2020).

Normalization plays an essential role in correcting data by removing biological variations. This step ensures that the results remain consistent and dependable across samples for analysis. The choice of normalization method is critical in proteomics, especially considering the practice of normalization prior to conducting liquid chromatography/mass spectrometry (LC/MS). Several normalization methods have been developed, each based on assumptions. Since protein abundance distribution tends to be heavily skewed toward zero, applying transformations to the intensity values is the practice before further normalization (L. Town et al. 2019).

Protein abundance data is prepared for statistical analysis and further exploration after completing the necessary cleaning, filtering, and normalization steps. The first step in analyzing the data is to compare the protein levels between two distinct conditions, which can be achieved through a t-test (Chen et al., 2020). However, the analysis can become more complex when the study needs to consider multiple factors. In such cases, ANOVA (analysis of variance) is a suitable alternative method that accounts for the variability in protein abundance levels across the various factors being studied. This can provide a more comprehensive understanding of the relationship between protein levels and the analyzed factors (Nanni et al. 2022).

7.2.3 Gene Editing Data Analysis

Every nucleated cell shares a similar genetic material, but what sets them apart is the distinctive genes they activate at particular times. To comprehend the genes involved in tissue-specific or developmental functions, researchers have traditionally developed libraries that contain all the expressed genes for a specific organ or stage of development (Li et al. 2020). These libraries, known as complementary DNA (cDNA) libraries, provide a snapshot of the actively expressed genes. They leverage that when mRNA is transcribed in eukaryotic cells, a poly(A) tail (a long sequence of adenine nucleotides) is added. This poly(A) tail distinguishes mRNA from other types of expressed RNAs and serves as a starting point for reverse transcription (Pulman 2014). Genomics can be broken down into three closely related domains:

1. Structural Genomics: The third deals with genome synthesizing and mapping.

2. Functional Genomics: In this area, analysts investigate mRNA expression involved in the study development, physiology, and sickness. Therefore, these gene differences can be used to explain variations in gene function.
3. Computational Genomics: These are techniques used to combat massive and intricate genomic data in this sector. This entails discovering methods that help one make meaning out of the vast volumes of genomic information today.

For large-scale genetic experiments, initial data requires processing to extract gene-specific information. Two algorithms based on deep neural networks were used to indicate the off-target mutations in the CRISPR-Cas9 genome editing. Two of these algorithms are a deep convolutional neural network and a deep feedforward neural network. These models are for training and testing purposes using newly released off-target datasets like the CRISPOR data set to measure their performance. To further evaluate, it can be profiled additionally in another off-target set designated by the GUIDE-seq (Lin and Wong 2018). The convolutional neural network performed best in this dataset, achieving an average AUC of approximately 97.2% after being tested using the five-fold cross-validation technique. The deep feedforward neural network likewise performed well for this condition, having a mean AUC of 97.0% (Aktas et al. 2019). Therefore, both deep neural network models are compared with current state-of-the-art off-target prediction approaches and traditional machine learning models for their respective AUC scores on both datasets (Sherkatghanad et al. 2023). However, one of the most significant issues for researchers working on gene editing technologies has been predicting off-target mutations as generated through CRISPR-Cas9. Such has happened since prediction models have been built about CRISPR-Cas9's mismatch scores. Therefore, many of these approaches are dominated by the vast amount of data produced during CRISPR-Cas9 and fail to meet the required precision (Lin and Wong 2018).

A well-formulated Java programming-based approach that provides robust and accurate sequencing analysis tools is provided through the Genome Analysis Toolkit (GATK). This software product includes the MapReduce approach, which divides the complicated hardware, whose task is to process vast quantities of next-generation sequential data, and the concrete logic, which is necessary for every instrument used for each of the analyses, into two separate parts (Roy et al. 2017). Researchers and developers can focus on improving the processing algorithms since the GATK engine is responsible for handling the complexities of accessing NGS data. Besides promoting higher efficiency and reliability to the created tools, it also increases the developers' productivity by enabling them to focus on analysis methodologies rather than on a standard data management engine, which is beneficial (Filgueira et al. 2022). Moreover, it allows us to get the best results in maximizing the accuracy and stability of the GATK engine concerning CPU and RAM utilization. In addition, it enables the automatic scaling of most analysis tools of shared-memory machines and distributed clusters. Still, having been in development for around a year, the GATK is the foundation of crucial instruments in projects such as the 1000 Genomes Project and The Cancer Genome Atlas. These include quality-score recalibration, indel discovery, as well as genotyping. They have proved highly efficient in processing Illumina, SOLiD, or 454 significant data volumes (Iakhnin 2019). The GATK shall be scaled up for extra data types such as local reference assembly, CNV detection, inversion-based research, and general structural variation programs (Ravenhall 2019).

Another widely employed tool is GAGE (Genome analysis and Genome editing). It encompasses the comprehensive use of the entire genome through bioinformatic analysis, primarily manifesting in different ways. GAGE is not limited to searching the whole genome by identifying the target sequences with confident PAM as their downstream end. It is usually one-third position out of every eighteen or thirty bases in these studies, which comprise different genomes. This translates into several target sequences, hence due to their large sizes. This wealth of information proves extremely valuable in assisting with the identification of plants and has great promise as a tool for GAGE (Hao et al. 2022).

In addition, GAGE moves another step forward by subjecting the selected target sequence in the plant of concern to all the genomes associated with the possible adulterants. It expressly confirms the targeted sequence and shows that it is done with an adequate GAGE on a fundamental level. More importantly, it enables the assessment of theoretical byproducts, limiting off-target mutations when using the CRISPR/Cas12a system (Centre and Düsseldorf 2021). The GAGE benefits from an ever-growing database that now exceeds 700 regarding published genomes, as these are being published rapidly. An ample supply of genomic data forms a sound base and has excellent prospects for the efficiency of GAGE (Hao et al. 2022).

7.3 Data Integration and Multi-Omics Approaches

Multi-omics datasets are utilized in conjunction with clinical data at the forerunner line to identify cellular functions. The multi-omics profile comprises genomic, proteomic, transcriptomic, metabolomic, and epigenetic data sets (Fig. 7.1). Moreover, it includes datasets comprising lipidome, phosphoproteome, and glycoproteome, among others. Integrating different omics data provides a systematic and comprehensive view of the complex biology (Eicher et al. 2020). The integrated approach in which individual omics datasets are either chained one after another or applied jointly offers an opportunity for unwrapping multi-faceted molecule interactions. Integrated approaches that evaluate the information passing through various omic categories provide a connection from genotype to phenotype. This quality improves the prediction accuracy for disease phenotypes, contributing to better treatment and preventive measures (Jendoubi 2021).

The emerging single-cell multimodal assays are potent tools to explore different aspects of cellular plasticity in development, tissue homeostasis, and diseases. One of the main difficulties in analyzing single-cell multimodal data is finding efficient ways to link and integrate data collected using various methods. This process has often been referred to as "data integration," encompassing techniques including removal of batch effects on single omics datasets and association of genetics, chromatin accessibility, and transcriptions (Vandereyken et al. 2023). Existing

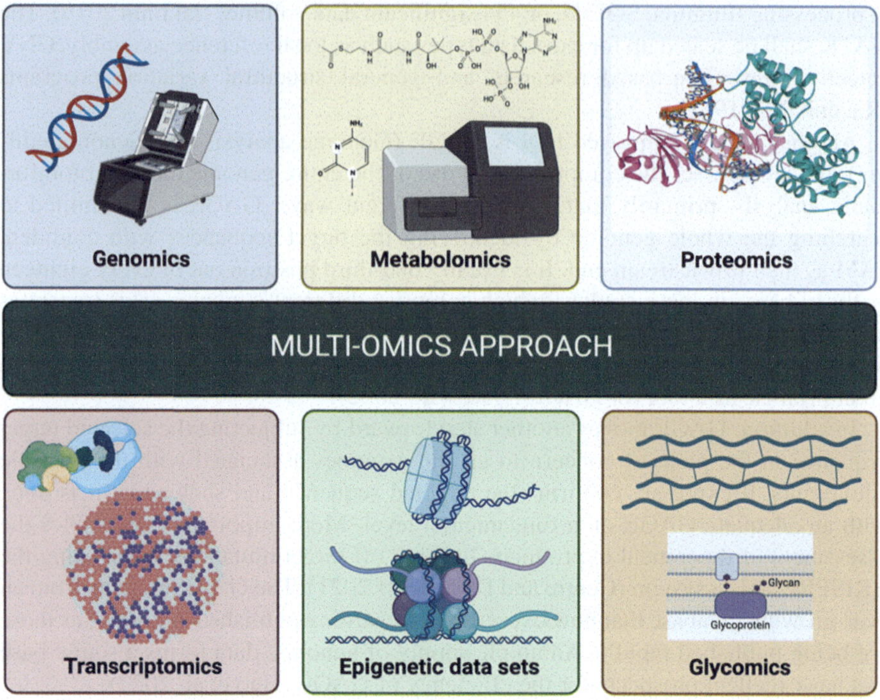

Fig. 7.1 Multi-omics approach

integration strategies may be based on a similar mathematical foundation but have different goals, theories, and assumptions. This calls for new definitions and conceptualizations of current practices to create newer strategies and approaches.

Recent investigations stress multi-omics analysis to understand better and clarify the studied system. For instance, in HNSCC cell lines, one instance could be analyzing ChIP-Seq and RNA-Seq data side by side. The analysis showed that particular H3K4me3 and H3K27ac cancer-associated histone marks were involved in transcriptional changes to some critical HNSCC driver genes such as EGFR, FGFR1, and FOXA1 (Kotipalli et al. 2021).

Data modality anchoring in the first stage of every data integration is critical to the process. However, during practice, experimental design typically dictates this choice and tremendously impacts downstream analysis. Selecting the anchor holds enormous undertones since selecting other anchors entails specific statistical and biological suppositions, calling for specialized research methodology approaches. Based on the chosen anchor, three distinct data integration approaches can be identified (Fig. 7.2) (Argelaguet et al. 2021).

- Vertical integration—the use of a matched assay approach in experimental designs in which many data modalities are simultaneously profiled from the same cell. Such techniques include single-cell analysis of genotype expression

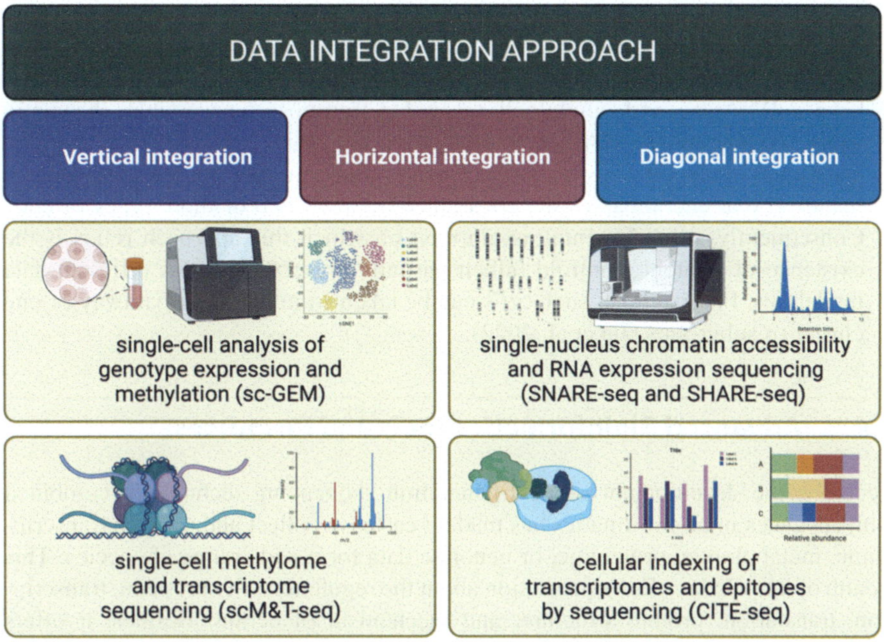

Fig. 7.2 Three distinct data integration approaches

and methylation (sc-GEM), single-nucleus chromatin accessibility and RNA expression sequencing (SNARE-seq and SHARE-seq), cellular indexing of transcriptomes and epitopes by sequencing (CITE-seq), single-cell methylome and transcriptome sequencing (scM&T-seq) (Hie et al. 2019). To illustrate, in a study involving mouse preimplantation embryos and data collected using scCAT-seq to measure chromatin accessibility and RNA expression simultaneously, iNMF supported the recovery of meaningful molecular features connected with diverse cell stages at the blastocyst stage. Likewise, on a data set of mouse postimplantation embryos where scNMT-seq simultaneously recorded both mRNA expressions, DNA methylation, and accessibilities of chromatins, an integrative analysis utilizing MOFA showed lineage-specific enhancer that is associated (Colomé-Tatché and Teis 2018).

- Horizontal integration—Unmatched assays are a case in point; for instance, single-cell RNA sequencing (scRNA-seq) involves profiling cells from the same tissue but obtained from different cohorts of donors. The other scenario involves combining data derived from several types of scRNA-seq assays, with their genes serving as connecting points. Horizontal integration is mostly considered noise that must be addressed whenever there are artificial technical differences among experiment batches (Argelaguet et al. 2021). Therefore, every experiment must have some actual biological fluctuation in this correction approach. This might well turn out to be one of the most critical steps in a cell analysis pipeline as more and more single-cell level atlas references become available—references such as the Human Cell Atlas referenced here (Regev et al. 2017).

- Diagonal integration—For instance, it appears separately on each cell that has been scRNA-seq and single-cell assay for transposase-accessible chromatin using sequencing (scATAC-seq) simultaneously (Luo et al. 2019). Generally, techniques of diagonal integration seek to reproduce an idealized low-dimensional manifold that encompasses the covariance between two or more types of data. Consequently, a fundamental assumption on which this approach relies is the existence of a latent manifold (albeit partially preserved) across different data modalities. For example, such cells can be taken from the same pathway or one cluster of subgroups (Jia et al. 2022).

7.4 Advanced Bioinformatics Tools and Resources

Recently, the development of next-generation sequencing technology combined with advances in bioinformatics has made it easier to collect and analyze transcriptomic, metabolomic, proteomic, or genomic data for a wide range of species. This wealth of knowledge allows prediction about the regulation of expression, transcription, translation, protein structure, and mechanism of action. Besides, it offers glimpses into factors like homology, mutations, and underlying events that cause structural and functional differences across generations. However, even as databases increase in information, they grow, and bioinformatics tools keep being modified such that the prediction of the functionality of a protein becomes more accurate (Manzoni et al. 2018). Such gradual evolution reveals the perpetual problem of bioinformatics studies. According to the National Center for Biotechnology Information (NCBI), bioinformatics is an area of science combining various aspects of various disciplines such as statistics, artificial intelligence, mathematical sciences, chemistry, computational engineering, biochemistry, and computer science (Can 2014).

Metabolomics utilizes mass spectroscopy and spectroscopy to generate high-throughput omics scale data. The mentioned platforms involve various procedures such as gas chromatography–mass spectrometry (GC–MS), liquid chromatography–mass spectrometry (LC–MS), capillary electrophoresis–mass spectrometry (CE–MS), and also other spectroscopic methods such as 13C-NMR, 1H-NMR, and Fourier-transform infrared (FTIR) (Misra 2021). Some of the notable metabolomics analysis tools have been discussed below.

- Signature mapping tool—SigMa is a signature mapping tool dependent on MATLAB. It converts the recorded raw urine spectra into a readable metabolic table. SigMa divides urine signals of unknowns spin system (SUS), NMR spectra in signature signals (SS) and bins. Applying this segmentation approach allows for the concurrent identification of urinary metabolites in large-scale NMR-based metabolomic investigations using a SigMA spectral database and an original automatic assignment algorithm (Khakimov et al., 2020).

- Global natural product social molecular networking (GNPS) tool—molecular networking analysis allows the storing, processing, sharing, as well as annotation and comparison of molecular networking results for both low-resolution data as well as GC-HRMS data (Aksenov et al., 2020). The GNPS MassIVE is a public hold which is dedicated to untargeted MS2 datasets and sample metadata. However, MSHub performs automated decomposition of compound fragmentation spectra by unsupervised non-negative matrix factorization. A further test of fragmentation pattern reproducibility across samples and molecular networking using GNPS analysis is embarked on (Aron et al. 2020).
- Automated spectral processing system for NMR (AlpsNMR)—There is a software application known as AlpsNMR that functions as an automatic spectral processing system in NMR. This package automates signal processing for untargeted NMR metabolomics data sets (Madrid-Gambin et al. 2020). It provides various functions such as loading spectra, data manipulation, automatic identification of anomalies, spectra alignment, peak picking, integration, and standardization. AlpsNMR is flexible; it can load Bruker or JDX samples for further data processing (Stanstrup et al. 2019).

Another significant component of the analysis is that the proteome is dynamic due to numerous complicated mechanisms regulating protein expression levels. However, its application is limited by the high costs associated with specialized personnel, dedicated facilities, equipment, and software for the equipment and database (Lill et al. 2021). While proteomics constantly changes, it poses problems through many stages of development, from preparing samples to collecting data. This provides much information concerning several protein folding models, 3-D structure prediction, structure–function prediction, and several methods such as 2-D gel electrophoresis, isoelectric focusing, 2-D protein picture and PMF (peptide mass fingerprinting), MS, and tandem MS. This richness of information leads to high performance in proteomics supported by bioinformatics tools introducing innovative algorithms for such comprehensive and unimodal data (Madrid-Gambin et al. 2020). The most widely used proteome analysis tools have been discussed below.

- Expert Protein Analysis System (ExPASy)—The users can easily access different databases and protein-specific analytical tools at ExPASy. It provides the correct entry for different databases and protein-related toolkits. The databases in ExPASy include SwissProt and TrEMBL, Swiss-2DPage, Prosite, Enzyme, and the Swiss Model repository (Paladhi et al. 2022).
- Basic Local Alignment Search tool (BLAST)—One of the most utilized platforms constantly revised is this platform. It uses simple yet strong approaches to analyze protein through amino acid sequence comparison and identification of protein homology. This process utilizes algorithms that provide the most appropriate fit for alignment (Misra 2021).
- Protein Data Bank (PDB) is a global collection center for macro-molecular structures updated weekly. It uses modeling software to incorporate more than

153,000 protein structures obtained through crystallography, X-ray, and nuclear magnetic resonance (NMR) methods. The servers these platforms use can group proteins according to their sequences, structures, and functions (Burley et al. 2017).

Similarly, genome editing, also known as gene editing, is a collection of technologies that allow scientists to change an organism's DNA. These technologies can add, remove, or alter genetic material at specific locations in the genome. The CRISPR/Cas system, initially identified in the adaptive immune system of bacteria and archaea, utilizes ~20-bp RNA CRISPR arrays to guide Cas nucleases for the specific recognition and cleavage of invader nucleic acid sequences (Li et al. 2023). Over the last decade, this system has evolved into a powerful genome editing tool, enabling sequence-specific mutagenesis at targeted genomic sites across various organisms, encompassing plants and animals (Li et al. 2021). The CRISPR/Cas genome editing tools are undergoing rapid modifications to expand their application potential further. Many laboratories have developed various computing packages to improve CRISPR/Cas with its research. Among them are tools to design single guide RNA (sgRNA) units and analyze off-and-on-target effects due to genome editing (Anzalone et al. 2020). Some of the widely used sgRNA web-based design tools have been discussed below.

- CRISPOR—CRISPOR provides an efficiency prediction, specificity prediction, and primer designing tool for building vectors and target and off-target recognition. It includes various empiric models of efficiency prediction whose outcome data is provided as a tabular presentation (Li et al. 2023). To predict specificity, CRISPOR uses both MIT and CFD, two often used tools in determining the specificity for a specific target sequence. CRISPOR also has two predictive models, namely out-of-frame scores and frameshift ratios, added to the CRISPR/ Cas outcome to boost the precision of the cutting process. In addition, there is information about some critical elements like GC content. Furthermore, essential elements such as GC composition or the number and type of mismatches (0–4 nt) are marked in the results (Chen et al. 2019a, b).
- CRISPR-P—It has been customized and used exclusively in making specific sgRNA for the plants from 75 of their genomes, with emphasis on cereal grains. In addition, CRISPR-P provides more variety of CRISPR/Cas PAM types such as St1Cas9, SpCas9, NmCas9, xCas9, and SaCas9 (Liu et al. 2017). Furthermore, CRISPR-P allows users to choose between a U3 and U6 sgRNA promoter-driven expression cassette. When users submit data to CRISPR-P, they can select from the gene ID/name, position on scaffold and chromosome, and the fasta format sequences. It uses rule sets 1/2 and CFD to predict and target effects (Lorenz et al. 2016).
- CRISPR RGEN—It is an advanced CRISPR/Cas library platform, which includes several sgRNA design tools. In particular, it uses Cas-designer for traditional CRISPR/Cas nucleases, BE-Designer for CRISPR base editing, and PE-Designer for CRISPR prime editing. In particular, PE-Designer works specifically

for SpCas9 while Cas-designer and BE-designer are universally adaptable regarding PAM (protospacer adjacent motif) support (Li et al. 2023). The out-of-frame score algorithm is applied in Microhomology-Predictor, a prediction tool that looks into possible results. The following algorithm evaluates the possible internal deletions from the MMEJ fixation method. However, Microhomology-Predictor goes further beyond CRISPR/Cas. Its usefulness is magnified if its recommended out-of-frame score exceeds 66 (Zhang et al. 2023). Hence, this instrument enables users to use it for different experiments on designing sgRNA with great precision.

Although different CRISPR/Cas systems will have similar patterns of genomic modifications, these differences are expressed at different levels. Such differences highly influence the choice of the best fit CRISPR/Cas systems for a particular experiment.

7.5 Conclusion

In conclusion, bioinformatics is critical in managing these omics data, including genomic, proteomic, and metabolomics. Combining data from these varied sources gives a comprehensive picture of the biological system and helps find new indicators. Nevertheless, problems such as uncontrolled variation and non-reproduction of results should be addressed while using open-access databases and conducting planned studies for accurate description and interpretation. Moreover, computational tools and modeled applications like predictive modeling can contribute significantly to the perception of biological systems. Currently, various publicly available ways and technologies can be used to integrate multi-omics data and reveal more profound insights. Hence, an integrative approach using multi-omics data provides a powerful means for revealing the mechanistic complexity of cellular signaling processes.

References

Aksenov AA, Laponogov I, Zhang Z, Doran SL, Belluomo I, Veselkov D et al (2020) Algorithmic learning for auto-deconvolution of GC-MS data to enable molecular networking within GNPS. BioRxiv:2020–2001

Aktas Ö, Dogan E, Ensari T (2019) Crispr/cas9 target prediction with deep learning. In: 2019 scientific meeting on electrical-electronics & biomedical engineering and computer science (EBBT). IEEE, pp 1–5

Alseekh S, Aharoni A, Brotman Y, Contrepois K, Ewald J, Fraser PD, Giavalisco P, Hall RD, Heinemann M, Link H, Luo J, Neumann S, Nielsen J, Saito K, Sauer U, Schroeder FC, Schuster S, Siuzdak G, Skirycz A, Sumner LW, Snyder MC, Tang H, Tohge T, Wang Y, Wen W, Wu S, Xu G, Zamboni N, Fernie AR (2021) Mass spectrometry-based metabolomics: a guide for annotation, quantification and best reporting practices. Nat Methods 18(7):747–756. https://doi.org/10.1038/s41592-021-01197-1

Anzalone AV, Koblan LW, Liu DR (2020) Genome editing with CRISPR-Cas nucleases, base editors, transposases and prime editors. Nat Biotechnol 38:824–844

Argelaguet R, Cuomo AS, Stegle O, Marioni JC (2021) Computational principles and challenges in single-cell data integration. Nat Biotechnol 39(10):1202–1215

Aron AT, Gentry EC, McPhail KL, Nothias LF, Nothias-Esposito M, Bouslimani A et al (2020) Reproducible molecular networking of untargeted mass spectrometry data using GNPS. Nat Protoc 15(6):1954–1991

Bartolec TK, Vázquez-Campos X, Norman A, Luong C, Johnson M, Payne RJ, Wilkins MR, Mackay JP, Low JKK (2023) Cross-linking mass spectrometry discovers, evaluates, and corroborates structures and protein-protein interactions in the human cell. Proc Natl Acad Sci USA 120(17):e2219418120

Burley SK, Berman HM, Kleywegt GJ, Markley JL, Nakamura H, Velankar S (2017) Protein Data Bank (PDB): the single global macromolecular structure archive. Methods Mol Biol (Clifton, N.J.) 1607:627–641

Can T (2014) Introduction to bioinformatics. Methods Mol Biol (Clifton, N.J.) 1107:51–71

Centre JR, Düsseldorf HHU (2021) Published plant genomes chronology (timeline). https://www.plabipd.de/timeline_view.ep

Chen C, Hou J, Tanner JJ, Cheng J (2019a) Bioinformatics methods for mass spectrometry-based proteomics data analysis. Int J Mol Sci 21(8):2873

Chen W, McKenna A, Schreiber J, Haeussler M, Yin Y, Agarwal V et al (2019b) Massively parallel profiling and predictive modeling of the outcomes of CRISPR/Cas9-mediated double-strand break repair. Nucleic Acids Res 47(15):7989–8003

Chen C, Hou J, Tanner JJ, Cheng J (2020) Bioinformatics methods for mass spectrometry-based proteomics data analysis. Int J Mol Sci 21(8):2873

Chen Y, Li M, Xu Y (2022) Guide to metabolomics analysis: a bioinformatics workflow. Meta 12:4

Chong J, Wishart DS, Xia J (2019) Using MetaboAnalyst 4.0 for comprehensive and integrative metabolomics data analysis. Curr Protoc Bioinformatics 68(1):e86

Colomé-Tatché M, Teis FJ (2018) Statistical single cell multi-omics integration. Curr Opin Syst Biol 7:54–59

Eicher T, Kinnebrew G, Patt A, Spencer K, Ying K, Ma Q, Machiraju R, Mathé EA (2020) Metabolomics and multi-omics integration: a survey of computational methods and resources. Meta 10:5

Fatumo S, Chikowore T, Choudhury A, Ayub M, Martin AR, Kuchenbaecker K (2022) A roadmap to increase diversity in genomic studies. Nat Med 28(2):243–250

Filgueira R, Awaysheh FM, Carter A, White DJ, Rana O (2022) SparkFlow: towards high-performance data analytics for spark-based genome analysis. In 2022 22nd IEEE International Symposium on Cluster, Cloud and Internet Computing (CCGrid). pp. 1007–1016. IEEE

Hao L, Xu W, Qi G, Xin T, Xu Z, Lei H, Song J (2022) GAGE is a method for identification of plant species based on whole genome analysis and genome editing. Commun Biol 5(1):1–7

Heiner M (2023) Predictive modeling in systems biology: from simulation to proteome validation. J Syst Bio Proteome Res 4(5):165

Hie B, Bryson B, Berger B (2019) Efcient integration of heterogeneous singlecell transcriptomes using Scanorama. Nat Biotechnol 37:685–691

Iakhnin S (2019) Modern systems for large-scale genomics data analysis in the cloud (Doctoral dissertation, Dissertation, Heidelberg, Universität Heidelberg, 2019)

Jendoubi T (2021) Approaches to integrating metabolomics and multi-omics data: a primer. Meta 11:3

Jia W, Sun M, Lian J, Hou S (2022) Feature dimensionality reduction: a review. Complex Intell Syst 8(3):2663–2693

Khakimov B, Mobaraki N, Trimigno A, Aru V, Engelsen SB (2020) Signature mapping (SigMa): an efficient approach for processing complex human urine 1H NMR metabolomics data. Anal Chim Acta 1108:142–151

Kotipalli A, Banerjee R, Kasibhatla SM, Joshi R (2021) Analysis of H3K4me3-ChIP-Seq and RNA-Seq data to understand the putative role of miRNAs and their target genes in breast cancer cell lines. Genomics Inform 19(2):e17

Li H, Yang Y, Hong W, Huang M, Wu M, Zhao X (2020) Applications of genome editing technology in the targeted therapy of human diseases: mechanisms, advances and prospects. Signal Transduct Target Ther 5(1):1–23

Li C, Brant E, Budak H (2021) Zhang B.CRISPR/Cas: a Nobel prize award-winning precise genome editing technology for gene therapy and crop improvement. J Zhejiang Univ Sci B 22:253–284

Li C, Chu W, Gill RA, Sang S, Shi Y, Hu X, Yang Y, Zaman QU, Zhang B (2023) Computational tools and resources for CRISPR/Cas genome editing. Genomics Proteomics Bioinformatics 21(1):108–126

Lill JR, Mathews WR, Rose CM, Schirle M (2021) Proteomics in the pharmaceutical and biotechnology industry: a look to the next decade. Expert Rev Proteomics 18(7):503–526

Lin J, Wong C (2018) Off-target predictions in CRISPR-Cas9 gene editing using deep learning. Bioinformatics 34(17):i656

Liu H, Ding Y, Zhou Y, Jin W, Xie K, Chen LL (2017) CRISPR-P 2.0: an improved CRISPR-Cas9 tool for genome editing in plants. Mol Plant 10(3):530–532

Lorenz R, Luntzer D, Hofacker IL, Stadler PF, Wolfinger MT (2016) SHAPE directed RNA folding. Bioinformatics 32(1):145–147

Luo C, Liu H, Xie F, Armand EJ, Siletti K, Bakken TE, Fang R, Doyle WI, Hodge RD, Hu L, Wang BA (2019) Single nucleus multi-omics links human cortical cell regulatory genome diversity to disease risk variants. Biorxiv:2019–2012

Madrid-Gambin F, Oller-Moreno S, Fernandez L, Bartova S, Giner MP, Joyce C, Ferraro F, Montoliu I, Moco S, Marco S (2020) AlpsNMR: an R package for signal processing of fully untargeted NMR-based metabolomics. Bioinformatics 36(9):2943–2945

Manzoni C, Kia DA, Vandrovcova J, Hardy J, Wood NW, Lewis PA, Ferrari R (2018) Genome, transcriptome and proteome: the rise of omics data and their integration in biomedical sciences. Brief Bioinform 19(2):286–302

Mattoli L, Gianni M, Burico M (2022) Mass spectrometry based metabolomic analysis as a tool for quality control of natural complex products. Mass Spectrom Rev 42:1358

Mayer G, Heinze G, Mischak H, Hellemons ME, Heerspink HJ, Bakker SJ, de Zeeuw D, Haiduk M, Rossing P, Oberbauer R (2011) Omics-bioinformatics in the context of clinical data. Methods in molecular biology (Clifton, N.J.) 719:479–497

Mehmood SS, Lu G, Luo D, Hussain MA, Raza A, Zafar Z, Zhang X, Cheng Y, Zou X, Lv Y (2021) Integrated analysis of transcriptomics and proteomics provides insights into the molecular regulation of cold response in Brassica napus. Environ Exp Bot 187:104480

Misra BB (2021) New software tools, databases, and resources in metabolomics: updates from 2020. Metabolomics 17:49

Mo Y, Jiao Y (2022) Advances and applications of single-cell omics technologies in plant research. Plant J 110(6):1551–1563. https://doi.org/10.1111/tpj.15772

Nagana Gowda GA, Raftery D (2019) Overview of NMR spectroscopy-based metabolomics: opportunities and challenges. NMR-Based Metabolomics: Methods and Protocols, 3–14

Nanni P, Gehrig P, Schlapbach R (2022) Mass spectrometry in proteomics: technologies, methods, and research applications for the life sciences. CHIMIA Int J Chem 76(1–2):73–80

Paladhi P, Dutta S, Pal S, Bose G, Ghosh P, Chattopadhyay R, Chakravarty B, Saha I, Ghosh S (2022) Novel mutations of TSPY1 gene associate spermatogenic failure among men. Reprod Sci 29(4):1241–1261

Perchey RT, Tonini L, Tosolini M, Fournié J-J, Lopez F, Besson A, Pont F (2019) PTMselect: optimization of protein modifications discovery by mass spectrometry. Sci Rep 9:4181

Pulman J (2014) A transcriptomics approach to understanding polymorphic and transcript level differences linked to isoquinoline alkaloid production in triploid varieties of Narcissus pseudonarcissus (Doctoral dissertation, University of Liverpool)

Ravenhall M (2019) A bioinformatic analysis of malaria host and pathogen genomics (Doctoral dissertation, London School of Hygiene & Tropical Medicine)
Reel PS, Reel S, Pearson E, Trucco E, Jefferson E (2021) Using machine learning approaches for multi-omics data analysis: a review. Biotechnol Adv 49:107739. Advance online publication
Regev A, Teichmann SA, Lander ES, Amit I, Benoist C, Birney E, Bodenmiller B, Campbell P, Carninci P, Clatworthy M, Clevers H (2017) The human cell atlas. elife 6:e27041
Reimand J, Isserlin R, Voisin V, Kucera M, Tannus-Lopes C, Rostamianfar A, Wadi L, Meyer M, Wong J, Xu C, Merico D, Bader GD (2019) Pathway enrichment analysis and visualization of omics data using g: Profiler, GSEA, Cytoscape and EnrichmentMap. Nat Protoc 14(2):482–517
Ren S, Hinzman AA, Kang EL, Szczesniak RD, Lu LJ (2015) Computational and statistical analysis of metabolomics data. Metabolomics 11:1492–1513
Roy A, Diao Y, Evani U, Abhyankar A, Howarth C, Le Priol R, Bloom T (2017) Massively parallel processing of whole genome sequence data: an in-depth performance study. In Proceedings of the 2017 ACM International Conference on Management of Data. pp. 187–202
Santa-Coloma TA (2022) Overlapping synthetic peptides as a tool to map protein-protein interactions– FSH as a model system of nonadditive interactions. Biochim Biophys Acta Gen Subj 1866(7):130153
Searle BC, Swearingen KE, Barnes CA, Schmidt T, Gessulat S, Küster B, Wilhelm M (2020) Generating high quality libraries for DIA MS with empirically corrected peptide predictions. Nat Commun 11(1):1548
Sherkatghanad Z, Abdar M, Charlier J, Makarenkov V (2023) Using traditional machine learning and deep learning methods for on- and off-target prediction in CRISPR/Cas9: a review. Brief Bioinform 24:3
Shreffler J, Huecker MR (2020) Type I and type II errors and statistical power. StatPearls. [internet]
Stanstrup J, Broeckling CD, Helmus R, Hoffmann N, Mathé E, Naake T et al (2019) The metaRbolomics toolbox in Bioconductor and beyond. Meta 9(10):200
Ting YS, Egertson JD, Bollinger JG, Searle BC, Payne SH, Noble WS, MacCoss MJ (2017) PECAN: Library-free peptide detection for data-independent acquisition tandem mass spectrometry data. Nat Methods 14:903–908
Town SEL, Dalla PV, Bicknell F, Belic NK, Violi JP, Steele JR, Padula MP (2019) What is normalization? The strategies employed in top-down and bottom-up proteome analysis workflows. Proteomes 7:3
Vandereyken K, Sifrim A, Thienpont B, Voet T (2023) Methods and applications for single-cell and spatial multi-omics. Nat Rev Genet 24(8):494–515
Veenstra TD (2021) Omics in systems biology: current progress and future outlook. Proteomics 21(3–4):2000235
Wishart DS (2019) Metabolomics for investigating physiological and pathophysiological processes. Physiol Rev 99:1819
Yang L, George J, Wang J (2020a) Deep profiling of cellular heterogeneity by emerging single-cell proteomic technologies. Proteomics 20(13):1900226
Yang Y, Liu X, Shen C, Lin Y, Yang P, Qiao L (2020b) In silico spectral libraries by deep learning facilitate data-independent acquisition proteomics. Nat Commun 11(1):146
Yang Y, Yang L, Zheng M, Cao D, Liu G (2023) Data acquisition methods for non-targeted screening in environmental analysis. TrAC Trends Anal Chem 116966
Zhang F, Ge W, Ruan G, Cai X, Guo T (2020) Data-independent acquisition mass spectrometry-based proteomics and software tools: a glimpse in 2020. Proteomics 20(17–18):1900276
Zhang XE, Liu C, Dai J, Yuan Y, Gao C, Feng Y, Wu B, Wei P, You C, Wang X, Si T (2023) Enabling technology and core theory of synthetic biology. Sci China Life Sci 66:1742–1785

The Role of Machine Learning in Biofertilizer Industry: From Data Analytics to Predictive Modelling

8

Gursharan Kaur, Palak Rana, Harleen Kaur Walia, and Vagish Dwibedi

Abstract

Agricultural production is an essential aspect of a country's economic development. Achieving sustainable crop production is consistently a challenge for agriculturists. Farmers have long faced challenges in generating maximum agricultural output due to the constantly shifting weather conditions. The incorporation of machine learning (ML) methods in the biofertilizer sector is a promising advancement in contemporary agriculture. Machine learning algorithms transform input data into novel solutions and diverse methods. Machine learning involves using gained information and rules to predict and make judgments about more data. This research examines the diverse use of machine learning in optimizing the utilization of biofertilizers, including data analytics and predictive modelling. Through the utilization of sophisticated algorithms and insights derived from data, machine learning can profoundly transform agricultural practices, leading to improved crop productivity and a greater focus on environmental responsibility. This chapter explores the important role of machine learning (ML) in determining the future of the biofertilizer sector, based on the current research and developing trends.

Keywords

Machine learning · Biofertilizer · Crop yield · Productivity · Predictive modelling

G. Kaur · P. Rana
Department of Biotechnology, UIBT, Chandigarh University, Mohali, Punjab, India

H. K. Walia
Department of Biotechnology, Thapar Institute of Engineering and Technology, Patiala, Punjab, India

V. Dwibedi (✉)
Department of Biotechnology, Chandigarh University, Mohali, Chandigarh, India
e-mail: vagish@volcani.agri.gov.il

S. Kaur et al. (eds.), *Metabolomics, Proteomics and Gene Editing Approaches in Biofertilizer Industry*, https://doi.org/10.1007/978-981-97-2910-4_8

8.1 Introduction

Human activities have contributed to ongoing soil erosion, resulting in crucial nutrient deficits in forest soils. Furthermore, it is projected that the global population will surpass 9 billion by the year 2050. Feeding the rising population poses a significant issue, and our main objective is to produce crops sustainably without causing harm to the ecology (Saha et al. 2023). In order to achieve optimal productivity, farmers excessively utilized artificial fertilizers to supply the primary essential nutrients for plants, including nitrogen, phosphorus, and potassium. The excessive utilization of chemical fertilizers leads to an elevation in soil acidity, a reduction in organic matter, humus, and beneficial organisms in the soil, hindered plant growth, alteration of soil pH, nourishment of pests, and the initiation of greenhouse gas emissions (Wu et al. 2021). In order to achieve agricultural sustainability, it is imperative to adopt a judicious and cost-effective utilization of nutrients that is also environmentally benign. Considering the circumstances, biofertilizer may be an appropriate choice (Kumar et al. 2021). The term "biological fertilizer," also referred to as "microinoculants," is used to describe a substance that contains living microorganisms. These microorganisms colonize the rhizosphere, which is the area around the roots of a plant, and enhance the plant's development by improving the availability and absorption of mineral nutrients (Saha et al. 2023; Chakraborty and Akhtar 2021a, b). Microorganisms primarily perform essential tasks such as nitrogen fixation in the soil, turning phosphorus into a plant-absorbable form, manufacturing chemicals that promote plant growth, and safeguarding plants from harmful organisms that might cause diseases. Furthermore, bacteria produce phytohormones that exert a substantial influence on plant growth regulation, significantly affect plant metabolism, and enhance the plant's response to stress (Patel et al. 2023). The challenges in agriculture encompass not only crop productivity, but also the quantities of fertilizer, insecticide, herbicide, and water required to achieve optimal yields for farmers. Machine learning is applied in agriculture to accurately address issues and provide recommendations for farmers, such as fertilizers and insecticides, based on their specific requirements. Various fields in our day-to-day existence extensively utilize several machine learning methods (Monika et al. 2022). The prediction analysis results are used to employ machine learning algorithms in order to forecast the optimal crop and its corresponding biofertilizer. A predictive analysis is conducted to suggest the three most appropriate crops considering soil temperature, nutrient levels, and potential crop revenue (Dey et al. 2024). The results can be utilized in two manners. The initial choice is the automated approach, where the farmer just chooses their location and a suitable crop is determined based on previous trials conducted in the vicinity. Additionally, you have the option to input soil data manually and receive a crop suggestion based on the specific value you provide (Kumar et al. 2023). The aim of this study is to investigate how machine learning approaches, from predictive modelling to data analytics, are being integrated into the biofertilizer sector to increase understanding and optimize biofertilizer usage for better agricultural results.

8.2 Biofertilizers: Types and Benefits

Biofertilizers are made up of live microbes that help plants grow better. These little living things use different ways to increase nutrient absorption, raise plant production and improve soil richness. These systems help plants by fixing nitrogen and making iron and phosphorus easier to use. They also defend against disease-causing bugs or harmful soil pollution while protecting the plant from environmental stressors (Mącik et al. 2019). Using tiny bugs from the soil as natural fertilizers is turning into a good and green way to lessen the bad effects of using too many chemicals on plants. The presence of these beneficial rhizobacteria contributes to the sustainability of farming. They are particularly effective in assisting plants in growing and developing in a way that benefits their productivity. Often called PGPR (plant growth-promoting bacteria), these are important for making soil better and richer. They do this in multiple ways, both directly and indirectly (Chakraborty and Akhtar 2021a, b). *Achromobacter*, a PGPR has been found to make root hairs longer and more numerous. This leads to higher amounts of nitrate and potassium being taken up by *Brassica napus*—also known as oilseed rape. This improvement is clear in the increased solid weight of both the shoot (from 22% to 33%) and root (ranging from 6% to upwards) (Thomas and Singh 2019). Nowadays, more and more organic food is being promoted with increased awareness about the harmful impact of chemical fertilizers. This has turned one kind of farming called "organic agriculture" into a fast-growing area within normal farming (Brenes-Muñoz et al. 2016). Land used for growing organic food has been increasing a lot. In 1999, it was about 53 billion square meters and grew to over four times that by 2016 (Willer and Lernoud 2018). Therefore, different biofertilizers are crucial for giving plant crops the right nutrients.

Biofertilizers come in various types, including those that fix nitrogen, sequester iron, solubilize phosphorus, solubilize potassium, and solubilize zinc (Fig. 8.1). Each of these biofertilizers serves a distinct purpose and contributes to specific functions in the growth and development of plants as discussed below. It is crucial to use each biofertilizer appropriately based on its intended role.

8.2.1 Nitrogen-Fixing Biofertilizers (NFB)

The NFB comprises nitrogen-fixing bacteria such as *Rhizobium* which belongs to the family *Rhizobiaceae*. These bacteria are important in making nitrogen from the atmosphere available to plants (Altaf 2021). Rhizobium functions symbiotically and can fix up to 50–100 kg of nitrogen per hectare each year. Rhizobium is also mainly found in leguminous plants where it penetrates and multiplies into root hairs and colonizes the roots of legumes leading to the formation of root nodules which are tumor-like growths that aid the production of ammonia. Furthermore, studies have also revealed that Rhizobium can fix atmospheric nitrogen through symbiotic relationships with legume plants and non-legumes (Abbey et al. 2019). Alongside the *Rhizobiaceae* family, other bacterial families are also considered nitrogen-fixers. Due to its heterotrophic and associative nature, the family *Spirillaceae* is beneficial for non-leguminous plants specifically those that utilize the C4-dicarboxylic

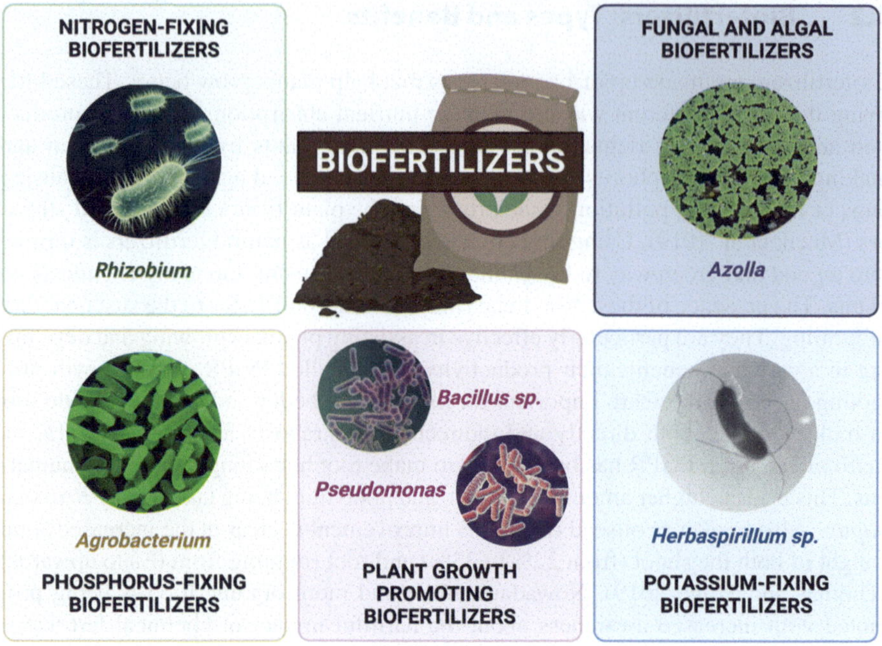

Fig. 8.1 Types of biofertilizers

photosynthesis pathway (Mahdi et al. 2010). *Azospirillum,* one of the members of *Spirillaceae* family, is a free-living, motile, and aerobic bacterium that can live in flood. It can fix up to 20–40 kg of nitrogen per hectare on an annual basis and improves the growth and development of a plant by assisting in the formation process of lateral roots as well as root hairs which often increase the surface area that is presented to roots for a better absorption rate. As a result, the plant's water status and nutrient profile improve (Pedraza et al. 2020).

8.2.2 Phosphorus Solubilizing Biofertilizers (PSB)

Phosphorus is the second most restrictive nutrient, fundamental for crop growth and productivity. With most phosphorus in the soil found to be insoluble in organic form, plants cannot utilize this material. Therefore, phosphorus must be soluble for plant absorption. PSBs are microorganisms capable of converting the insoluble and unavailable type of phosphorus to a dissolvable solvate form (orthophosphates) (Kaur et al. 2024). For instance, phospho-bacterin a PSB contributes significantly to solubilizing the insoluble forms of phosphates such as di and tricalcium phosphate hydroxyapatites rock Phosphas (Mahdi et al. 2010). These bacteria are usually present in large amounts in the soil and rhizosphere of plants. Phospho-bacterin produces organic acids that lower the soil's pH, which facilitates the dissolution of other phosphate compounds and releases available amounts of phosphorus for plant use (Tian et al. 2021). Phosphatase enzymes, organic acids, and soil pH reduction are other mechanisms PSBs enable in the transformational process from phosphorus of soils into soluble and readily available orthophosphate forms plants cannot directly utilize it, so it is converted into soluble forms by bacteria such as *Rhizobium, Pseudomonas, Bacillus, and Agrobacterium* (Billah et al. 2019).

8.2.3 Plant Growth-Promoting Biofertilizers (PGPB)

Plant growth and yield are promoted by a collection of PGPR (plant growth-promoting rhizobacteria) bacteria through several mechanisms, including phytohormone production, phosphate solubilization, and nitrogen fixation among others. A wide range of benefits in promoting plant growth while sustaining natural flora have been accrued from the use of PGPR that synthetic fertilizers fail to achieve. In the rhizosphere, there is a large variety of PGPR communities that actively contribute to crop productivity. Some of the PGPR genera capable of plant growth-promoting activity are *Pseudomonas, Burkholderia Rhizobium Enterobacter Mesorhizobium, Erwinia, Mycobacterium, Flavobacterium, Bacillus,* and many others (Kaur et al. 2024).

The inoculation of plants with PGPR has consistently been shown to stimulate plant growth, increase yield significantly, and promote P or K solubilization and N absorption along with other important elements through numerous studies as well as reviews. In addition, the studies show that inoculation of PGPR leads to better development for roots that have a bigger surface area and more root hairs than before. Furthermore, PGPR takes on a defensive role for these plants against several pathogens either via direct interaction or by promoting host resistance (Basu et al. 2021). Recent studies show that inoculation of PGPR in agricultural soils increases their productivity much more than those without such an injection. Using PGPR is an environmentally friendly and cost-saving approach that improves plant development and reduces stress conditions.

8.2.4 Fungal and Algal Biofertilizers

Unquestionably fungal species are an important part of the soil community as they play significant roles in decomposition, ecosystem regulation, and biological control. They contribute immensely to the phosphorus macronutrient supply. Many species of fungi are specialized in phosphorus solubilization which converts unavailable insoluble (organic) forms to available soluble (inorganic). These fungi are called phosphate-solubilizing mycelia (Yang et al. 2022). For instance, mycorrhizal fungi increase soil quality through positive changes in its porosity, stability, aggregate formation, aeration properties, dynamics of water, and resistance to abiotic stress. Such mycorrhizal fungi promote the growth of a host plant and protect it from pathogenic attack (Kaur et al. 2024).

Similarly, the major two types of algae used as biofertilizers are blue-green algae and *Azolla* (nonsymbiotic). *Azolla* with scaly leaves and floating roots are tiny plants that fix nitrogen. Importantly, *Azolla* is characterized by a symbiotic relationship with nitrogen-fixing organisms like *Nitella* and *Anabaena* in particular (Kaur and Purewal 2019).

8.2.5 Potassium Solubilizing Biofertilizers (PFB)

Potassium is an important nutrient for crop growth since it has a role in regulating the opening and closing of stomata to maintain normal water balance. Even though the whole supply of total potassium available in the soil is great, only small quantities are easily accessible for consumption by plants. Potassium exists in three simultaneous forms in the soil system: unavailable forms slowly available and readily available (Abbey et al. 2019).

Herbspirillum sp. is a member of the *Spirillaceae* family, forming a symbiotic association that contributes to atmospheric nitrogen fixation in sugarcane plants. The inoculation of plants with *Herbspirillum* improves nitrogen availability, promotes nitrate uptake, and significantly, enables the absorption of potassium and phosphate. Additionally, this symbiotic association promotes the production of growth-promoting substances such as kinetin, auxin, and gibberellic acid (Chaudhary et al. 2022).

8.2.6 Application and Challenges in Biofertilizer Production

Biofertilizers are applied through seeds, seedlings or directly to the soil. Whichever method is applied, it has its advantages and disadvantages depending on factors like characteristics of the inoculated material, crop type cultivated as well as environmental conditions found at a particular site together with the technical farmer's knowledge. Nevertheless, some pre-application tricks should be followed, such as sun exposure avoidance or overnight storage after use. Secondly, biofertilizers should be kept in proper temperature conditions, not below 0 °C but less than 35 °C (Mącik et al. 2019). Soil inoculation is the suggested method for directing a huge quantum of microbial strains into the soil. In this method, granules (0.5–1.5 cm) are used as carrier materials; commonly here granular forms of peat, perlite talcum powder or soil aggregates are applied (Asif et al. 2018). Soil inoculation usually consists of depositing these granules beneath, above, and close to the seed. Seeds in furrows can be sprayed with the liquid forms of biofertilizers and hydroponic systems could also serve to introduce these microorganisms (Fasusi et al. 2021).

Soil treatment has all those benefits for inoculants as it gives control of concentration and location, and protection from adverse effects caused by pesticides or fungicides, besides saving seed coats. This approach lowers the likelihood of spending part of an inoculant from seeding machinery. Soil inoculation increases the probability of seeds contacting a much higher concentration of biofertilizer than is achieved by seed treatment (Ballard and Wright 2022). Nevertheless, this technique presents some disadvantages mainly technical that include the need for specialized equipment, larger quantities of biofertilizer requiring more storage space and transport difficulties (Mącik et al. 2019).

8.3 Machine Learning in Biofertilizer Industry

The maximum capabilities of technologies for precision agriculture can be fully utilized when combined with advanced data processing methods, such as machine learning, to extract valuable information that affects crop output (Abbas et al. 2020). Machine learning algorithms transform input data into novel solutions and diverse methods. It involves utilizing gained information and algorithms to predict and make judgments about other parts of the data. Common supervised learning techniques in machine learning are artificial neural networks, linear regression models, and random forests (Osman et al. 2023). Various empirical and analytical yield modelling techniques have been utilized for different crops. Implementing these strategies in diverse regions is challenging due to the need for extensive understanding of crop and soil illustrated in Fig. 8.2(Yin et al. 2021). Predictive yield modelling incorporates many remote sensing approaches using satellites. Nevertheless, these techniques are insufficient in providing adequate spatial information about small-scale farms for the purpose of optimizing crop inputs through site-specific management (Kasampalis et al. 2018; Abbas et al. 2020). The modelling industry has increasingly embraced recent breakthroughs in machine learning and data-driven modelling, which have proven effective in solving and comprehending intricate interactions. Various machine learning approaches have demonstrated their promise for application in crop prediction (Van Klompenburg et al. 2020).

For example, the study showed that the application of digestate composed of a combination of CW–BF: NPK (3:1:1) led to the maximum grain output of 7521 kg/hectare, higher panicle length, test weight, and a greater number of full grains compared to the other combinations. A gradient-boosting machine learning model revealed that NPK was identified as the first principal component, with significant loading values and digestate was identified as the second principal component, suggesting its important involvement in the manufacture of fertilizers. Thus, the utilization of these hybridized fertilizers, coupled with accurate statistical analysis and machine learning techniques, has the potential to enhance rice production (Show et al. 2023). The study conducted by Ranjan and Parida (2019) suggests a reliable machine learning random forest (RF) classification method used to mapping of paddy acreage. 1.60 tons of paddy yield per hectare was predicted by a basic linear regression yield model (Ranjan and Parida 2019). Moreover, another paper proposes a technique for enhancing the recommendations of biofertilizers in coffee crops by utilizing deep collaborative matrix factorization. In order to achieve optimal results, the complex interconnections among the health of coffee plants, weather conditions, soil characteristics, and different types of biofertilizers are carefully examined and customized. The proposed model attained an accuracy of 97.88%

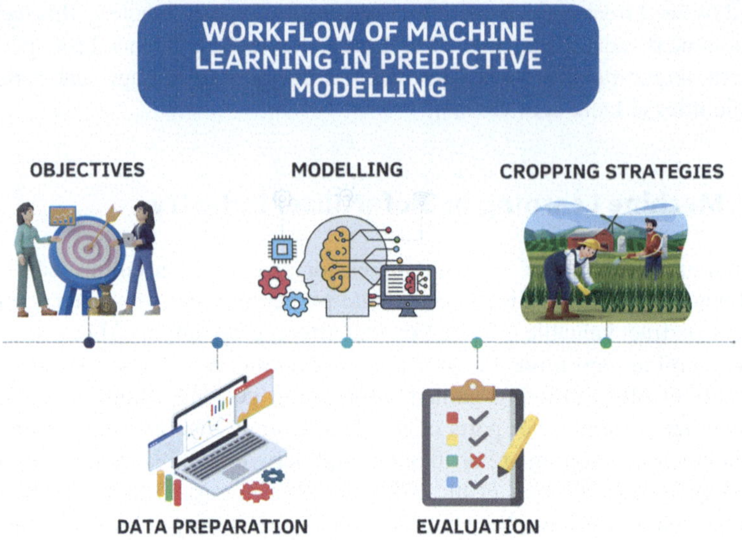

Fig. 8.2 Workflow of machine learning in predictive modelling of crops

indicating that the system had the ability to accurately and precisely suggest biofertilizers with exceptional accuracy and precision (Raveena and Surendran 2023). Four machine learning algorithms, specifically support vector regression (SVR), elastic net (EN), k-nearest neighbor (k-NN), and linear regression (LR), were employed to predict the yield of potato (Solanum tuberosum) tubers using data obtained from soil and crop properties gathered from six fields in Atlantic Canada through proximal sensing techniques. The SVR models exhibited superior performance compared to all other models across all datasets, as measured by the root mean square error (RMSE). However, the performance of k-NN was unsatisfactory in three out of four datasets. Additionally, the study demonstrated that substantial datasets are necessary to produce valuable outcomes using either methodology (Abbas et al. 2020). One of the research projects introduces several machine learning techniques, including linear discriminant analysis, k-NN, and SVM, for evaluating genomes and predicting rhizosphere molecular mechanisms to identify growth-boosting bacteria and propose a secondary metabolic model for future growth. The NCBI Dataset was utilized for experimental purposes. The SVM model achieved a Precision of 0.97, a Recall of 0.91, and an F-measure of 0.95. SVM achieved the highest accuracy among the various machine learning algorithms when predicting the presence of plant promoting rhizobacteria (Indumathi et al. 2021).

Weather conditions have a significant impact on agricultural productivity. Multiple studies have aimed to assess the effect of climate change on crop yields, but the year-to-year impact in various places is still uncertain (Waldhoff et al. 2020). Numerous studies on the effects of climate change use agro-meteorological variables, like wind speed, air temperature, sun radiation, and precipitation to determine whether or not a certain region's climate is suitable for growing a particular crop

(Wang et al. 2020). The process of constructing the indicator involves developing statistical agricultural models based on meteorological factors and determining the indicator values through classification. When comparing two machine learning algorithms, random forest (RF) and support vector machine (SVM), it was found that RF outperformed SVM in the independent test set. Therefore, RF was chosen to create ICAI, which has three values: yield loss, normal, and yield increment. The ICAI findings indicate that the over the past three decades, wheat farming in Southern Jiangsu has significantly enhanced its ability to adapt to the local surroundings. Furthermore, when the annual temperature rises, the likelihood of experiencing decreased crop yields in Northern Jiangsu and increased crop yields in Southern Jiangsu will also grow (Xu et al. 2019). Hence, integrating advanced data processing techniques, particularly machine learning, unlocks the full potential of precision agriculture technologies. Machine learning algorithms convert unprocessed data into practical insights, facilitating the prediction of crop performance and enhanced efficiency and production.

8.4 Software Tools and Resources

Machine learning algorithms are not dependent on rules established by human experts. A machine learning framework is considered a genuine learning system when it is not explicitly built to execute a certain task, but rather programmed to acquire the ability to perform that task through learning (Kumar and Sharma 2022). Machine learning (ML) is primarily focused on making predictions, while statistical modelling tends to prioritize interpretation. Using excessive computer capacity to train intricate models of rapidly changing network environments is not a cost-efficient strategy, unlike conventional methods (Singh et al. 2022). Machine learning can be classified into three primary types: supervised learning, unsupervised learning, and regression learning illustrated in Fig. 8.3. Supervised learning is creating a model that captures the connection between measured data features and corresponding labels (Singh et al. 2023). Once the model is established, it can be utilized to assign labels to fresh, unfamiliar data. These problems can be further separated into classification tasks and regression tasks. In classification, the labels are discrete categories, whereas in regression, the labels are continuous values. One significant distinction is that classical technique employs a more comprehensive mathematical approach, whereas machine learning methods focus more on information processing (Sen et al. 2020). The five machine learning algorithms include random forest (RF), support vector machine (SVM), k-nearest neighbors (k-NN), decision tree (DT), and eXtreme gradient boosting (XGBoost) (Sudhan Reddy and Neerugatti 2023).

Random forest is an ensemble learning technique, employed to categorize data by creating a forest composed of a variable number of trees and then averaging the predictions made by each decision tree. For example, to predict soybean output in three agroecological zones in northern Nigeria. The results showed that the combination of rhizobium inoculant and phosphorus had the average yield increased by

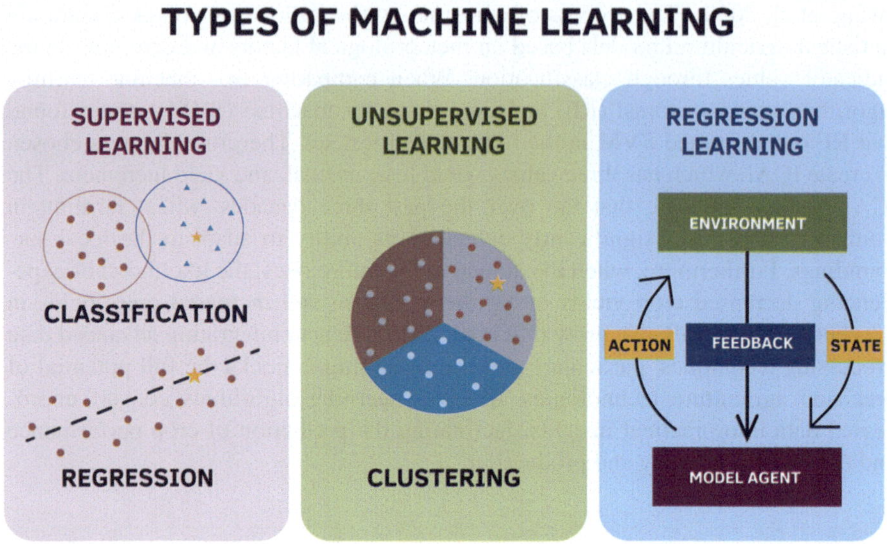

Fig. 8.3 Types of machine learning

111%, especially in tough environmental situations (Jemo et al. 2023). The support vector machine (SVM) is a classification technique that efficiently minimizes model complexity while precisely fitting the training set. The method reduces the possibility of generalization mistakes by finding linear hyperplanes that maximize the distance between two sides or edges of multiple hyperplanes (Kok et al. 2021). For example, by using SVM model the significance of the microbial genera in charge of the variation among fertilizers was assessed. The highest ranking bacterial genera in terms of relative relevance between biofertilizer and urea application were *Anaerolineace, Vulgatibacter,* and *Paenibacillus*, whereas the highest ranking fungal genera were *Cochliobolus, Sordariales,* and *Dothideomycetes* (Liu et al. 2021). The K-nearest neighbor algorithm is a widely used technique for finding the k-most comparable training samples in a given training set to a target item. This technique is not well adapted for processing high-dimensional data and is only suitable to a very small set of surrounding samples for category determinations (Dey et al. 2024). A decision tree (DT) is a classification technique that uses a tree structure to categorize data based on a sequence of decisions. It is extensively utilized in managing organized data. For example, a decision tree was randomly chosen from the Cowpea dataset using the RF model. The decision tree demonstrates that the criteria for dividing each node are based on the selection of features that are closer to the root node and are more important and have a significant impact on how the classification results are divided (Li et al. 2024). XGBoost is highly esteemed in the field of machine learning algorithms due to the use of a gradient-boosted decision tree that allocates the model with exceptional adaptability, velocity, and efficiency in comparison to alternative models. The study conducted to identify microbial strains that can alleviate the impacts of drought is essential for improving crop resilience and yield revealed that the gradient-boosted trees model demonstrated the highest level of accuracy among the models, achieving an accuracy rate of 87% with a standard deviation of 4%. The results showed the highest improvement (68.0), however, it required a substantial amount of computational resources, with a total computation time of 3,381,260.0 (Miller et al. 2023). The study was conducted to assess the performance of five different machine learning models using a dataset taken from the Kaggle repository to select or identify necessary nutrients at a particular location. The yields of individual data sets of 10 horticultural and 11 agricultural crops, as well as the combined yield of the two Agri-horticultural crops, were used to train the models support vector machine, XGBoost, random forest, k-NN, and decision tree. When comparing the accuracy of the five machine learning models, XGBoost showed the greatest level with precision rate of 99.09% (Dey et al. 2024). A machine learning model is developed to accurately determine the level of moisture in the soil. This model is then implemented utilizing Internet of Things (IoT) devices (Goap et al. 2018). The user provides information on the utilization of various technologies such as ML's crop and soil monitoring systems, remote sensing, satellite imaging, drones, and precision technology. Autonomous robots are being developed to accomplish labor-intensive tasks, such harvesting crops, faster and more efficiently than human laborers (Javaid et al. 2023). The significance of machine learning is in its ability to facilitate informed decision-making in the field of precision agriculture, as crop output is influenced by an extensive range of external influences. Trying to understand the collective influence of these aspects without the assistance of modelling poses a significant problem (Shaikh et al. 2022). Various biotic and abiotic factors significantly affect agricultural yield. Utilizing machine learning algorithms to analyze the public dataset of a specific country, the recommendations for growing appropriate crops can offer a wide range of benefits to users, ultimately leading to the highest possible crop production (Abioye et al. 2022).

8.5 Challenges and Future Directions

Although this study provides vital insights into the implementation of machine learning models, it is crucial to recognize its limitations to fully understand the context and extent of the findings. Initially, the models undergo training and are evaluated using a robust dataset that has been acquired. Nevertheless, the efficacy of the models may change when employed on diverse datasets. This is a frequent limitation in machine learning applications, as the effectiveness of the models is typically strongly influenced by the particular attributes of the training data (Murdoch et al. 2019; Miller et al. 2023). Although machine learning models can generate precise predictions, they do not inherently offer a detailed understanding of the underlying biological mechanisms. Thus, it is imperative to verify predictions under experimental conditions to authenticate their biological significance and practicality (Lopatkin and Collins 2020). Although there are certain limits, the study is a notable achievement in the incorporation of machine learning in agriculture and paves the way for further progress in this promising subject.

Future directions of machine learning have great potential in the investigation of ensemble methods. Ensemble approaches, which involve integrating predictions of numerous models, have demonstrated their efficacy in enhancing accuracy and resilience across various machine learning projects (Ganaie et al. 2022). By harnessing the advantages of many models, ensemble approaches have the potential to produce more precise and dependable predictions in the field of microbial strain selection. Subsequent research efforts could explore different ensemble methodologies, such as bagging, boosting, and stacking, and evaluate their efficacy to standalone models (Guindo et al. 2021). Another promising area for future research involves exploring more sophisticated machine learning approaches, such as deep learning, that have the ability to learn intricate patterns from data with many dimensions. Within the framework of microbial strain selection, these models have the ability to reveal complex connections between microbial characteristics, environmental variables, and plant reactions, resulting in more precise forecasts (Miller et al. 2023). Furthermore, future research endeavors could encompass the utilization of the developed machine learning approaches on other datasets. Genomic data and soil microbiome data have the potential to offer a substantial amount of information regarding the attributes of various microbial strains and their interactions with plants and the environment (Pirttilä et al. 2021). Furthermore, future research should explore the suitability of machine learning models in various crops, soil types, and meteorological situations. This has the potential to result in the creation of more adaptable and resilient models that can accommodate a diverse array of agricultural situations.

8.6 Conclusion

The extensive use of dangerous chemicals, specifically pesticides employed as fertilizers, is presenting a significant threat to the surrounding agricultural sector. Biofertilizers offer a viable alternative to these dangerous chemicals since they contribute to sustainable agriculture by enhancing crop productivity and preserving long-term soil fertility. In the agricultural industry, precision agriculture has become a game-changer by providing accurate information on the viability of crops based on dominant nutrient, climate, and soil pH variables through the use of machine learning-based modelling. The implementation of machine learning to identify microbial strains represents an enormous shift from traditional approaches, which could increase the project's efficiency. To get the most accurate outcomes, certain activities require the use of certain approaches and models. Even with this study's positive findings, there are still issues that need to be resolved. For instance, the accuracy of models can be significantly impacted by the type and quantity of accessible data. Furthermore, the properties of microbes can vary greatly, which makes it challenging to correctly identify and classify them. To overcome these obstacles and raise the accuracy rates of, more developments in ML algorithms are required.

References

Abbas F, Afzaal H, Farooque AA, Tang S (2020) Crop yield prediction through proximal sensing and machine learning algorithms. Agronomy 10(7):1046

Abbey L, Abbey J, Leke-Aladekoba A, Iheshiulo EMA, Ijenyo M (2019) Biopesticides and biofertilizers: types, production, benefits, and utilization. In: Byproducts from agriculture and fisheries: adding value for food, feed, pharma, and fuels. Wiley, pp 479–500

Abioye EA, Hensel O, Esau TJ, Elijah O, Abidin MSZ, Ayobami AS et al (2022) Precision irrigation management using machine learning and digital farming solutions. Agri Eng 4(1):70–103

Altaf MM (2021) Functional diversity of nitrogen-fixing plant growth-promoting Rhizobacteria: the story so far. In: Soil nitrogen ecology. Springer International Publishing, Cham, pp 327–348

Asif M, Mughal AH, Bisma R, Mehdi Z, Saima S, Ajaz M, Masood A, Malik MA, Sidique S (2018) Application of different strains of biofertilizers for raising quality forest nursery. Int J Curr Microbiol App Sci 7:3680–3686. https://doi.org/10.20546/ijcmas.2018.710.425

Ballard RA, Wright D (2022) Soil microbial inoculants for sustainable agriculture: limitations and opportunities. Soil Use Manag 38(3):1340–1369. https://doi.org/10.1111/sum.12811

Basu A, Prasad P, Das SN, Kalam S, Sayyed RZ, Reddy MS, El Enshasy H (2021) Plant growth promoting rhizobacteria (PGPR) as green bioinoculants: recent developments, constraints, and prospects. Sustain For 13(3):1140

Billah M, Khan M, Bano A, Hassan TU, Munir A, Gurmani AR (2019) Phosphorus and phosphate solubilizing bacteria: keys for sustainable agriculture. Geomicrobiol J 36:904–916

Brenes-Muñoz T, Lakner S, Brummer B (2016) What influences the growth of organic farms? Evidence from a panel of organic farms in Germany. Ger J Agric Econ 65:1–15

Chakraborty T, Akhtar N (2021a) Biofertilizers: characteristic features and applications. In: Inamuddin, Ahamed, MI, Boddula R, Rezakazemi M (eds) Biofertilizers. Wiley https://doi.org/10.1002/9781119724995.ch15

Chakraborty T, Akhtar N (2021b) Biofertilizers: prospects and challenges for future. In: Biofertilizers. Wiley, pp 575–590

Chaudhary P, Singh S, Chaudhary A, Sharma A, Kumar G (2022) Overview of biofertilizers in crop production and stress management for sustainable agriculture. Front Plant Sci 13:930340

Dey B, Ferdous J, Ahmed R (2024) Machine learning based recommendation of agricultural and horticultural crop farming in India under the regime of NPK, soil pH and three climatic variables. Heliyon 10(3):e25112

Fasusi OA, Cruz C, Babalola OO (2021) Agricultural sustainability: microbial biofertilizers in rhizosphere management. Agriculture 11(2):163. https://doi.org/10.3390/agriculture11020163

Ganaie MA, Hu M, Malik AK, Tanveer M, Suganthan PN (2022) Ensemble deep learning: a review. Eng Appl Artif Intell 115:105151

Goap A, Sharma D, Shukla AK, Krishna CR (2018) An IoT based smart irrigation management system using machine learning and open source technologies. Comput Electron Agric 155:41–49

Guindo ML, Kabir MH, Chen R, Liu F (2021) Particle swarm optimization and multiple stacked generalizations to detect nitrogen and organic-matter in organic-fertilizer using Vis-NIR. Sensors 21(14):4882

Indumathi V, Megala SS, Padmapriya R, Jayanthi MSB (2021) Prediction and analysis of plant growth promoting bacteria using machine learning for millet crops. Ann Rom Soc Cell Biol 25(6):1826–1833

Javaid M, Haleem A, Khan IH, Suman R (2023) Understanding the potential applications of artificial intelligence in agriculture sector. Adv Agrochem 2(1):15–30

Jemo M, Devkota KP, Epule TE, Chfadi T, Moutiq R, Hafidi M et al (2023) Exploring the potential of mapped soil properties, rhizobium inoculation, and phosphorus supplementation for predicting soybean yield in the savanna areas of Nigeria. Front Plant Sci 14:1120826

Kasampalis DA, Alexandridis TK, Deva C, Challinor A, Moshou D, Zalidis G (2018) Contribution of remote sensing on crop models: a review. J Imaging 4(4):52

Kaur S, Dwibedi V, Sahu PK, Kocher GS (eds) (2024) Metabolomics, proteomes and Gene Editing Approaches in Biofertilizer Industry. Springer Nature
Kaur P, Purewal SS (2019) Biofertilizers and their role in sustainable agriculture. In: Biofertilizers for sustainable agriculture and environment, pp 285–300
Kok ZH, Shariff ARM, Alfatni MSM, Khairunniza-Bejo S (2021) Support vector machine in precision agriculture: a review. Comput Electron Agric 191:106546
Kumar D, Dutt S, Raigond P, Changan SS, Lal MK, Tiwari RK et al (2021) Different biofertilizers and their application for sustainable development. In: Microbial technology for sustainable environment, pp 31–48
Kumar P, Sharma M (2022) Data, machine learning, and human domain experts: none is better than their collaboration. Int J Hum Comput Interact 38(14):1307–1320
Kumar JA, Parimala N, Pitchai R (2023) Crop selection and yield prediction using machine learning algorithms. In: 2023 Second International Conference on Augmented Intelligence and Sustainable Systems (ICAISS). IEEE, pp 669–673
Li Y, Li R, Ji R, Wu Y, Chen J, Wu M, Yang J (2024) Research on factors affecting global grain legume yield based on explainable artificial intelligence. Agriculture 14(3):438
Liu Q, Pang Z, Yang Z, Nyumah F, Hu C, Lin W, Yuan Z (2021) Bio-fertilizer affects structural dynamics, function, and network patterns of the sugarcane rhizospheric microbiota. Microb Ecol 84:1195–1211
Lopatkin AJ, Collins JJ (2020) Predictive biology: modelling, understanding and harnessing microbial complexity. Nat Rev Microbiol 18(9):507–520
Mącik M, Gryta A, Frąc M (2019) Biofertilizers in agriculture: an overview on concepts, strategies and effects on soil microorganisms. Adv Agron 162:31–87. https://doi.org/10.1016/bs.agron.2020.02.001
Mahdi SS, Hassan GI, Samoon SA, Rather HA, Dar SA, Zehra B (2010) Bio-fertilizers in organic agriculture. J Phytology 2(10):42–54
Miller T, Mikiciuk G, Kisiel A, Mikiciuk M, Paliwoda D, Sas-Paszt L et al (2023) Machine learning approaches for forecasting the best microbial strains to alleviate drought impact in agriculture. Agriculture 13(8):1622
Monika K, Ramprakash B, Muthuramalingam S, Mirdula K (2022) Crop fertilizer prediction using regression analysis and machine learning algorithms. In: In 2022 5th International Conference on Contemporary Computing and Informatics (IC3I). IEEE, pp 1261–1266
Murdoch WJ, Singh C, Kumbier K, Abbasi-Asl R, Yu B (2019) Definitions, methods, and applications in interpretable machine learning. Proc Natl Acad Sci 116(44):22071–22080
Osman AI, Zhang Y, Lai ZY, Rashwan AK, Farghali M, Ahmed AA et al (2023) Machine learning and computational chemistry to improve biochar fertilizers: a review. Environ Chem Lett 21(6):3159–3244
Patel C, Singh J, Karunakaran A, Ramakrishna W (2023) Evolution of nano-biofertilizer as a green technology for agriculture. Agriculture 13(10):1865
Pedraza RO, Filippone MP, Fontana C, Salazar SM, Ramírez-Mata A, Sierra-Cacho D, Baca BE (2020) Azospirillum. In: Beneficial microbes in agro-ecology. Academic Press, pp 73–105
Pirttilä AM, Mohammad Parast Tabas H, Baruah N, Koskimäki JJ (2021) Biofertilizers and biocontrol agents for agriculture: how to identify and develop new potent microbial strains and traits. Microorganisms 9(4):817
Ranjan AK, Parida BR (2019) Paddy acreage mapping and yield prediction using sentinel-based optical and SAR data in Sahibganj district, Jharkhand (India). Spat Inf Res 27(4):399–410
Raveena S, Surendran R (2023) Recommending the right biofertilizer using deep collaborative matrix factorization in the coffee plantation. In: In 2023 7th International Conference on Electronics, Communication and Aerospace Technology (ICECA). IEEE, pp 476–482
Saha S, Paul D, Poudel TR, Basunia NM, Hasan T, Hasan M et al (2023) Biofertilizer science and practice for agriculture and forestry: a review. J Appl Biol Biotechnol 11:31–44
Sen PC, Hajra M, Ghosh M (2020) Supervised classification algorithms in machine learning: a survey and review. In: Emerging technology in modelling and graphics: proceedings of IEM graph 2018. Springer, Singapore, pp 99–111

Shaikh TA, Rasool T, Lone FR (2022) Towards leveraging the role of machine learning and artificial intelligence in precision agriculture and smart farming. Comput Electron Agric 198:107119

Show BK, Panja S, Ghosh Thakur R, Basu A, Koley A, Ghosh A et al (2023) Optimisation of anaerobic digestate and chemical fertiliser application to enhance rice yield—a machine-learning approach. Sustain For 15(18):13706

Singh A, Janu N, Trivedi S, Jain M (2022) Precision agriculture and machine learning. In: In 2022 IEEE world conference on Applied Intelligence and Computing (AIC). IEEE, pp 659–664

Singh P, FNU K, Encarnação T (2023) Marine bacteria for biofertilizers. In: Marine organisms: a solution to environmental pollution? Uses in bioremediation and in biorefinery. Springer International Publishing, Cham, pp 189–203

Sudhan Reddy DM, Neerugatti UR (2023) A comparative analysis of machine learning models for crop recommendation in India. Revue d'Intelligence Artificielle 37(4):1181

Thomas L, Singh I (2019) Microbial biofertilizers: types and applications. In: Biofertilizers for sustainable agriculture and environment. Springer, pp 1–19

Tian J, Ge F, Zhang D, Deng S, Liu X (2021) Roles of phosphate solubilizing microorganisms from managing soil phosphorus deficiency to mediating biogeochemical P cycle. Biology 10(2):158. https://doi.org/10.3390/biology10020158

Van Klompenburg T, Kassahun A, Catal C (2020) Crop yield prediction using machine learning: a systematic literature review. Comput Electron Agric 177:105709

Waldhoff ST, Wing IS, Edmonds J, Leng G, Zhang X (2020) Future climate impacts on global agricultural yields over the 21st century. Environ Res Lett 15(11):114010

Wang Q, Li W, Xiao C, Ai W (2020) Evaluation of high-resolution crop model meteorological forcing datasets at regional scale: air temperature and precipitation over major land areas of China. Atmos 11(9):1011

Willer H, Lernoud J (2018) The World of Organic Agriculture Statistics and Emerging Trends 2018. Research Institute of Organic Agriculture (FiBL), Frick, and IFOAM—Organics International, Bonn. https://doi.org/10.4324/9781849775991

Wu L, Wang Y, Zhang S, Wei W, Kuzyakov Y, Ding X (2021) Fertilization effects on microbial community composition and aggregate formation in saline-alkaline soil. Plant Soil 463:523–535

Xu X, Gao P, Zhu X, Guo W, Ding J, Li C et al (2019) Design of an integrated climatic assessment indicator (ICAI) for wheat production: a case study in Jiangsu Province, China. Ecol Indic 101:943–953

Yang T, Li L, Wang B, Tian J, Shi F, Zhang S, Wu Z (2022) Isolation, mutagenesis, and organic acid secretion of a highly efficient phosphate-solubilizing fungus. Front Microbiol 13:793122

Yin X, Struik PC, Goudriaan J (2021) On the needs for combining physiological principles and mathematics to improve crop models. Field Crop Res 271:108254

The Role of Artificial Intelligence in Biofertilizer Development

9

Babita Thakur and Sukhminderjit Kaur

Abstract

In the field of developing biofertilizers, the integration of artificial intelligence (AI), machine learning (ML), and data analytics has become a shifting approach that presents previously unheard chances to improve sustainable agriculture practices. The contribution of AI, ML, and data analytics to the advancement of biofertilizer research and application is thoroughly examined in this chapter. This chapter examines the most current developments, approaches, difficulties, and prospects for using these technologies to improve agricultural yield, soil fertility, and environmental sustainability. The complementary relationship between technology and biology in opening up new vistas in microbial ecology, soil science, and precision agriculture through an investigation of AI-driven decision support systems, ML algorithms for predictive modeling, and data analytics for data-driven insights. This study highlights the need for ongoing research, cooperation, and the use of new technologies to solve global concerns in food security, environmental degradation, and agricultural sustainability. It does this by illuminating the revolutionary potential of AI, ML, and data analytics.

Keywords

Artificial intelligence · Biofertilizers · Data analytics · Machine language

B. Thakur
Department of Biotechnology, Chandigarh University, Mohali, Punjab, India

S. Kaur (✉)
Department of Biotechnology, Chandigarh University, Mohali, Chandigarh, India
e-mail: sukhminderjit.uibt@cumail.in

S. Kaur et al. (eds.), *Metabolomics, Proteomics and Gene Editing Approaches in Biofertilizer Industry*, https://doi.org/10.1007/978-981-97-2910-4_9

9.1 Introduction

Technology integration has become a revolutionary factor in the search for sustainable agricultural practices, altering conventional farming methods and managing land. The integration of artificial intelligence (AI), machine learning (ML), and data analytics has attracted considerable interest among the many uses of technology in agriculture, especially in the field of biofertilizer development. The need for sustainable alternatives to chemical fertilizers is greater than ever, as the world's population continues to rise and food production demands rise. In this context, biofertilizers have emerged as a possible alternative since they make use of the symbiotic interactions between plants and microorganisms to improve soil fertility and nutrient absorption (Rani et al. 2023; Subeesh and Mehta 2021). The integration of AI, ML, and data analytics has unparalleled prospects to transform the identification, enhancement, and use of biofertilizers. Researchers and practitioners may handle the complexity of microbial ecology, soil dynamics, and crop physiology with accuracy and efficiency by utilizing sophisticated algorithms, computational models, and data-driven insights. This integration not only quickens the development of biofertilizers but also provides farmers with customized solutions to improve soil health, maximize crop output, and minimize environmental impact (Jiang et al. 2022).

AI is essential to understand the complex relationships between microbial populations, soil characteristics, and plant–microbe interactions because of its capacity to handle large datasets and imitate human cognitive capabilities. AI algorithms help in the identification of new microbial strains with desirable characteristics, the forecasting of microbial behavior in different environments, and the streamlining of fermentation processes for the generation of biofertilizers. Furthermore, data-driven insights for precision agriculture are made possible by AI-driven decision support systems, which allow for real-time monitoring of soil health indicators, nutrient dynamics, and crop responses to biofertilizer applications (Miller et al. 2023).

In the field of biofertilizer research, ML, a subset of AI, provides strong tools for data analysis, pattern identification, and predictive modeling. ML algorithms can anticipate microbial interactions, identify trends in microbial diversity, and optimize formulation parameters for optimal performance by utilizing supervised and unsupervised learning approaches. Additionally, the creation of customized biofertilizer solutions based on particular crop kinds, soil types, and agroecological circumstances is made possible by ML-driven predictive analytics, which improves resource efficiency and lessens environmental damage (Puavualoaia and Necula 2023).

In the creation of biofertilizers, data analytics, a collection of statistical, computational, and visual aids, is the foundation of evidence-based decision-making. Data analytics platforms allow researchers to find patterns, make sense of data, and derive practical recommendations for improving agronomic practices, streamlining production procedures, and formulating biofertilizers. These recommendations are made possible by merging multi-dimensional datasets covering genomics, metabolomics, soil physicochemical properties, and agronomic variables (Yadav et al. 2021).

This chapter included information about how the creation of biofertilizers is entering a new age of innovation and sustainability with the integration of AI, ML, and data analytics. Researchers and practitioners may open new avenues in soil health management, nutrients, and crop resilience by combining the power of technology, biology, and data science. This will pave the way for a future that is more sustainable and food secure.

9.2 AI in Biofertilizer Development: Unleashing the Power of Computational Intelligence in Microbial Ecology and Formulation Optimization

AI has become a disruptive force across a range of industries, changing our understanding of, and ability to process, and application of complicated datasets to solve challenging issues. AI has great potential to improve resource efficiency, reduce environmental impact, and promote sustainable agriculture practices. Using computational intelligence to better understand the complex dynamics of microbial communities, optimize formulation parameters, and improve soil–plant interactions is a particularly promising use of AI in agriculture when it comes to the creation of biofertilizers (Fasusi et al. 2021).

Utilizing the mutually beneficial interactions between plants and beneficial microbes, biofertilizers provide a sustainable substitute for chemical fertilizers. Biofertilizers boost crop output, lower pollution levels, and increase agricultural sustainability by improving nutrient availability, supporting soil health, and inhibiting plant diseases. However, the effectiveness of biofertilizers depends on a sophisticated knowledge of soil dynamics, microbial ecology, and plant–microbe interactions, which poses significant difficulties for both researchers and practitioners (Ghannam and Techtmann 2021). Advancements in microbial ecology and formulation optimization have been made possible in recent years by the use of AI in biofertilizer production. The way we investigate and utilize the potential of microbial communities in agricultural systems has changed dramatically as a result of AI-powered algorithms. These algorithms are distinguished by their capacity to evaluate enormous datasets, identify trends, and forecast results. The intricacies of microbial diversity, functional dynamics, and ecological niches may be untangled by researchers through the use of ML techniques, neural networks, and deep learning architectures. This allows them to find novel microbial strains with advantageous characteristics for use as biofertilizers (Ibáñez et al. 2023).

AI also helps optimize biofertilizer formulations by anticipating microbial activity in a variety of environmental situations, simulating fermentation processes, and discovering synergistic microbial consortia for improved nutrient cycling and encouraging plant development. Employing computational simulations, optimization algorithms, and predictive analytics, AI-driven decision support systems enable researchers to create customized biofertilizer formulations that are optimized for particular soil types, crop varieties, and agroecological conditions. This maximizes efficacy and minimizes effects on the environment.

AI further permits real-time monitoring of nutrient dynamics, crop responses to biofertilizer treatments, and indicators of soil health, which facilitates precision agricultural techniques and adaptive management tactics. AI-driven monitoring systems enable real-time agronomic decision-making, minimize resource waste, and optimize nutrient management through the integration of sensor technologies, Internet of Things devices, and remote sensing platforms (Dhanaraju et al. 2022).

By enabling the power of computational intelligence to open up new vistas in microbial ecology, soil science, and precision agriculture, the incorporation of AI into the biofertilizer production process signifies a paradigm change in sustainable agriculture. Researchers and practitioners may expedite innovation, boost agricultural output, and encourage environmental stewardship with the goal of a more resilient and sustainable food system for coming generations by utilizing AI's collective potential (Araújo et al. 2023).

9.3 Harnessing Machine Learning for Innovative Solutions in Biofertilizer Research

ML has emerged as a transformative tool in various scientific disciplines, revolutionizing the way we analyze data, extract patterns, and make predictions. In the context of agriculture, ML holds tremendous potential for advancing biofertilizer research, offering innovative approaches to deciphering complex biological systems, optimizing microbial formulations, and enhancing soil fertility. As the global demand for sustainable agricultural practices continues to rise, there is a growing need to leverage ML techniques to unlock the full potential of biofertilizers in enhancing crop productivity, mitigating environmental degradation, and promoting agricultural sustainability. Biofertilizers, which harness the beneficial interactions between microorganisms and plants to enhance nutrient uptake and soil health, represent a promising alternative to chemical fertilizers (Kumar et al. 2022).

Biofertilizers provide several benefits by promoting symbiotic connections between nitrogen-fixing bacteria, phosphate-solubilizing fungi, and plant roots. These benefits include a decreased need for synthetic inputs, better soil structure, and increased crop resistance to stressors. Nonetheless, the creation and refinement of biofertilizers need a profound comprehension of microbial ecology, formulation dynamics, and ecosystem interactions, posing significant obstacles to scholars and professionals. To overcome these obstacles and realize the full potential of biofertilizers in sustainable agriculture, ML techniques provide creative solutions. ML methods allow researchers to evaluate large datasets including microbial genomes, metabolomics, soil physicochemical parameters, and agronomic factors using supervised and unsupervised learning approaches. ML enables the extraction of actionable insights, pattern recognition, and correlation analysis from heterogeneous data sources. This leads to the identification of novel microbial strains, clarification of microbial community dynamics, and optimization of biofertilizer formulations that are customized for different soil types, crop varieties, and environmental conditions. Predictive modeling of soil nutrient dynamics, crop responses to biofertilizer treatments, and microbial behavior is made possible by supervised learning techniques such as random forests, support vector machines (SVM), and neural networks. Researchers may create precise and reliable models for forecasting nutrient availability, optimizing biofertilizer formulations, and improving agronomic decision-making in actual agricultural systems by training models on historical data and cross-referencing predictions with experimental results (Guardiola-Márquez et al. 2023).

Unsupervised learning methods such as association rule mining, dimensionality reduction, and clustering allow scientists to uncover important microbial taxa, investigate intricate interactions within microbial communities, and clarify ecological niches that are essential to the effectiveness of biofertilizers. By combining ML with omics technologies, scientists may study the genetic, metabolic, and functional variety of microbial consortia. This will help create customized biofertilizer formulations that are best suited for certain soil-plant systems. Furthermore, real-time monitoring of crop performance, nutrient dynamics, and indicators of soil health is made possible by ML-driven predictive analytics, which makes precision agriculture and adaptive management techniques possible. Agronomic decision-making in dynamic agricultural contexts may be improved, nutrient management can be optimized, and resource waste can be reduced, all thanks to the integration of sensor technologies, Internet of Things (IoT) devices, and remote sensing platforms offered by ML-driven monitoring systems (Algren et al. 2021).

With the use of ML techniques, biofertilizer research can advance and new perspectives on soil dynamics, crop nutrition, and microbial ecology can be gained. Researchers and practitioners may expedite the development and deployment of sustainable biofertilizer technologies and pave the path for a more resilient, productive, and ecologically sustainable agriculture system by using the combined capabilities of ML algorithms (Salih and Nassief 2021).

9.3.1 Machine Learning Approaches in Biofertilizer Research

The study of biofertilizers heavily relies on ML techniques, especially when it comes to microbial community analysis, dynamics, and formulation optimization. Biofertilizer effectiveness is increased by employing both supervised and unsupervised learning approaches to glean valuable insights from microbiological data. We explore each strategy below:

9.3.1.1 Supervised Learning Techniques

When the desired output or target variable is known ahead of time, supervised learning techniques are a type of ML algorithms that are used to map input characteristics to the desired output using labeled training data. To forecast different outcomes on microbial populations, soil health, and crop performance, these strategies are commonly used in biofertilizer research (Bhardwaj et al. 2014). The following are some supervised learning strategies that are frequently applied in biofertilizer research:

9.3.1.1.1 Support Vector Machines

Support vector machines, or SVMs, are strong supervised learning algorithms that perform well in regression and classification applications. Finding the hyperplane in the feature space that optimally divides data points into various classes is how SVM operate. The main goal of SVM is to increase the margin that separates the nearest data points, or support vectors, from the hyperplane. SVM can handle datasets with high dimensionality and nonlinear feature interactions using this method, which also helps it generalize effectively to new data. In the field of biofertilizer research, SVM are utilized to categorize microbial taxa according to their functional functions, forecast microbial interactions, and pinpoint important microbial markers linked to soil fertility and the enhancement of plant development (Garg et al. 2023). Small-to-medium-sized datasets work particularly well for SVMs, which can handle classification problems that are both linear and nonlinear by utilizing kernel functions. In big datasets, SVMs might need a lot of processing power and be sensitive to the selection of hyperparameters. SVM, in general, are useful instruments in the study of biofertilizers because they can analyze intricate microbiological data and provide insights that improve soil health and biofertilizer formulations (Sreemoyee et al. 2021).

9.3.1.1.2 Random Forests

Within the family of ensemble learning algorithms, random forests are flexible and strong supervised learning technique. The way it works is that, to increase accuracy and resilience, it builds many decision trees throughout the training phase. To boost the model's capacity for generalization and minimize overfitting, each decision tree is constructed by randomly choosing portions of the training dataset's characteristics and data points. In biofertilizer research, random forests are frequently used to forecast soil health indicators based on the composition of the microbial community, find microbial biomarkers linked with certain soil conditions, and analyze microbial diversity. Large datasets and high-dimensional data handling are two of random forests' main benefits, which make them ideal for complicated biological and environmental datasets (Liu et al. 2012). Additionally, random forests offer insights on feature relevance and are resistant to noise and outliers in the data, which can aid researchers in identifying the critical variables influencing the dynamics of the microbial community and the health of the soil. On the other hand, random forests may be computationally demanding for training on very large datasets and may need precise parameter adjustment to maximize performance. To analyze complicated microbiological data and make predictions to improve soil fertility and crop yield in sustainable agricultural systems, random forests are a useful tool in biofertilizer research (Lee et al. 2020).

9.3.1.1.3 Neural Networks

In the field of biofertilizer research, neural networks, a class of computer models inspired by the composition and operation of the human brain, have shown to be indispensable instruments. Neural networks are essential in this setting for simulating complex interactions between microbial populations and plant hosts, anticipating microbial responses to environmental stimuli, and improving biofertilizer formulations. Neural networks are composed of linked layers of artificial neurons that receive input signals, apply activation functions, and transmit output signals to successive layers. This architecture makes it easier to extract complicated patterns from high-dimensional biological data. Neural networks such as convolutional neural networks (CNNs) and recurrent neural networks (RNNs) are especially good at processing large datasets that include metabolomic profiles, genetic sequences, and microbial community dynamics in the context of biofertilizers. CNNs are particularly good at analyzing spatial data; for example, they can recognize images in studies of microbial ecology and help identify different species of microorganisms and their functions within ecosystems (Dalgaty et al. 2021). RNNs, on the other hand, are experts at modeling temporal sequences, which makes it possible to forecast how bacteria will develop over time and evaluate the dynamic connections that exist within microbial communities. Neural networks are incredibly flexible in capturing nonlinear relationships between input features and output labels, despite their computational demands and overfitting susceptibility. This flexibility enables researchers to gain new insights into microbial ecology, optimize biofertilizer formulations, and improve soil fertility and crop productivity in sustainable agricultural systems. The integration of neural networks promises to spur innovation and propel revolutionary ideas toward attaining global food security as biofertilizer research progresses (Garg et al. 2023).

9.3.1.1.4 Linear Regression

In order to predict correlations between many parameters impacting soil health, microbial populations, and crop yield, researchers studying biofertilizers frequently use linear regression, a fundamental supervised learning approach. When it comes to biofertilizers, linear regression is a useful technique for estimating soil nutrient levels based on environmental factors, evaluating the effect of applying biofertilizer on crop yield, and measuring the correlation between microbial diversity and soil health indicators. Researchers may determine and measure the contributions of several parameters, including soil pH, moisture content, and nutrient concentrations, to total soil fertility and the makeup of the microbial community through the use of linear regression analysis (Lee et al. 2020). Additionally, using linear regression to estimate the impacts of biofertilizer treatments on important agronomic parameters enables farmers to make well-informed decisions about the choice and use of biofertilizer formulations. Despite assuming a linear relationship between the independent and dependent variables, linear regression offers findings that are easy to understand and shed light on the strength and direction of relationships between variables. It is imperative to acknowledge the constraints of linear regression, specifically its incapacity to represent intricate nonlinear correlations seen in biological systems. However, when used wisely and in concert with other statistical and ML methods, linear regression greatly advances our knowledge of soil–plant interactions, sustainable agricultural practices, and the effectiveness of biofertilizers, spurring innovation and encouraging environmental stewardship in contemporary agriculture (Dhanaraju et al. 2022).

9.3.1.2 Unsupervised Learning Techniques

In the study of biofertilizers, unsupervised learning methods are essential for revealing hidden structures, correlations, and patterns in microbial communities and soil ecosystems. Researchers may examine the intricacy of microbial diversity, pinpoint ecological niches, and evaluate the effects of biofertilizers on soil health and plant–microbe interactions by utilizing these methodologies, which include clustering algorithms, dimensionality reduction approaches, and association rule mining. Algorithms for clustering and hierarchical clustering divide data into groups or clusters according to similarity metrics. This makes it easier to identify microbial communities with comparable functional profiles and to track their distribution in various soil types and environmental settings. To effectively visualize and interpret the dynamics of microbial communities, dimensionality reduction techniques like principal component analysis (PCA) and t-distributed stochastic neighbor embedding (t-SNE) help reduce the dimensionality of high-dimensional datasets while maintaining significant relationships among data points (Salih and Nassief 2021). In contrast, association rule mining finds intriguing correlations, associations, or patterns in big datasets that aid in identifying co-occurrence patterns between different microbial taxa, microbial interactions, and microbial consortia linked to particular soil properties or plant–microbe interactions. Researchers can optimize soil management practices, develop targeted biofertilizer formulations, and advance sustainable agricultural systems by utilizing unsupervised learning techniques to gain valuable insights into the structural and functional aspects of microbial communities. To get relevant conclusions and useful insights for improving soil fertility, crop production, and environmental sustainability in agriculture, it is crucial to carefully evaluate the data and combine them with domain expertise (Sujatha and Jaidhar 2023).

9.3.1.2.1 Clustering Algorithms

Clustering algorithms are critical tools in biofertilizer development, allowing researchers to understand the complicated dynamics of microbial communities and uncover microbial consortia with potential uses in soil fertility and plant health. Clustering methods such as k-means clustering and hierarchical clustering help to categorize microbial species into groups with similar functional profiles or ecological niches by partitioning data into discrete clusters based on similarity metrics. Clustering algorithms help identify microbial consortia associated with certain soil types, crop kinds, or environmental circumstances, influencing the selection and manufacture of biofertilizer products that are customized to target habitats. Furthermore, clustering algorithms are critical for revealing microbial community structures, diversity patterns, and interactions within soil ecosystems, providing insights into the intricate links between microbial diversity, soil health indicators, and plant–microbe interactions (Amaral et al. 2023). Researchers may use clustering algorithms to identify new microbial strains, assess their potential for biofertilizer applications, and optimize biofertilizer formulations to improve soil fertility, nutrient cycling, and crop yield in sustainable agricultural systems. To draw meaningful conclusions and drive decision-making in biofertilizer research and development, clustering results must be carefully interpreted, findings validated using complementary analytical approaches, and domain knowledge integrated (Hernández-Álvarez et al. 2023).

9.3.1.2.2 Dimensionality Reduction Techniques

Dimensionality reduction techniques are critical in biofertilizer research because they overcome the constraints provided by large biological datasets and make it easier to extract relevant insights from complicated microbial community data. Principal component analysis (PCA), t-distributed stochastic neighbor embedding (t-SNE), and other dimensionality reduction technologies let researchers display, evaluate, and study microbial community dynamics. Dimensionality reduction approaches help to identify underlying patterns, structures, and interactions within microbial communities and soil ecosystems by lowering the dimensionality of multi-dimensional datasets while keeping significant linkages between data points. Dimensionality reduction techniques aid in the identification of key variables driving microbial diversity, soil health indicators, and plant–microbe interactions, guiding the selection and formulation of biofertilizer products that are optimized for specific soil types, crop varieties, and environmental conditions (Oyewole and Thopil 2023). Furthermore, dimensionality reduction techniques make it easier to integrate data from many sources, such as genomic sequences, metabolomic profiles, and environmental factors, allowing researchers to acquire a more complete knowledge of microbial ecology and ecosystem dynamics. However, to draw meaningful conclusions and drive decision-making in biofertilizer research and development, dimensionality reduction results must be carefully interpreted, findings validated using complementary analytical approaches, and domain knowledge integrated (Garma and Osório 2023).

9.3.1.2.3 Association Rule Mining

Association rule mining is a useful tool in biofertilizer research because it allows researchers to discover relevant relationships, correlations, and patterns in big and complicated datasets containing microbial populations, soil characteristics, and agronomic factors. Association rule mining provides insights into the functional roles of microorganisms, nutrient cycling processes, and soil health indicators relevant to biofertilizer development by analyzing co-occurrence patterns among microbial taxa, identifying microbial interactions, and elucidating ecological relationships within soil ecosystems. Researchers can use association rule mining to find microbial consortia linked with certain soil conditions, crop types, or agricultural techniques, leading to the development of biofertilizer products customized to target habitats (Mishra et al. 2019). Additionally, association rule mining helps identify new symbiotic connections, microbial interactions, and ecological niches that are essential for improving soil fertility, and nutrient availability, and promoting plant development. To design tailored biofertilizer formulations that are suited for sustainable agriculture, researchers may better comprehend the intricate relationships between microorganisms, soil health indicators, and plant–microbe interactions by utilizing association rule mining. To get relevant results and direct decision-making in the creation of biofertilizers, it is crucial to carefully evaluate association rule mining results, validate findings using complementary analytical approaches, and incorporate domain expertise (Pinu et al. 2019).

9.4 Predictive Modeling of Soil Health Indicators and Crop Responses to Biofertilizer

A complex process combining a variety of datasets, analytical methods, and domain expertise, predictive modeling of soil health indicators and crop responses to biofertilizer treatments helps improve resource management strategies and support agricultural decision-making. To estimate crop yields, evaluate the effectiveness of biofertilizer treatments, and monitor soil fertility, this method entails creating prediction models that make use of ML algorithms, statistical methods, and agronomic concepts. The first step in predictive modeling is the creation of extensive datasets that include microbial populations, environmental conditions, soil parameters, and agronomic techniques. Predictive models are trained on these datasets, which also help to capture the intricate relationships between crop responses, biofertilizer formulations, and soil health indicators (Kharnaior and Tamang 2022). Nutrient levels, pH, organic matter content, microbial diversity, and soil structure are important markers of soil health that affect plant development, nutrient absorption, and total crop yield.

To create prediction models for soil health indicators and crop responses to biofertilizers, supervised learning techniques including support vector machines (SVM), random forests, neural networks, and linear regression are frequently used. Researchers can estimate soil nutrient levels, evaluate soil fertility, and project crop yields under various environmental circumstances and biofertilizer treatments thanks to these algorithms, which use the previous data to predict future outcomes. To verify model predictions and evaluate their accuracy, dependability, and suitability for use in a variety of agroecosystems, predictive models can also include data from controlled settings, experimental plots, and field trials (Wilhelm et al. 2022).

Researchers may enhance biofertilizer formulations, adjust agronomic practices, and improve predictive models to maximize agricultural output and sustainability by comparing observed outcomes with model projections. Farmers and other agricultural professionals may also utilize predictive modeling to make well-informed decisions on crop rotation plans, nutrient management, and the use of biofertilizer by predicting crop responses to biofertilizer treatments (Chaudhary et al. 2022). Farmers may improve soil health, increase nutrient use efficiency, and reduce the environmental effects of chemical fertilizers by optimizing biofertilizer application rates, timing, and techniques. In agricultural management and biofertilizer research, predictive modeling is a useful technique that helps optimize soil fertility, crop production, and environmental sustainability. Predictive modeling provides stakeholders with the ability to make evidence-based decisions, adjust to changing environmental conditions, and develop resilient agricultural systems that can address the challenges of climate change and global food security by integrating diverse datasets, analytical techniques, and stakeholder engagement (Mitter et al. 2021).

9.5 Optimization of Fermentation Processes and Formulation Parameters Using AI Algorithms for Biofertilizers

A cutting-edge strategy in the creation of biofertilizers is the optimization of fermentation processes and formulation parameters using AI algorithms. This technique has substantial prospects for improving product efficacy, consistency, and scalability. Through the use of AI algorithms, such as ML and deep learning techniques, researchers can identify critical process parameters, analyze large datasets produced during fermentation, and optimize formulation compositions to maximize the production of microbial biomass, nutrient content, and bioactivity (Pisoschi et al. 2018). Complex interactions between fermentation conditions, microbial growth kinetics, and product quality parameters may be modeled by ML methods, including supervised and unsupervised learning approaches. The best fermentation conditions, including temperature, pH, oxygenation, and nutrient availability, may be found by researchers using prediction models trained on historical fermentation data. These parameters encourage the development and metabolic activity of advantageous bacteria. Furthermore, by using ML algorithms to find patterns and correlations in fermentation datasets, researchers may increase product consistency and reproducibility, troubleshoot inefficient processes, and optimize fermentation procedures. Advanced tools for assessing high-dimensional fermentation data, including genomic sequences, metabolomic profiles, and fermentation kinetics, are provided by deep learning techniques like convolutional neural networks (CNNs) and artificial neural networks (ANNs) (Harfouche et al. 2019).

To improve the performance and efficacy of products, these methods can reveal hidden patterns, forecast the results of fermentation, and improve formulation factors including microbial strains, nutrient sources, and carrier materials. Moreover, researchers may dynamically modify fermentation settings in response to shifting ambient conditions, substrate availability, and microbial activity thanks to AI algorithms, which provide real-time monitoring and adaptive control of fermentation processes. Through the integration of AI-powered process control systems with fermentation bioreactors, scientists may increase product yields, reduce production costs, and optimize fermentation kinetics all while preserving product quality standards (Subeesh and Mehta 2021). All things considered, the research and commercialization of biofertilizers might be greatly advanced by employing AI algorithms to optimize fermentation processes and formulation parameters. Through the utilization of data analytics, predictive modelling, and process optimization techniques, researchers can expedite the identification and expansion of innovative biofertilizer products, tackle sustainability issues in agriculture, and facilitate the advancement of more robust and efficient agricultural systems. The future of sustainable agriculture will be shaped by the incorporation of AI-driven techniques into biofertilizer development processes, which will be essential in unlocking new insights and spurring innovation as AI technologies continue to grow (Nath et al. 2021).

9.6 Data Analytics for Biofertilizer Optimization

To maximize the efficacy of biofertilizer formulations and promote sustainable agriculture, data analytics is a valuable tool. Data analytics approaches allow researchers to pinpoint the main parameters affecting the effectiveness of biofertilizers and create focused formulation optimization strategies by examining a variety of information including soil characteristics, microbial populations, plant responses, and environmental conditions. Data analytics enhances biological interpretation by identifying beneficial microbial consortia, predicting soil nutrient dynamics, and predicting crop responses to biofertilizer applications. It does this by combining statistical analysis, ML algorithms, and bioinformatics tools. Data analytics also helps researchers find hidden relationships, patterns, and interactions in agricultural datasets (Akila et al. 2022). These findings inform the best combinations of nutrients, crop varieties, and application rates for biofertilizers that are customized for a given soil type, agroecological conditions, and application rates. Biofertilizer producers and agricultural practitioners may boost soil fertility, improve nutrient management techniques, and support sustainable crop production systems by utilizing the insights obtained from data analytics. There is a lot of potential for improving global food security, reducing environmental effects, and promoting innovation in agricultural sustainability when data analytics technologies are integrated into biofertilizer optimization processes as they develop and become more complex (den Broeck et al. 2005).

9.6.1 Utilization of Multi-Dimensional Datasets for Uncovering Correlations Between Microbial Diversity and Soil Health

The development and effectiveness of biofertilizers, which are becoming more widely acknowledged as sustainable substitutes for conventional fertilizers in contemporary agriculture, are significantly influenced by soil health and microbial diversity. Understanding and using microbial variety is crucial to maximizing soil fertility and agricultural yield because of the complex interactions that exist between microbial communities, soil characteristics, and plant health. To get the best results in biofertilizer production, a comprehensive strategy that takes into account the intricate interactions among many elements is necessary. Thus, integrating multi-dimensional information is essential to get a thorough knowledge of microbial diversity and soil health (Antle et al. 2017). By employing multi-dimensional information, scientists may investigate the complex relationships between soil health indices and microbial diversity. These databases cover a wide range of parameters, such as the physicochemical characteristics of the soil, the makeup of the microbial community, environmental influences, and agronomic techniques. Researchers can develop more efficient biofertilizer formulations and soil management techniques by combining data from several sources to better understand the dynamic interactions between microbial populations and their surroundings.

Integrating different datasets is crucial for thorough research since it can provide a full picture of complicated occurrences. Integrated datasets allow researchers to find hidden patterns, find correlations, and acquire insights that would not be obvious when evaluating individual datasets separately since they include data from different sources and disciplines. This methodology improves the precision and dependability of studies, facilitates well-informed decision-making, and fosters creativity and exploration. When tackling complex issues such as climate change, public health emergencies, and sustainable development—where a sophisticated comprehension of interrelated components is crucial—integrated datasets are especially helpful (Zaheer and Humayoun Akhtar 2017). All things considered, combining various datasets enables a deeper investigation of phenomena, promotes multidisciplinary cooperation, and advances knowledge and practice in a variety of sectors. Several key factors are responsible for why integrating diverse datasets is crucial:

9.6.1.1 Holistic Understanding

Researchers can obtain a thorough grasp of complicated systems by combining data from several sources. Researchers can get a fuller view of how diverse components interact and affect results by integrating data on multiple facets of a phenomenon or process (Sharpley et al. 2015).

9.6.1.2 Finding Patterns and Trends

A variety of datasets offer a plethora of information that may be used to identify patterns, trends, and correlations that are not immediately obvious when looking at individual datasets. Deeper knowledge of underlying systems and events can result from analyzing data across several dimensions, which can reveal hidden links and insights (Alzoubi et al. 2023).

9.6.1.3 Increased Reliability and Accuracy

Combining several datasets can increase the predictability and accuracy of analysis. Through cross-referencing data from many sources, researchers may verify their findings, spot flaws or discrepancies, and lessen any potential biases (Venkidasamy et al. 2019).

9.6.1.4 Facilitating Well-Informed Decision-Making

In-depth analysis made possible by integrated datasets offers insightful information to aid in decision-making. Decision-makers can profit from a more comprehensive knowledge of the factors driving outcomes and the possible repercussions of certain choices, whether in policy formation, resource allocation, or company strategy (Datta et al. 2014).

9.6.1.5 Promoting Creativity and Exploration

Combined datasets have the power to generate new ideas and discoveries. Through the integration of data from several disciplines or domains, scholars can reveal new perspectives, spot developing patterns, and formulate theories for more research. Breakthroughs and developments that would not have been achievable with isolated datasets may result from this multidisciplinary approach (Bala et al. 2023).

9.6.1.6 Optimizing Resource Allocation

Funding, manpower, and infrastructure are only a few of the resources that are scarce in many sectors. By concentrating on areas where they are most likely to have an impact or produce significant findings, researchers may more efficiently prioritize their efforts and allocate resources when varied datasets are integrated.

9.6.1.7 Handling Complex Challenges

A lot of the most important issues of the day, such as food security, public health emergencies, and climate change, have many facets and connections. Researchers may approach these problems from a variety of perspectives by integrating varied datasets and taking into account the intricate relationships between social, environmental, and economic issues (Hurley et al. 2013).

9.6.1.8 Encouraging Transparency and Accountability

In research and decision-making procedures, transparency and accountability are fundamental concepts. Transparency, cooperation, and stakeholder trust may all be promoted by academics by integrating varied datasets and providing stakeholders with easy access to findings.

Integrating various datasets for thorough analysis is crucial for increasing understanding, spurring creativity, and tackling the intricate problems that society faces. Researchers can get a more detailed knowledge of events, make better judgments, and ultimately contribute to favorable social, economic, and environmental outcomes by merging data from several sources and disciplines.

9.7 Spatial Data Analysis Techniques

Understanding soil health, microbial diversity, and agricultural landscapes is made possible by the application of spatial data analysis techniques, especially Geographic Information Systems (GIS) and remote sensing, in the field of biofertilizers. By using these tools, researchers and practitioners may monitor changes over time, assess geographical trends, and make well-informed decisions about soil management, crop yield, and the use of biofertilizers. A strong software tool for capturing, storing, analyzing, and visualizing geographical data is Geographic Information Systems (GIS). Researchers using GIS may combine many information, including soil characteristics, crop distribution, land cover, and climatic factors, into a single spatial framework while studying biofertilizers (Babbar et al. 2022). They may evaluate the geographical distribution of soil nutrients, pinpoint regions that would be good candidates for biofertilizer application, and improve fertilizer management techniques to increase crop production and quality by overlaying and evaluating these spatial layers. On the other hand, remote sensing entails gathering data about the surface of the Earth from a distance, usually through the use of satellite or aerial images. Remote sensing data can offer important insights into vegetation health, soil moisture levels, and land cover dynamics in biofertilizer studies. Through the analysis of remote sensing photos, scientists may spot shifts in crop growth patterns, pinpoint areas under stress or nutrient deficit, and track the efficacy of biofertilizer treatments over time. There are several benefits to integrating GIS and remote sensing methods in biofertilizer research (Nungula et al. 2023).

The fields of biofertilizer research and innovation should benefit greatly from the application of GIS and remote sensing, two potent spatial data analysis approaches. Researchers can enhance their understanding of soil–plant interactions, improve fertilizer management strategies, and help develop sustainable agriculture solutions that support crop productivity, soil health, and environmental stewardship by utilizing the capabilities of these tools. In the next years, GIS and remote sensing will continue to be crucial parts of agricultural management plans and biofertilizer research as technology advances (Adejumo and Adebiyi 2020).

9.8 Integration of Multi-Dimensional Datasets in Biofertilizer Research

Understanding the intricate relationships between soil, bacteria, plants, and fertilizers requires integrating multi-dimensional information into biofertilizer research. To have a thorough knowledge of soil health, microbial ecology, and agricultural sustainability, this integration entails merging heterogeneous datasets from numerous sources and disciplines. By combining multi-dimensional information, scientists may learn more about the dynamics of ecosystems, nutrient cycling, and interactions between microbes, plants, and soil. With the help of this comprehensive understanding of agricultural systems, tailored biofertilizer formulations, precision farming techniques, and sustainable soil management methods have been developed.

In many domains, including biofertilizer research, methods for data preparation, standardization, and quality control are crucial parts of the data analysis pipeline. By ensuring that datasets are clear, consistent, and trustworthy for further analysis, these tactics seek to improve the validity and interpretability of study findings. Preprocessing data entails several procedures, such as variable encoding, transformation, and cleaning. Finding and addressing outliers, inconsistent data, and missing values from the dataset are all part of cleaning (Zhao et al. 2024). This might entail eliminating outliers that drastically diverge from the rest of the sample or using imputation techniques to restore missing data. To stabilize variance and adhere to statistical analysis presumptions, transformation techniques like logarithmic or square root transformations might be used. Quantitative studies can incorporate categorical variables by encoding them into numerical representations using techniques like label encoding or one-hot encoding (Wang et al. 2016).

To standardize data to a similar scale, normalization is yet another crucial step in the preparation of data. Z-score normalization and min-max scaling are two popular normalizing methods. For datasets with roughly normal distributions, Z-score normalization rescales the data to have a mean of 0 and a standard deviation of 1. When used in datasets with non-normal distributions or known ranges of values, min–max scaling rescales the data to a specified range, usually between 0 and 1. The validity and dependability of the dataset are ensured by quality control measures. Data points that significantly differ from the rest of the dataset are identified using outlier detection techniques, which may also hint at measurement or data-collecting mistakes. Error checking entails confirming the dataset's correctness by performing internal consistency tests or cross-referencing with other sources. To guarantee the dataset's dependability for analysis, validation processes compare the outcomes with data that has been gathered independently or with pre-established standards. The act of recording metadata, encompassing data sources, collecting techniques, and preparation stages, improves data analysis accountability, repeatability, and transparency (Bargaz et al. 2018).

Researchers may make sure that datasets are ready for further analysis by putting these techniques for data preparation, normalization, and quality control into practice. This will result in more solid and trustworthy research findings in biofertilizer research and other domains.

9.9 Real-Time Monitoring of Nutrient Dynamics, Environmental Factors, and Crop Performance

An essential component of contemporary agriculture and biofertilizer research is the real-time monitoring of crop performance, environmental variables, and nutrient dynamics. Through the ongoing gathering, evaluation, and interpretation of data on crop health, weather patterns, nutrient levels, and soil conditions, farmers and researchers are better equipped to make timely decisions that maximize agricultural operations. The availability, absorption, and cycling of vital nutrients within the soil–plant system are referred to as nutrient dynamics. With real-time monitoring, nutrient levels, including those of nitrogen, phosphorus, potassium, and micronutrients, can be continuously assessed, assisting in the prevention of excesses or deficiencies that may hurt crop development and output. Farmers may ensure optimal nutrient availability for plant uptake by adjusting irrigation timings and fertilizer inputs by tracking nutrient dynamics in real time. Environmental elements that affect crop growth and development include temperature, humidity, rainfall, and sun radiation. Farmers are better equipped to predict and manage possible hazards like drought, frost, or insect outbreaks when environmental conditions are monitored in real time. This is because real-time monitoring offers insightful information on weather patterns and climatic trends. Farmers can reduce the negative effects of unfavorable weather conditions on crop yields by implementing adaptive management systems that combine environmental data with crop performance parameters (Kwak and Kim 2017). Monitoring crop performance entails keeping tabs on important metrics related to plant growth, health, and production during the growing season. Monitoring plant growth phases, the leaf area index, the amount of chlorophyll, and yield measures including grain quality and biomass buildup are all part of this process. To maximize agricultural production, resource efficiency, and sustainability, real-time monitoring of nutrient dynamics, environmental conditions, and crop performance is crucial. Through the utilization of developments in sensor technology, data analytics, and decision support systems, farmers and researchers may obtain timely and useful information to optimize crop management techniques, reduce input expenses, and augment overall farm profitability (Nungula et al. 2023).

9.10 Challenges and Limitations

In the context of biofertilizer research and agriculture, integrating strong bioinformatics infrastructure and data integration platforms is fraught with difficulties and constraints. First off, the complexity and sheer amount of data produced by contemporary agricultural systems need the development of such infrastructure. This contains information on a variety of topics, such as crop performance, environmental factors, microbial populations, and soil characteristics. Investing much in technology, software, and people resources is necessary to build and maintain a strong bioinformatics infrastructure. Furthermore, technical issues with data formats, interoperability, and the scalability of computing resources arise when integrating disparate information. Furthermore, a major problem is validating prediction models across a range of agricultural circumstances.

The temperature, kinds of soil, crop species, and management techniques of agricultural systems vary greatly, which makes it difficult to create models that effectively generalize across many settings. To guarantee the dependability and resilience of forecast models, validation research has to cover an extensive spectrum of climatic circumstances and cropping schemes. In addition, ethical issues about intellectual property rights, data privacy, and fair access to technology are crucial. To foster innovation and protect the interests of stakeholders, matters about data security, ownership, and intellectual property rights concerning algorithms and software tools must be properly handled.

For agricultural research and development to be collaborative, knowledge-sharing, and capacity-building, equal access to technology and data is crucial. Researchers, legislators, industry stakeholders, and the larger agricultural community must work together to address these obstacles and limitations. Standardized procedures, best practices, and legal frameworks that support openness, justice, and moral behavior in biofertilizer research and agriculture must also be developed.

References

Adejumo IO, Adebiyi OA (2020) Agricultural solid wastes: causes, effects, and effective management. In: Strategies of sustainable solid waste management, vol 8. IntechOpen

Akila A, Parameswari R, Jayakumari C (2022) Big data in healthcare: management, analysis, and future prospects. In: Handbook of intelligent healthcare analytics: knowledge engineering with big data analytics. Wiley, pp 309–326

Algren M, Fisher W, Landis AE (2021) Machine learning in life cycle assessment. In: Data science applied to sustainability analysis. Elsevier, pp 167–190

Alzoubi S, Jawarneh M, Bsoul Q, Keshta I, Soni M, Khan MA (2023) An advanced approach for fig leaf disease detection and classification: leveraging image processing and enhanced support vector machine methodology. Open Life Sci 18(1):20220764

Amaral L d O, Miranda GV, Souza J d S, Moitinho ACR, Cristeli DS, Silva HK d, Anjos RSR d, Alliprandini LF, Unêda-Trevisoli SH (2023) Application of artificial neural networks to predict genotypic values of soybean derived from wide and restricted crosses for relative maturity groups. Agronomy 13(10):2476

Antle JM, Basso B, Conant RT, Godfray HCJ, Jones JW, Herrero M, Howitt RE, Keating BA, Munoz-Carpena R, Rosenzweig C et al (2017) Towards a new generation of agricultural system data, models and knowledge products: design and improvement. Agric Syst 155:255–268

Araújo SO, Peres RS, Ramalho JC, Lidon F, Barata J (2023) Machine learning applications in agriculture: current trends, challenges, and future perspectives. Agronomy 13(12):2976

Babbar D, Chauhan SK, Sharma D, Upadhyay K, Dwivedi MD, Sahana M, Kumar S (2022) Spatial analysis of soil quality using geospatial techniques in Botanic Garden of Indian Republic, Noida, Uttar Pradesh, India. Environ Sustain 5(4):471–492

Bala S, Garg D, Sridhar K, Inbaraj BS, Singh R, Kamma S, Tripathi M, Sharma M (2023) Transformation of agro-waste into value-added bioproducts and bioactive compounds: micro/nano formulations and application in the agri-food-pharma sector. Bioengineering 10(2):152

Bargaz A, Lyamlouli K, Chtouki M, Zeroual Y, Dhiba D (2018) Soil microbial resources for improving fertilizers efficiency in an integrated plant nutrient management system. Front Microbiol 9:1606

Bhardwaj D, Ansari MW, Sahoo RK, Tuteja N (2014) Biofertilizers function as key player in sustainable agriculture by improving soil fertility, plant tolerance and crop productivity. Microb Cell Factories 13:1–10

Chaudhary P, Singh S, Chaudhary A, Sharma A, Kumar G (2022) Overview of biofertilizers in crop production and stress management for sustainable agriculture. Front Plant Sci 13:930340

Dalgaty T, Miller JP, Vianello E, Casas J (2021) Bio-inspired architectures substantially reduce the memory requirements of neural network models. Front Neurosci 15:612359

Datta N, Pal M, Roy U, Mitra R, Pradhan A (2014) World Journal of Pharmaceutical Research. Infection 13(11):15. https://doi.org/10.20959/wjpr201911-15964

den Broeck J, Argeseanu Cunningham S, Eeckels R, Herbst K (2005) Data cleaning: detecting, diagnosing, and editing data abnormalities. PLoS Med 2(10):e267

Dhanaraju M, Chenniappan P, Ramalingam K, Pazhanivelan S, Kaliaperumal R (2022) Smart farming: Internet of Things (IoT)-based sustainable agriculture. Agriculture 12(10):1745

Fasusi OA, Cruz C, Babalola OO (2021) Agricultural sustainability: microbial biofertilizers in rhizosphere management. Agriculture 11(2):163

Garg D, Sridhar K, Stephen Inbaraj B, Chawla P, Tripathi M, Sharma M (2023) Nano-biofertilizer formulations for agriculture: A systematic review on recent advances and prospective applications. Bioengineering 10(9):1010

Garma LD, Osório NS (2023) Demystifying dimensionality reduction techniques in the 'omics' era: a practical approach for biological science students. Biochem Mol Biol Educ 52:165

Ghannam RB, Techtmann SM (2021) Machine learning applications in microbial ecology, human microbiome studies, and environmental monitoring. Comput Struct Biotechnol J 19:1092–1107

Guardiola-Márquez CE, Santos-Ramirez MT, Figueroa-Montes ML, Valencia-de Los Cobos EO, Stamatis-Félix IJ, Navarro-López DE, Jacobo-Velázquez DA (2023) Identification and characterization of beneficial soil microbial strains for the formulation of biofertilizers based on native plant growth-promoting microorganisms isolated from northern Mexico. Plan Theory 12(18):3262

Harfouche AL, Jacobson DA, Kainer D, Romero JC, Harfouche AH, Mugnozza GS, Moshelion M, Tuskan GA, Keurentjes JJB, Altman A (2019) Accelerating climate resilient plant breeding by applying next-generation artificial intelligence. Trends Biotechnol 37(11):1217–1235

Hernández-Álvarez C, Peimbert M, Rodriguez-Martin P, Trejo-Aguilar D, Alcaraz LD (2023) A study of microbial diversity in a biofertilizer consortium. PLoS One 18(8):e0286285

Hurley BRA, Ouzts A, Fischer J, Gomes T (2013) Paper presented at IAPRI world conference 2012: effects of private and public label packaging on consumer purchase patterns. Packag Technol Sci 29:399–412. https://doi.org/10.1002/pts

Ibáñez A, Garrido-Chamorro S, Vasco-Cárdenas MF, Barreiro C (2023) From lab to field: biofertilizers in the 21st century. Horticulturae 9(12):1306

Jiang Y, Luo J, Huang D, Liu Y, Li D (2022) Machine learning advances in microbiology: a review of methods and applications. Front Microbiol 13:925454

Kharnaior P, Tamang JP (2022) Metagenomic-metabolomic mining of kinema, a naturally fermented soybean food of the Eastern Himalayas. Front Microbiol 13:1–20. https://doi.org/10.3389/fmicb.2022.868383

Kumar S, Sindhu SS, Kumar R et al (2022) Biofertilizers: an ecofriendly technology for nutrient recycling and environmental sustainability. Curr Res Micro Sci 3:100094

Kwak SK, Kim JH (2017) Statistical data preparation: management of missing values and outliers. Korean J Anesthesiol 70(4):407–411

Lee H, Wang J, Leblon B (2020) Using linear regression, random forests, and support vector machine with unmanned aerial vehicle multispectral images to predict canopy nitrogen weight in corn. Remote Sens 12(13):2071

Liu Y, Wang Y, Zhang J (2012) New machine learning algorithm: random forest. In: Information Computing and Applications: Third International Conference, ICICA 2012, Chengde, China, September 14–16, 2012. Proceedings. Springer, Berlin, pp 246–252

Miller T, Mikiciuk G, Kisiel A, Mikiciuk M, Paliwoda D, Sas-Paszt L, Cembrowska-Lech D, Krzemińska A, Kozioł A, Brysiewicz A (2023) Machine learning approaches for forecasting the best microbial strains to alleviate drought impact in agriculture. Agriculture 13(8):1622

Mishra BK, Hati S, Das S (2019) Bio-nutritional aspects of Tungrymbai, an ethnic functional fermented soy food of Khasi Hills, Meghalaya, India. Clin Nutr Exp 26:8–22. https://doi.org/10.1016/j.yclnex.2019.05.004

Mitter EK, Tosi M, Obregón D, Dunfield KE, Germida JJ (2021) Rethinking crop nutrition in times of modern microbiology: innovative biofertilizer technologies. Front Sustain Food Sys 5:606815

Nath S, Roy M, Sikidar J, Deb B, Sharma I, Guha A (2021) Characterization and in-vitro screening of probiotic potential of novel Weissella confusa strain GCC_19R1 isolated from fermented sour rice. Curr Res Biotechnol 3:99–108. https://doi.org/10.1016/j.crbiot.2021.04.001

Nungula EZ, Mugwe J, Nasar J, Massawe BHJ, Karuma AN, Maitra S, Seleiman MF, Dindaroglu T, Khan N, Gitari HI (2023) Land degradation unmasked as the key constraint in sunflower (*Helianthus annus*) production: Role of GIS in Revitalizing this vital sector. Cogent Food Agric 9(2):2267863

Oyewole GJ, Thopil GA (2023) Data clustering: application and trends. Artif Intell Rev 56(7):6439–6475

Pinu FR, Beale DJ, Paten AM, Kouremenos K, Swarup S, Schirra HJ, Wishart D (2019) Systems biology and multi-omics integration: viewpoints from the metabolomics research community. Metabolites 9(4):76

Pisoschi AM, Pop A, Cimpeanu C, Turcuş V, Predoi G, Iordache F (2018) Nanoencapsulation techniques for compounds and products with antioxidant and antimicrobial activity—a critical view. Eur J Med Chem 157:1326–1345. https://doi.org/10.1016/j.ejmech.2018.08.076

Puavualoaia V-D, Necula S-C (2023) Artificial intelligence as a disruptive technology—a systematic literature review. Electronics 12(5):1102

Rani P, Rajput S, Thakur B, Kaur S (2023) Immobilization and co-mobilization: an unexploited biotechnological tool for enhancing efficiency of biofertilizers. In: Metabolomics, proteomes and gene editing approaches in biofertilizer industry. Springer Nature Singapore, Singapore, pp 219–236

Salih SR, Nassief AM (2021) Applying negative association rule utilizing unsupervised machine learning models. In: Proceedings of First International Conference on Mathematical Modeling and Computational Science: ICMMCS, vol 2020. Springer, pp 73–81

Sharpley AN, Bergström L, Aronsson H, Bechmann M, Bolster CH, Börling K, Djodjic F, Jarvie HP, Schoumans OF, Stamm C et al (2015) Future agriculture with minimized phosphorus losses to waters: research needs and direction. Ambio 44:163–179

Sreemoyee B, Nilay K, Pragati A, Priyank J (2021) Machine learning concepts for correlated Big Data privacy. J Big Data 8(1):157

Subeesh A, Mehta CR (2021) Automation and digitization of agriculture using artificial intelligence and internet of things. Artif Intell Agric 5:278–291

Sujatha M, Jaidhar CD (2023) Machine learning-based approaches to enhance the soil fertility—a review. Expert Syst Appl 240:122557

Venkidasamy B, Selvaraj D, Nile AS, Ramalingam S, Kai G, Nile SH (2019) Indian pulses: a review on nutritional, functional and biochemical properties with future perspectives. Trends Food Sci Technol 88:228–242. https://doi.org/10.1016/j.tifs.2019.03.012

Wang YH, Dong J, Zhang JX, Zhai J, Ge B (2016) Effects of mimic of manganese superoxide dismutase on 2,4,6-trinitrobenzene sulfonic acid-induced colitis in rats. Arch Pharm Res 39(9):1296–1306. https://doi.org/10.1007/s12272-016-0811-z

Wilhelm RC, van Es HM, Buckley DH (2022) Predicting measures of soil health using the microbiome and supervised machine learning. Soil Biol Biochem 164:108472

Yadav SK, Patel JS, Singh BN, Bajpai R, Teli B, Rajawat MVS, Sarma BK (2021) Biofertilizers as microbial consortium for sustainability in agriculture. In: Plant, soil and microbes in tropical ecosystems. Springer, pp 349–368

Zaheer K, Humayoun Akhtar M (2017) An updated review of dietary isoflavones: nutrition, processing, bioavailability and impacts on human health. Crit Rev Food Sci Nutr 57(6):1280–1293. https://doi.org/10.1080/10408398.2014.989958

Zhao G, Zhu X, Zheng G, Meng G, Dong Z, Baek JH, Jeon CO, Yao Y, Xuan YH, Zhang J et al (2024) Development of biofertilizers for sustainable agriculture over four decades (1980–2022). Geog Sustain 5(1):19–28

Algal Nanobiofertilizers: Prospects and Challenges

10

K. Sowmiya, K. Praveen, S. Krishna Kumar, and M. Priya

Abstract

Sustainable agriculture is a global priority, as it aims to ensure food production while protecting our planet's resources for future generations emphasizing the alternative for chemical fertilizers. Chemicals are replaced with the use of biofertilizers including bacteria, fungi, and algae for plant growth promotion that certainly enriches the soil. Besides, nanofertilizers turned out to be an alternate version of chemical fertilizers that overcome the residual effect of toxic chemicals, where the coated nanoparticles aid in targeted and timely release of nutrients to the soil. Together the components, nanoparticles and biofertilizer yielded the promising approach of nanobiofertilizers as ecofriendly alternative in agricultural sector by overcoming the demerits of conventional fertilization. As ubiquitous inhabitant, algae already journeying as biofertilizer and biostimulant that imparts plant health positively as well as helps in maintaining soil health with their exudates. Apart from being a soil and crop booster, algal cells itself act as biological factories that manufacture nanoparticles that imposes algae as an

K. Sowmiya (✉)
Department of Agricultural Microbiology, University of Agricultural Sciences, Bengaluru, India

Agronomy section, Central Sericultural Research and Training Institute, Mysore, India

K. Praveen
Department of Agricultural Microbiology, University of Agricultural Sciences, Bengaluru, India

S. Krishna Kumar
Department of Agricultural Entomology, University of Agricultural Sciences, Bengaluru, India

M. Priya
Division of Microbiology, Indian Agricultural Research Institute, New Delhi, India

S. Kaur et al. (eds.), *Metabolomics, Proteomics and Gene Editing Approaches in Biofertilizer Industry*, https://doi.org/10.1007/978-981-97-2910-4_10

eligible candidate for nanobiofertilizer (NBF) synthesis. This chapter summarizes the application of algal-based nanofertilizers in the different streams of agriculture like crop production and protection. The prospects of algal NBFs and the challenges involved in the synthesis and use are also mentioned.

Keywords

Nanobiofertilizer · Algae · Plant health · Nanoparticles · Fertilizers

10.1 Introduction

Globally around 800 million people suffer from appalling starvation, and by 2050, the world's population is expected to exceed 9.7 billion (FAO 2018). Food instability that is quickly spreading might cause conflict, environmental change, and economic uncertainty. The biggest challenges facing agricultural scientists today are declining farm productivity, nutritional scarcity, and climate change. In order to fulfill the need for 2–3 billion tonnes of grains, or a 50–70% increase in yearly output, it is necessary to expand agriculture sustainably (Ayub et al. 2022). The extensive utilization of fertilized agricultural soils, tillage methods, fossil fuels, and livestock dung are the main sources of greenhouse gas emissions, significantly impacting agriculture. The degradation of soil fertility, food security, is also a result of intensive plant cultivation, excessive fertilizer and pesticide treatments, and other factors. Crop yields significantly decrease due to climate instability and climate change-related variables, including rising temperatures, increased atmospheric CO_2 levels, and altered rainfall patterns (Verma et al. 2022). In addition, an unstable and poor-quality diet contributes to several health issues. Furthermore, unexpected seasonal and geographic changes disturb plant life cycles; climate change affects the nutritional content of crops. Because of this enormous problem, it will be necessary to work together to protect natural resources to support increased agriculture and reduce its detrimental environmental effects (Li et al. 2017).

In addition, abiotic and biotic stresses are major constraints that adversely affect crops' growth, development, and yield. There are different types of abiotic stressors such as drought, salinity, temperature stress, excessive water, heavy metal stress, UV stress, etc. (Shinozaki and Yamaguchi-Shinozaki 2007). Synthetic chemical fertilizers have been a crucial component in efforts over the past 60 years to increase agricultural production and meet the world's substantial food demand (Kumawat et al. 2022). According to scientific and agricultural sources, extensive use of agrochemicals over an extended period can have disastrous effects on the environment. This has emerged as the main cause of soil and water pollution, soil erosion, nutritional imbalance, depressed soil fertility, poor water retention, and disruption of intrinsic soil biodiversity. These agricultural chemicals enter the earth's hydrosphere and lead to eutrophication (Sharma et al. 2023).

Most robust and sustainable ways to boost agricultural productivity has been suggested, and it involves using plant growth-promoting rhizobacteria (PGPR) as biofertilizers. Through different direct and indirect ways, they provide plant nutrients, hence fostering plant development. N_2 fixation, siderophore, phytohormone, exopolysaccharide, and nutrient solubilization are examples of direct mechanisms. Their rhizosphere inoculation makes the plants stress-tolerant or stress-resilient (Riaz et al. 2021). The efficient use of these natural resources without negative repercussions to the environment has encouraged the incursion of new technologies such as nanotechnology, precision farming, and smart nano-interventions, and the revolution in farming's circular bio-economies is what makes this feasible for sustainable and climate-smart agriculture (Mahapatra et al. 2022).

10.2 Nanofertilizer Intervention: Opportunity for Modern-Day Agriculture

Nanotechnology has emerged as the sixth most revolutionary technology after the green revolution of the 1960s and the biotechnology revolution of the 1990s. Modern agriculture is intended to infuse nanotechnology, biotechnology, and other disciplines of science into agricultural sciences to transform traditional farming practices into precision agriculture ensures food security for the country's growing population in a cost-effective and eco-sustainable way (El-Ghamry et al. 2018). As better alternatives, nano- and biofertilizers have gained momentum and become increasingly popular in the agricultural sector. Nanobiofertilizer is a step ahead in the innovative field of agriculture, toward the joint formulation of biofertilizer and nanoparticles into nanobiofertilizer, a more effective alternative for eco-sustainable agriculture (Duhan et al. 2017). "Nanobiofertilizer" is created by encasing biofertilizers that have been reduced to nanosize (1–100 nm) under a coating made of a suitable nanomaterial (Ma et al. 2010). They control how nutrients are delivered to the desired location and reduce the effects of environmental stressors. Their key characteristics include a diminished reliance on chemical fertilizers, increased nutrient usage efficacy, enhanced nutrient availability and absorption, mass output in a shorter amount of time, environmental sustainability, economic effectiveness, and renewable fertilization (Thirugnanasambandan 2019). NPs can affect plant–microbe interactions in direct and indirect ways. Direct effects include nutrient availability in the rhizosphere, while indirect effects include microbial stimulation (Abdelghany et al. 2022).

Applications of nanotechnology in agriculture have focused on the research for the activation of the secondary metabolism in plants, a mechanism of adaptation and evolution as a defense to harsh environmental conditions (biotic and abiotic) (Aguirre-Becerra et al. 2022). The influence of NPs on plant protection against abiotic stressors has been documented in several previous studies. Several applications of NPs and NMs have been proposed and studied to achieve this goal. For example, producing nanocomposites containing fertilizers, pesticides, and microorganisms is the most common application (Usman et al. 2020) and nanoencapsulation of macronutrients, micronutrients, microorganisms, and organic compounds (Sambangi et al. 2022). However, because of our incomplete understanding of the interactions between biofertilizers, nanoparticles, and plants, microbe-based nanobiofertilizers have not yet received widespread endorsement or commercialization. On either physically or chemically synthesized nanofertilizers, recent research has concentrated. This will enable researchers to study and create more practical nanobiofertilizers, enabling their commercialization and distribution to farmers.

10.3 A New Member of the Successive Plant Microbiome: Algae

Algae are a class of prehistoric photosynthetic creatures that range from prokaryotic cyanobacteria to eukaryotic microalgae. Around 800,000 distinct species of algae are found in nature, and only 5000 have been discovered so far (Lee and Ryu 2021). Furthermore, only a small portion of these algal species have been selected in order to study their potential utility in plant development under certain growth conditions. Most often, size is used to distinguish between micro- and macroalgae. Rhodophyta, brown algae, and green algae (Chlorophyta) are the three classes of big, photosynthetic plants known as macroalgae that are inhabited by water. *Arthrospira, Chlorella,* and *Nostoc* are examples of common microalgal taxa (Elster 2002). A significant component of the interactions between microorganisms and plants in the natural environment is cyanobacteria, a kind of prokaryotic microalga. In order to guard against biotic and abiotic challenges, the algal community builds a biological soil crust with the help of other microorganisms (Krug et al. 2020).

Most microbiologists used to think that algae only lived in freshwater and saltwater, which were thought to be their natural environments. They did not think algae could thrive in soil or on plant surfaces. However, more than 30 years ago, scientists studied how algae propagated through soil (Davey and Clarke 1991). In prior research, microscopic examination and chlorophyll content were used to identify the cryptogrammic flora in Antarctic soil. According to Davey and Clarke (1991), the Oscillatoriaceae family controlled the soil up to 7–9 cm. With the aid of other bacteria, the algae build a biological soil crust in warm deserts, polar regions, and semi-arid conditions to defend itself from biotic and abiotic pressures (Krug et al. 2020). Previous studies (Liu and Chen 2016; Treves et al. 2016; Zhu et al. 2018; Gantar and Elhai 1999; Gantar 2000) documented the presence of chlorella in the soil and on the surface of plant leaves and, *Nostoc* and *Anabaena* spp., on the surface of plant roots. Furthermore, Leach et al. (2017) reported microalgae in plant tissues and soil. In a recent study, the microalgae were analyzed using metagenome sequencing and the findings indicated that microalgae, which include eukaryotic and prokaryotic algae species, must be classified as members of the microbiome (Xu et al. 2018).

10.4 Algae as a Biofertilizer

Soil microflora has been found to promote biomass productivity and improve soil fertility. However, algal microflora has also been found as a true bio-based fertilizer for agricultural techniques that are both environmentally favorable and

Table 10.1 List of algal biofertilizers and their impact on plant health

Group	Algal Species	Effects
Cyanobacteria	*Scytonema hofmanni*	Salinity tolerance
	Anabaena cylindrica	Increases the common bean's nodulation, plant growth, and yield.
	Consortia of Nostoc ellipsosporum and *N. Punctiforme*	Cyanobacteria thrive in salt-affected, semi-arid soils that can be bio-improved
	Oscillatoria sp., *Microcoleus vaginatus, Nostoc commune*	Cyanobacteria release extracellular polymeric compounds that bind soil particles, enhancing surface stability, decreasing clay dispersion, and the establishment of biological soil crusts and soil loss prevention.
Eukaryotic microalgae	*Chlorella ellipsoida*	Increased antioxidant activity and resistance to salt stress
	Chlorella vulgaris	Improves water usage efficiency, stomatal closure effectiveness, ROS generation, and drought tolerance
	Dunaliella salina	Salt stress tolerance
	Chlorella fusca	Delayed plant senescence
	Chlorella sp.	Shortens the time it takes for seeds to germinate, boosts the development of roots and leaves, and speeds up the photosynthesis of maize.

pollution-free. Although photosynthetic, these algae are adapted to extremes of environments and have been observed to thrive under extreme light settings with limited nutrients such as C and N levels and a low water need. They are extremely independent of alerts in terrestrial N pools that generate primary necessary amino acids, growth promoters, and hormones that are required for good crop growth in cropping systems because they are excellent N-fixing agents. Algal biofertilizers and their impact on plant growth with special characteristics are enlisted in Table 10.1.

10.4.1 Algal Nano Biofertilizers

The agriculture industry is struggling with a fundamental problem: reduced soil fertility due to the excessive use of chemical fertilizers. NPs can be employed to create safe, efficient, and environmentally friendly nanofertilizers (Subramaniyam et al. 2015). Poor surface topography may harm the chemical and physical characteristics of nanoparticles. Wet-chemical pyrolysis methods are particularly used for the synthesis of nanoparticles, with the NPs generated in liquid media with numerous stabilizing, capping, reducing, and reactant agents. Chemical reduction, with organic or inorganic reducing chemicals either colloidally or via the sol-gel method, is commonly used to produce NPs (Salem and Fouda 2021). Although these approaches produce a higher volume of NPs at a lower cost, they have some drawbacks (Sharma et al. 2018). In the physical synthesis of NPs, ultrasonication, spray pyrolysis, laser radiation, electron beam, ion implantation, and vapor phase are all utilized. These physical methods have disadvantages that make them unsuitable for nanoparticle syntheses, such as being costly, having a poor production rate, and requiring a large quantity of energy to sustain the high temperatures and pressures needed (Salem and Fouda 2021; Sharma et al. 2018).

Algal nanobiofertilizers utilize extracts derived from algae, which contain a range of essential nutrients, growth-promoting substances, and bioactive compounds that contribute to plant growth and development. Algal extracts used in nanobiofertilizers contain bioactive compounds that have beneficial effects on plant growth. These compounds can include plant hormones, enzymes, amino acids, vitamins, and antioxidants. They contribute to various physiological processes in plants, such as root development, nutrient absorption, photosynthesis, and stress tolerance. Nanoparticles can protect the nutrients from leaching or volatilization, increasing their availability for plant uptake. This results in improved nutrient utilization and reduced nutrient losses, leading to enhanced plant growth and productivity. Algal nanobiofertilizers offer a sustainable and environmentally friendly approach to agriculture. They reduce reliance on synthetic chemical fertilizers, which can have negative impacts on ecosystems and human health. Algal extracts are biodegradable and pose minimal risks to the environment. Additionally, the targeted nutrient delivery and enhanced nutrient uptake reduce nutrient runoff and minimize environmental pollution (Chojnacka et al. 2020; Singh et al. 2019).

10.4.2 Algae as a Factory for Nanoparticle Synthesis

Numerous techniques, including physical, chemical, and biological can be used to synthesize nanoparticles. Greener production of nanoparticles is more environmentally friendly than other approaches (Kim et al. 2007). Biological agents, including bacteria, viruses, yeast, fungi, plants, and algae, are involved in synthesis of nanoparticles. (Gilaki 2010; Schneidewind et al. 2012). Among them, algae have been considered a potential candidate for nanoparticle synthesis due to their fast growth, high metal accumulation capacity, ability to produce diverse secondary metabolites, high CO_2 fixation efficiency (Behera et al. 2023). They can withstand a wide range of climatic conditions and are extensively dispersed over a variety of habitats (Beetul et al. 2016). The branch phyconanotechnology deals with the study of algae mediated biosynthesis of nanometals (Sharma et al. 2016). Many different types of algae, including macro- and microalgae, have been utilized for nanoparticle synthesis. Microalgae that are utilized for nanoparticles production include *Spirulina platensis*, *Lyngbya majuscule*, and *Chlorella vulgaris.* Diatoms such as *Navicula atomus* and *Diadesmis gallica* synthesize gold and silica gold nanocomposites (LewisOscar et al. 2016; Schröfel et al. 2011). Macroalgae including green algae (Chlorophyta), brown algae (Phaeophyceae), and red algae (Rhodophyta) which are found in brackish or freshwater resources synthesize nanoparticles (Kokabi and Yousefzadi 2015). There are several reports related to marine algae-based nanoparticle synthesis. Marine algae such as *Turbinaria conoides* (Sheeba and Thambidurai 2009), *Gelidiella acerosa* (Vivek et al. 2011), and *Ulva fasciata* (Rajesh et al. 2012) have the ability to produce silver nanoparticles. *Sargassum muticum* synthesize zinc oxide (Azizi et al. 2014) and silver nanoparticles (Govindaraju et al. 2015).

10.4.2.1 Advantages of Algae Mediated Nanoparticles over Chemical Synthesis of Nanoparticles

Algal nanoparticles offer some key advantages over chemical nanoparticles. Algae are photosynthetic organisms that can be grown year-round using simple and cost-effective methods, such as open pond systems or photobioreactors as well as the synthesis of a significant quantity of nanoparticles by algae is economically feasible as compared to chemical synthesized nanoparticles which require high energy intensive process (Khan et al. 2022; Michael et al. 2022; Chan et al. 2022). Algal nanoparticles can be produced by greener method by using various biomolecules (pigments, polysaccharides, proteins) in the algal extract whereas the chemical synthesis involves the usage of hazardous solvents in the process which pose an adverse impact to the environment (Khanna et al. 2019; Nahvi et al. 2021). Due to its low toxicity, environmental friendliness, and greener synthesis process, algal-capped and stabilized nanoparticles are widely considered.

10.4.2.2 Mechanism of Nanoparticle Synthesis by Algae

Algae can be cultivated, and algal extract is made by heating or boiling the algae in either water or an organic solvent for a certain period. Subsequently, molar solutions of ionic metal compounds are prepared and incubated with the algal extract with continuous stirring under controlled conditions. Various factors such as pH, temperature, pressure, concentration of algal extracts, incubation period, size, and other reaction conditions influence the characteristics and applications of biosynthesized nanoparticles (Jacob et al. 2021). Algal mediated synthesis of nanoparticles can be carried out by two different ways either by extracellulary or intracellularly which may vary depending upon the dosage of metal ions and the type of algae used (Dahoumane et al. 2012).

10.4.2.2.1 Extracellular Synthesis of Algal Nanoparticles

Algal extracts contain various biomolecules such as polysaccharides, peptides, reducing sugars, proteins, pigments or other reducing factors that act as reducing, capping, and stabilizing agents of the metal ions which precipitate the metal resulting in the extracellular formation of nanoparticles in aqueous phase (Mahdavi et al. 2013).

10.4.2.2.2 Intracellular Synthesis of Algal Nanoparticles

Nanoparticles may accumulate intracellularly due to algal metabolism, such as photosynthesis and respiration. As a result of energy-generating reactions within the photosynthetic or by respiratory ETS (Electron Transport System) and by redox reactions at the thylakoids membrane, cell membrane and cytoplasm, NADPH and NADPH-dependent reductase are recycled which act as reducing agents, resulting in the intracellular reduction of metallic ions to nanoparticles by Cyanobacteria (Sharma et al. 2016; Dahoumane et al. 2014).

10.4.2.3 Characteristics of Metallic Nanoparticles Synthesized by Algae

The properties of nanoparticles synthesized by algae/algal extracts can be ensured by various methods as summarized below.

10.4.2.3.1 Color Change

Indication of the color change of the reaction mixture from green, colorless, or light yellow to dark yellowish or brown color (Selvaraj et al. 2020).

10.4.2.3.2 Measuring Absorbance by UV-Vis Spectroscopy

Different types of NP can be characterized by plasmon resonance and measuring the collective oscillations of conduction band electrons according to electromagnetic waves. UV-Vis spectroscopy absorbance ranging 390–450 nm confirms the formation of nanoparticles (Fatima et al. 2020).

10.4.2.3.3 Detection of Reductant by Fourier Transform Infrared (FTIR) Spectroscopy

Different types of functional groups such as Amino, amide, and carbonyl groups in algal extract interact with metal particles resulting in agglomeration and stabilization of nanoparticles. FTIR is involved in the identification of active group which is responsible for the reduction of metal ions into nanoparticles and detect the modification in absorption peaks (Fatima et al. 2020). The extract of green microalga *Chlorella vulgaris* contains carboxyl and hydroxyl group which plays a role in biosynthesis of AgNPs (Xie et al. 2007).

10.4.2.3.4 Determination of Zeta Potential

The stability and the charge of the nanoparticles can be analyzed by Zetasizer. The negative value of the zeta potential shows the efficiency of the chemical (reducing/capping) compounds found in algal extracts (Haider and Mehdi 2014).

10.5 Algal Species Used for Nanobiofertilizer

1. *Chlorella* sp.: Chlorella is a genus of green algae that is widely used in nanobiofertilizer production. It is known for its high protein content, amino acids, vitamins, minerals, and growth regulators. *Chlorella* extracts are rich in nutrients and bioactive compounds that promote plant growth and development.
2. *Spirulina* sp.: a filamentous blue-green algae that is often utilized in nanobiofertilizer production. It is an excellent source of proteins, essential amino acids, vitamins, minerals, antioxidants, and growth-promoting substances. *Spirulina* extracts provide a range of nutrients that enhance plant growth and productivity.
3. *Dunaliella* sp.: *Dunaliella* is a type of green microalgae that thrives in high-salinity environments. It is rich in beta-carotene, a precursor of vitamin A, as well as other pigments, amino acids, and minerals. *Dunaliella* extracts are known to have beneficial effects on plant growth, stress tolerance, and nutrient uptake.

4. *Ascophyllum nodosum*: *Ascophyllum nodosum* is a brown seaweed that is commonly used in nanobiofertilizer production. It is a rich source of bioactive compounds, including plant hormones (such as auxins, cytokinins, and gibberellins), amino acids, carbohydrates, vitamins, minerals, and antioxidants. *Ascophyllum* nodosum extracts promote root development, nutrient absorption, and overall plant growth.
5. *Gracilaria* sp.: *Gracilaria* is a red seaweed that is often employed in nanobiofertilizer production. It contains polysaccharides, amino acids, vitamins, minerals, and bioactive compounds that contribute to plant growth and stress tolerance. *Gracilaria* extracts have been shown to improve nutrient uptake, enhance crop yield, and confer resistance against various biotic and abiotic stresses (Ammar et al. 2022; Asimakis et al. 2022; Zou et al. 2021; Alam et al. 2020). Various algal species that synthesize different nanoparticles are enlisted in the Table 10.2.

Table 10.2 Synthesis and application of various algal-mediated nanoparticles in agriculture

Synthesized Nanoparticles	Algae	Features of nanoparticles	Reference
Silver	Filamentous cyanobacteria *Plectonema boryanum* UTEX 485	Size: Intracellularly synthesized nanoparticles –10 nm Extracellularly synthesized –1 to 200 nm	Thakkar et al. (2010)
	Spirulina plantensis	Size: 30–50 nm Antibacterial property	Sharma et al. (2015)
	Oscillatoria willie NTDM01	Size—100–200 nm	Ali et al. (2011)
	Cylindrospermum, Microcheate, Lyngbya, Anabaena,	Size—38–40 nm	Husain et al. (2015)
	Chlorella vulgaris	Size—10 nm Antibacterial efficacy against *Staphylococcus aureus*	Soleimani and Habibi-Pirkoohi (2017)
	Nannochloropsis oculata and *Chlorella vulgaris*	Size—15 nm	Mohseniazar et al. (2011)
	Chlorella pyrenoidosa	Size—5–10 nm	Elumalai et al. (2013)
	Chlamydomonas reinhardtii	Size—10 nm Inhibitory effect on *Listeria monocytogenes*	Ahmadi et al. (2016)
	Ulva fasciata	Size—40.5 nm and spherical in shape	Hamouda et al. (2018)
	Codium capitatum	Size—30 nm	Kannan et al. (2013)
	Padina pavonica,	Size—45–64 nm	Abdel-Raouf et al. (2019)
	Chaetomorpha antennina	Size—256.2 nm (diameter) Antibacterial activity	Roy and Anantharaman (2017)
	Sargassum ilicifolium	Antibacterial activity	Roy and Anantharaman (2018)

(continued)

Table 10.2 (continued)

Synthesized Nanoparticles	Algae	Features of nanoparticles	Reference
Gold	*Plectonema boryanum*	Size—20 nm	Lengke et al. (2006)
	Lyngbya majuscula	Size—41 nm and had antibacterial activities against *S. aureus* and *B. Subtilis*	Bakir et al. (2018)
	Sargassum wightii	Size—8–12 nm	Singaravelu et al. (2007)
	Ulva intestinalis, Phormidium valderianum, and Rhizoclonium fontinales	Size—43 nm 8–24 nm 10–30 nm, respectively	Parial et al. (2012)
	Chlorella vulgaris	Spherical Size—2–10 nm Anti-pathogenic	Annamalai and Nallamuthu (2015)
	Klebsormidium flaccidum, and Tetraselmis kochinensis	Size—15–20 nm	Dahoumane et al. (2012)
	Sargassum muticum	Size—3–8 nm	Namvar et al. (2015)
Zinc oxide	*Sargassum muticum*	Size—30–57 nm	Azizi et al. (2014)
	Ulva lactuna	Size—10–50 nm Antibiofilm and insecticidal activity	Ishwarya et al. (2018)

10.6 Role of Algal Nanoparticles in Agriculture

The application of nanoparticles could revolutionize agriculture by improving crop growth, nutrient uptake, and disease resistance. Using Nanosensors in agriculture can detect the disease symptoms in plants before it is evidently seen by the farmers (Chhipa and Joshi 2016). Through the use of nanofertilizers and nanopesticides, excessive fertilizer and pesticide use can be reduced and nutrients are prevented from being lost, and plants are able to absorb nutrients more effectively through targeted delivery and slow release of nutrients.

10.6.1 Algal Nanoparticles as Nanofertilizer

Nanobioferilizer has several unique advantages over conventional fertilizers. Due to their small particle size and high surface area, nanoparticles are able to assist different metabolic processes in plants and enhance the reactivity of the nutrient with other compound and improve nutrient use efficiency of plants (Singh 2017). Nanobiofertilizers are highly soluble in various solvents and thereby increase the bioavailability of nutrients. The penetration capacity of nanofertilizers is much higher than that of bulk materials (Prasad et al. 2012). The use of nanofertilizers facilitates the controlled and effective release of nutrients throughout the crop duration, whereas bulk fertilizers are only available for a short timeframe (Manjunatha et al. 2016).

Silver nanoparticles synthesized by brown seaweed aqueous extract of Sargassum ilicifolium increased the seed germination of Abelmoschus esculentus and Raphanus sativus var. Longipinnatus and could be utilized as a nanobiofertilizer to increase crop growth (Roy and Anantharaman 2018). Application of Silver nanoparticles synthesized by *Sargassum plagiophyllum* induced the seed germination of *Vigna mungo, Vigna radiate*, and *Arachis hypogaea* (Mary et al. 2015). Similarly seed germination of cucumber was induced by the application of silver nanoparticles produced by *Caulepa racemosa* and *Ulva lactuca* (Aravindan et al. 2014).

10.6.2 Nanoparticles-Induced Plant Stress Tolerance/Alleviation

The beneficial role of nanoparticles in plant stress tolerance several field studies and controlled experiments have considered NPs for enhancing crop productivity under normal as well as stressful conditions. NPs can efficiently deliver fertilizers, herbicides, and pesticides and improve the physicochemical properties of soil, including soil water-holding capacity. In addition, NPs can deliver specific materials, such as nucleotides, proteins, and chemicals under stressed conditions, mitigating those stress effects and improving crop growth and yield. Several studies have reported the potential role of various NPs, e.g., Ag, Ca, Fe, K, CuO, SiO_2, Se, Ti, and Zn, under controlled growth conditions: boosting crop production and resistance to abiotic stresses such as drought salinity and flood and other abiotic stresses (Zulfiqar and Ashraf 2021). Plant adaptation triggers biochemical, morphological, physiological, and molecular mechanisms to combat abiotic stresses. The application of nanoparticles (NPs) can help plants to withstand abiotic stresses by a concentration-dependent effect on growth and development in the plants (Rastogi et al. 2017). The application of nanoparticles helps a plant's drought resistance, as observed in a study by using silica NPs in hawthorns to increase the growth of seedlings and the physiological response to drought (Ashkavand et al. 2015). Similar results were documented in *Triticum aestivum*, where increased starch and gluten levels in plants ultimately improved their growth under drought conditions (Sarraf et al. 2022). Similarly, Jute seed treated with hydroxyapatite nanoparticles (CaNPs) increased drought tolerance by increasing proline synthesis (Jaberzadeh et al. 2013). One notable finding of proteomics was to assess the influence of Ag-NPs on the synthesis of glyoxalase II, regarded as an indication of cytotoxicity. When nicotinic acid and potassium nitrate (KNO_3) are combined with Ag-NPs, it improves plant flood tolerance (Das et al. 2016). It has been documented that the application of different types of NPs can alleviate the harmful effects of salt-stressed plants. For example, Zn-NPs alleviated the adverse effect of salinity in canola (Elsheery et al. 2020), rapeseed and mustard (Djanaguiraman et al. 2018). Similarly, Ag-NPs alleviated the salinity effects during seed germination of pearl millet. Excessive salinity results in an osmotic and ionic imbalance in plants, leading to ROS overproduction that damages biological molecules and their metabolic activities. Further application of different kinds of nanoparticles and their potential role in elevating various stressors are enlisted in Table 10.3.

Table 10.3 Nanoparticles and their potential role in elevating various stressors

Nanoparticles	Size of nanoparticles	Type of stress	Test plant	Inference	Reference
SiO_2 NPs	15–30 nm	Salinity	Tomato	Elevated salt stress genes' expression patterns.	Almutairi (2016)
TiNPs	20–30 nm	Drought	Wheat	Enhances wheat's agronomic characteristics, seed gluten content, and starch content	Jaberzadeh et al. (2013)
Chitosan–PVA and CuNPs	20–40 nm	Salinity	Tomato	Enhanced the water content in plants, photosynthetic rate, fresh weight, and chlorophyll concentration.	Hernández-Hernández et al. (2018)
CuNPs	30–40 nm	Drought	Maize	Increased carotenoid, chlorophyll, levels and decreased reactive oxygen species (ROS) due to the actions of scavenging enzymes in maize cultivated under drought stress	Van Nguyen et al. (2021)
TiNPs	21 nm	Drought	Onion	Improved seedling development. Additionally, it boosted the activity hydrolytic and antioxidant of enzymes.	Laware and Raskar (2014)
TiNPs	7–40 nm	Cold	Chick pea	Enhanced plant redox state and increased cold tolerance	Mohammadi et al. (2013)
CeNPs	10 nm	Salinity	Rice	Increased the synthesis of nitric oxide by promoting the NIA2-encoding nitrate reductase's transcription and regulating the protein's dephosphorylation, which led to NO production and increased salt tolerance in plants	Zou et al. (2021)

Table 10.3 (continued)

Nanoparticles	Size of nanoparticles	Type of stress	Test plant	Inference	Reference
FeNPs	18–94 nm	Drought	Rice	Enhanced plant physiological characteristics and a reduced ROS production	Ahmed et al. (2021)
CeO_2	10–40 nm	Salinity	Soya bean Rapeseed	Increased the rate of photosynthesis, water usage efficiency, and plant development Enhanced photosynthetic apparatus performance and altered salt stress responses in plants by restriction of salt absorption	Cao et al. (2017) Rossi et al. (2016)

10.6.3 Algal Nanobiofertilizers and Their Potential as Bioinsecticides

Algal nanobiofertilizers have shown potential as bioinsecticides, offering a sustainable and environmentally friendly alternative to chemical insecticides. Algal extracts contain bioactive compounds that possess insecticidal properties, including deterrent, repellent, and toxic effects on pests. By incorporating these algal extracts into nanobiofertilizers, the insecticidal activity can be enhanced and targeted specifically toward pests, minimizing harm to beneficial organisms.

The potential of algal nanobiofertilizers as bioinsecticides lies in their ability to disrupt key biological processes in pests, such as feeding, growth, development, and reproduction. They can act on the nervous system, disrupt hormonal regulation, or interfere with essential biochemical pathways, ultimately leading to pest mortality or reduced population growth.

10.6.3.1 Modes of Action and Specific Targeting Mechanisms Against Insect Pests

Algal nanobiofertilizers can exhibit modes of action and specific targeting mechanisms against insect pests, providing an alternative approach to chemical insecticides. Here are some of the common ways in which algal nanobiofertilizer can act against insect pests:

1. Insecticidal Compounds: Algal extracts used in nanobiofertilizers can contain natural compounds with insecticidal properties. These compounds may include secondary metabolites, alkaloids, terpenoids, or other bioactive substances that are toxic or repellent to insects. When applied to plants, these compounds can deter or harm pests, reducing their feeding activity, growth, or survival.

2. Antifeedant Effects: Some algal extracts incorporated in nanobiofertilizers can possess antifeedant properties, making plants less attractive or palatable to insects. The bioactive compounds present in the algal extracts can interfere with the feeding behavior of pests, reducing their damage to crops and inhibiting their population growth.
3. Growth Disruption: Algal nanobiofertilizers may contain bioactive compounds that disrupt the growth and development of insect pests. These compounds can interfere with insect hormone regulation, molting, or metamorphosis, leading to abnormal growth patterns, sterility, or mortality of the pests. By targeting specific life stages or physiological processes, the nanobiofertilizers can effectively control pest populations.
4. Repellent Effects: Algal extracts can possess natural repellent properties, deterring insects from approaching or landing on plants. These repellent effects can disrupt the host-seeking behavior of pests, reducing their ability to locate and infest crops. By creating a barrier or undesirable sensory response, algal nanobiofertilizers can effectively repel insect pests.
5. Systemic Effects: Algal nanobiofertilizers have the potential to induce systemic responses in plants, making them more resistant or less favorable to insect pests. The bioactive compounds present in the algal extracts can trigger plant defense mechanisms, such as the production of defensive chemicals, phytoalexins, or the reinforcement of cell walls. These systemic effects can enhance the plant's ability to withstand pest attacks and reduce the extent of damage.
6. Integrated Pest Management (IPM) Approach: Algal nanobiofertilizers can be integrated into IPM strategies, which the aim to combine multiple pest management techniques for sustainable control. By incorporating insecticidal properties into the fertilizers, pests can be targeted while minimizing harm to beneficial organisms and the environment. This approach promotes a balanced ecosystem and reduces reliance on chemical insecticides.

It is worth noting that the efficacy of algal nanobiofertilizers in pest control can vary depending on the specific pest species, crop, application method, and environmental factors. Further research and field testing are needed to optimize the formulation and application strategies to achieve effective and targeted pest control with algal nanobiofertilizers.

10.7 Challenges and Limitations in the Adoption and Commercialization of Algal Nanobiofertilizers

The adoption and commercialization of algal nanobiofertilizers face several challenges and limitations. Here are some key factors that can hinder their widespread implementation:

1. Limited Research and Development: Algal nanobiofertilizers are still in the early stages of research and development. Further studies are needed to optimize for-

mulations, understand their modes of action, determine dosage recommendations, and evaluate their long-term effects on crops, soils, and the environment. More research is also required to assess their efficacy across different crop species and pest management scenarios.

2. Regulatory Considerations: The registration and regulatory approval process for new agricultural products, including algal nanobiofertilizers, can be complex and time-consuming. These products need to meet specific safety and efficacy requirements set by regulatory authorities. The cost and time associated with obtaining regulatory approvals can pose challenges for small-scale manufacturers or hinder the commercialization of algal nanobiofertilizers.
3. Standardization and Quality Control: Ensuring consistent quality, stability, and reproducibility of algal nanobiofertilizers can be challenging. Variations in algal species, growth conditions, extraction methods, and formulation techniques can lead to variability in the composition and efficacy of the final product. Establishing standardized production protocols and quality control measures is essential to ensure reliable and effective algal nanobiofertilizers.
4. Cost and Scalability: The production cost of algal nanobiofertilizers can be relatively high compared to conventional chemical fertilizers. Factors such as algal cultivation, processing, formulation, and scale-up can contribute to the overall cost. Cost-effective production methods need to be developed to make algal nanobiofertilizers economically viable and competitive in the market. Additionally, scaling up production to meet commercial demands may require significant investments in infrastructure and technology.
5. Knowledge and Awareness: Limited knowledge and awareness about algal nanobiofertilizers among farmers, agronomists, and other stakeholders can hinder their adoption. Education and awareness programs are necessary to disseminate information about the benefits, efficacy, and proper application of algal nanobiofertilizers. Demonstrating their effectiveness through field trials and sharing success stories can help build confidence and promote their uptake.
6. Market Acceptance and Competition: Algal nanobiofertilizers face competition from established chemical fertilizers and conventional pest control products. Market acceptance may be influenced by factors such as farmer preferences, trust in traditional practices, and cost-effectiveness compared to existing alternatives. Demonstrating the economic benefits, environmental sustainability, and long-term value of algal nanobiofertilizers is crucial to gain market acceptance.

10.8 Environmental Impacts and Safety Considerations

10.8.1 Agricultural Regulations

Algal nanobiofertilizers are typically regulated under agricultural or crop protection regulations in many countries. These regulations govern the registration, labeling, and use of agricultural products, including fertilizers and biopesticides. Regulatory agencies such as the United States Environmental Protection Agency (EPA) and the

European Commission's Directorate-General for Health and Food Safety (DG SANTE) provide guidelines and requirements for the registration and approval of agricultural products.

10.8.2 Product Registration

Algal nanobiofertilizers may need to undergo a registration process before they can be marketed and used commercially. This process typically involves providing scientific data on product safety, efficacy, and environmental impacts. Registration requirements may include data on product composition, toxicology, ecotoxicology, residue levels, and environmental fate. The registration process ensures that the product meets safety standards and is suitable for its intended use.

10.8.3 Labeling and Safety Data Sheets (SDS)

Regulations often require clear and accurate labeling of algal nanobiofertilizers. Labels should include product name, active ingredients, directions for use, application rates, safety precautions, and storage recommendations. Safety Data Sheets (SDS) provide comprehensive information on the potential hazards, handling, storage, and emergency procedures associated with the product.

1. Maximum Residue Limits (MRLs): MRLs may be established for algal nanobiofertilizer components to ensure the safety of crops and food products. MRLs define the maximum allowable concentration of residues on crops or food items resulting from the use of the product. Compliance with MRLs helps ensure consumer safety and facilitates international trade of agricultural commodities.

Environmental Impact:

   (a) It reduces the risk of long-term environmental persistence compared to synthetic chemical fertilizers and pesticides due to having biodegradability mechanism.
   (b) It is essential to assess their potential impacts on beneficial insects, pollinators, and other non-target organisms to ensure they do not disrupt ecological balance.
   (c) Careful application practices and adherence to recommended dosages can help minimize the risk of water contamination.

2. Human Health and Safety: During the handling and application of algal nanobiofertilizers, it is important to follow appropriate safety measures to prevent direct contact or inhalation of the product. Personal protective equipment (PPE) and safe handling practices should be employed. It generally have low risks of leaving harmful residues on crops and food products. However, it is crucial to establish maximum residue limits (MRLs) for algal nanobiofertilizer components to ensure consumer safety and comply with regulatory standards. Alongside, assessment of the potential allergenic properties or their components

of algal nanobiofertilizers and provide appropriate labeling and warnings much needed.

3. Occupational Safety and Health Regulations: Occupational safety and health regulations aim to protect workers involved in the production, handling, and application of algal nanobiofertilizers. These regulations may include guidelines on personal protective equipment (PPE), training requirements, exposure limits, and hazard communication.

10.8.4 Ecotoxicological Studies

Ecotoxicological studies are necessary to evaluate the effects of algal nanobiofertilizers on various organisms and ecosystems. These studies assess acute and chronic toxicity, effects on reproduction, and other relevant endpoints to understand the potential risks to the environment. Ecotoxicological evaluations play a vital role in assessing the potential impacts of algal nanobiofertilizers on non-target organisms and the overall ecosystem. These evaluations help determine the toxicity and potential risks associated with the use of these fertilizers. Here are some key aspects of ecotoxicological evaluations and their impact on non-target organisms:

10.9 Future Directions and Research Opportunities

1. The field of algal nanobiofertilizers is still evolving, and there are several promising future directions and research opportunities that can further advance their development and application. Here are some areas that hold potential for future research:
2. Optimization of Formulation and Delivery Systems: Continued research can focus on optimizing the formulation and delivery systems of algal nanobiofertilizers. This includes exploring innovative encapsulation techniques, controlled release mechanisms, and novel carriers to enhance stability, efficacy, and targeted delivery of active compounds to plants.
3. Understanding Mechanisms of Action: Further studies can delve into understanding the specific mechanisms of action of algal nanobiofertilizers against pests and pathogens. Investigating the interactions between algal components and target organisms at the molecular level can provide insights into their modes of action, resistance management, and potential synergistic effects with other biocontrol agents.
4. Expansion of Targeted Pest and Disease Control: Research can focus on expanding the range of pests and diseases that can be effectively controlled using algal nanobiofertilizers. This includes exploring their efficacy against a broader spectrum of insect pests, plant diseases, and weed species. Additionally, investigations into their potential as plant growth-promoting agents can contribute to enhanced crop productivity and stress tolerance.

5. Integration with Other Pest Management Strategies: There is an opportunity to explore the integration of algal nanobiofertilizers with other pest management strategies, such as biological control agents, cultural practices, and precision agriculture technologies. Evaluating the compatibility and synergistic effects of these approaches can enhance overall pest management effectiveness and sustainability.
6. Environmental Impact Assessments: Further studies can focus on conducting comprehensive environmental impact assessments of algal nanobiofertilizers. This includes evaluating their potential long-term effects on soil health, microbial communities, non-target organisms, and ecosystem functioning. Understanding their ecological impacts can help develop guidelines for responsible and sustainable use.
7. Scale-up and Commercialization: As algal nanobiofertilizers transition from the laboratory to commercial applications, research can focus on scaling up production methods and optimizing cost-effectiveness. This includes exploring large-scale cultivation techniques for algal biomass production, refining extraction and processing methods, and developing scalable manufacturing processes.
8. Regulatory Considerations: Future research can contribute to the development of robust regulatory frameworks and guidelines specific to algal nanobiofertilizers. This includes addressing safety considerations, efficacy testing protocols, standardized evaluation criteria, and risk assessment methodologies. Collaboration between researchers, industry stakeholders, and regulatory authorities can ensure responsible commercialization and regulatory compliance.

10.10 Conclusion

The application of nanoparticles in agriculture has immense potential to increase the yield and quality of crops by smart and sustained delivery of nutrients/pesticides to the target site. Exploring the potential of algal nanoparticles as biofertilizers, nanobiostimulants or nanopesticides would reduce the hazardous impact of excessive application of agrochemicals. However, till date only few studies are reported regarding the use of algal synthesized nanoparticles in agriculture. Research and development toward synthesis and application of algal-based nanobiofertilizers and their interaction studies with plants for nutrient uptake and disease control would improve crop production.

References

Abdelghany AM, El-Banna AA, Salama EA, Ali MM, Al-Huqail AA, Ali HM, Paszt LS, El-Sorady GA, Lamlom SF (2022) The individual and combined effect of nanoparticles and biofertilizers on growth, yield, and biochemical attributes of peanuts (*Arachis hypogea* L.). Agronomy 12(2):398

Abdel-Raouf N, Al-Enazi NM, Ibraheem IBM, Alharbi RM, Alkhulaifi MM (2019) Biosynthesis of silver nanoparticles by using of the marine brown alga *Padina pavonia* and their characterization. Saudi J Biol Sci 26(6):1207–1215

Aguirre-Becerra H, Feregrino-Perez AA, Esquivel K, Perez-Garcia CE, Vazquez-Hernandez MC, Mariana-Alvarado A (2022) Nanomaterials as an alternative to increase plant resistance to abiotic stresses. Front Plant Sci 13:1023636

Ahmadi FS, Tanhaeian A, Pirkohi MH (2016) Biosynthesis of silver nanoparticles using *Chlamydomonas reinhardtii* and its inhibitory effect on growth and virulence of *Listeria monocytogenes*. Iran J Biotechnol 14(3):163

Ahmed T, Noman M, Manzoor N, Shahid M, Abdullah M, Ali L, Wang G. Hashem A, Al-Arjani ABF, Alqarawi AA, Abd_Allah EF, 2021. Nanoparticle-based amelioration of drought stress and cadmium toxicity in rice via triggering the stress-responsive genetic mechanisms and nutrient acquisition. Ecotoxicology and Environmental Safety, 209, p.111829.

Alam MA, Xu JL, Wang Z (eds) (2020) Microalgae biotechnology for food, health and high value products. Springer, Singapore

Ali DM, Sasikala M, Gunasekaran M, Thajuddin N (2011) Biosynthesis and characterization of silver nanoparticles using marine cyanobacterium, *Oscillatoria willei* NTDM01. Dig J Nanomater Biostruct 6(2):385–390

Almutairi ZM (2016) Effect of nano-silicon application on the expression of salt tolerance genes in germinating tomato ('Solanum lycopersicum 'L.) seedlings under salt stress. Plant Omics 9(1), pp.106–114

Ammar EE, Aioub AA, Elesawy AE, Karkour AM, Mouhamed MS, Amer AA, El-Shershaby NA (2022) Algae as bio-fertilizers: between current situation and future prospective Saudi. J Biol Sci 29(5):3083–3096

Annamalai J, Nallamuthu T (2015) Characterization of biosynthesized gold nanoparticles from aqueous extract of *Chlorella vulgaris* and their anti-pathogenic properties. Appl Nanosci 5:603–607

Aravindan D, Azeez R, Nallamuthu T (2014) Biosynthesis of silver nanoparticles using two seaweeds and their potential towards environment. Nano Biomed Eng 6(4):94–103

Ashkavand P, Tabari M, Zarafshar M, Tomaskova I, Struve D (2015) Effect of SiO_2 nanoparticles on drought resistance in hawthorn seedlings. Leśne Pr Badaw 76(4):350

Asimakis E, Shehata AA, Eisenreich W, Acheuk F, Lasram S, Basiouni S, Emekci M, Ntougias S, Taner G, May-Simera H, Yilmaz M (2022) Algae and their metabolites as potential biopesticides. Microorganisms 10(2):307

Ayub MA, Naeem A, Rehman MZ, Farooqi ZUR, Umar W, Fatima H, Nadeem M, Shabaan M (2022) Role of nanotechnology in enhancing crop production and produce quality. In: Sustainable nanotechnology for environmental remediation. Elsevier, pp 703–764

Azizi S, Ahmad MB, Namvar F, Mohamad R (2014) Green biosynthesis and characterization of zinc oxide nanoparticles using brown marine macroalga *Sargassum muticum* aqueous extract. Mater Lett 116:275–277

Bakir EM, Younis NS, Mohamed ME, El Semary NA (2018) Cyanobacteria as nanogold factories: chemical and anti-myocardial infarction properties of gold nanoparticles synthesized by *Lyngbya majuscula*. Mar Drugs 16(6):217

Beetul K, Gopeechund A, Kaullysing D (2016) Challenges and opportunities in the present era of marine algal applications. In: Thajuddin N, Dhanasekaran D (eds) Algae-organisms for imminent biotechnology. IntechOpen, pp 237–276

Behera M, Behera PR, Bhuyan PP, Singh L, Pradhan B (2023) Algal nanoparticles and their antibacterial activity: current research status and future prospectives. Drugs Drug Candid 2(3):554–570

Cao Z, Stowers C, Rossi L, Zhang W, Lombardini L, Ma X (2017) Physiological effects of cerium oxide nanoparticles on the photosynthesis and water use efficiency of soybean (Glycine max (L.) Merr.). Environmental Science: Nano 4(5), pp.1086–1094.

Chan SS, Low SS, Chew KW, Ling TC, Rinklebe J, Juan JC, Ng EP, Show PL (2022) Prospects and environmental sustainability of phyconanotechnology: a review on algae-mediated metal nanoparticles synthesis and mechanism. Environ Res 212:113140

Chhipa H, Joshi P (2016) Nanofertilisers, nanopesticides and nanosensors in agriculture. In: Nanoscience in food and agriculture, vol 1. Springer, pp 247–282

Chojnacka K, Moustakas K, Witek-Krowiak A (2020) Bio-based fertilizers: a practical approach towards circular economy. Bioresour Technol 295:122223

Dahoumane SA, Djediat C, Yepremian C, Coute A, Fievet F, Coradin T, Brayner R (2012) Recycling and adaptation of *Klebsormidium flaccidum* microalgae for the sustained production of gold nanoparticles. Biotechnol Bioeng 109(1):284–288

Dahoumane SA, Yéprémian C, Djédiat C, Couté A, Fiévet F, Coradin T, Brayner R (2014) A global approach of the mechanism involved in the biosynthesis of gold colloids using micro-algae. J Naopart Res 16:1–12

Das A, Ray R, Mandal N, Chakrabarti K (2016) An analysis of transcripts and enzyme profiles in drought-stressed jute (*Corchorus capsularis*) and rice (*Oryza sativa*) seedlings treated with $CaCl_2$, hydroxyapatite nano-particle and β-amino butyric acid. Plant Growth Regul 79: 401–412

Davey MC, Clarke KJ (1991) The spatial distribution of microalgae on Antarctic fellfield soils. Antarct Sci 3(3):257–263

Djanaguiraman M, Nair R, Giraldo JP, Prasad PVV (2018) Cerium oxide nanoparticles decrease drought-induced oxidative damage in sorghum leading to higher photosynthesis and grain yield. ACS Omega 3(10):14406–14416

Duhan JS, Kumar R, Kumar N, Kaur P, Nehra K, Duhan S (2017) Nanotechnology: the new perspective in precision agriculture. Biotechnol Rep 15:11–23

El-Ghamry AM, Mosa AA, Alshaal TA, El-Ramady HR (2018) Nanofertilizers vs. biofertilizers: new insights. Environ Biodivers Soil Secur 2:51–72

Elsheery NI, Helaly MN, El-Hoseiny HM, Alam-Eldein SM (2020) Zinc oxide and silicon nanoparticles to improve the resistance mechanism and annual productivity of salt-stressed mango trees. Agronomy 10(4):558

Elster J (2002) Ecological classification of terrestrial algal communities in polar environments: dedicated to professor Josef Svoboda on the occasion of his 70th birthday. In: Geoecology of Antarctic ice-free coastal landscapes. Springer, pp 303–326

Elumalai S, Santhose BI, Devika R, Revathy S (2013) Collection, isolation, identification, and biosynthesis of silver nanoparticles using microalga *Chlorella pyrenoidosa*. Nanosci Technol 4(1):59

FAO (2018) The future of food and agriculture: alternative pathways to 2050. Food and Agriculture Organization of the United Nations, Rome

Fatima R, Priya M, Indurthi L, Radhakrishnan V, Sudhakaran R (2020) Biosynthesis of silver nanoparticles using red algae *Portieria hornemannii* and its antibacterial activity against fish pathogens. Microb Pathog 138:103780

Gantar M (2000) Mechanical damage of roots provides enhanced colonization of the wheat endo rhizosphere by the dinitrogen-fixing cyanobacterium *Nostoc* sp. strain 2S9B. In: Biology and fertility of soils. Springer, pp 250–255

Gantar M, Elhai J (1999) Colonization of wheat para-nodules by the N_2-fixing cyanobacterium *Nostoc* sp. strain 2S9B. New Phytol 141(3):373–379

Gilaki M (2010) Biosynthesis of silver nanoparticles using plant extracts. J Biol Sci 10(5):465–467

Govindaraju K, Krishnamoorthy K, Alsagaby SA, Singaravelu G, Premanathan M (2015) Green synthesis of silver nanoparticles for selective toxicity towards cancer cells. IET Nanobiotechnol 9(6):325–330

Haider MJ, Mehdi MS (2014) Study of morphology and zeta potential analyzer for the silver nanoparticles. Int J Sci Eng Res 5(7):381–387

Hamouda R, Abd El-Mongy M, Eid KF (2018) Antibacterial activity of silver nanoparticles using *Ulva fasciata* extracts as reducing agent and sodium dodecyl sulfate as stabilizer. Int J Pharmacol 14(3):359–368

Hernández-Hernández H, González-Morales S, Benavides-Mendoza A, Ortega-Ortiz H, Cadenas-Pliego G, Juárez-Maldonado A (2018) Effects of chitosan–PVA and Cu nanoparticles on the growth and antioxidant capacity of tomato under saline stress. Molecules, 23(1), p.178

Husain S, Sardar M, Fatma T (2015) Screening of cyanobacterial extracts for synthesis of silver nanoparticles. World J Microbiol Biotechnol 31:1279–1283

Ishwarya R, Vaseeharan B, Kalyani S, Banumathi B, Govindarajan M, Alharbi NS, Kadaikunnan S, Al-Anbr MN, Khaled JM, Benelli G (2018) Facile green synthesis of zinc oxide nanoparticles using *Ulva lactuca* seaweed extract and evaluating their photocatalytic, antibiofilm and insecticidal activity. J Photochem Photobiol B Biol 178:249–258

Jaberzadeh A, Moaveni P, Moghadam HRT, Zahedi H (2013) Influence of bulk and nanoparticles titanium foliar application on some agronomic traits, seed gluten and starch contents of wheat subjected to water deficit stress. Not Bot Horti Agrobot Cluj Napoca 1:201–207

Jacob RH, Shanab SM, Shalaby EA (2021) Algal biomass nanoparticles: chemical characteristics, biological actions, and applications. Biomass Convers Biorefinery 13:1–15

Kannan RRR, Stirk WA, Van Staden J (2013) Synthesis of silver nanoparticles using the seaweed *Codium capitatum* PC Silva (Chlorophyceae). S Afr J Bot 86:1–4

Khan F, Shahid A, Zhu H, Wang N, Javed MR, Ahmad N, Xu J, Alam MA, Mehmood MA (2022) Prospects of algae-based green synthesis of nanoparticles for environmental applications. Chemosphere 293:133571

Khanna P, Kaur A, Goyal D (2019) Algae-based metallic nanoparticles: synthesis, characterization and applications. J Microbiol Methods 163:105656

Kim JS, Kuk E, Yu KN, Kim JH, Park SJ, Lee HJ, Kim SH, Park YK, Park YH, Hwang CY, Kim YK (2007) Antimicrobial effects of silver nanoparticles. Nanomedicine 3(1):95–101

Kokabi M, Yousefzadi M (2015) Checklist of the marine macroalgae of Iran. Bot Mar 58(4):307–320

Krug L, Erlacher A, Markut K, Berg G, Cernava T (2020) The microbiome of alpine snow algae shows a specific inter-kingdom connectivity and algae-bacteria interactions with supportive capacities. ISME J 14(9):2197–2210

Kumawat KC, Razdan N, Saharan K (2022) Rhizospheric microbiome: bio-based emerging strategies for sustainable agriculture development and future perspectives. Microbiol Res 254:126901

Laware SL, Raskar S (2014) Effect of titanium dioxide nanoparticles on hydrolytic and antioxidant enzymes during seed germination in onion. Int. J. Curr. Microbiol. App. Sci 3(7), pp.749–760

Leach JE, Triplett LR, Argueso CT, Trivedi P (2017) Communication in the phytobiome. Cell 169(4):587–596

Lee SM, Ryu CM (2021) Algae as new kids in the beneficial plant microbiome. Front Plant Sci 12:599742

Lengke MF, Ravel B, Fleet ME, Wanger G, Gordon RA, Southam G (2006) Mechanisms of gold bioaccumulation by filamentous cyanobacteria from gold (III)–chloride complex. Environ Sci Technol 40(20):6304–6309

LewisOscar F, Vismaya S, Arunkumar M, Thajuddin N, Dhanasekaran D, Nithya C (2016) Algal nanoparticles: synthesis and biotechnological potentials. In: Algae-organisms for imminent biotechnology, vol 7. IntechOpen, pp 157–182

Li R, Tao R, Ling N, Chu G (2017) Chemical, organic and bio-fertilizer management practices effect on soil physicochemical property and antagonistic bacteria abundance of a cotton field: implications for soil biological quality. Soil Tillage Res 167:30–38

Liu J, Chen F (2016) Biology and industrial applications of *Chlorella*: advances and prospects. Adv Biochem Eng Biotechnol 153:1–35

Ma X, Geiser-Lee J, Deng Y, Kolmakov A (2010) Interactions between engineered nanoparticles (ENPs) and plants: phytotoxicity, uptake and accumulation. Sci Total Environ 408(16):3053–3061

Mahapatra DM, Satapathy KC, Panda B (2022) Biofertilizers and nanofertilizers for sustainable agriculture: phycoprospects and challenges. Sci Total Environ 803:149990

Mahdavi M, Namvar F, Ahmad MB, Mohamad R (2013) Green biosynthesis and characterization of magnetic iron oxide (Fe_3O_4) nanoparticles using seaweed (*Sargassum muticum*) aqueous extract. Molecules 18(5):5954–5964

Manjunatha SB, Biradar DP, Aladakatti YR (2016) Nanotechnology and its applications in agriculture: a review. J Farm Sci 29(1):1–13

Mary XA, Mathuran T, Saji M, Kumar SA (2015) *In-vitro* germination effect of biologically synthesized silver nanoparticles from *Sargassum plagiophyllum* on *Arachis hypogaea*, *Vigna mungo*, and *Vigna radiate*. IRJNAS 12:278–284

Michael A, Singh A, Roy A, Islam MR (2022) Fungal-and algal-derived synthesis of various nanoparticles and their applications. Bioinorg Chem Appl 2022:1

Mohammadi R, Maali-Amiri R, Abbasi A (2013) Effect of TiO 2 nanoparticles on chickpea response to cold stress. Biological trace element research, 152, pp.403-410.

Mohseniazar M, Barin M, Zarredar H, Alizadeh S, Shanehbandi D (2011) Potential of microalgae and lactobacilli in biosynthesis of silver nanoparticles. Bio Impacts 1(3):149

Nahvi I, Belkahla S, Asiri SM, Rehman S (2021) Overview and prospectus of algal biogenesis of nanoparticles. In: Microbial nanotechnology: green synthesis and applications. Springer, pp 121–134

Namvar F, Azizi S, Ahmad MB, Shameli K, Mohamad R, Mahdavi M, Tahir PM (2015) Green synthesis and characterization of gold nanoparticles using the marine macroalgae *Sargassum muticum*. Res Chem Inter 41:5723–5730

Parial D, Patra HK, Dasgupta AK, Pal R (2012) Screening of different algae for green synthesis of gold nanoparticles. Eur J Pharmacol 47(1):22–29

Prasad TNVKV, Sudhakar P, Sreenivasulu Y, Latha P, Munaswamy V, Reddy KR, Sreeprasad TS, Sajanlal PR, Pradeep T (2012) Effect of nanoscale zinc oxide particles on the germination, growth and yield of peanut. J Plant Nutr 35(6):905–927

Rajesh S, Raja DP, Rathi JM, Sahayaraj K (2012) Biosynthesis of silver nanoparticles using *Ulva fasciata* (Delile) ethyl acetate extract and its activity against *Xanthomonas campestris* pv. *alvacearum*. J Biopest 5:119

Rastogi A, Zivcak M, Sytar O, Kalaji HM, He X, Mbarki S, Brestic M (2017) Impact of metal and metal oxide nanoparticles on plant: a critical review. Fron Chem 5:78

Riaz U, Murtaza G, Anum W, Samreen T, Sarfraz M, Nazir MZ (2021) Plant Growth-Promoting Rhizobacteria (PGPR) as biofertilizers and biopesticides. In: Microbiota and biofertilizers: a sustainable continuum for plant and soil health. Springer, pp 181–196

Rossi L, Zhang W, Lombardini L, Ma X, (2016) The impact of cerium oxide nanoparticles on the salt stress responses of Brassica napus L. Environmental Pollution 219:28–36

Roy S, Anantharaman P (2017) Biosynthesis of silver nanoparticles by *Chaetomorpha antennina* (Bory de Saint-Vincent) Kutzing with its antibacterial activity and ecological implication. J Nanomed Nanotechnol 8(467):2

Roy S, Anantharaman P (2018) Biosynthesis of silver nanoparticles by *Sargassum Ilicifolium* (Turner) C. Agardh with their antimicrobial activity and potential for seed germination, p 2

Salem SS, Fouda A (2021) Green synthesis of metallic nanoparticles and their prospective biotechnological applications: an overview. Biol Trace Elem Res 199:344–370

Sambangi P, Gopalakrishnan S, Pebam M, Rengan AK (2022) Nano-biofertilizers on soil health, chemistry, and microbial community: benefits and risks. Proc Indian Natl Sci Acad 88(3):357–368

Sarraf M, Vishwakarma K, Kumar V, Arif N, Das S, Johnson R, Janeeshma E, Puthur JT, Aliniaeifard S, Chauhan DK, Fujita M (2022) Metal/metalloid-based nanomaterials for plant abiotic stress tolerance: an overview of the mechanisms. Plan Theory 11(3):316

Schneidewind H, Schuler T, Strelau KK, Weber K, Cialla D, Diegel M, Mattheis R, Berger A, Möller R, Popp J (2012) The morphology of silver nanoparticles prepared by enzyme-induced reduction. Beilstein J Nanotechnol 3(1):404–414

Schröfel A, Kratošová G, Bohunicka M, Dobrocka E, Vávra I (2011) Biosynthesis of gold nanoparticles using diatoms—silica-gold and EPS-gold bionanocomposite formation. J Nanopart Res 13:3207–3216

Selvaraj P, Neethu E, Rathika P, Jayaseeli JPR, Jermy BR, Abdul Azeez S, Borgio JF, Dhas TS (2020) Antibacterial potentials of methanolic extract and silver nanoparticles from marine algae. Biocatal Agric Biotechnol 28:101719

Sharma G, Jasuja ND, Kumar M, Ali MI (2015) Biological synthesis of silver nanoparticles by cell-free extract of *Spirulina platensis*. J Nanotechnol 2015:1–6

Sharma A, Sharma S, Sharma K, Chetri SP, Vashishtha A, Singh P, Kumar R, Rathi B, Agrawal V (2016) Algae as crucial organisms in advancing nanotechnology: a systematic review. J Appl Phycol 28:1759–1774

Sharma G, Pandey S, Ghatak S, Watal G, Rai PK (2018) Potential of spectroscopic techniques in the characterization of "green nanomaterials". In: Nanomaterials in plants, algae, and microorganisms. Academic Press, In, pp 59–77

Sharma B, Tiwari S, Kumawat KC, Cardinale M (2023) Nano-biofertilizers as bio-emerging strategies for sustainable agriculture development: potentiality and their limitations. Sci Total Environ 860:160476

Sheeba JM, Thambidurai S (2009) Extraction, characterization, and application of seaweed nanoparticles on cotton fabrics. J Appl Polym Sci 113(4):2287–2292

Shinozaki K, Yamaguchi-Shinozaki K (2007) Gene networks involved in drought stress response and tolerance. J Exp Bot 58(2):221–227

Singaravelu G, Arockiamary J, Ganesh K, Govindaraju K (2007) A novel extracellular synthesis of monodisperse gold nanoparticles using marine alga, *Sargassum wightii* Greville. Colloids Surf B Biointerfaces 57:97–101

Singh MD (2017) Nano-fertilizers is a new way to increase nutrients use efficiency in crop production. IJAS ISSN 9(7):0975–3710

Singh JS, Kumar A, Singh M (2019) Cyanobacteria: a sustainable and commercial bio-resource in production of bio-fertilizer and bio-fuel from waste waters. Environ Sustain 3-4:100008

Soleimani M, Habibi-Pirkoohi M (2017) Biosynthesis of silver nanoparticles using *Chlorella vulgaris* and evaluation of the antibacterial efficacy against *Staphylococcus aureus*. Avicenna J Med Biotechnol 9(3):120

Subramaniyam V, Subashchandrabose SR, Thavamani P, Megharaj M, Chen Z, Naidu R (2015) *Chlorococcum* sp. MM11—a novel phyco-nanofactory for the synthesis of iron nanoparticles. J Appl Phycol 27:1861–1869

Thakkar KN, Mhatre SS, Parikh RY (2010) Biological synthesis of metallic nanoparticles. Nanomedicine 6(2):257–262

Thirugnanasambandan T (2019) Advances and trends in nano-biofertilizers. SSRN:59

Treves H, Raanan H, Kedem I, Murik O, Keren N, Zer H, Berkowicz SM, Giordano M, Norici A, Shotland Y, Ohad I (2016) The mechanisms whereby the green alga *Chlorella ohadii*, isolated from desert soil crust, exhibits unparalleled photodamage resistance. New Phytol 210(4):1229–1243

Usman M, Farooq M, Wakeel A, Nawaz A, Cheema SA, ur Rehman H, Ashraf I, Sanaullah M (2020) Nanotechnology in agriculture: current status, challenges and future opportunities. Sci Total Environ 721:137778

Van Nguyen D, Nguyen HM, Le NT, Nguyen KH, Nguyen, HT, Le HM, Nguyen AT, Dinh NTT, Hoang SA, Van Ha C (2021) Copper nanoparticle application enhances plant growth and grain yield in maize under drought stress conditions. Journal of Plant Growth Regulation, pp.1–12

Verma KK, Song XP, Joshi A, Tian DD, Rajput VD, Singh M, Arora J, Minkina T, Li YR (2022) Recent trends in nano-fertilizers for sustainable agriculture under climate change for global food security. Nanomater 12(1):173

Vivek M, Kumar PS, Steff S, Sudha S (2011) Biogenic silver nanoparticles by *Gelidiella acerosa* extract and their antifungal effects. Avicenna J Med Biotechnol 3(3):143

Xie J, Lee JY, Wang DI, Ting YP (2007) Identification of active biomolecules in the highyield synthesis of single crystalline gold nanoplates in algal solutions. Small 3(4):672–682

Xu J, Zhang Y, Zhang P, Trivedi P, Riera N, Wang Y, Liu X, Fan G, Tang J, Coletta-Filho HD, Cubero J (2018) The structure and function of the global citrus rhizosphere microbiome. Nat Commun 9(1):4894

Zhu H, Li S, Hu Z, Liu G (2018) Molecular characterization of eukaryotic algal communities in the tropical phyllosphere based on real-time sequencing of the 18S rDNA gene. BMC Plant Biol 18(1):1–14

Zou Y, Zeng Q, Li H, Liu H, Lu Q (2021) Emerging technologies of algae based wastewater remediation for biofertilizer production: a promising pathway to sustainable agriculture. J Chem Technol Biotechnol 96(3):551–563

Zulfiqar F, Ashraf M (2021) Nanoparticles potentially mediate salt stress tolerance in plants. Plant Physiol Biochem 160:257–268

Nanobiofertilizers: A New Paradigm for Sustainable Agriculture

11

Yashika Yadav, Babita Thakur, and Sukhminderjit Kaur

Abstract

Nanotechnology refers to any nanoscale technology with multiple practical applications. Their broad applications are in the development of material science, drug delivery, cosmetics, biology, biotechnology, agriculture, and food. Nanobiotechnology is used in the creation of fertilizers to enhance the dynamics of soil plant nutrition for long-term crop management. With the potential properties of nanomaterials and biofertilizers, nanobiotechnology aids in the creation of new, inexpensive, and eco-friendly nanobiofertilizers. Nanoparticles are used in agriculture because of their unique physiochemical properties. The effectiveness of farming is increased by the delivery of agrochemicals using nanoparticles. By being nano-encapsulated, conventional and alternative fertilizers can be released gradually and precisely, supporting sustainable agriculture. Both plant production and plant pollution can be reduced with the use of nanobiofertilizers. In this chapter, we focused on the production of nanobiofertilizers and their mechanism of action and also its advantages and disadvantages in the field of agriculture.

Keywords

Nanotechnology · Nanobiofertilizers · Nanomaterials

Y. Yadav · B. Thakur
Department of Biotechnology, Chandigarh University, Mohali, Punjab, India

S. Kaur (✉)
Department of Biotechnology, Chandigarh University, Mohali, Chandigarh, India
e-mail: sukhminderjit.uibt@cumail.in

S. Kaur et al. (eds.), *Metabolomics, Proteomics and Gene Editing Approaches in Biofertilizer Industry*, https://doi.org/10.1007/978-981-97-2910-4_11

11.1 Introduction

The field of nanobiotechnology combines the fields of nanotechnology and biotechnology. Nanomaterial-based strategies are being developed to enhance current biotechnological procedures and get beyond their drawbacks. Nanotechnology is based on the use of nanoscale materials (such as nanowires, nanotubes, nanomembrane, and nanoparticles) in various fields (Shahcheraghi et al. 2022). Objects with a size range of 1–100 nm are called nanoparticles. Currently, titanium, gold, magnesium, zinc, alginate, copper, and silver are used to create several metallic nanostructures (Hasan 2015). Metallic inorganic nanomaterials include silver, gold, aluminium, cadmium, copper, iron, zinc, and lead nanoparticles. Graphene, fullerene, single and multiple-wall carbon nanotubes, activated carbon, and carbon black are examples of carbon-based nanomaterials. Carbon-free organic compounds including liposomes, cyclodextrin, dendrimers, and micelles are used to make the organic-based nanomaterials. Any mixture of metal, metal oxide, carbon, and/or organic nanoparticles, as well as carbon and/or organic-based nanomaterials, makes up composite nanomaterials. These substances possess intricate configurations, like an organic-metal lattice (Majhi and Yadav 2021). Nanoparticles are produced by microorganisms through a green and environmentally benign process called biosynthesis. Numerous microorganisms, including eukaryotes and prokaryotes, are used to create metallic nanoparticles. Actinomycetes, fungi, bacteria, and algae are some of these microorganisms. The particles may have been synthesized intracellularly or extracellularly. It is dependent upon the placement of the nanoparticle (Markus et al. 2016). However, extracellular synthesis has more applications than intracellular. Most fungi produce nanoparticles extracellularly. To create nanoparticles, intracellular synthesis entails transferring ions into microbial cells while enzymes are present. Extracellular synthesis involves adjoining cellular components like cell walls that bind to metal ion foam nanoparticles (Spadari et al. 2017). Distinct bacteria and fungi are capable of forming nanoparticles (Table 11.1) extracellularly and intracellularly (Matei et al. 2021).

There are many applications for nanoparticles in the medical industry, water treatment, agriculture, human health, and environmental safety. Hazardous organic compounds and industrial pollutants can be degraded and changed into safe forms by using nanoparticles as a photocatalyst (Garg et al. 2023) (Fig. 11.1).

Nanobiofertilizer development is one of the most exciting uses of nanotechnology in agriculture. Utilizing helpful microbes in conjunction with nanomaterials, these fertilizers can improve soil health and plant nutrient delivery more effectively. Since they can improve crop yield, advance sustainable agricultural methods, and completely change nutrient management, researchers, legislators, and agricultural practitioners have been paying close attention to nanobiofertilizers in recent years. By combining biofertilizers with nanotechnology, nanobiofertilizers capitalize on the special qualities of nanomaterials and the biological activity of microorganisms. It has long been known that biological nitrogen fixation, phosphorus solubilization, and nutrient mobilization by nitrogen-fixing bacteria, phosphate-solubilizing bacteria, and mycorrhizal fungi are how biofertilizers enhance soil fertility and

Table 11.1 Nanoparticle production using different bacteria and fungi

Bacteria	Nanoparticles produced	Synthesis location	References
Thermomonospora sp.	Au	Extracellular	Kasthuri et al. (2009)
Rhodococcus sp.	Au	Intracellular	Park et al. (2021)
Bacillus sp.	Nanoselenium	Extracellular	Bharathi et al. (2020)
Bacillus licheniformis	Ag	Intracellular	Schlüter et al. (2014)
Lactobacillus kimchicus	Au	Intracellular	Mathivanan et al. (2019)
Bacillus subtilis	ZnO	Extracellular	Dhandapani et al. (2020)
Fungi			
Pediococcus pentosaceus	Ag	Extracellular	Garcia-Ruiz et al. (2015)
Aspergillus niger BSC-1	Fe_2O_3	Extracellular	Kumari et al. (2020)
Fusarium oxysporum	CdS	Extracellular	Rai et al. (2009)
Aspergillus fumigatus	Ag	Extracellular	Rajan et al. (2016)
Fusarium oxysporum	Platinum	Extracellular	Chatterjee et al. (2020)
Verticillium sp.	Au	Intracellular	Ramanathan et al. (2013)

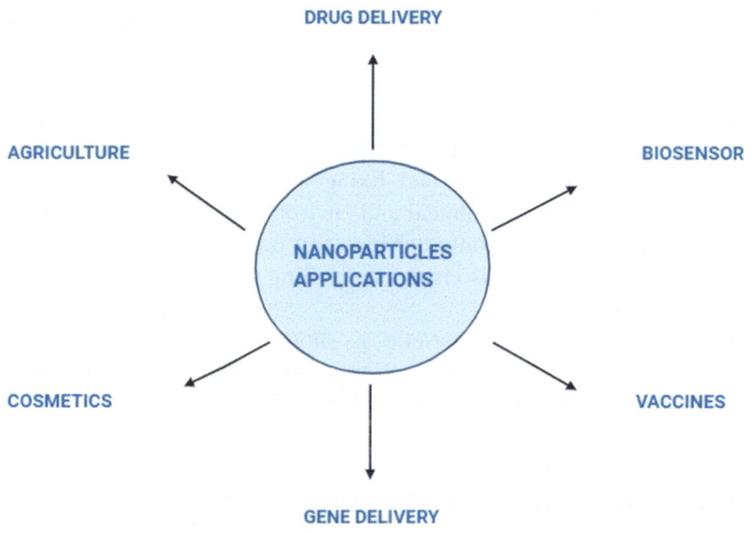

Fig. 11.1 Nanoparticles applications

encourage plant development. Through the use of nanomaterials like nanoparticles and nanocomposites, nanobiofertilizers improve the stability, efficacy, and delivery of biofertilizers. This presents previously unheard-of possibilities for addressing nutrient deficiencies, lowering reliance on chemical fertilizers, and lessening the negative environmental effects of conventional fertilizers (Rani et al. 2023).

In the field of agriculture, the use of biofertilizers also has some drawbacks, including short shelf lives, specific fertilizers for particular crops, instability in the field due to the absence of required soil and environmental conditions (temperature and pH sensitivity), requiring specific storage conditions, high dosages, etc. So, there is a need for advanced biofertilizers that are appropriate for sustainable agriculture and effective in enhancing soil fertility and plant tolerance. To get over all these problems, nanoparticle-based biofertilizer formulations have been created. Nanoparticles have an advantage over other materials when it comes to protecting and promoting plant development because of their unique optical qualities, high surface-to-volume ratio, high absorption capacity, and controlled-release rate (Pati et al. 2021a). Nano-bio technologically modified fertilizers as well as pesticides to significantly reduce crop loss. Several nanofertilizers, nanobiosensors, and nanopesticides have been created. Processes based on nanotechnology have been created to improve agricultural and environmental practices by reducing soil pollution (Usman et al. 2020).

11.2 Biofertilizers

Live microorganisms are incorporated into materials to create biofertilizers. According to Kumar et al. (2018), they settle down in the rhizosphere of the host plant, where they enhance the nutrient supply and encourage plant development. Creating a dynamic and sustainable soil environment for crop growth needs microorganisms that promote plant growth and engage in a variety of biological activities. Nutrient availability that is simple for plants to absorb is a feature of biofertilizers. They create components that promote plant development and boost soil fertility by solubilizing insoluble phosphates and fixing atmospheric nitrogen in the soil. Biofertilizers are naturally economical and environmentally beneficial (Mazid and Khan 2015). Biofertilizers not only enhance soil fertility but also shield the plant from pests and diseases. They also increase seedling survival and the life of the root system (Youssef and Eissa 2014).

Biofertilizers come in several varieties as shown in Table 11.2. Nitrogen fixers, potassium, phosphorus, and plant growth promoting rhizobacteria (PGPR) are the most widely used types of biofertilizers (Shafiq et al. 2022).

Biofertilizer production consists of six major steps. Active organisms must be determined first. Then there is the isolation of target microbes to separate them from their habitat. The microorganisms that have been separated will be cultured in glasshouses, shaking flasks, and petri plates. Next, it is decided which carrier material will work best for creating biofertilizers. There are various methods of propagation available to determine the ideal conditions for an organism. After that, a prototype is created and tested. Finally, biofertilizers are being tested extensively in various environments to determine their applicability and limitations. Sterilization of carrier material is necessary. The carrier should be easy to process, have a sufficient pH buffering capacity, and be gamma or autoclave sterilized (Mohammadi et al. 2012).

Table 11.2 Various types of biofertilizers and their examples

Biofertilizers	Groups	Examples	References
Nitrogen—Fixing	Free-living and symbiotic	*Azotobacter, Azospirillum* spp., *Rhizobium, Frankia, Anabaena, Clostridium,* and *Anabaena*	Bhat et al. (2015)
Potassium—Solubilizing	Bacteria and fungi	*Bacillus* spp., *Arthrobacter* spp., *Aspergillus niger*	Etesami et al. (2017)
Phosphate—Solubilizing	Bacteria and fungi	*Pseudomonas, Rhizobium, Bacillus, Enterobacter, Penicillium,* and *Aspergillus*	Anand et al. (2016)
Plant growth promoting	Plant growth promoting rhizobacteria	*Agrobacterium, Frankia, Bacillus, Arthrobacter, Pseudomonas,* and *Rhizobium*	Yadav et al. (2017)

11.3 Mechanism of Activity of Plant Growth Promoting Rhizobia

Rhizobacteria with the ability to promote plant development do so in two different ways. They are divided into two groups: mechanisms that are direct and indirect (Kundan et al. 2015). Directly, PGPR increases the development of plants by either promoting resource recovery (nitrogen and phosphorus) or modifying plant hormones or indirectly, PGPR increases plant growth by mitigating the impact of several diseases on plant development and fostering the creation of the biocontrol agent foam (Glick et al. 2012).

11.3.1 Direct Mechanisms

11.3.1.1 Fixation of Nitrogen

For plant development and output, nitrogen is the most important nutrient. However since they are unable to consume atmospheric nitrogen, plants do not utilize it. Even though 80 percent of gaseous nitrogen is present in our atmosphere, nitrogen-fixing bacteria and the nitrogenase enzyme system transform atmospheric nitrogen into ammonia (Tairo and Ndakidemi 2013). Nitrogen-fixing organisms often fall into two categories: (a) symbiotic nitrogen-fixing bacteria and (b) non-symbiotic nitrogen-fixing forms, including cyanobacteria; Rhizobiaceae family members grow symbiotically with leguminous plants (like rhizobia) and non-leguminous trees (like Frankia). Nitrogen-fixing bacteria that are not symbiotic hardly supply a trace of the fixed nitrogen required by the bacterially attached host plant (Bhattacharyya and Jha 2012). Rhizobia colonize as intracellular symbionts as a result of the intricate interactions between the symbiont and host that lead to the development of nodules during the symbiotic relationship. Another name for nitrogen fixers is diazotrophs (Glick et al. 2012).

11.3.1.2 Phosphorus Solubilization

Even though soil has a lot of phosphorus, the majority of it is insoluble. Phosphorus cannot promote plant development because it is only soluble in two forms that plants may absorb: monobasic and dibasic. Examples of inorganic and organic phosphorus include phosphotryesters, phosphomonoesters, and inositol phosphate (soil phytate) (Rahaiee et al. 2020). Chemical fertilizers contain soluble inorganic phosphorus, but it is put on the field and rapidly becomes immobile. This led researchers to focus on economical and eco-friendly solutions to increase crop production in low-phosphorus soils. Under such circumstances, microorganisms with the ability to solubilize inorganic phosphorus are crucial as a possible supply of phosphorus for plants (Zaidi et al. 2009). Low-molecular-weight organic acids like citric and gluconic acid, which are foamed by different soil microbes, are the source of inorganic phosphorus solubilization. These low-molecular-weight organic acids have carboxyl and hydroxyl groups that can chelate the phosphate-attached cations, converting insoluble phosphorous to soluble form (McComb et al. 2013). However, organic phosphorus is mineralized as a result of the creation of several phosphatases, which catalyse the hydrolysis of phosphoric esters. Above all, according to Guang-Can et al. (2008), the same bacterial strain may mineralize and solubilize phosphate. *Micrococcus, Bacillus, Azotobacter,* and *Pseudomonas* are among the bacteria that can solubilize insoluble inorganic phosphorus microorganisms (Shafiq et al. 2022).

11.3.2 Indirect Mechanisms

The majority of rhizobacteria's indirect plant growth promoting mechanisms also function as biocontrol agents. The most common way that microorganisms that promote plant development inhibit the spread of plant diseases is by synthesizing different antibiotics. A large number of compounds are being studied, and some are being sold. The majority of commercialized rhizobacterial products function as bioinoculants to combat plant diseases rather than improving plant nutrition or reducing abiotic stressors (Hermosa et al. 2012). Some pathogens cause plant illnesses, like *Fusarium, Pythium, Rhizoctonia,* and *Sclerotium*. And some biofertilizers like *Bacillus subtilis, P. fluorescens, and Trichoderma harzianum* address this. *Several* antifungal compounds, such as pyoluteorin, tensin, 2,4-diacetyl phloroglucinol, phenazines, and pyrrolnitrin, are produced by rhizobacteria (Bhattacharyya and Jha 2012). The host plant can be protected from pathogenic fungi, bacteria, and viruses when some rhizobacteria and plant roots come into contact which is known as induced systemic resistance (ISR) (Pieterse et al. 2014).

11.4 Nanobiofertilizers

Nanobiofertilizers are a combination of biofertilizers and nanoparticles. Biofertilizers are compounds that include live microorganisms, while nanoparticles are objects with sizes ranging from 1 to 100 nm. Biofertilizers encapsulated in nanoparticles make up nanobiofertilizers. They are economical and environmentally beneficial (Akhtar et al. 2022). Nanobiofertilizers are more efficient than other traditional methods because they produce superior and longer-lasting outcomes. By gradually releasing nutrients, they can improve the efficiency with which plants utilize nutrients (Du et al. 2018).

Nanomaterials like polymers, chitosan, and zeolite are used for nano-encapsulation of organic materials (Qureshi et al. 2018). Nutrient stability is increased when biofertilizers are coated with nanoparticles on their surface. For crop fertilization, nanobiofertilizers may enable more contact and active absorption of these nutrients. This fertilizer may also sustainably supply plants with the bioavailable nutrients they require at various growth stages (Vejan et al. 2016). Nanobiofertilizers contain microorganisms (such as fungus and bacteria). Through a variety of mechanisms, including siderophore production, phosphorus solubilization, and nitrogen fixation, they support plant development (Mala et al. 2017).

Nanobiofertilizers not only promote plant growth but also further provide resistance against various disease pathogens. In research, it has been investigated that nanobiofertilizers play a pest-resistant role against diseases like wilt in tomato crops caused by *Ralstonia solanacearum* species (Gatahi et al. 2016). Other investigations by researchers are about the role of nanobiofertilizers in shielding crops from pathogens. The crop is protected from pathogens by fertilizers that contain plant growth-regulating rhizobacteria (*Pseudomonas putida*, *Pseudomonas fluorescens*, *Paenibacillus elgii*, and *Bacillus subtilis*) that fight harmful bacterial and fungal strains in the rhizosphere of legumes (Gouda et al. 2018).

There are three processes in the manufacturing of nanobiofertilizers: The biofertilizer culture must first develop; next, it must be encapsulated with nanoparticles; finally, its purity, shelf life, efficacy, and quality must be evaluated (Panichikkal et al. 2021). It is made by mixing PGPR suspension in a 2:1 ratio with a solution containing 1.5% sodium alginate, 3% starch, and 4% bentonite. Following this mixture's encapsulation, the microcapsules are thoroughly cleaned with sterile distilled water using a crosslinking calcium chloride solution (Saberi-Rise and Moradi-Pour 2020). Salicylic acid and nanoparticles have also been used to generate nanobiofertilizer. ZnONPs (1 g/mL), salicylic acid (1.5 mM), and sodium alginate (2% w/w) are added to the biofertilizer using this approach (Table 11.3).

The coating solution is made of calcium chloride (3% solution), and 1-mm beads are made, air-dried, and incubated at 4 °C (Panichikkal et al. 2021). A research claims that by combining organic waste from plants, animals, and food scraps with nanoparticles, nanobiofertilizer may be produced. Either pyrolyzed or allowed to decompose naturally, the organic waste was cleansed and broken down into little pieces before being rinsed with water to eliminate any impurities. According to

Table 11.3 Effects of nanobiofertilizers on various plants

Nanobiofertilizers	Plants	Composition	Mechanism reported	Response	Disadvantages	References
Nanosilicab fertilizers	Triticum aestivum	SiO2 NPs	Penetrates cell wall of root and enter extracellular and endothermal space	Root length 16.88% Shoot length 34.77%	Higher concentration of nanosilicab can reduce plant growth by affecting enzymatic and nonenzymatic activities	Akhtar and Ilyas (2022)
Zinc oxide nanobiofertilizers	Vicia faba L.	ZnO + CA + NPs	Via stomata or absorbed by roots	Root length 127.17% Shoot length 124.44% Chl a 206.66% and Chl b 162.69%	–	Ragab et al. (2022)
Iron and zinc nanobiofertilizers	CHICKPEA (Cicer arietinum)	Fe3O4 + ZnO nanoparticles	–	Shoot length 33.66 (cm) Chl a 129.17 and Chl b 202.33	It can be potentially toxic to soil microbes and their associated functions	Tului et al. (2021)
Titanium oxide and zinc oxide nanobiofertilizers	Zea mays L.	PGPR, TiO2, and ZnONPs	Reduce cd toxicity, resulting in defence responses and improved stress metabolism	Root length 9.1 AB ± 0.06 (cm) Shoot length 22.5 AB ± 0.02 (cm) increases of 502%, 238%, and 127% in the content of carotenoids and chlorophyll a and b	–	Shafiq et al. (2022)

Nanobiofertilizers	Plants	Composition	Mechanism reported	Response	Disadvantages	References
Nanozinc fertilizers	Cotton plant (Gossypium herbaceum)	NZn	–	Root length NZnO 2.16, NZn1 3.70, NZn2 2.54 Shoot length NZnO 6.50, NZn1 9.18, NZn2 12.17 dry weight g/plant	–	Hussein and Abou-Baker (2018)
Lithovit nanobiofertilizers	Date palm (Phoenix dactylifera L)	Inorganic- nano-CaCO3	Lithovit particles penetrate into the intercellular compartment and dock with their negatively charged surface on the outside of the cell membrane	Shoot length 7.34 and 7.22 (cm)	–	Ghazzawy (2020)
Nano-iron biofertilizers	Lemon (Citrus Limon L.)	Nano-iron+ *Bacillus subtilis*	–	Shoot length A 36.74, B 30.05, C 33.69 dry weight in g Chlorophyll A 960.29, B 829.71, C 600.0 µg.m-2	They can be potentially toxic to soil microbes and their associated functions, such as nitrogen (N) mineralization and decomposition of organic materials	Al-Zuhairi et al. (2021)
Nanopotassium biofertilizers	Eggplant (Solanum melongena L.)	Nanopotassium + *Azotobacter chroccocum*	–	Root length 21.73% Chlorophyll 56.12, 53.03, 51.29%	–	Al-Fahdawi and Allawi (2019)

(continued)

Table 11.3 (continued)

Nanobiofertilizers	Plants	Composition	Mechanism reported	Response	Disadvantages	References
Iron nano-oxide biofertilizers	Maize	Nano-iron oxide + *mycorrhiza* and *pseudomonas*	–	Protein content 0.044 0.011 increase in grain yield	They can be toxic	Eliaspour et al. (2020)
Nanosilicon biofertilizers	Wheat	Nanosilicon + rhizobacteria and *mycorrhiza*	–	Chlorophyll 2.87 mg	–	Nouraldinvand et al. (2021)
Nano-iron and molybdenum biofertilizers	Phaseolus vulgaris	Nano-iron and molybdenum + biofertilizers	–	Shoot length Pole 113.6, bush 50.5	They can be toxic	Al-Burki and Al-Ajeel (2021)
Nano Ca + B biofertilizers	Peanuts (*Arachis hypogea* L.)	Nanocalcium and boron + mycorrhiza and phosphorine	–	Shoot length 0.23 cm Protein content 0.02	Plants may suffer Ca toxicity. This may prevent the germination of seeds and reduce plant growth rates	Abdelghany et al. (2022)
Iron nanochelate biofertililizers	Chickpea (*Cicer arietinum* L.)	Phosphate Barvar-2 biofertilizer, iron nano-chelate and superabsorbent	Superabsorbent via maintaining soil moisture and preventing nitrogen leaching, iron nanochelate via providing iron required for nitrogen metabolism and phosphate Barvar-2 through affecting nitrogen fixation caused increase in grain protein and chickpea chlorophyll index	Protein content 21.86%	–	Olorunsogbon et al. (2022)

Nanobiofertilizers	Plants	Composition	Mechanism reported	Response	Disadvantages	References
Nano Fe-Si biofertilizers	Triticale (Triticosecale Wittmack)	Nano Fe-Si + *Azospirillum lipoferum* and *Pseudomonas putida*	–	Increased chlorophyll index (50.23%)	–	Aghaei et al. (2022)
Zinc nanobiofertilizers	Sunflower (*Helianthus annuus* L.)	Nano-Zn + Azotovit and phosphate	–	Shoot length 30.53 cm	–	Alamery and Ahmed (2020)

Markus et al. (2016), this partially degraded or pyrolyzed trash was mixed with nanoparticles to create nanobiofertilizer.

There are two methods for applying nanobiofertilizers to plants: either independently as biofertilizers and nanofertilizers or jointly as a nanobiofertilizer. When utilized independently, nanoparticles can directly impact plant growth by enhancing photosynthesis, carbon sequestration, seed germination, enzyme activity, and nitrogen fixation.

Nanobiofertilizers applied to soil, leaves, or seeds have a special capacity to penetrate plants (Pateiro et al. 2021). Nanobiofertilizers can easily pass through plant tissue and quickly reach their destination since they are so little (Garg et al. 2023). When injected into the rhizosphere, nanobiofertilizers enter the vascular tissue through the apoplastic route, which allows them to reach the centre of the root as well as its core. In the meantime, the symplastic route is utilized in this pathway to get beyond the Casparian strip barrier. Conversely, phyllospheric application of nanobiofertilizers causes them to follow the route of phloem translocation. By enlarging the chloroplast and grana size and amount of chlorophyll, the silicon nanocomposites help the plants become more photosynthetically capable (Yadav et al. 2021) (Fig. 11.2).

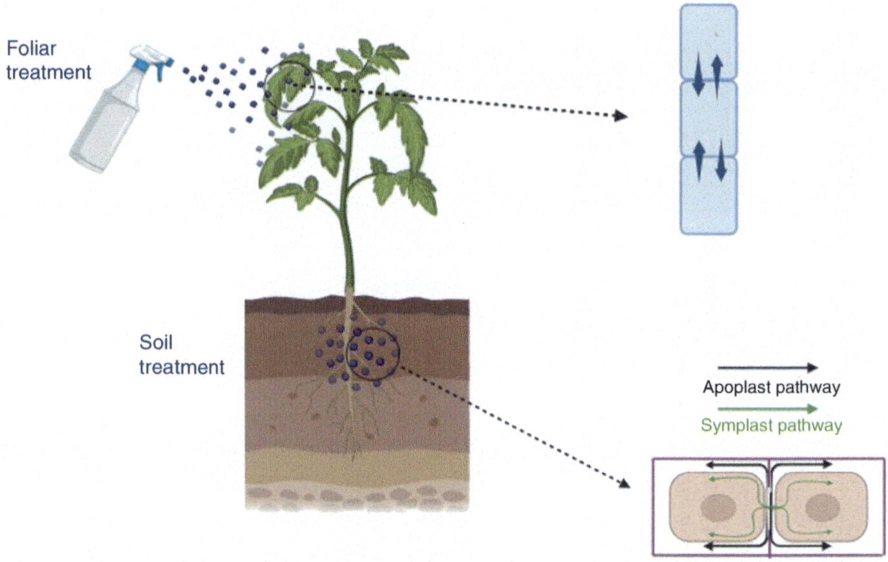

Fig. 11.2 Uptake of nanobiofertilizers by plant cells

11.5 Mechanisms of Action of Nanobiofertilizers

The intricate interaction between nanomaterials and biofertilizers, which is designed to maximize nutrient absorption, encourage plant development, and improve soil fertility, is the mechanism of action behind nanobiofertilizers. These strategies enhance soil health, ease nutrient transport, and reduce environmental stresses by using the distinct physicochemical features of nanoparticles (NPs) and the biological activities of biofertilizers. To fully utilize nanobiofertilizers in agricultural activities, it is imperative to comprehend these mechanisms. Nanomaterials' ability to modify soil characteristics is one of the main processes. Nanoparticles can affect the structure, texture, and porosity of soil because of their high surface area-to-volume ratio and reactivity. By improving soil aggregation, increasing soil aeration, and improving water retention capacity, they create an environment that is favourable for root development and nutrient uptake. Furthermore, since nanoparticles interact with soil constituents including organic matter and clay minerals, they facilitate the sorption, desorption, and mineralization of nutrients, hence improving plant nutrient availability (Pati et al. 2021b).

Nanomaterials assist in the regulated release of nutrients, which is another important technique. Through the encapsulation of biofertilizers inside matrices of nanomaterials, nanobiofertilizers enable the prolonged release of nutrients. By limiting nutrient losses from leaching and runoff and lowering the frequency of fertilizer treatments, this controlled-release method guarantees a steady supply of vital nutrients to plants. Additionally, biofertilizers are shielded by nanomaterials from environmental deterioration and unfavourable soil conditions, extending their effectiveness and improving their stability in the soil environment. The physiology and metabolism of plants are directly impacted by nanobiofertilizers, which enhance nutrient absorption and stress tolerance. Nanoparticles, especially those made of metal and metal oxide, have the ability to activate processes in plants that promote development and cause physiological reactions in plants, such as longer roots, better photosynthesis, and strengthened antioxidant defence systems. The bioactive qualities of nanoparticles, which include their capacity to control hormone signalling pathways, scavenge reactive oxygen species, and alter gene expression linked to stress reactions and nutrition transport, are thought to be responsible for these reactions (Rahaiee et al. 2020).

Additionally, the inherent qualities of nanoparticles provide nanobiofertilizers extra advantages. Because of their antibacterial qualities, some nanoparticles, for example, can inhibit soil-borne diseases and encourage the growth of advantageous microbes linked to biofertilizers. By lowering the frequency of plant illnesses and preserving soil health, its antibacterial action supports sustainable farming methods.

The modes of action of nanobiofertilizers raise questions about their possible safety and potential effects on the environment, notwithstanding any potential advantages. Comprehensive evaluation and risk management techniques are necessary due to the interaction between nanoparticles and soil microorganisms as well as their long-term impacts on soil health and ecosystem dynamics. Furthermore, assessing the possible effects of nanoparticles on soil and water quality requires a knowledge of their fate and movement in the environment. The several activities that together increase nutrient availability, encourage plant development, and boost soil health are all part of the mechanisms of action of nanobiofertilizers. Nanobiofertilizers provide creative answers to the problems of contemporary agriculture while fostering sustainability and environmental stewardship by utilizing the special qualities of nanomaterials and the biological activities of biofertilizers. To completely comprehend and improve these mechanisms for useful use in agricultural systems, further study and cooperation are nonetheless required (Akhtar et al. 2022).

11.6 Advantages of Nanobiofertilizers

The plethora of benefits that nanobiofertilizers provide position them as a viable agricultural breakthrough, bolstering soil health, crop production, and conservation measures. To start with, plants can receive vital nutrients from nanobiofertilizers in a targeted and effective manner. These formulations prevent environmental degradation of the active components and enable regulated release by encasing biofertilizers within nanomaterials, such as nanoparticles or nanocomposites. Nutrient losses due to leaching or volatilization are minimized by this targeted delivery method, which guarantees that nutrients are supplied progressively over time in accordance with the plant's development phases. In turn, this improves crop yields and quality since nanobiofertilizers increase the efficiency of nutrient absorption.

By limiting the negative environmental effects of chemical fertilizers and lowering dependency on them, nanobiofertilizers also contribute to environmental sustainability. The health of ecosystems and biodiversity is severely hampered by the use of traditional fertilizers, which frequently cause nutrient runoff, soil erosion, and water contamination. By using the special qualities of nanoparticles and the biological activity of microorganisms, nanobiofertilizers, however, provide a more environmentally friendly option for nutrient management. Nanobiofertilizers support soil health, water quality, and ecosystem resilience by encouraging biological nutrient cycling, enhancing soil structure, and reducing nutrient losses. Furthermore, because nanobiofertilizers have controlled-release characteristics, fewer fertilizer applications are needed, conserving resources and lessening the environmental impact of conventional agricultural methods (Singh et al. 2019).

Additionally, by raising crop yields, lowering input costs, and enhancing overall farm profitability, nanobiofertilizers contribute to farmers' financial gains. Nanobiofertilizers assist farmers to produce better-quality, higher-yield food by boosting plant growth and nutrient absorption efficiency. This increases market competitiveness and returns on investment and benefits farmers. Sustainable agricultural methods and long-term profitability are further enhanced by the long-term soil health advantages of nanobiofertilizers, which include increased soil fertility and less reliance on chemical inputs. Because they provide agronomic, environmental, and financial benefits that facilitate the shift to more resilient and sustainable agricultural systems, farmers can benefit from the use of nanobiofertilizers.

11.7 Challenges and Limitations

Despite their many potential benefits, nanobiofertilizers have several issues that need to be resolved before they can be widely used and commercialized in the agricultural industry.

11.7.1 Toxicity

The possible toxicity of nanomaterials to plants, soil organisms, and the environment is one of the main issues with nanobiofertilizers. Even while nanoparticles have special qualities that improve plant development and nutrient delivery, their unforeseen consequences on biological systems need to be carefully considered. Research has indicated that specific nanoparticles, especially those based on metals, may have phytotoxic or cytotoxic effects on soil microorganisms and plants, hence affecting the dynamics of ecosystems and soil fertility. Concerns have also been raised concerning the possible bioaccumulation and transmission of nanoparticles via the food chain, which raises issues with their long-term implications on the environment and human health. To guarantee the safe and responsible application of nanobiofertilizers in agriculture, thorough risk evaluations and toxicity studies are thus crucial (Khan et al. 2019).

11.7.2 Regulatory Issues

The creation, authorization, and marketing of nanobiofertilizers are severely hampered by regulatory concerns. Varied countries have varied rules, norms, and criteria when it comes to nanotechnology in agriculture, making the regulatory environment complicated and always changing. The licencing process is frequently hampered by regulatory authorities' lack of precise policies and procedures that are specialized to nanobiofertilizers. Standardized procedures for characterisation and risk assessment are necessary since nanoparticles have unique features and behaviours that make it difficult to classify and identify them. To overcome regulatory obstacles and provide market access for nanobiofertilizers, regulatory framework harmonization, stakeholder involvement, and open communication are required (Salata 2004).

11.7.3 Commercialization

The commercialization of nanobiofertilizers is hindered by several issues, such as scalability, cost-effectiveness, and market acceptability. Technological issues with manufacturing uniformity, quality control, and production costs arise when nanobiofertilizer production systems are scaled up from laboratory to commercial scale. The production of nanomaterials, which are frequently costly and resource-intensive, may restrict the affordability and availability of nanobiofertilizers for smallholder farmers and other agricultural sectors with low resources. Furthermore, elements like trust, transparency, and the perceived advantages vs. hazards of nanobiofertilizers affect their market acceptability and customer perception. To overcome commercialization barriers and encourage the use of nanobiofertilizers in agriculture, it is crucial to educate stakeholders, form alliances with key industry players, and highlight the agronomic and environmental advantages of these products.

Innovation in technology, multidisciplinary cooperation, and regulatory alignment are necessary to address the obstacles and constraints related to nanobiofertilizers. Nanobiofertilizers have the potential to change nutrient management, increase crop yield, and support sustainable agricultural practices by addressing toxicity concerns, negotiating regulatory challenges, and overcoming commercialization obstacles. To fully utilize nanobiofertilizers and realize their promise in tackling the global concerns of food security and environmental sustainability, more research, stakeholder involvement, and policy support are required (Aghaei et al. 2022).

11.8 Current Scenario of Nanobiofertilizer Development and Research

Since nanoparticles yield results quickly, researchers worldwide are investigating the function and application of nanoparticles in agriculture. Recently, some studies have been conducted that have shown the use and formulation of nanobiofertilizers. Iranian researchers have reported the majority of research on the evolution of ecologically friendly nanobiofertilizers. This accelerates the production of improved crop-care products (Bibi et al. 2022). In India, *Pseudomonas taiwanensis* and nano-gypsum have various applications; they increase soil moisture, enhance nutrient availability, and promote bacterial diversity and growth. Using this nanobiofertilizer enhanced the activity of soil enzymes as well. Nano-gypsum and *Pseudomonas taiwanensis* are combined to enhance the germination time, shoot length, root length, and chlorophyll content of *Zea mays* (Chaudhary et al. 2021). A 7-day application of nano-Zn and biofertilizer, according to a group of Indian experts, markedly improved grain output. Chinese and Pakistani researchers have worked together and found that the extreme amount of Zn is hazardous to plants and pollutes the soil. But nanobiofertilizers with Zn nanoparticles reduce zinc's negative effects and work well as a fertilizer. It raises the weight of the cob, the quantity of leaves, the weight of a thousand grains, the height of the plant, and the biological weight (Saboor et al. 2021).

11.9 Future Prospective

With the advancement of agriculture, it is expected that the use of nanobiofertilizers will rise. The promise of a novel approach to biotechnology and nanotechnology for sustainable crop management will always be welcomed to achieve nutritional security with constrained resources. Nanoparticle utilization is still in its experimental stages and is not yet widespread. It will take a lot of work to advance future research to reduce the hazards associated with employing nanobiofertilizers while keeping in mind both the beneficial and detrimental aspects of nano-technological interventions. Results from laboratory studies cannot be taken as absolute confirmation of the efficacy of a nanobiofertilizer method in actual agricultural practices. To accurately describe the environmental effects of nanobiofertilizers, testing should be conducted in natural environments using a suitable experimental methodology. To create nanobiofertilizers with high quality, ease of use, low cost, and long shelf life, modern tools and procedures must be used. Farmers need to be made aware of the negative consequences of chemical fertilizers as well as how nanobiofertilizers will be more affordable and have longer-lasting effects. It is necessary to create a multipurpose nanobiofertilizer that is suitable for many crops.

References

Abdelghany AM, El-Banna AAA, Salama EAA, Ali MM, Al-Huqail AA, Ali HM, Paszt LS, El-Sorady GA, Lamlom SF (2022) The individual and combined effect of nanoparticles and biofertilizers on growth, yield, and biochemical attributes of peanuts (*Arachis hypogea* L.). Agronomy 12(2):398

Aghaei F, Seyed Sharifi R, Farzaneh S (2022) The effect of some nanoparticles and biofertilizers on chlorophyll fluorescence components and some physiological traits of Triticale (Triticosecale Wittmack) at different irrigation levels. IJPB 14(3):13–40

Akhtar N, Ilyas N (2022) Role of nanosilicab to boost the activities of metabolites in Triticum aestivum facing drought stress. Plant Soil 477(1):99–115

Akhtar N, Ilyas N, Meraj TA, Pour-Aboughadareh A, Sayyed RZ, Mashwani Z-R, Poczai P (2022) Improvement of plant responses by nanobiofertilizer: a step towards sustainable agriculture. Nano 12(6):965

Alamery AA, Ahmed N A (2020) Effect of biofertilizers and zinc nano particles on growth, yield and oil percentage of sunflower (*Helianthus annuus* L.)

Al-Burki HAH, Al-Ajeel SAHS (2021) Effect of bio-fertilizer and nano-elements on growth and yield of two *Phaseolus vulgaris* L. varieties. Plant Archives 21(1):1191–1194

Al-Fahdawi AJJ, Allawi MM (2019) Impact of biofertilizers and nano potassium on growth and yield of eggplant (*Solanum melongena* L.). Plant Archives 19(2):1809–1815

Al-Zuhairi FFA, Al-Aareji JMA, Al-Taae AK (2021) Effect of nano and regular iron spraying and bio-fertilization on growth of local lemon transplants *Citrus limon* L. budded on sour orange. IOP Conf Series Earth Environ Sci 735(1):12045

Anand K, Kumari B, Mallick MA et al (2016) Phosphate solubilizing microbes: an effective and alternative approach as biofertilizers. Int J Pharm Pharm Sci 8(2):37–40

Bharathi S, Kumaran S, Suresh G, Ramesh M, Thangamani V, Pugazhvendan SR, Sathiyamurthy K (2020) Extracellular synthesis of nanoselenium from fresh water bacteria Bacillus sp., and its validation of antibacterial and cytotoxic potential. Biocatal Agric Biotechnol 27:101655

Bhat TA, Ahmad L, Ganai MA, Khan OA (2015) Nitrogen fixing biofertilizers; mechanism and growth promotion: a review. J Pure Appl Microbiol 9(2):1675–1690

Bhattacharyya PN, Jha DK (2012) Plant growth-promoting rhizobacteria (PGPR): emergence in agriculture. World J Microbiol Biotechnol 28:1327–1350

Bibi F, Ilyas N, Arshad M, Khalid A, Saeed M, Ansar S, Batley J (2022) Formulation and efficacy testing of bio-organic fertilizer produced through solid-state fermentation of agro-waste by *Burkholderia cenocepacia*. Chemosphere 291:132762

Chatterjee S, Mahanty S, Das P, Chaudhuri P, Das S (2020) Biofabrication of iron oxide nanoparticles using manglicolous fungus *Aspergillus niger* BSC-1 and removal of Cr (VI) from aqueous solution. Chem Eng J 385:123790

Chaudhary P, Khati P, Chaudhary A, Maithani D, Kumar G, Sharma A (2021) Cultivable and metagenomic approach to study the combined impact of nanogypsum and *Pseudomonas taiwanensis* on maize plant health and its rhizospheric microbiome. PLoS One 16(4):e0250574

Dhandapani P, Prakash AA, AlSalhi MS, Maruthamuthu S, Devanesan S, Rajasekar A (2020) Ureolytic bacteria mediated synthesis of hairy ZnO nanostructure as photocatalyst for decolorization of dyes. Mater Chem Phys 243:122619

Du C, Abdullah JJ, Greetham D, Fu D, Yu M, Ren L, Li S, Lu D (2018) Valorization of food waste into biofertiliser and its field application. J Clean Prod 187:273–284

Eliaspour S, Seyed Sharifi R, Shirkhani A, Farzaneh S (2020) Effects of biofertilizers and iron nano-oxide on maize yield and physiological properties under optimal irrigation and drought stress conditions. Food Sci Nutr 8(11):5985–5998

Etesami H, Emami S, Alikhani HA (2017) Potassium solubilizing bacteria (KSB): mechanisms, promotion of plant growth, and future prospects–a review. J Soil Sci Plant Nutr 17(4):897–911

Garcia-Ruiz A, Crespo J, López-de-Luzuriaga JM, Olmos ME, Monge M, Rodriguez-Álfaro MP, Martin-Álvarez PJ, Bartolome B, Moreno-Arribas MV (2015) Novel biocompatible silver nanoparticles for controlling the growth of lactic acid bacteria and acetic acid bacteria in wines. Food Control 50:613–619

Garg D, Sridhar K, Stephen Inbaraj B, Chawla P, Tripathi M, Sharma M (2023) Nano-biofertilizer formulations for agriculture: a systematic review on recent advances and prospective applications. Bioengineering 10(9):1010

Gatahi DM, Wanyika HN, Kavoo A, Kihurani A, Ateka EM (2016) Enhancement of bacterial wilt resistance and rhizosphere health in tomato using bionanocomposites. Int J Hort Sci Technol 3(2):129–144

Ghazzawy HS (2020) Enhance in vitro propagation of date palm (*Phoenix dactylifera* L.) Barhee cv. by using nano-bio fertilizers (Lithovit). Middle East J Appl Sci:572–582. https://doi.org/10.36632/mejas/2020.10.3.49

Glick BR et al (2012) Plant growth-promoting bacteria: mechanisms and applications. Scientifica 2012:1

Gouda S, Kerry RG, Das G, Paramithiotis S, Shin H-S, Patra JK (2018) Revitalization of plant growth promoting rhizobacteria for sustainable development in agriculture. Microbiol Res 206:131–140

Guang-Can TAO, Shu-Jun T, Miao-Ying CAI, Guang-Hui XIE (2008) Phosphate-solubilizing and-mineralizing abilities of bacteria isolated from soils. Pedosphere 18(4):515–523

Hasan S (2015) A review on nanoparticles: their synthesis and types. Res J Recent Sci 2277:2502

Hermosa R, Viterbo A, Chet I, Monte E (2012) Plant-beneficial effects of Trichoderma and of its genes. Microbiology 158(1):17–25

Hussein MM, Abou-Baker NH (2018) The contribution of nano-zinc to alleviate salinity stress on cotton plants. R Soc Open Sci 5(8):171809

Kasthuri J, Kathiravan K, Rajendiran N (2009) Phyllanthin-assisted biosynthesis of silver and gold nanoparticles: a novel biological approach. J Nanopart Res 11:1075–1085

Khan I, Saeed K, Khan I (2019) Nanoparticles: properties, applications and toxicities. Arab J Chem 12(7):908–931. https://doi.org/10.1016/j.arabjc.2017.05.011

Kumar S, Paritosh K, Pareek N, Chawade A, Vivekanand V (2018) De-construction of major Indian cereal crop residues through chemical pretreatment for improved biogas production: an overview. Renew Sust Energ Rev 90:160–170

Kumari RM, Kumar V, Kumar M, Agrawal A, Pareek N, Nimesh S (2020) WITHDRAWN: extracellular biosynthesis of silver nanoparticles using *Aspergillus terreus*: evaluation of its antibacterial and anticancer potential. Elsevier

Kundan R, Pant G, Jadon N, Agrawal PK et al (2015) Plant growth promoting rhizobacteria: mechanism and current prospective. J Fertil Pestic 6(2):9

Majhi KC, Yadav M (2021) Synthesis of inorganic nanomaterials using carbohydrates. In: Green sustainable process for chemical and environmental engineering and science. Elsevier, pp 109–135

Mala R, Celsia Arul Selvaraj R, Barathi Sundaram V, Blessina Siva Shanmuga Rajan R, Maheswari Gurusamy U (2017) Evaluation of nano structured slow release fertilizer on the soil fertility, yield and nutritional profile of *Vigna radiata*. Recent Pat Nanotechnol 11(1):50–62

Markus J, Mathiyalagan R, Kim Y, Abbai R, Singh P, Ahn S, Elizabeth Z, Perez J, Hurh J, Yang DC (2016) Intracellular synthesis of gold nanoparticles with antioxidant activity by probiotic lactobacillus kimchicus DCY51T isolated from Korean kimchi. In: Enzyme and microbial technology. Elsevier Inc. https://doi.org/10.1016/j.enzmictec.2016.08.018

Matei E, Gaidau C, Râpă M, Stefan LM, Ditu LM, Predescu AM, Stanca M, Pantilimon MC, Berechet MD, Predescu C, Mosutiu A (2021) Sustainable coated nanostructures based on alginate and electrospun collagen loaded with antimicrobial agents. Coatings 11(2):1–16. https://doi.org/10.3390/coatings11020121

Mathivanan K, Selva R, Chandirika JU, Govindarajan RK, Srinivasan R, Annadurai G, Duc PA (2019) Biologically synthesized silver nanoparticles against pathogenic bacteria: synthesis, calcination and characterization. Biocatal Agric Biotechnol 22:101373

Mazid M, Khan TA (2015) Future of bio-fertilizers in Indian agriculture: an overview. Int J Agric Food Res 3(3):10–23

McComb RB, Bowers GN Jr, Posen S (2013) Alkaline phosphatase. Springer Science & Business Media, pp 215–216

Mohammadi K, Sohrabi Y et al (2012) Bacterial biofertilizers for sustainable crop production: a review. J Agric Biol Sci 7(5):307–316

Nouraldinvand FA, Sharifi RS, Siadat SA, Khalilzadeh R (2021) Effects of nano silicon concentrations and bio-fertilizer on yield and grain filling components of wheat in different irrigation regimes

Olorunsogbon T, Adesanya Y, Atiyeh HK, Okonkwo CC, Ujor VC, Ezeji TC (2022) Effects of clostridium beijerinckii and medium modifications on acetone—butanol—ethanol production from switchgrass. Front Bioeng Biotechnol 10:942701

Panichikkal J, Puthiyattil N, Raveendran A, Nair RA, Krishnankutty RE (2021) Application of encapsulated *Bacillus licheniformis* supplemented with chitosan nanoparticles and rice starch for the control of *Sclerotium rolfsii* in *Capsicum annuum* (L.) seedlings. Curr Microbiol 78:911–919

Park S, Classen A, Gohou HM, Maldonado R, Kretschmann E, Duvernay C, Kim GJ, Ronholm J (2021) A new, reliable, and high-throughput strategy to screen bacteria for antagonistic activity against *Staphylococcus aureus*. BMC Microbiol 21(1):1–10. https://doi.org/10.1186/s12866-021-02265-4

Pateiro M, Gómez B, Munekata PES, Barba FJ, Putnik P, Kovačević DB, Lorenzo JM (2021) Nanoencapsulation of promising bioactive compounds to improve their absorption, stability, functionality and the appearance of the final food products. Molecules 26(6):1547. https://doi.org/10.3390/molecules26061547

Pati S, Dash A, Maity S, Pattnaik S, Mohapatra S, Samantaray DP (2021a) Nano biofertilizer for sustainable agriculture. In: Agriculturally important microorganisms: mechanisms and applications for sustainable agriculture, vol 1(12). Routledge, pp 201–208. https://doi.org/10.1201/9781003245841-10

Pati S, Dash A, Maity S, Pattnaik S, Mohapatra S, Samantaray DP (2021b) Nano biofertilizer for sustainable agriculture. In: Agriculturally important microorganisms. CRC Press, pp 201–208

Pieterse CMJ, Zamioudis C, Berendsen RL, Weller DM, Van Wees SCM, Bakker PAHM (2014) Induced systemic resistance by beneficial microbes. Annu Rev Phytopathol 52:347–375

Qureshi A, Singh DK, Dwivedi S (2018) Nano-fertilizers: a novel way for enhancing nutrient use efficiency and crop productivity. Int J Curr Microbiol App Sci 7(2):3325–3335

Ragab SM, Turoop L, Runo S, Nyanjom S (2022) Nanoparticle treatments based on zinc oxide and *Moringa oleifera* leaf extracts alleviate salinity stress in faba bean (*Vicia faba* L.). JACEN 11(1):42–65

Rahaiee S, Assadpour E, Faridi Esfanjani A, Silva AS, Jafari SM (2020) Application of nano/microencapsulated phenolic compounds against cancer. Adv Colloid Interf Sci 279:102153. https://doi.org/10.1016/j.cis.2020.102153

Rai M, Yadav A, Gade A (2009) Silver nanoparticles as a new generation of antimicrobials. Biotechnol Adv 27(1):76–83

Rajan A, Cherian E, Baskar G (2016) Biosynthesis of zinc oxide nanoparticles using *Aspergillus fumigatus* JCF and its antibacterial activity. Int J Mod Sci Technol 1(2):52–57

Ramanathan R, Field MR, O'Mullane AP, Smooker PM, Bhargava SK, Bansal V (2013) Aqueous phase synthesis of copper nanoparticles: a link between heavy metal resistance and nanoparticle synthesis ability in bacterial systems. Nanoscale 5(6):2300–2306

Rani P, Rajput S, Thakur B, Kaur S (2023) Immobilization and co-mobilization: an unexploited biotechnological tool for enhancing efficiency of biofertilizers. In: Metabolomics, proteomes and gene editing approaches in biofertilizer industry. Springer Nature Singapore, Singapore, pp 219–236

Saberi-Rise R, Moradi-Pour M (2020) The effect of Bacillus subtilis Vru1 encapsulated in alginate—bentonite coating enriched with titanium nanoparticles against *Rhizoctonia solani* on bean. Int J Biol Macromol 152:1089–1097

Saboor A, Ali MA, Ahmed N, Skalicky M, Danish S, Fahad S, Hassan F, Hassan MM, Brestic M, El Sabagh A et al (2021) Biofertilizer-based zinc application enhances maize growth, gas exchange attributes, and yield in zinc-deficient soil. Agriculture 11(4):310

Salata OV (2004) Applications of nanoparticles in biology and medicine. J Nanobiotechnol 2:1–6. https://doi.org/10.1186/1477-3155-2-3

Schlüter M, Hentzel T, Suarez C, Koch M, Lorenz WG, Böhm L, Düring R-A, Koinig KA, Bunge M (2014) Synthesis of novel palladium (0) nanocatalysts by microorganisms from heavy-metal-influenced high-alpine sites for dehalogenation of polychlorinated dioxins. Chemosphere 117:462–470

Shafiq T, Yasmin H, Shah ZA, Nosheen A, Ahmad P, Kaushik P, Ahmad A (2022) Titanium oxide and zinc oxide nanoparticles in combination with cadmium tolerant bacillus pumilus ameliorates the cadmium toxicity in maize. Antioxidants 11(11):2156

Shahcheraghi N, Golchin H, Sadri Z, Tabari Y, Borhanifar F, Makani S (2022) Nano-biotechnology, an applicable approach for sustainable future. 3 Biotech 12(3):65

Singh P, Ghosh D, Manyapu V, Yadav M, Majumder S (2019) Synergistic impact of iron (III) oxide nanoparticles and organic waste on growth and development of *Solanum lycopersicum* plants: new paradigm in nanobiofertilizer. Plant Archives 19:339–344

Spadari C d C, Lopes LB, Ishida K (2017) Potential use of alginate-based carriers as antifungal delivery system. Front Microbiol 8:1–11. https://doi.org/10.3389/fmicb.2017.00097

Tairo EV, Ndakidemi PA (2013) Possible benefits of rhizobial inoculation and phosphorus supplementation on nutrition, growth and economic sustainability in grain legumes. Am J Res Comm 1(12):532–556

Tului V, Janmohammadi M, Abbasi A, Vahdati-Khajeh S, Nouraein M (2021) Influence of iron, zinc and bimetallic Zn-Fe nanoparticles on growth and biochemical characteristics in chickpea (*Cicer arietinum*) cultivars. Agri Forest 67:179–193

Usman M, Farooq M, Wakeel A, Nawaz A, Cheema SA, ur Rehman, H., Ashraf, I., & Sanaullah, M. (2020) Nanotechnology in agriculture: current status, challenges and future opportunities. Sci Total Environ 721:137778

Vejan P, Abdullah R, Khadiran T, Ismail S, Nasrulhaq Boyce A (2016) Role of plant growth promoting rhizobacteria in agricultural sustainability—a review. Molecules 21(5):573

Yadav AN, Verma P, Singh B, Chauhan VS, Suman A, Saxena AK (2017) Plant growth promoting bacteria: biodiversity and multifunctional attributes for sustainable agriculture. Adv Biotechnol Microbiol 5(5):1–16

Yadav SK, Patel JS, Singh BN, Bajpai R, Teli B, Rajawat MVS, Sarma BK (2021) Biofertilizers as microbial consortium for sustainability in agriculture. In: Plant, soil and microbes in tropical ecosystems. Springer, pp 349–368

Youssef MMA, Eissa MFM (2014) Biofertilizers and their role in management of plant parasitic nematodes. A review. J Biotechnol Pharm Res 5(1):1–6
Zaidi A, Khan M, Ahemad M, Oves M (2009) Plant growth promotion by phosphate solubilizing bacteria. Acta Microbiol Immunol Hung 56(3):263–284

Foliar-Based Nanoformulations: Leads and Flaws

12

Gajendra B. Singh, Adarsh Sharma, Joythi Thapa, and Nidhi

Abstract

Nanotechnology presents significant opportunities to enhance the quality of life in various fields, including agriculture and food systems. Researchers are actively exploring the innovative ways to develop various nano-based agro products like nanofertilizers, nanopesticides, etc. for foliar application. Foliar application is a method of treating plants by administering liquid formulations directly onto their leaves. Numerous nanoformulations have recently developed to strengthen foliar application's efficacy by overcoming many inherent challenges like precise targeting of active compounds to specific plant tissues, improved penetration, and controlled release and reduced environmental impact. Nanoencapsulation provides protection against degradation, prolonging the shelf life of formulations. Plants absorb essential nutrients via stomata, cuticle, and trichomes during foliar application. Foliar nanoformulations offer several advantages over conventional counterparts by providing a direct route for nutrient delivery to plants without relying on the root system or requiring adequate topsoil moisture, under adverse conditions such as root diseases and drought. Notably, these formulations require only small quantities for application while delivering nutrients in a controlled and sustainable manner leading to improved and increased yield quality standards. Timing of foliar spray is pivotal for effectiveness and minimal plant stress. Ideal times are early morning or late evening when temperatures are mild, light is low, and wind speed is minimal. This promotes optimal absorption and minimizes heat stress. This chapter delves into the pros and cons of foliar-based nanoformulations and highlights further research and development in this area.

Keywords

G. B. Singh (✉) · A. Sharma · J. Thapa · Nidhi
Department of Biotechnology, UIBT, Chandigarh University, Mohali, Punjab, India
e-mail: gajendra.singh@cumail.in

S. Kaur et al. (eds.), *Metabolomics, Proteomics and Gene Editing Approaches in Biofertilizer Industry*, https://doi.org/10.1007/978-981-97-2910-4_12

Nanoformulations · Foliar · Nanofertilizer · Nanopesticide · Agriculture · Nanofarming · Nanoparticles

12.1 Nanoformulation for Foliar Application

12.1.1 Nanoformulation

The latest engine for the advancement of environmentally friendly farming is nanotechnology. It makes smart nano-based formulations for minerals and insecticides using nanomaterials, which can help tackle some of the challenges that traditional farming faces. Various special characteristic has been shown by nanomaterials because of their large surface area, small structure, and the ability to alter the properties depending on their intended application. These qualities make them appropriate for use in formulations of pesticides based on nanotechnology. The features and activity of pesticides, such as their ability to disperse, spread, stay stable, migrate to their targeted location, can be improved through nano-based formulations. By making traditional pesticides more effective, control release of active ingredient (AI), durable, low requirement, and less susceptible to environmental breakdown, they can also significantly boost efficiency, safety, and financial benefits. By doing this, fewer pesticides may leak into the environment and leave behind dangerous residues. Additionally, in response to biological and external stimuli, nanoformulations can disperse their AI and split them more accurately through regulated dispersion mechanisms or targeted distribution. They can become more environmentally friendly and effective as a result. One of the most popular subjects in the field of nanotechnology for agriculture is the invention of new, sophisticated nanoformulations that can enter and deliver to the desired site, remain steady and active under spray conditions (sun, heat, and rain), extend the period of action, and minimize waste into the environment (Zhao et al. 2017).

12.1.2 Different Varieties of Nanoformulation

Nanoformulations come in various types, such as nanoemulsion, nanocapsule, nanosphere, and nanogel. Figure 12.1 represents water-based dispersion nanoformulation.

12.1.2.1 Nanoemulsion

Nanoemulsion refers to emulsions with droplet sizes of less than 100 nm. They are made up of water, oil, and an emulsifier (Dey and Singh 2023). For the formation of tiny droplets, emulsifier is essential because it lowers the interfacial tension between oil and water phases. Through electrostatic repulsion and steric hindrance, the emulsifier also stabilizes nanoemulsions by preventing droplet coalescence. The kinetic stability of nanoemulsion is because they do not split instantly instead require a while to do so (Gupta et al. 2016). Fertilizer or pesticide nanoemulsion

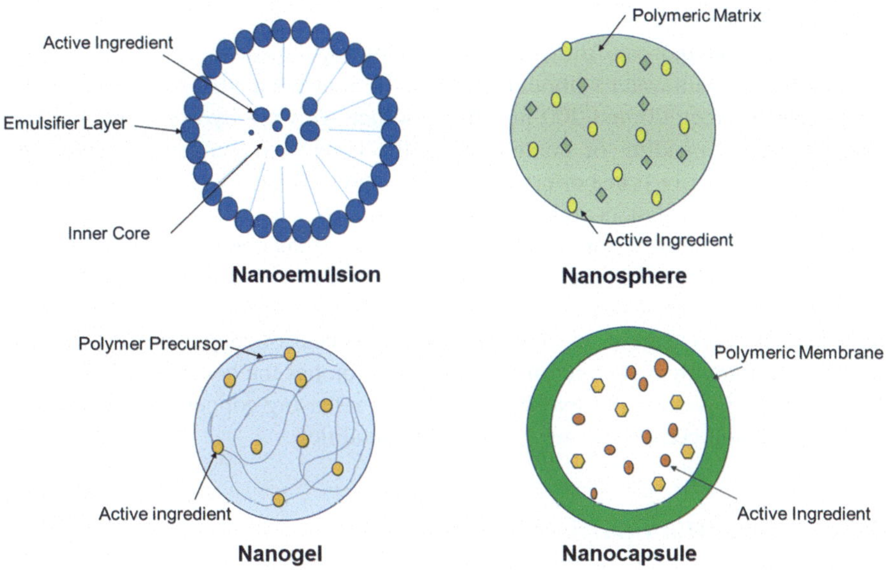

Fig. 12.1 Various types of water-based dispersion nanoformulation

made and distributed in a biphasic media dissolves the substance and significantly increases its absorption and effectiveness (Feng et al. 2018).

12.1.2.2 Nanosphere

Nanosphere is spherical polymeric diameters ranging from 10 to 200 nm (Mehndiratta et al. 2015). It is an aggregate with a uniform distribution of the active ingredient throughout the matrix of polymers. The matrix either dissolves, encapsulates, entraps, or gets attached to the AI in a uniform manner and shields it from the chemical and enzymatic deterioration. Nanosphere has hollow structure with lower mass, this allows the better loading of functional ingredient and makes it suitable for the targeted delivery of the AI. The composition of the polymer and its fluid-absorbing capacity determine the rate of release of AI at the target. Nanosphere releases the AI through diffusion in a controlled and sustainable manner (Pathak et al. 2019).

12.1.2.3 Nanocapsule

Nanocapsule is a colloidal particle created through the polymerization of premade polymers, such as PEG, PLGA, PCL, and PLA. Although their size varies between 5 and 1000 nm, they are typically between 100 and 500 nm. They are composed of an oil-filled core which contains AI and has a polymeric shield. It is further encapsulated with nanomaterial which controls the monitored delivery in a response to the environmental, chemical, biological, and thermal stimuli (Pathak et al. 2019).

12.1.2.4 Nanogel

Nanogel is small nanoscale, three-dimensional hydrogel made up of swellable polymer which can retain water without melting in the aqueous media. Nanogels usually have a size between 20 and 200 nm. Because nanogels are hydrophobic and less susceptible to expanding or contracting with changes in temperature, they are thought to be superior transporters than nanospheres (Bhagat et al. 2013). In addition to preventing outbursts and possible breaches, they can greatly enhance packing and discharge ability (Paula et al. 2011). In accordance with the guidelines for sustainable agriculture (Dwivedi et al. 2016), the use of nanogels is advised for the distribution of essential oils and signals. According to Brunel et al. (2013), pure chitosan nanogels can be used to boost the effectiveness of copper-based fungicide therapies. Employing a nanogel instead of a solution has the benefits of being simpler to handle, having better leaf dispersion, and releasing copper onto leaf or soil gradually without compromising its fungicidal qualities.

12.1.3 Carrier Nanoparticle

12.1.3.1 Silica Nanoparticle

Abamectin, an insecticide, was enclosed with scratched silica nanoparticles produced with NaOH. By enclosing it, the abamectin was protected against photolysis. In another work, nanoporous silica nanoparticles were used for encasing the insecticide avermectin. The greatest quantity of avermectin which was dispersed at pH 10 in 30% ethanol was almost 42% w/w. By using 0.4 mg/mL nanomaterials in 40% ethanol, an alternate encapsulating technique that made use of supercritical CO_2 fluid was capable of to increase the release duration of avermectin by 1.87 days to almost 27 days. Additionally, by using this method, the avermectin was protected from UV degradation (Pestovsky and Martínez-Antonio 2017).

12.1.3.2 Gold Nanoparticles

The *Camellia sinensis* leaf after pretreatment with 250 mg/L of gold nanoparticles was treated with ferbam, both chemicals were able to enter inside, this phenomenon took place despite ferbam not being widespread effective. Ferbam was absorbed by the plant more quickly after tiny number of nanoparticles were added (Hou et al. 2016).

12.1.3.3 Zinc Oxide Nanoparticles

Zinc oxide nanoparticles were used to indirectly disable the pesticide thiram. The pesticide showed dependent on dose action toward the infectious fungi *Phytophthora capsici*, with concentrations between 0.05 and 2 g/L. It was a procedure that produced species of oxygen that were reactive. It is fascinating to note that the nanoparticles helped the pesticide photodegradation, which successfully removed any leftover thiram after the fungicidal treatment (Xue et al. 2014).

12.1.3.4 Calcium Alginate Nanoparticles

At doses ranging from 5 and 10 mg/L, the pesticide imidacloprid showed less cytotoxicity toward Vero cell populations when it was enclosed in calcium alginate nanoparticles as opposed to its permitted form. Conversely, it showed improved effectiveness toward the leafhoppers 11–15 days after spraying at a pesticide dosage of 0.145 mg/L (Pestovsky and Martínez-Antonio 2017).

12.1.3.5 Chitosan Nanoparticles

When enclosed in sodium tripolyphosphate-crosslinked chitosan nanoparticles, the volatile oil of *Zataria multiflora* exhibited improved effectiveness toward the plant's infectious fungus *Botrytis cinerea*. It was seen once the oil was added to the strawberries (Fragaria × ananassa) at an amount of 1500 ppm and in the laboratory at amounts that varied from 188 to 1500 ppm. It is important to remember that inside the exact same concentration range, the untreated chitosan nanoparticles itself showed some antifungal properties (Mohammadi et al. 2015).

12.1.3.6 Polyethylene Glycol Nanoparticles

Acephate, an organic phosphate pesticide, was successfully prevented from breaking down by encasing it in polyethylene glycol nanoparticles. There was neither cytotoxicity nor oral toxicity in the final composition. In contrast with standard acephate, administered *Spodoptera litura* exposed to 180–300 ppm of enclosed acephate showed a concentration-related reduction in acetylcholinesterase activity. As a result, *Spodoptera litura* and tea spider mites (*Oligonychus coffeae*) had a high death rate. Under field circumstances, *Spodoptera litura* and *Lipaphis erysimi* reemerged on the seventh day, while *Bemisia tabaci* reemerged after 5 days. In a natural setting, the nanoparticles were sprayed one time (Pradhan et al. 2013). Table 12.1 outlines a variety of carrier nanoparticles.

12.2 Application Methods

Nanoformulation (NFs) must be administered in a manner that preserves their robust properties, including efficacy, controlled time-release, solubility, targeted delivery, stability, and reduced toxicity. The efficacy plus influence of these NFs are predominantly determined by their application methods for safe delivery and disposal. Below are enumerated several approaches for delivering nanofertilizers to plants.

12.2.1 In Vitro Methods

Aeroponics and hydroponics are methodologies for cultivating plants in soilless culture. In aeroponics, the root of plant is situated in an environment devoid of solid medium, receiving nutrients through a consistent misting application. This technique offers precise control over the root zone environment. However, its less

Table 12.1 Various carrier nanoparticles

Active ingredient (AI)	Nanoparticle	Quantity	Effect	References
Avermectin	Silica	0.4 mg/mL in 40% ethanol	Protects avermectin from photolysis, UV degradation, and increases the release duration of Avermectin	Pestovsky and Martínez-Antonio (2017)
Ferbam	Au	250 mg/L	Increased the absorption rate of ferbam by plant leaf	Hou et al. (2016)
Thiram	ZnO	0.05–2 g/L	Thiram shows dose dependent activity and disabled thiram by photodegradation	Xue et al. (2014)
Imidacloprid	Calcium alginate	5–10 mg/L	Decreased the cytotoxicity of imidacloprid toward Vero cell	Pestovsky and Martínez-Antonio (2017)
Sodium tripolyphosphate	Chitosan	188–1500 ppm	Increases antifungal property of *Zataria multiflora* oil	Mohammadi et al. (2015)
Acephate	Polyethylene glycol	180–330 ppm	Concentration-related reduction in acetylcholinesterase activity of *Spodoptera litura*	Pradhan et al. (2013)

widespread adoption is due to its substantial nutrient demand necessary for sustained and rapid plant development (Kalra et al. 2020). In contrast, hydroponics, too referred to "solution culture," utilizes natural liquid chemical salts to submerge the root of plant in a macronutrients and micronutrients solution. Factors such as pH, oxygen availability, and solution volume are meticulously managed, often incorporating supporting materials like sand or gravel in commercial setups. The system operates by replacing used solution with fresh nutrient solutions, yet it faces challenges such as common pathogen intrusion and high moistness levels that can induce excessive wilting in grounded plants (Solanki et al. 2015).

12.2.2 In Vivo Methods

In vivo methodologies encompass distinct approaches, notably foliar and soil application, for administering micronutrients to plants. The foliar method involves the direct delivery of micronutrients to plant leaves through the application of liquid fertilizers via spraying. An advantage of this technique is the expedited nutrient uptake by plants during their rapid growth phase due to the decreased duration between fertilizer application and absorption.

Conversely, soil application entails the deposition of micronutrients like iron (Fe), manganese (Mn), and copper (Cu) onto soil particles, enhancing the uptake of these specific micronutrients (Taiz and Zeiger 2010). This method largely involves the absorption of nutrients by leaf epidermal cells and stomata. Implementing nanofertilizers through this approach may offer significant agronomic advantages in terms of nutrient assimilation.

Consideration of the duration for stomatal pore opening is essential due to its involvement in the process. The study focused on the application of nitrogen, phosphorus, and potassium with using a mixture of nanoparticles as foliar treatments. Significantly positive results were observed upon utilizing nanochitosan-NPK fertilizer on wheat, as it notably increased the mobilization, crop, and yield index. This phenomenon was replicated in watermelon plants through nanocomposite foliar uptake (Wang et al. 2013). Moreover, foliar practices of gold nanoparticle (NPs) showed promising results in terms of enhancing branches height, branch number, width, and seed production (Marzouk et al. 2019). Optimal foliar uptake depends not only on particle size but also on various factors like plant species, environmental conditions (including gas, temperature, water), and nanoparticle delivery techniques (Birbaum et al. 2010).

12.3 Mechanism of Foliar Uptake

The underlying premise centers on the solubility of specific fertilizer nutrients in water, facilitating their direct application to the aerial tissues of plants. The nutrient assimilation occurs through either cuticular penetration or stomatal entry, subsequently traversing into the plant cell where it becomes available for metabolic processes.

These nutrients (nanoparticles) applied through foliage can infiltrate plants via the leaf epidermis or stomata, subsequently traversing either the apoplastic or symplastic pathway (Ullah et al. 2020; Lv et al. 2019). The xylem and phloem systems are pivotal in the passage of these nanoparticles (Raliya et al. 2016), while accumulation primarily occurs within the vacuoles and cell walls (Shahid et al. 2017; Li et al. 2020; Singh et al. 2011). Figure 12.2 shows several variables affect the assimilation, translocation, and aggregation of nanoparticles (Su et al. 2019), including abiotic elements such as humidity and temperature, as well as the physical characteristics of the nanoparticles (size and shape), and the physiological traits of the plants. Encompassing abiotic elements such as humidity and temperature, and physical characteristics of the nanoparticles themselves like (size and shape), and the physiological characters of the plants. Notably, foliar application of NPs demonstrates increased efficacy and environmental friendliness compared to exposure through roots or soil (Achari and Kowshik 2018; Clarke et al. 2020).

Figure 12.3 depicts the diverse determinants, encompassing plant attributes, ambient circumstances, and nanoparticle properties, which collectively exert influence on the absorption and translocation processes associated with foliar-applied nanoparticles (Rana et al. 2021).

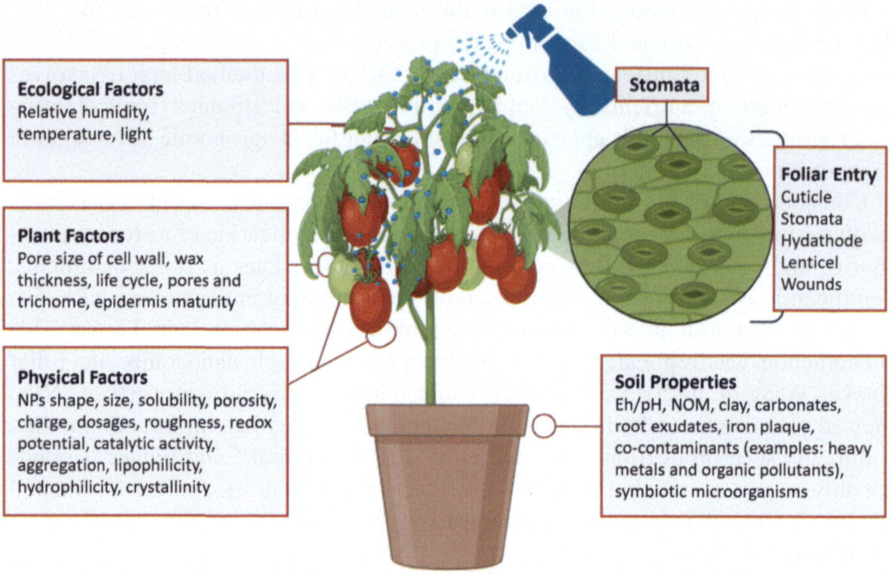

Fig. 12.2 Variables influencing the absorption and movement of engineered nanoparticles in plants through foliar applications. (Image created by BioRender.com)

In order for foliar-based nanofertilizers to facilitate plant growth through assimilation, the nutrients must initially traverse the cuticle and epidermal layers of the leaf, subsequently gaining entry into the cytoplasm of the leaf's cellular structure. The primary hurdle in this process is believed to be the cuticle, which presents the greatest resistance to nutrient penetration. Once this penetration is accomplished, the uptake of nutrients by leaves is expected to be quite similar to the absorption process of the same nutrients by roots. The key discrepancy lies in the distinct environments within which these plant components operate.

Figure 12.4 illustrates there exist two primary pathways for the absorption of foliar-based nanofertilizers into leaves, namely via the stomata and the external cuticle. While the predominant mode of nutrient uptake is commonly believed to be through the cuticle, soluble nutrients also gain entry indirectly through the stomata. However, a contentious discussion persists regarding the significance of stomatal absorption into the internal structure of the leaves. Before 1970, substantial discussions surrounded the significance of stomata intake of foliar-based nanoformulations (Oosterhuis 2009). It was shown that because of the surface tension of water, hydrophobic characteristics of leaf surfaces, and the configuration of the stomata, it was not possible for a droplet to penetrate the large plant leaf. Additionally, the rates of ions intake of foliar sprays typically show high efficiency during the night, a period when stomata are closed, compared to the daytime when they are open. New evidence has recently emerged suggesting the potential uptake of large anions through stomata, indicating the possibility of these structures serving as a pathway for a limited nutrient entry into leaves. Conventionally, the prevailing notion is that

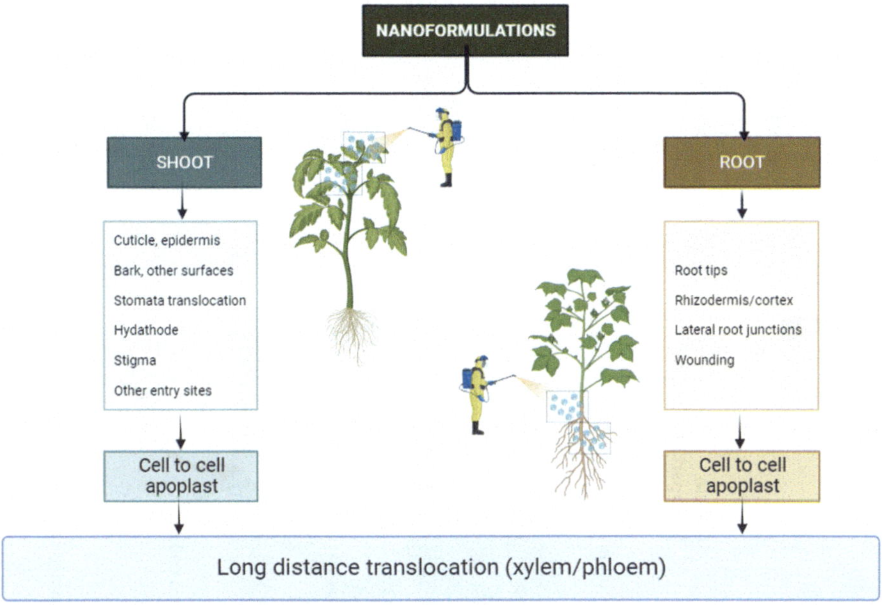

Fig. 12.3 Proposed routes for nanoparticle internalization and movement within plant structures. (Image created by BioRender.com)

the intake of water and soluble nutrients occurs predominantly through the cuticle of leaf unless surfactants are present. It is important to note that this stomatal penetration is viable only in the immediate aftermath of application when the spray deposits are in a liquid state. Subsequently, the cuticle remains the primary pathway for uptake.

Stomatal entry serves as a primary route for absorption on the leaf exterior, as the epidermis restricts absorption because of minute opening size (Xiong et al. 2016). Research indicates that particles exceeding a size of 20 nm encounter impediments during translocation through cellular membrane apertures (Alshaal and El-Ramady 2017). Although stomata are commonly found on the lower surface of leaf, specific plant species possess stomata on the upper surface, potentially enhancing the prospect of nanoparticle absorption on the leaf's exterior. Furthermore, younger leaves demonstrate higher nutrient absorption capabilities compared to more mature leaves, attributed to their thinner wax layer and less developed physiological barriers that impede metal absorption (Shahid et al. 2017). This phenomenon mirrors the nutrient absorption process witnessed in maturing seeds, primarily relying on the utilization of new leaves because of the lack of leaf wax barriers formation (Rawat et al. 2017).

Nanoparticles exhibit an inclination to deposit and accumulate on the surface of leaves, possibly leading to their absorption by plants over a period of time. Larger particles typically adhere to the waxy leaf surface, while smaller particles can permeate through leaf trichomes, entering the leaf and being delivered to various plant

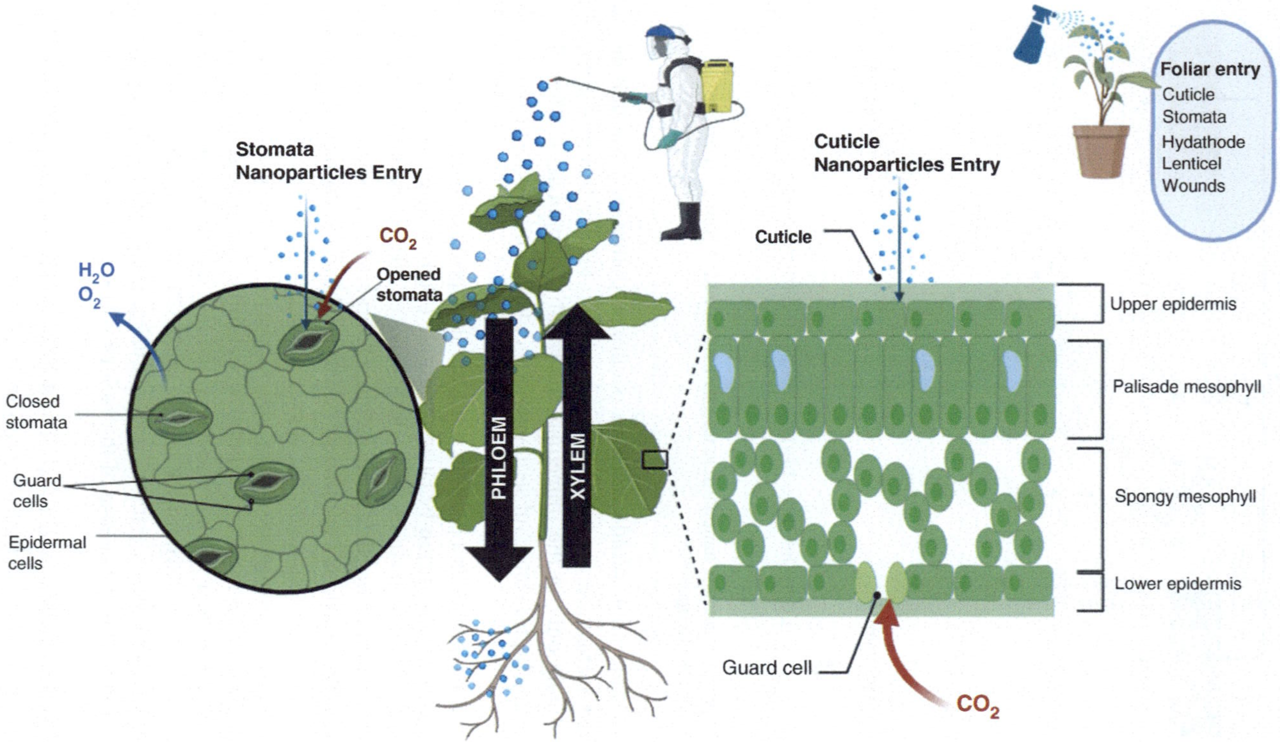

Fig. 12.4 Schematic diagram of the uptake and translocation pathways of NPs in plants. (Image created by BioRender.com)

parts. Research has confirmed the detection of undamaged nanoparticles within the mesophyll, epidermis, and vascular tissues of leaves that were sprayed with silver nanoparticles. Studies have elucidated the transfer phenomenon of nanoparticles from foliar tissues to diverse anatomical regions within plants, including roots and nascent leaves (Xiong et al. 2019; Li et al. 2017; Sharma et al. 2017). Furthermore, observations have revealed that aerosolized gold nanoparticles (spherical, 30–90 nm) are capable of entering plant stomata through gaseous absorption, although the magnitude of this uptake is limited (Raliya et al. 2015). A relative study analyzed the application of NPs through foliar spray in suspension and aerosol form. It was found that foliar spraying of NP suspension resulted in the delivery of 60–70%. However, the analogous absorption rate of these NPs necessitates further investigation (Su et al. 2019).

Internalization of nanoparticles within plant cells predominantly occurs via stomata, with their subsequent movement contingent on particle size. Stomatal pores in leaves typically measure around 100 nm in diameter, although the waxy and hydrophobic nature of stomata can limit the entry of larger particles (Su et al. 2019). Larger particles ranging from 50 to 200 nm primarily traverse through the apoplast, while smaller particles sized between 10 and 50 nm are chiefly transported through the symplast. After being absorbed by plants, NPs (nanoparticles) move through the plant's vascular system, specifically the phloem (Achari and Kowshik 2018; Raliya et al. 2016). Zinc is taken in by the outer layer cells of plants, then moved to the vascular bundle, and finally transported to the grain organ through the phloem. Research indicates that smaller-sized NPs exhibit enhanced absorption through pores of leaf, resulting in a larger influx of nanoparticles into the plants (Deshpande et al. 2017).

Nanoparticle exhibit various pathways for entry into plant cells, encompassing processes such as endocytosis, utilization of ion or water molecular pathways, formation of carrier protein complexes, induction of physical damage, and engagement with organic constituents (Zahedi et al. 2020). Upon uptake by plants, these NPs can undergo redox reactions and other transformative processes leading to alterations in their morphology. Studies have demonstrated that specific foliar nanoparticle (NP) types possess the capability to make the development of novel openings in the cellular wall, thereby enhancing their penetration into cells.

Plants via pre-existing pores or fissures in their foliage can easily take up iron nanoparticles. This uptake process may be facilitated by enzymes and chelating agent that participate in the transportation of Fe to all parts of plant (Jalali et al. 2017). Studies have suggested that foliar-applied NPs are primarily absorbed through hydrophilic pathways, as opposed to lipophilic routes, possibly due to the encapsulation and absorption of NPs by the epidermal wax before penetrating the plant tissue (Hu et al. 2017). Endocytosis represents an additional mechanism by which nanoparticles (NPs) are internalized by plant cells. NPs that traverse the cell wall can access the phloem through endocytosis, potentially facilitating their transportation to the plant roots (Wang et al. 2016; Kranjc et al. 2018). Observations have shown that nanoparticles applied to leaves can enter plants via trichomes and subsequently spread to whole plant (Anusuya and Sathiyabama 2015). Research has demonstrated that NPs sprayed onto foliage can move in two directions: downwards toward the roots via phloem or upwards to different plant parts through the xylem (Ma et al. 2018).

The process through which aqueous solutions penetrate plant leaves via stomata is intricate and subject to diverse influencing factors, including the surface tension properties of the liquid, the angle of incidence upon the leaf surface, and the structural characteristics of the opening walls (Greene and Bukovac 1974). Moreover, the penetration of stomata takes place exclusively during the transient phase following application, when the sprayed deposits are in a liquid state. Subsequently, the cuticle continues to be the exclusive pathway for further uptake.

12.4 Comparative Analysis of Foliar and Basal Uptake Mechanisms

Physiological roles of leaf and root in plants involve distinct roles that influence the absorption of nanoparticles (NPs) within these plant organs. Roots produce particular exudates that hinder the uptake of NPs, while leaves have the ability to capture NPs within their cell walls after being applied to the foliage. When plants are displayed to metallic nanoparticles like silver and zinc oxide nanoparticles via either leaf or root, the levels of intake and resulting toxicity vary significantly (Thunugunta et al. 2018). Plant nutrient application to foliage results in downward transport and circulation throughout the plants (Kannan and Charnel 1986). Likewise, the application of nanoparticles (NPs) to leaves elevates total nanoparticles in the root system, aiding nutrient absorption by root cells and bolstering their resilience against diseases (Elmer and White 2016). Conflicting results indicate that the translocation of metallic nanoparticles to plant roots through foliar application or their transfer to shoots via root absorption may not consistently occur. This variability could contribute to significant differences in toxicity levels.

Stomata in plants exhibit a higher amount of nutrient uptake compared to root cell. As a result, employing nanoformulations via foliar application offers a swifter correction of nutrient deficiencies in plants by enhancing the aggregation of the applied nutrients (Alshaal and El-Ramady 2017). This method is also proven effective in supplementing specific nutrients crucial for plant growth without exerting an impact on the soil environment, as mentioned in source. It is crucial to emphasize that employing identical nanoparticles (NPs) via root or foliar techniques may lead to divergent outcomes owing to the variations in response times. Specifically, soil-applied NPs demonstrate slower responses compared to foliar application. Moreover, these methods differ in terms of bioavailability, primarily due to the environmental constraints that affect root absorption, including soil surface assimilation, desorption, transportation (Raliya et al. 2015).

12.5 Foliage Application of Nanoparticles

Nanoparticles applied through the foliage are frequently employed for agricultural purposes as biological sensors, after harvest additives for preservation, fertilizers as well as plant nutritional supplements (Tighe-Neira et al. 2018; Sun et al. 2020). Rain and absorption on earth compounds cause crops to use ground-applied chemical fertilizers at a low level, which results in a spike in the usage of artificial chemical fertilizers and a process known as eutrophication (Raliya et al. 2016). Crops that absorb chemical fertilizers with poor solubility grow weakly because the plants are incapable to take in adequate traces of elements (Li et al. 2016).

Foliar-applied nanoformulation provides numerous benefits over typical ground-applied fertilizers, including faster plant uptake, greater cost-effectiveness, and less negative effects on soil health (Meier et al. 2020). Application of foliage can replenish deficient soils with vitamins and other minerals. Twenty-one studies have demonstrated that slow-release nanoformulation improved plant uptake of potassium, phosphate, and nitrogen (Salem et al. 2016). Mesoporous silica nanoformulations inhibited the early combustion of vital nutrients and lessened the destructive outcomes of nanopesticides on the habitat (Wang et al. 2016.) Additionally, foliar nanoformulation can boost crop nutrition deficiencies, quickly accomplish biological fortification, and lessen crop deficiencies in specific vital nutrients (Deshpande et al. 2017). Investigations also demonstrate foliar nanoformulation antimicrobial action. Because of its antibacterial qualities, silver can be a key ingredient in detergents and spray solutions that treat illnesses in plants brought on by mold and fungus (El-Batal et al. 2016). Silver formulation of NPs is commonly employed as fungicides (Li et al. 2017). Additionally, certain biologically generated NPs, such as those produced by soil-isolated bacteria, have strong antibacterial action (Mishra et al. 2017). Due to their antibacterial qualities, artificial metal-based nanoparticles, including Cu, Zn bimetallic small particles, are frequently employed as both pesticides and antibacterial substances (Antonoglou et al. 2018). Research on artificial antimicrobial substances, such as mixed ions and bimetallic form agents, has been promising over the past few years. Studies have demonstrated that the combined antibacterial activities of many metals are more effective than those of individual metals.

12.6 Consequences of Foliar-Based Nanoformulation

Despite the fact that the need for food has gone up drastically, fertilizer use has generally been inefficient. While nanoformulation can benefit plants by improving resistance to disease, reducing stress caused by oxidation, and boosting both quality and yield (Sturikova et al. 2018; Lu et al. 2020; Li et al. 2020), large amounts appropriate to nanoformulation can occasionally harm crops (Tighe-Neira et al. 2018). Zinc shortage, as long as, can result in proliferation retardation (Deshpande et al. 2017) but zinc overabundance prevents development of seedlings. Crop fertilization using Zn nanoformulation sprayed on leaves can be successful, but practical

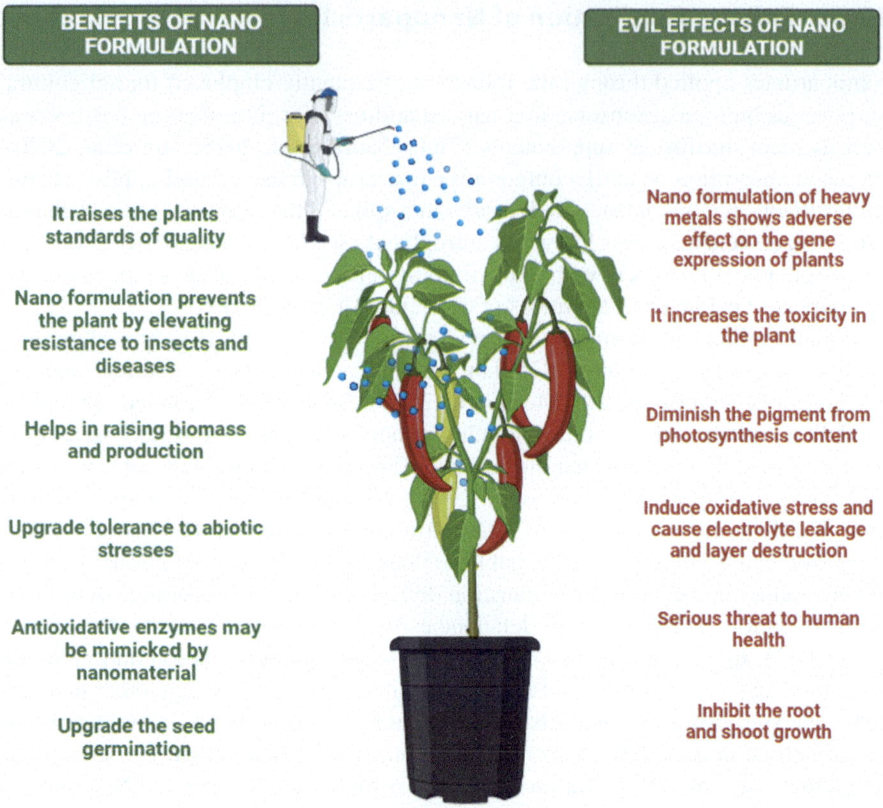

Fig. 12.5 Adversarial and beneficial effects of foliar spraying of NPs on plants. (Image created by BioRender.com)

application requires careful consideration of a number of factors (Alshaal and El-Ramady 2017). Figure 12.5 illustrates the adversarial and beneficial effects of foliar spraying of NPs on plants.

12.6.1 Advantages of Foliar Nanoformulation

12.6.1.1 Crops Quality

It takes high-quality nutrients to raise the plant's standards of quality. According to certain research investigations, applying nanofertilizer through foliage resulted in agricultural goods of higher quality than using conventional fertilizer. The use of nanoparticles of metal oxide boosted cotton fiber quality indicators such as homogeneity proportion and fiber toughness compared to control. According to Prasad et al. (2012), fertilizer applied by nanoformulation is totally regulated and has boosted peanut production and amount of protein. When zinc and iron nanoformulation were applied foliar, fodder the grain's phosphorus quantity, the plant matter, protein content in crude form, plus accessible carbohydrate content all enhanced relative to chemical fertilizers (Sharifi et al. 2016). According to Sham (2017), using nanoparticles made of ZnO topically improved sunflower oil nutritional value and other quality indicators. Zinc chemical fertilizers augmented the soluble carbohydrate levels, likely due to zinc's involvement in photosynthesis, chlorophyll formation, starch generation, and the action of carbonic anhydrase, an enzyme that accelerates carbohydrate synthesis (Singh and Kumar 2012). Zinc is beneficial for root expansion because it facilitates in the consumption of essential health-giving nutritious elements via crops, particularly nitrogen-containing compounds, which is needed for the production of proteins. Zinc also participates in the formation of carbohydrates, germ maturity is associated with the cellular breakdown in carbohydrates, polypeptides, and messenger molecules from plants, particularly auxin (El-Metwally et al. 2018). When compared to ordinary Fe, nano-Fe significantly affected the seed amount of protein in black-eyed peas by a factor of two (Delfani et al. 2014). According to Jamadar (2016), foliar application of nanoparticles' usage in foliage has risen both in activity of enzymes, zinc consumption within upland paddy by 53% and 48% above control.

12.6.1.2 The Massive Increase in Yield

By elevating resistance to insects and diseases, raising biomass and production, boosting the amount in dietary, and adding to the quality in herb, foliar nanoformulations have a favorable impact on crops (Sharifi et al. 2016; Lu et al. 2020). For instance, C-nanotubes boosted the development and extension of seeds in a range of agricultural products, which may be connected with their capacity to improve the water intake. In *Lycopersicon esculentum* (tomato), foliar usage of released in aerosol form ZnO, CuO, and nanoparticles made of TiO_2 proved an effective method to boost immunity against pathogens, raise the level of chlorophyll, and improve the rate of photosynthetic activity (Raliya et al. 2015; Elmer and White 2016).

Several studies find the possibility by using nanofertilizers to boost crop productivity in recent years. The production metrics of wheat crops were enhanced as a result of foliar applications of nanofertilizer (Abdel-Aziz et al. 2018). Because nitrogen nanofertilizers enhance in metabolism process and enhance growth hormone activity, applying them on chickpeas enhanced both the yield and its components and tended to promote flowering and production of grains (Drostkar et al. 2016). In addition to cutting the price of fertilizer and minimizing the risk of contamination, the usage of nanoformulations has a larger impact on enhancing cotton crop production. According to Singh and Kumar's (Singh and Kumar 2017) research, using 500 ppm of nanoscale ZnS for 55 days after seeding dramatically boosted sunflower seed output. As it has a significant area as per volume, nanoscale ZnO has shown to be far more efficient at improving efficiency and absorption of Zn. According to Dapkekar et al. (2018), the dosage needed for nano-scopic zinc manure was 10 times lower than that of ordinary zinc SO_4. According to Afshar et al. (2014), there was a proposal that the usage of nanoformulation rose the amount of zinc of cereals yet did not alter their grain yield, amount of protein, number of spikelet's per spike, weight of 1000 kernels, etc. This was likely due to a rise in enzyme activity and improved breakdown of carbohydrates. With the foliar spray of nanoformulation of silver, potato tuber yield enhanced presumably because of its antibacterial impact that may have enabled the seedlings remain fresher for longer periods of time within the soil and consequently generated more strong plants. Plant length, the pods quantity, the ripe pods number per plant, the number of 100 germ volume, the seeds number in each plant, seeds length, the seeds and pods yield, branches on each side, and also the biological efficiency in peanuts are some characteristics that are significantly influence by nano usage of foliar exercise in chelate molybdenum.

12.6.1.3 Suppression of Abiotic Stress

Salinity, dryness, pollutants, floods, extreme temperatures, the ozone layer, and UV radiation are a few instances of abiotic stressors. Drought and salinity are two abiotic stressors that pose significant dangers to agricultural productivity worldwide. Dry and salt are two examples of these abiotic factors that can reduce agricultural yield by up to 50% (Wang et al. 2003). Due to osmotic, ionic tension, saltiness promotes nutritional disparity, injury in membrane, and enzyme prohibition in crops. The growth of roots is impacted by salinity, which lowers their intake of nutrients. Nanomaterials may mimic antioxidative enzymes including catalase, superoxide dismutase, and peroxidase. These enzymes and antioxidants (Upadhyaya et al. 2015) are constantly obtaining oxygen species that are reactive. Salt existence into the ground reduces growth factors, like moisture content, intake of N, Ca, P, K, Na, and cotton yield. Salty circumstances have an impact on root growth, which lowers nutrient intake. According to Hussein and Abou-Baker (2018), applying zinc NPs improved radicle absorption, along with increased absorption of nutrients. It also significantly altered the dry and fresh rice weight (Upadhyaya et al. 2015), sunflower generation of organic material (Torabian et al. 2018), wheat production in salty conditions (Babaei et al. 2017), maize production during the lack of moisture (Farnia et al. 2015). As stated by Soliman et al. (2015), foliar spraying of a Hoagland solution containing ZnO and Fe3O4 nanoparticles can mitigate the effects of salt stress on Moringa plants by enhancing enzyme activity associated with salt tolerance. Plants' ability to withstand salt is significantly indicated by the build-up of less Na. By promoting development of root systems, minimizing nutritional diseases, and boosting the absorption of nutrients by the roots, potassium (K) level indicates tolerance to salt in crops and mitigates the adverse impacts of NaCl on nutritional absorption (El-Fouly et al. 2000). The bigger significant surface area or high reactive regions in the NPs facilitated the increased activity of enzymes associated with tolerance to salt. The presence of iron ad zinc content, that are essential to the forbearance to salt procedure, is increased by the nanoparticles (NP), which also facilitate the easier intake of fertilizer and improve the effects of Hoagland solution.

12.6.2 Adverse Effect of Foliar Nanoformulation

The application of foliar nanoformulations in agriculture is restricted due to their possibility of undiscovered damage (Lu et al. 2020). Crops subjected to increased levels of metal nanoparticles (NPs) through foliar application exhibit notable alterations in gene expression (Dappe et al. 2019), reduction in photosynthetic pigment content, induction of oxidative stress (OS), and the occurrence of electrolyte leakage and membrane damage (Larue et al. 2014). CuO and silver nanoparticles have the potential to negatively impact the composition and operation of crops' photosynthesis machinery. Foliage NPs made from copper oxide have the potential to inhibit root and shoot growth, and following application to the foliage, closure of stomata signs may occur. Applying TiO_2 nanoparticles through the foliage to plants causes DNA damage, diminished the plant matter, and lowered activity of enzymes (Teszlák et al. 2018). Using nanoformulation directly can result in leaf burning, that lowers the efficiency of photosynthetic processes, or pore blockage from NP accumulation, that obstructs gas exchange and photosynthesis. Reduced intake of several crucial components can also be the outcome of pore blockage (Kannan and Charnel 1986).

12.6.2.1 Plant Toxicity

Adverse impacts of nanoformulations on higher plants have not been extensively studied. By stimulating oxidative stress signaling, the accumulation of nanoparticles such as TiO, ZnO, CeO_2, and Ag NPs on outer layer of cells along with the organelles themselves induce the free radical damage in germ cell (Buzea et al. 2007). The nanoparticles of Fe_3O_4 sprayed on the foliage to grow maize in either small or large amount interact with ion channels that reside in the plasma barrier, affecting the capacity for photosynthesis (Racuciu and Creanga 2007; Tan et al. 2009). Based on certain research (Roh et al. 2012; Gopinath et al. 2010), presence of silver nanoparticles interfered with membrane polypeptides, thus activating signaling pathways and impeding cellular growth.

12.6.2.2 Detrimental Effect on People

Particles of greater size are unable to infiltrate systems of life with the ease that nanoparticles can. Nanoparticles can potentially cross cell barriers, and their ability to do so could be influenced by their size. It is possible to breathe in the nanostructures by agricultural laborers during the application process, which may reach the blood and other target areas like the cerebral cortex, liver function, or heart (Suppan 2017). These tiny particles could have an impact on regulation processes on other enzymes and polypeptides (Boulares et al. 2007). Additionally, a potential is there that nanomaterials will find their way into the supply chains of food. This in the end impacts the healthcare system for people. Thus, there is imperative that human resources be developed through an awareness of the complexity involved in the cultivation of crops system to support uses for nanotechnologies in agriculture with effectiveness (Mukhopadhyay 2014).

12.7 Conclusion

In conclusion, the utilization of nanoformulations in agriculture, particularly through foliar application, represents a promising avenue for addressing various challenges in traditional farming practices. Nanoformulations, including nanoemulsions, nanospheres, nanocapsules, and nanogels, offer a versatile toolkit for enhancing the delivery and efficacy of minerals and insecticides. The encapsulation of agricultural agents in carrier nanoparticles, such as silica nanoparticles, gold nanoparticles, zinc oxide nanoparticles, and calcium alginate nanoparticles, demonstrates potential benefits in terms of protection against photolysis, improved absorption, and controlled release. The mechanisms involved in the application of nanoparticle-based fertilizers through foliage are complex, involving pathways such as stomatal entry and cuticular penetration. The assimilation, translocation, and aggregation of nanoparticles within plants are influenced by various factors, including the size of nanoparticles, physiological barriers, and plant age. Understanding these processes is crucial for optimizing the effectiveness of nanoformulations in promoting plant growth and nutrient uptake. Advantages of foliar nanoformulations are diverse, encompassing faster plant uptake, cost-effectiveness, reduced environmental impact, and improved crop quality and yield. These nanoformulations have demonstrated antimicrobial properties, aiding in disease resistance and overall plant health. Additionally, the potential for suppressing abiotic stressors, such as salinity and drought, further highlights the versatility of nanoformulations in mitigating adverse environmental conditions. Despite these promising benefits, it is essential to consider potential challenges and adverse effects associated with nanoformulations, including plant toxicity and potential impacts on human health. Further research and careful monitoring of these nanotechnological applications are crucial to ensure their safe and sustainable integration into agricultural practices. In summary, the integration of nanoformulations in agriculture presents a transformative approach with the potential to revolutionize crop management, improve resource efficiency, and contribute to global food security. Continued research, coupled with responsible deployment and monitoring, will be essential to unlock the full potential of nanotechnology in sustainable and environmentally friendly farming practices.

References

Abdel-Aziz HMM, Hasaneen MNA, Omer AM (2018) Foliar application of nano chitosan NPK fertilizer improves the yield of wheat plants grown on two different soils. Egypt J Exp Biol(Bot) 14(1):63–72

Achari GA, Kowshik M (2018) Recent developments on nanotechnology in agriculture: plant mineral nutrition, health, and interactions with soil microflora. J Agric Food Chem 66(33):8647–8661

Afshar I, Haghighi AR, Shirazi M (2014) Comparison the effects of spraying different amounts of nano zinc oxide and zinc oxide on, wheat. Int J Plant Animal Environ Sci 4(3):688–693

Alshaal T, El-Ramady H (2017) Foliar application: from plant nutrition to biofortification. EBSS 1:71–83

Antonoglou O, Moustaka J, Adamakis IDS, Sperdouli I, Pantazaki AA, Moustakas M, Dendrinou-Samara C (2018) Nanobrass CuZn nanoparticles as foliar spray nonphytotoxic fungicides. ACS Appl Mater Interfaces 10(5):4450–4461

Anusuya S, Sathiyabama M (2015) Foliar application of β-D-glucan nanoparticles to control rhizome rot disease of turmeric. Int J Biol Macromol 72:1205–1212

Babaei K, Seyed Sharifi R, Pirzad A, Khalilzadeh R (2017) Effects of bio fertilizer and nano Zn-Fe oxide on physiological traits, antioxidant enzymes activity and yield of wheat (*Triticum aestivum* L.) under salinity stress. J Plant Interact 12(1):381–389

Bhagat D, Samanta SK, Bhattacharya S (2013) Efficient management of fruit pests by pheromone nanogels. Sci Rep 3(1):1294

Birbaum K, Brogioli R, Schellenberg M, Martinoia E, Stark WJ, Günther D, Limbach LK (2010) No evidence for cerium dioxide nanoparticle translocation in maize plants. Environ Sci Technol 44(22):8718–8723

Boulares N, Guergouri K, Tabet N, Monty C (2007) Nanotechnology and its applications first Sharjah international conference on nanotechnology and its applications. In: AIP Conf Proc, vol 929, pp 172–176

Brunel F, El Gueddari NE, Moerschbacher BM (2013) Complexation of copper (II) with chitosan nanogels: toward control of microbial growth. Carbohydr Polym 92(2):1348–1356

Buzea C, Pacheco II, Robbie K (2007) Nanomaterials and nanoparticles: sources and toxicity. Biointerphases 2(4):MR17–MR71

Clarke CR, Tegg RS, Thompson HK, Frederick C, Haynes KG, Kramer M, Wilson CR (2020) Low-dose foliar treatments of the auxin analog 2, 4-D reduce potato common scab and powdery scab for multiple potato cultivars and enhance root development. Crop Prot 136:105208

Dapkekar A, Deshpande P, Oak MD, Paknikar KM, Rajwade JM (2018) Zinc use efficiency is enhanced in wheat through nanofertilization. Sci Rep 8(1):6832

Dappe V, Dumez S, Bernard F, Hanoune B, Cuny D, Dumat C, Sobanska S (2019) The role of epicuticular waxes on foliar metal transfer and phytotoxicity in edible vegetables: case of Brassica oleracea species exposed to manufactured particles. Environ Sci Pollut Res 26:20092–20106

Delfani M, Baradarn Firouzabadi M, Farrokhi N, Makarian H (2014) Some physiological responses of black-eyed pea to iron and magnesium nanofertilizers. Commun Soil Sci Plant Anal 45(4):530–540

Deshpande P, Dapkekar A, Oak MD, Paknikar KM, Rajwade JM (2017) Zinc complexed chitosan/TPP nanoparticles: a promising micronutrient nanocarrier suited for foliar application. Carbohydr Polym 165:394–401

Dey S, Singh GB (2023) Green nanotechnology approaches in vaccinology: advantages and disadvantages in biomedical sciences. In: Nanovaccinology: clinical application of nanostructured materials research to translational medicine. Springer International Publishing, Cham, pp 281–299

Drostkar E, Talebi R, Kanouni H (2016) Foliar application of Fe, Zn and NPK nano-fertilizers on seed yield and morphological traits in chickpea under rainfed condition. J Resour Ecol 4(1):221–228

Dwivedi S, Saquib Q, Al-Khedhairy AA, Musarrat J (2016) Understanding the role of nanomaterials in agriculture. In: Microbial inoculants in sustainable agricultural productivity: vol 2: Functional applications, pp 271–288

El-Batal AI, Gharib FAEL, Ghazi SM, Hegazi AZ, Hafz AGMAE (2016) Physiological responses of two varieties of common bean (*Phaseolus vulgaris* L.) to foliar application of silver nanoparticles. Nanomater Nanotechnol 6:13

El-Fouly MM, Moubarak ZM, Salama ZA (2000, November) Micronutrient foliar application increases salt tolerance of tomato seedlings. In: International symposium on techniques to control salination for horticultural productivity, vol 573, pp 467–474

Elmer WH, White JC (2016) The use of metallic oxide nanoparticles to enhance growth of tomatoes and eggplants in disease infested soil or soilless medium. Environ Sci Nano 3(5):1072–1079

El-Metwally IM, Doaa MR, Abo-Basha AEAM, Abd El-Aziz M (2018) Response of peanut plants to different foliar applications of nano-iron, manganese and zinc under sandy soil conditions. Middle East J Appl Sci 8(2):474–482

Farnia A, Omidi MM, Farnia A (2015) Effect of nano-zinc chelate and nano-biofertilizer on yield and yield components of maize (*Zea mays* L.), under water stress condition. Indian J Nat Sci 5(29):4614–4624

Feng J, Zhang QI, Liu QI, Zhu Z, McClements DJ, Jafari SM (2018) Application of nanoemulsions in formulation of pesticides. In: Nanoemulsions. Academic Press, pp 379–413

Gopinath P, Gogoi SK, Sanpui P, Paul A, Chattopadhyay A, Ghosh SS (2010) Signaling gene cascade in silver nanoparticle induced apoptosis. Colloids Surf B: Biointerfaces 77(2):240–245

Greene DW, Bukovac MJ (1974) Stomatal penetration: effect of surfactants and role in foliar absorption. Am J Bot 61(1):100–106

Gupta A, Eral HB, Hatton TA, Doyle PS (2016) Nanoemulsions: formation, properties and applications. Soft Matter 12(11):2826–2841

Hou R, Zhang Z, Pang S, Yang T, Clark JM, He L (2016) Alteration of the nonsystemic behavior of the pesticide ferbam on tea leaves by engineered gold nanoparticles. Environ Sci Technol 50(12):6216–6223

Hu J, Guo H, Li J, Wang Y, Xiao L, Xing B (2017) Interaction of γ-Fe 2 O 3 nanoparticles with *Citrus maxima* leaves and the corresponding physiological effects via foliar application. J Nanobiotechnol 15:1–12

Hussein MM, Abou-Baker NH (2018) The contribution of nano-zinc to alleviate salinity stress on cotton plants. R Soc Open Sci 5(8):171809

Jalali M, Ghanati F, Modarres-Sanavi AM, Khoshgoftarmanesh AH (2017) Physiological effects of repeated foliar application of magnetite nanoparticles on maize plants. J Agron Crop Sci 203(6):593–602

Jamadar A (2016) Phosphorus use efficiency in upland paddy through use of nanoparticles and PSB. M.Sc.(Agri.) Thesis

Kalra T, Tomar PC, Arora K (2020) Micronutrient encapsulation using nanotechnology: nanofertilizers. Plant Arch 20(2):1748–1753

Kannan S, Charnel A (1986) Foliar absorption and transport of inorganic nutrients. Crit Rev Plant Sci 4(4):341–375

Kranjc E, Mazej D, Regvar M, Drobne D, Remškar M (2018) Foliar surface free energy affects platinum nanoparticle adhesion, uptake, and translocation from leaves to roots in arugula and escarole. Environ Sci Nano 5(2):520–532

Larue C, Castillo-Michel H, Sobanska S, Trcera N, Sorieul S, Cécillon L, Sarret G (2014) Fate of pristine TiO2 nanoparticles and aged paint-containing TiO2 nanoparticles in lettuce crop after foliar exposure. J Hazard Mater 273:17–26

Li P, Du Y, Huang L, Mitter N, Xu ZP (2016) Nanotechnology promotes the R&D of new-generation micronutrient foliar fertilizers. RSC Adv 6(73):69465–69478

Li CC, Dang F, Li M, Zhu M, Zhong H, Hintelmann H, Zhou DM (2017) Effects of exposure pathways on the accumulation and phytotoxicity of silver nanoparticles in soybean and rice. Nanotoxicology 11(5):699–709

Li WQ, Qing T, Li CC, Li F, Ge F, Fei JJ, Peijnenburg WJ (2020) Integration of subcellular partitioning and chemical forms to understand silver nanoparticles toxicity to lettuce (*Lactuca sativa* L.) under different exposure pathways. Chemosphere 258:127349

Lu L, Huang M, Huang Y, Corvini PFX, Ji R, Zhao L (2020) Mn3O4 nanozymes boost endogenous antioxidant metabolites in cucumber (*Cucumis sativus*) plant and enhance resistance to salinity stress. Environ Sci Nano 7(6):1692–1703

Lv J, Christie P, Zhang S (2019) Uptake, translocation, and transformation of metal-based nanoparticles in plants: recent advances and methodological challenges. Environ Sci Nano 6(1):41–59

Ma C, White JC, Zhao J, Zhao Q, Xing B (2018) Uptake of engineered nanoparticles by food crops: characterization, mechanisms, and implications. Annu Rev Food Sci Technol 9:129–153

Marzouk NM, Abd-Alrahman HA, EL-Tanahy AMM, Mahmoud SH (2019) Impact of foliar spraying of nano micronutrient fertilizers on the growth, yield, physical quality, and nutritional value of two snap bean cultivars in sandy soils. Bull Natl Res Cent 43(1):1–9

Mehndiratta P, Jain A, Singh GB, Sharma S, Srivastava S, Gupta S, Gupta N (2015) Magnetite nanoparticle aided immobilization of *Pseudomonas* sp. GBS. 5 for carbazole degradation. J Biochem Technol 5(4):823–825

Meier S, Moore F, Morales A, González ME, Seguel A, Meriño-Gergichevich C et al (2020) Synthesis of calcium borate nanoparticles and its use as a potential foliar fertilizer in lettuce (*Lactuca sativa*) and zucchini (*Cucurbita pepo*). Plant Physiol Biochem 151:673–680

Mishra S, Singh BR, Naqvi AH, Singh HB (2017) Potential of biosynthesized silver nanoparticles using *Stenotrophomonas* sp. BHU-S7 (MTCC 5978) for management of soil-borne and foliar phytopathogens. Sci Rep 7(1):45154

Mohammadi A, Hashemi M, Hosseini SM (2015) Nanoencapsulation of Zataria multiflora essential oil preparation and characterization with enhanced antifungal activity for controlling *Botrytis cinerea*, the causal agent of gray mould disease. Innovative Food Sci Emerg Technol 28:73–80

Mukhopadhyay SS (2014) Nanotechnology in agriculture: prospects and constraints. Nanotechnol Sci Appl 7:63–71

Oosterhuis D (2009) Foliar fertilization: mechanisms and magnitude of nutrient uptake. In: Proceedings of the fluid forum, pp 15–17

Pathak C, Vaidya FU, Pandey SM (2019) Mechanism for development of nanobased drug delivery system. In: Applications of targeted nano drugs and delivery systems. Elsevier, pp 35–67

Paula HC, Sombra FM, de Freitas Cavalcante R, Abreu FO, de Paula RC (2011) Preparation and characterization of chitosan/cashew gum beads loaded with *Lippia sidoides* essential oil. Mater Sci Eng C 31(2):173–178

Pestovsky YS, Martínez-Antonio A (2017) The use of nanoparticles and nanoformulations in agriculture. J Nanosci Nanotechnol 17(12):8699–8730

Pradhan S, Roy I, Lodh G, Patra P, Choudhury SR, Samanta A, Goswami A (2013) Entomotoxicity and biosafety assessment of PEGylated acephate nanoparticles: a biologically safe alternative to neurotoxic pesticides. J Environ Sci Health B 48(7):559–569

Prasad TNVKV, Sudhakar P, Sreenivasulu Y, Latha P, Munaswamy V, Reddy KR, Pradeep T (2012) Effect of nanoscale zinc oxide particles on the germination, growth and yield of peanut. J Plant Nutr 35(6):905–927

Racuciu M, Creanga D (2007) TMA-OH coated magnetic nanoparticles internalized in vegetal tissue. Rom J Phys 52(3/4):395

Raliya R, Nair R, Chavalmane S, Wang WN, Biswas P (2015) Mechanistic evaluation of translocation and physiological impact of titanium dioxide and zinc oxide nanoparticles on the tomato (*Solanum lycopersicum* L.) plant. Metallomics 7(12):1584–1594

Raliya R, Franke C, Chavalmane S, Nair R, Reed N, Biswas P (2016) Quantitative understanding of nanoparticle uptake in watermelon plants. Front Plant Sci 7:1288

Rana RA, Siddiqui MN, Skalicky M, Brestic M, Hossain A, Kayesh E et al (2021) Prospects of nanotechnology in improving the productivity and quality of horticultural crops. Horticulturae 7(10):332

Rawat M, Nayan R, Negi B, Zaidi MGH, Arora S (2017) Physio-biochemical basis of iron-sulfide nanoparticle induced growth and seed yield enhancement in *B. juncea*. Plant Physiol Biochem 118:274–284

Roh JY, Eom HJ, Choi J (2012) Involvement of Caenohabditis elegans MAPK signaling pathways in oxidative stress response induced by silver nanoparticles exposure. Toxicol Res 28:19–24

Salem NM, Albanna LS, Awwad AM (2016) Green synthesis of sulfur nanoparticles using *Punica granatum* peels and the effects on the growth of tomato by foliar spray applications. Environ Nanotechnol Monit Manag 6:83–87

Shahid M, Dumat C, Khalid S, Schreck E, Xiong T, Niazi NK (2017) Foliar heavy metal uptake, toxicity and detoxification in plants: a comparison of foliar and root metal uptake. J Hazard Mater 325:36–58

Sham SP (2017) Effect of foliar application of nano zinc particles on growth, yield and qualities of sunflower (*Helianthus annus* L.). Sc.(Agri.) Thesis. University of Agricultural Science, Dharwad, Karnataka, India

Sharifi R, Mohammadi K, Rokhzadi A (2016) Effect of seed priming and foliar application with micronutrients on quality of forage corn (*Zea mays*). Environ Exp Biol 14(4):151

Sharma G, Rout DP, Singh G (2017) Characterization of goat milk protein and comparison of milk proteins using polyacrylamide gel electrophoresis. Int J Sci Eng Appl Sci 4:2395–3470

Singh O, Kumar S (2012) Productivity and profitability of rice (*Oryza sativa*) as influenced by high fertility levels and their residual effect on wheat (*Triticum aestivum*). Indian J Agron 57(2):143–147

Singh MD, Kumar B (2017) Bio efficacy of nano zinc sulphide (ZnS) on growth and yield of sunflower (*Helianthus annuus* L.) and nutrient status in the soil. Int J Agric Sci 9:0975–3710

Singh GB, Srivastava A, Saigal A, Aggarwal S, Bisht S, Gupta SS, Gupta N (2011) Biodegradation of carbazole and dibenzothiophene by bacteria isolated from petroleum-contaminated sites. Biorem J 15(4):189–195

Solanki P, Bhargava A, Chhipa H, Jain N, Panwar J (2015) Nano-fertilizers and their smart delivery system. In: Nanotechnologies in food and agriculture. Springer, pp 81–101

Soliman AS, El-feky SA, Darwish E (2015) Alleviation of salt stress on *Moringa peregrina* using foliar application of nanofertilizers. J Hortic Forest 7(2):36–47

Sturikova H, Krystofova O, Huska D, Adam V (2018) Zinc, zinc nanoparticles and plants. J Hazard Mater 349:101–110

Su Y, Ashworth V, Kim C, Adeleye AS, Rolshausen P, Roper C, Jassby D (2019) Delivery, uptake, fate, and transport of engineered nanoparticles in plants: a critical review and data analysis. Environ Sci Nano 6(8):2311–2331

Sun H, Du W, Peng Q, Lv Z, Mao H, Kopittke PM (2020) Development of ZnO nanoparticles as an efficient Zn fertilizer: using synchrotron-based techniques and laser ablation to examine elemental distribution in wheat grain. J Agric Food Chem 68(18):5068–5075

Suppan S (2017) Applying nanotechnology to fertilizer: rationales, research, risks and regulatory challenges. The Institute for Agriculture and Trade Policy works locally and globally, Brazil

Taiz L, Zeiger E (2010) Plant physiology, 5th edn. Sinauer Associates Inc. Publishers, Sunderland

Tan XM, Lin C, Fugetsu B (2009) Studies on toxicity of multi-walled carbon nanotubes on suspension rice cells. Carbon 47(15):3479–3487

Teszlák P, Kocsis M, Scarpellini A, Jakab G, Kőrösi L (2018) Foliar exposure of grapevine (*Vitis vinifera* L.) to TiO2 nanoparticles under field conditions: photosynthetic response and flavonol profile. Photosynthetica 56(4):1378–1386

Thunugunta T, Channa Reddy A, Kodthalu Seetharamaiah S, Ramanna Hunashikatti L, Gowdra Chandrappa S, Cherukatu Kalathil N, Dhoranapalli Chinnappa Reddy LR (2018) Impact of zinc oxide nanoparticles on eggplant (*S. melongena*): studies on growth and the accumulation of nanoparticles. IET Nanobiotechnol 12(6):706–713

Tighe-Neira R, Carmora E, Recio G, Nunes-Nesi A, Reyes-Diaz M, Alberdi M et al (2018) Metallic nanoparticles influence the structure and function of the photosynthetic apparatus in plants. Plant Physiol Biochem 130:408–417

Torabian S, Zahedi M, Khoshgoftarmanesh A (2018) Effect of foliar spray of zinc oxide on some antioxidant enzymes activity of sunflower under salt stress, J Agric Sci Technol:9446313

Ullah H, Li X, Peng L, Cai Y, Mielke HW (2020) In vivo phytotoxicity, uptake, and translocation of PbS nanoparticles in maize (*Zea mays* L.) plants. Sci Total Environ 737:139558

Upadhyaya H, Shome S, Tewari S, Bhattacharya MK, Panda SK (2015) Effect of Zn nano-particles on growth responses of rice. In: Nanotechnology: novel perspectives and prospects, pp 508–512

Wang W, Vinocur B, Altman A (2003) Plant responses to drought, salinity and extreme temperatures: towards genetic engineering for stress tolerance. Planta 218:1–14

Wang P, Menzies NW, Lombi E, McKenna BA, Johannessen B, Glover CJ et al (2013) Fate of ZnO nanoparticles in soils and cowpea (*Vigna unguiculata*). Environ Sci Technol 47(23):13822–13830

Wang P, Lombi E, Zhao FJ, Kopittke PM (2016) Nanotechnology: a new opportunity in plant sciences. Trends Plant Sci 21(8):699–712

Xiong T, Austruy A, Pierart A, Shahid M, Schreck E, Mombo S, Dumat C (2016) Kinetic study of phytotoxicity induced by foliar lead uptake for vegetables exposed to fine particles and implications for sustainable urban agriculture. J Environ Sci 46:16–27

Xiong T, Zhang T, Dumat C, Sobanska S, Dappe V, Shahid M et al (2019) Airborne foliar transfer of particular metals in *Lactuca sativa* L.: translocation, phytotoxicity, and bioaccessibility. Environ Sci Pollut Res 26:20064–20078

Xue J, Luo Z, Li P, Ding Y, Cui Y, Wu Q (2014) A residue-free green synergistic antifungal nanotechnology for pesticide thiram by ZnO nanoparticles. Sci Rep 4:5408

Zahedi SM, Karimi M, Teixeira da Silva JA (2020) The use of nanotechnology to increase quality and yield of fruit crops. J Sci Food Agric 100(1):25–31

Zhao X, Cui H, Wang Y, Sun C, Cui B, Zeng Z (2017) Development strategies and prospects of nano-based smart pesticide formulation. J Agric Food Chem 66(26):6504–6512

Supplementing Micronutrients to Plants Through Microbes

13

Yalavarthi Nagaraju, Duppala Manoj Kumar, Praveen S. Patted, Pramod Kumar Sahu, and Savani Ajit Kumar

Abstract

Micronutrients play a vital role in plants, and though they are required in minor quantities, their importance is immense. The deficiency of micronutrients hinders the growth and quality of produce, and the end user also gets affected. Micronutrients can be applied artificially or managed through agronomic practices or microbe-mediated bioavailability is also explored. Among all the methods, microbe-based methods are feasible and cost-effective. Several microbes are being explored for micronutrient supplements and the development of biofortified varieties. However, more focus is required on the microbiome-based approaches to deliver the unanswered areas of scope. Lack of information on the importance and role of microbes in making micronutrient availability hinders the growth in this area. Hence, efforts were taken to delineate some of the more focused issues in this chapter.

Keywords

Y. Nagaraju (✉)
Central Sericultural Research and Training Institute, Berhampore, West Bengal, India

D. M. Kumar
Agricultural College, Bapatla, ANGRAU, Guntur, India

P. S. Patted
ICAR-central potato research Institute, Shimla, India

P. K. Sahu
Agricultural Microbiology, ICAR National Bureau of Agriculturally Important Microorganisms, Mau Nath Bhanjan, Uttar Pradesh, India

S. A. Kumar
ICAR-NAARM, Hyderabad, India

S. Kaur et al. (eds.), *Metabolomics, Proteomics and Gene Editing Approaches in Biofertilizer Industry*, https://doi.org/10.1007/978-981-97-2910-4_13

Micronutrients · Microbes · Biofortification · Microbiome

13.1 Introduction

Nutrients that are required in trace levels but essentially found in all plant tissues are generally referred to as micronutrients; without them, the plant is undernourished and potentially declines in plant productivity. Micronutrients are present in very minimal concentrations in the soil. In lesser proportions than macronutrients, micronutrients are critical building blocks for plant growth and development. Metal micronutrients such as molybdenum (Mo), iron (Fe), manganese (Mn), copper (Cu), nickel (Ni), and zinc (Zn) take part in several processes in plant cells or help to form proteins. While non-metal chlorine (Cl) helps maintain the electrical and osmotic equilibrium, metalloid micronutrient boron (B) plays structural functions in cellular metabolism (Williams 2017).

Micronutrient fertilization is gaining popularity among the agricultural community for several reasons. The removal of micronutrients from soils through soil erosion and long-term cropping, crop yields directly related to the micronutrient removal from soil, and the widespread substitution of micronutrient-rich manures with inorganic fertilizers decreased the soils micronutrient capacity. As a result of all these elements, farmers are currently debating whether micronutrient fertilization is still necessary to address the shifting nutritional needs of crops. Micronutrients play a key role in photosynthesis, nitrogen fixation, redox reactions, enzyme activation, cell division, and seed formation. Deficiency of one or few micronutrients can significantly impact plant growth and quality. As a result, micronutrient deficits are less frequent than macronutrient deficiencies, even though they are crucial for specific cell processes (Mallarino et al. 2017). There is a widespread issue with the micronutrients deficiency in arable soil (Monreal et al. 2016). Because of the high yields produced by the widespread application of nitrogen fertilizers, particularly after World War II, to small grains like corn and wheat (*Triticum aestivum* L.), researchers and academics have been inspired to investigate the possibility of increasing yields by utilizing other nutrients (Bruns 2017).

Many environmental factors, such as disease and insect infestations, poor soil fertility, and insufficient water availability, negatively impact crop productivity (Adisa et al. 2019). Only fertile soil that offers all the vital elements needed by plants can sustain good crop production because, as the saying goes, "If we feed the soil, it will feed us." To grow healthy plants, improve agricultural output, and maintain soil health, soil fertility must be kept at an ideal level. Since soils provide the basis for quality food production, water conservation, nutrient cycling through oxidoreduction, mitigating climate change, and biodiversity conservation thus in recent years, soil quality and health gained tremendous attention from the scientific community. It is important to gain the knowledge on soil processes for the successful soil health management and to produce higher yields and maintain the agricultural economy. Legumes have several advantages, such as boosting SOM, enhancing soil porosity, structure, nutrients cycling, pH buffering, flora and fauna dynamics in the soil, and restricting the pests and diseases (Dhaliwal et al. 2019).

13.2 Global Micronutrient Scenario in Soils

Alkaline soils, as seen in African and Asian countries like Lebanon, Syria, Egypt, Pakistan, Italy, and India, have substantial Mn deficiencies. Zinc (Zn) insufficiency has been reported throughout the world. Nevertheless, majority of Indian soil contains inappropriate levels of micronutrients resulting in their poor bioavailability for plant absorption. Boron deficiency was commonly observed in Indian soils, it has been reported that 68% deficiency in the red soils of Bihar to a 2% deficiency in the alluvial soils of Gujarat. Alfisol soils in West Bengal and Assam had the highest B deficit (54–86%) because of heavy rainfall, which causes water-soluble B levels to drop. A greater concentration of B was found in the Indo-Gangetic plains with salty soil, whereas Madhya Pradesh and Rajasthan had a reasonable amount of B. Copper-deficient soil was found in Kerala, the Himalayan Tarai zone, Bihar, Uttar Pradesh, and north Madhya Pradesh (Singh 2001).

Despite the presence of iron in the earth's crust, alkali soils of the Indo-Gangetic areas have low quantities of plant-available iron. According to statistics on soil studies, over 12% of Indian soils are low in iron. About 1–5% of soils in India were found to be Mn deficient. The states of Bihar, Punjab, Haryana, and Madhya Pradesh were the ones where the manganese shortage was most prevalent. Furthermore, excessive lime application in Orissa's red lateritic soils led to Mn shortage. There is an 11% molybdenum shortage in Indian soils, including the hill region soils of Andhra Pradesh, the Konkan and Malabar areas, and the northern and northeastern Himalaya regions (Bhupalraj et al. 2002).

Conversely, crops in the Punjab region's calcareous alkaline soils might be poisonous due to their high accessible Mo amounts. Because of the high Mo content identified, wheat straw used as feed was discovered poisonous to cattle. 51.2% of the 65,000 soil samples from India were low in zinc, making Indian soils the most zinc-poor in the world, with a widespread deficit throughout the Indo-Gangetic plains. States like Andhra Pradesh, Punjab, Haryana, Uttar Pradesh, Bihar, and Madya Pradesh have shown a decrease in Zn deficiency (Singh et al. 2006).

13.3 Impact of Micronutrients on Plants

Improving soil fertility and fulfilling the nutritional needs of developing crops are the two main goals of soil nutrient management. In order to produce lucrative crops and maintain soil health, it is crucial to improve soil fertility. Healthy soils must have a diversified community of soil flora and fauna that reduces disease and pests, fosters positive symbiotic interactions, decomposes and recycles crucial plant nutrients, and decomposes organic matter (Siqueira Freitas et al. 2018). The increased decomposition rate of organic matter added to the soil with a high C:N ratio also affects soil fertility and health (Ladha et al. 2011). Although some studies claimed that additional fertilizer had little impact on the breakdown of high C:N materials, fertilizer administration stimulates microbial activity and improves the decomposition of organic waste (Sutradhar et al. 2017). Nevertheless, judicious fertilizer usage over time might increase SOM in the soil profile, enhancing soil fertility, health, and crop productivity. The enhanced availability of certain micronutrients, such as Mn and Zn, was demonstrated in studies to be closely connected to greater SOM application rates.

13.4 Factors Determining the Bioavailability of Micronutrients

13.4.1 pH

Small quantities of trace elements are necessary for plant development and are essential cofactors in metabolic activities. The process of nitrogen fixation, respiration, carbon and light trapping are only a few examples of how boron (B) and zinc (Zn) are crucial components in higher plants functions. The substrate pH directly influences the availability of nutrients near roots and plant absorption. At a pH range of 6.0–6.5, macronutrients are readily accessible, but micronutrients become less so at an pH >7.0 (Ferrarezi et al. 2022). For example, more micronutrients are readily accessible to plants at low pH conditions, which typically favor plant development (Lončarić et al. 2008). However, several of these micronutrients and non-essential substances might become hazardous when present in excessive concentrations. Soil pH rises with depth while keeping the root zone pH in near neutral range for healthy root growth (Atta et al. 2020). Lower soil pH can boost root activity and activation of genes involved in systemic acquired resistance (SAR), ion transporter genes, and easing physiological problems brought on by phloem sieve tube obstruction (Li et al. 2020). Phosphate combines with aluminum (Al) and iron (Fe) at acidic pH to generate precipitates (sparingly soluble to insoluble), but it reacts with calcium (Ca) and magnesium (Mg) at pH >7.5 to form soluble complexes. Additionally, the soil's redox conditions impact the availability of iron (Fe), manganese (Mn), aluminum (Al), and sulfur (S) (Fig. 13.1).

13.4.2 Microbial Diversity

The soil surrounding the root is the richest source of nutrients, water and offers a feasible habitat for expanding microbial activities in return for different nutrient and metabolic processes. There are several plant-derived substances in the rhizosphere that the microorganisms that improve plant defenses against pests and diseases would probably employ as food sources or elicitors. Proteobacteria (Gram negative), Actinobacteria (Gram positive), Acidobacteria, and Bacteroidetes are the most prevalent taxa in the root zone. Among the microbial diversity, proteobacteria

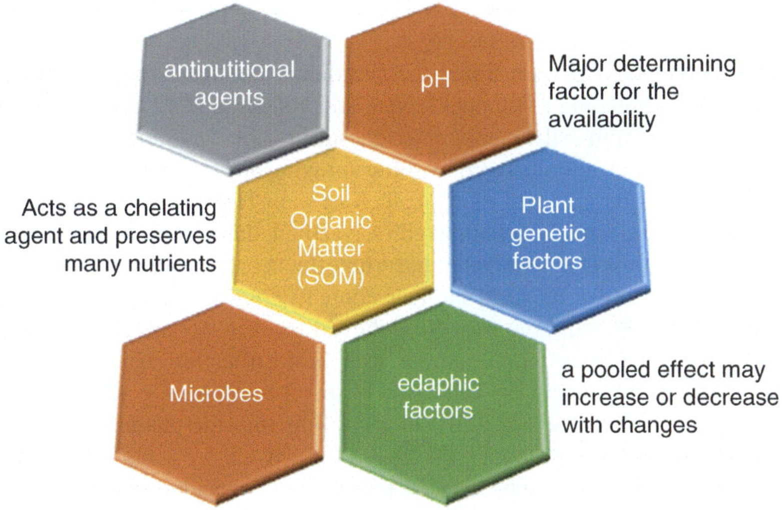

Fig. 13.1 Factors effecting the micronutrients availability in soil

dominate the plant surfaces, followed by the firmicutes, Bacteroides (Khatoon et al. 2020). The PGPR increase root development (root interception) by several mechanisms, allowing the root system to have greater contact with and access to the nutrients in the soil solution. The most crucial process is root elongation by PGPR, which produces IAA and breaking the ethylene cycle (using ACC deaminase). It has been demonstrated that PGPR promotes root hair elongation, increases lateral root length and number, and inhibits main root development (Santoyo et al. 2021).

There are endophytic microorganisms everywhere, and they have been isolated from every plant tissue that has been studied. Of these, seed-endophytes are considered major symbionts and may be important in the growth of seedlings; these microorganisms may be transferred through vertical or horizontal transmission. A significant study has been done on various rhizospheric microorganisms and how they affect plant health (Berendsen et al. 2012). The root's influence on the biodiversity of microorganisms can be observed in the rhizosphere (Gottel et al. 2011).

13.4.3 Fertilization

Due to its many advantages, applying micronutrients through soil is the most adaptable and successful strategy, especially for addressing B deficiency. Zinc sulfate heptahydrate (21% Zn) at 62.5 kg ha^{-1} or $ZnSO_4{\cdot}H_2O$ (33% Zn) at 40 kg per hectare has been proven to be equally practical and cost-effective for treating Zn deficiency in various crops (Rasheed et al. 2020). The desired salts of micronutrients and a single or combination are sprayed onto the leaves, where the stomata and epidermis allow absorption. Foliar feeding has been proven more successful in some circumstances than soil treatments for using nutrients and quickly reducing problems with visual and soil insufficiency. Numerous research has been published on applying foliar mineral fertilizers to various crops.

Recent studies have demonstrated the significance of nano foliar treatment for the protection and feeding of plants (Triveni and Nagaraju 2022). Numerous nutrients, including selenium, silica, copper oxide, sulfur, zinc, and ferric oxide nanoparticles (NPs), have been administered by foliar spraying and using silver nanoparticles and TiO_2. The unique qualities of NPs, such as their high surface reactivity, high surface area, and electron exchangeability, verified their significance in various demanding situations. Due to reduced harmful effects on wheat crops at lower dilutions, besides boosting grain yield and zinc content, ZnO-NPs have potential advantages over $ZnSO_4$. By utilizing iron NPs, the Cd level in wheat has been reduced, and FeO_3-NPs have been employed for iron biofortification. The outcomes are explicitly explained by using the activation of metal transporter genes, such as iron transporters, to regulate the entry of cadmium ions into the plants. Therefore, the external supply of Fe may hinder the absorption of Cd by roots (Hussain et al. 2020). The size of the NPs to be applied is calculated considering plant leaf structure (pore diameters in cell walls). Greater than the size of leaf pores, NPs will not transfer as well. When various NPs are applied to leaf surfaces, they enter in to plants through stomatal holes or trichrome bases for translocation in different plant tissues.

13.4.4 Soil Organic Matter (SOM)

Using organic inputs increases the soil's physical, chemical, and biological activity, making plants' micronutrients available for absorption. Hence crop residue management at the field level plays a key role in micronutrient availability. Soil organic matter addition not only improves soil health but also helps in ecosystem functions. In general practice, farmers apply farm yard manure and vermicompost as a source of nutrients and soil conditioner. In the present conditions, the availability of organic matter is limited, and non-traditional practices completely separate cattle rearing from agriculture. The increasing demand for the products pushes the integrated framing toward independent farming.

13.5 Antinutrient Compounds in Plants

Plants contain diverse antinutrient compounds such as glucosinolates or goitrogens, lectins, oxalates, phytates or myo-inositol hexa-phosphate, saponins, and tannins, which reduce the availability of micronutrients. The most prevalent type of phosphorus stored in seeds is myo-inositol hexa-phosphate. Because it decreases the bioavailability of critical micronutrients, PA is regarded as an antinutrient. The cations Ca^{2+}, Mg^{2+}, K^{+}, Fe^{3+}, and Zn^{2+} are highly chelated by PA because it is negatively charged. Phytate or phytin is the typical form of PA mixed salts (Raboy 2003). According to Mitchikpe et al. (2008), these insoluble salts stop vital nutrients from being absorbed in the human gut, which may result in micronutrient shortages. Majorly, inositol hexa-phosphate (InsP6) and inositol pentaphosphate (insP5) are inhibitory to micronutrient absorption, notably linked to Fe and Zn absorption (Al Hasan et al. 2016). It works as a potent scavenger of positively charged metal ions to create phytate. Rice is a poor supply of vital micronutrients while being the main source of nutrition for >350 billion people around the globe. Enhancing the bioavailability of micronutrients in rice, novel techniques like biofortification and lowering the myo-inositol hexa-phosphate or PA content of rice have been developed.

Additionally, consequences of global climate change, notably growing atmospheric CO_2 concentration, are anticipated to increase myo-inositol hexa-phosphate content and decrease concentrations of the majority of vital micronutrients in rice grain. In crops, many PA biosynthesis-related genes have been found and described. The basis for adjusting the PA content of rice grains should come from a thorough identification of putative genes involved in myo-inositol hexa-phosphate synthesis, deposition during seed development, and the development of methods to decrease their expression. The growth of low-PA rice mutants with much-decreased grain PA concentration was accompanied by lower yields and subpar agronomic performance, which make it difficult to employ these mutants effectively in breeding programs. However, the creation of low-PA rice has been successful, thanks to transgenic technology without interfering with seed or plant growth. Additionally, improving micronutrient levels, decreasing the PA concentration in the endosperm, and changing the distribution of micronutrients in rice grains are all potential methods for promoting mineral bioavailability. Overexpression of phytase enzyme in Arabidopsis reduced the quantity of phytic acid in the seeds without affecting the germination and subsequent growth.

13.6 Microbes Involved in the Micronutrients Acquisition

Lorenz Hiltner introduced the word "rhizosphere" to describe the abundance of miniature creatures (bacteria, fungus, and archaea) near and inside plant roots, since then, many studies have identified the multifaceted interactions between soil biota and plant hosts through plant secretions and root exudates (Fitzpatrick et al. 2020). The "soil microbial loop" hypothesis states that carbon and nutrients are cycled between the microorganisms and soil, and micronutrients become available through microbial oxidoreduction and decomposition process of soil organic matter and are subsequently "scavenged" by the plant (Paungfoo-Lonhienne et al. 2010). The significance of plant–microbe via soil interactions controls important nutrient cycles, plant inflicting properties of microbes have been demonstrated via numerous cultivation-dependent and metagenomics research. The rhizosphere microbiome has acted as a secondary genome bank based on microbial diversity, supplying a plant with substances and features obtained from microbes. The host genotype directs the general makeup of the microbial communities in the rhizosphere. Root exudates, metabolites generated from plants, primarily serve as nutrients and signaling agents for microorganisms in the rhizosphere (Mohapatra et al. 2021). Nutrients

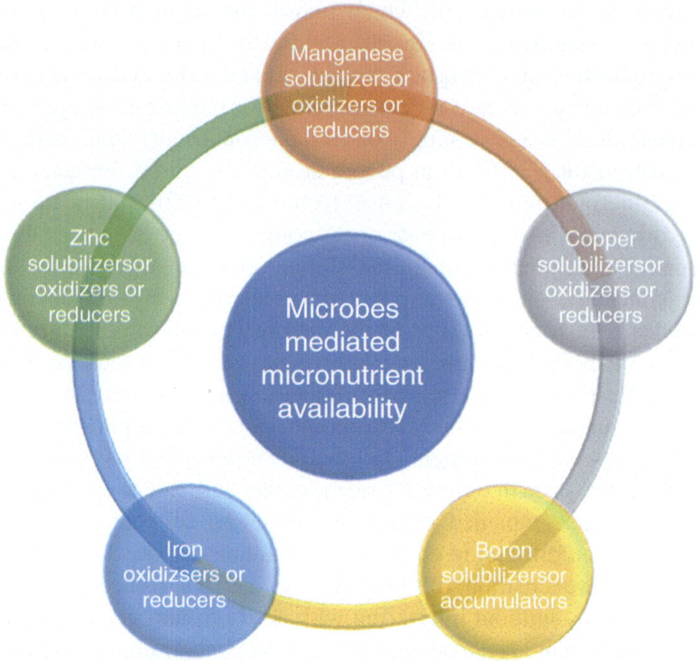

Fig. 13.2 Microbes mediated micronutrients availability to plants in soils

are soil-associated or insoluble precipitates, and root absorption of these nutrients is challenging (Fig. 13.2). In this scenario, PGPR bacteria are essential to liberate insoluble nutrients for their cellular metabolism and facilitate plant root absorption (Etesami and Adl 2020). In the view of reducing the use of inorganic straight or complex fertilizers, increasing the availability of nutrients to different field crops, and bioremediation (metal remediation), the use of PGPR as inputs has emerged as a quick and reliable source (Etesami and Adl 2020).

13.7 Zinc (Zn) Transforming Microorganisms

The productivity and quality of basic food crops are commonly hampered by Zn shortage. According to estimates, soils used for cultivating rice are Zn deficient in roughly 50% of cases. Due to Zn immobilization in the soil, soluble $ZnSO_4$ often results in inferior fertilizer usage efficiency (1–5%) (Krithika and Balachandar 2016). Alternative strategies for improving plant Zn uptake are Zn-mobilizing microorganisms. In rice (*O. sativa*) seedlings, fertilization with $ZnSO_4$ and application of zinc oxide along with zinc-mobilizing bacteria had equal effects on plant Zn accumulation (Krithika and Balachandar 2016). OsZIP1 and OsZIP5 gene expression was elevated in rice infected with the zinc-mobilizing bacterium *Enterobacter cloacae*, but the bacteria suppressed OsZIP4 gene expression in the presence of zinc oxide (Krithika and Balachandar 2016). The microbial obtrusion of ZIP genes may result from the microbe-mediated zinc management, which is also reported in *Hordeum vulgare* and *Medicago sativa* with the inoculation of *R. irregularis* (Nguyen et al. 2021).

13.8 Manganese (Mn) Transforming Microorganisms

Plants may absorb only the divalent form of manganese (Mn^{2+}). The highly saline and oxidative stress stimulates the development of insoluble forms of manganese oxides, making manganese unavailable in well-aerated calcareous soils for plant uptake; however, Mn^{2+} soil dressing is ineffectual in treating shortage because of quick oxidation of Mn^{2+} (Andresen et al. 2018). Rhizosphere acidification caused by microbes would aid plant uptake of Mn since soil pH affects the availability of Mn. Acidophilic bacteria strains employed for Mn bioleaching from ores or mining wastewater, including *Acinetobacter* sp. and *Lysinibacillus* sp., are examples of bacterial mobilization of Mn. *Aspergillus terreus* and *Penicillium daleae*, two fungal strains capable of solubilizing Mn, have been isolated from manganese mine tailings. Their capacity to do so has been linked to mycelia's production of organic acids like gluconic acid, oxalic acid, maleic acid, and citric acid (Mohanty et al. 2017). IRT1, a Fe^{2+} importer, is also one of the plants' Mn transporters due to its poor selectivity. The pH and redox potential of the soil also affects the availability of both Fe and Mn. As a result, microorganisms that can cause plants to exhibit iron deficiency symptoms may also increase Mn absorption by plants.

13.9 Copper (Cu) Transforming Microorganisms

Under toxic and Cu-deficient environments, AM fungus can improve plant fitness. Cu content in cucumber (*Cucumis sativus*) shoots rose due to *G. mosseae* colonization because of extraradical mycelia-dependent nutrient absorption (Singh et al. 2022). Although contemporary application of phosphorous changed the distribution patterns of the copper contents contributed by *G. mosseae*. The mycorrhizal contribution to plant copper uptake in the mycorrhizal white clover was independent of P availability. Microbial siderophores or manipulating certain plant genes have been linked to bacterial advances in plant Cu feeding. Siderophores secreting bacterial strains, *P. fluorescens* and *Rhizobium leguminosarum* bv *phaseoli*, improved the absorption of Cu and Fe in *Medicago sativa* (Alfalfa) seedlings (Carrillo-Castañeda et al. 2002). The plant-beneficial bacterium *A. brasilense* reduced plant stress symptoms in cucumber plants with Cu deficit or dual Cu and Fe deficiency through CsFRO gene expression, which take up Cu following FRO-mediated decrease, which is consistent with the bacteria-enhanced plant nutrient absorption by improved root growth (Marastoni et al. 2020).

13.10 Iron (Fe) Transforming Microorganisms

Iron (Fe) is a crucial component of cellular metabolism. It has role in producing chlorophyll, preserving the structure and function of chloroplast, nucleic acids synthesis, and respiration. Complexation is a fundamental approach used by plants and related bacteria to deal with extremely limited Fe. Siderophores and organic acids interact directly with iron ore surfaces, such as silicates and oxyhydroxides, which helps disintegrate and translocate the micronutrients. Additionally, using a ligand exchange mechanism, these complexation agents may remove iron from proteins, phenols, humic compounds, and Fe-organic acids (Zhang et al. 2019). Rhizospheric microorganisms, primarily bacteria and fungi, secrete siderophores of low molecular weights (400–1500 Da) that are more effective in scavenging Fe (binding constant $K_f > 10^{30}$). In addition to iron, siderophores are discovered to interact with trace metals, including cadmium, copper, aluminum, In, lead, and zinc, assisting in the availability for plants for their improved growth (Yu et al. 2017). Currently, 500 distinct siderophores are known, of which 270 have been structurally described (Kramer et al. 2020).

Gram-positive and Gram-negative members have various mechanisms for absorbing iron using the TonB-ExbBD complex, the p-ATPase system, and periplasmic SBPs. According to Fukushima et al. (2014), lipoprotein SBPs tethered to the plasma membrane of Gram-positive species bind the ferrous ions using catechol or citrate type of siderophores and moved into the cell through the p-ATPase system. Nevertheless, various redox (oxidation–reduction) processes are used by the soil/phytomicrobiome to move the Fe in the soil solution close to the roots. In particular, ferric ions liberated from Fe minerals are quickly oxidized to ferrous ions under aerobic conditions and at pH 5–8. Conversely, under anaerobic conditions, ferrous ions are reduced by inorganic ions or by iron reducing bacteria in soil or water (Tang et al. 2019). Soil microbes use numerous biogeochemical processes to make them available and absorb iron (II/III), which eventually aids in improved Fe absorption by plants. Ferrous can be reduced directly or indirectly. Direct reduction of Fe(III) to Fe(II) requires energy (a respiratory substrate). Dissimilatory Fe reduction (DFeR), membrane-bound reductases, helps in using a variety of ferrous minerals (Lovley et al. 2004). Microorganisms from the Bacillales family, including Geobacter, Shewanella, and others, prefer hydrogen and organic carbon as electron donors (oxidation).

13.11 Mycorrhizae

The availability of micronutrients in the soil is positively impacted by the production of siderophores, acidity near roots, and the transport of nutrients to plants by external mycelium. Additionally, through the overexpression of HvZIP13, the colonization of *Rhizophagus irregularis* in Zn-deficient circumstances resulted in enhanced grain Zn contents (Watts-Williams and Cavagnaro 2018). Wheat grain concentrations of zinc and iron were higher when PGPB and AM fungi were combined than when they were used separately, indicating improved yield-related characteristics. The PGPB improved the numerous metabolic processes that affect plant development metrics. However, AM takes part in several functions to effectively absorb micronutrients in root tissues and promote plant development by producing signaling molecules and other substances that cause alteration in the root architecture of agricultural plants. In addition, using PGPB and AM fungi together results in more significant weed management, pathogen protection, and stress relief. Endophyte inoculation dramatically lowered the myo-inositol hexa-phosphate content in the grains of wheat by around 26.00% over the RDF, which led to rise in iron and zinc content (Yadav et al. 2020).

13.12 Prospects of Microbe-Mediated Biofortification of Food Grains

Globally, a quiet pandemic of nutrient shortages in humans and animals is affecting several developing nations. Nutritional quality is also threatened by a lack of variety in food, which includes cereal-based products that lack essential minerals. The importance of biofortification as a method to increase nutritional quality and security is being considered to meet the existing and future demands of the ever-increasing population. For the nutritional enrichment of field crops, oilseeds, and fodder crops, a variety of novel and focused biofortification techniques have been proposed in this effort. The soil and foliar fertilizer treatment, including soil application and agronomic approach, raises the micronutrient status of crops. The traditional method of biofortification is by selecting elite plants or genotypes with desirable nutritional features and breeding them without losing agricultural or commercial output.

Nevertheless, the transgenic/biotechnological method entails the creation of transgenes to re-translocate micronutrients across tissues to increase their bioavailability. Through various methods, including synthesis, mobilization, transformations, and siderophore generation, soil microbes increase the nutrient content in the rhizosphere, increasing the number of minerals absorbed by plants. In biofortification, various micronutrient sources, such as mineral solutions, chelates, and nanoparticles, are crucial because they control the rates and methods of plant absorption. A sustainable and economical strategy, biofortification increased agricultural productivity and quality characteristics to reduce hidden hunger.

When *Pseudomonas* and *Bacillus* sp. were applied to the soil, the pH fell, converting the insoluble complex chemicals such as ZnS, ZnO, and $ZnCO_3$ into ions in the broth. In wheat (DS-68 var.), inoculation of *Arthrobacter* sp. and *A. sulfonivorans* increased the zinc and iron concentration. In another study, endophytes such as *A. sulfonivorans* and *E. hirae* increased average root number and root surface area by 1.6- and 2-fold than the control, respectively, in soil with less iron content. However, the average number of root tips and the surface area changed over the control by 1.2 and 1.5 times, respectively, in cases where the soil's Fe concentration was high. These improved root characteristics immediately aided in the wheat grains' Fe fortification (Singh et al. 2018). Wheat production and micronutrient status were further increased with the co-inoculation of *B. subtilis* and *Arthrobacter* sp., a potential zinc solubilizer, and endophytes (*E. hirae* and *A. sulfonivorans*).

13.13 Conclusions

Many microorganisms can improve plant nutrition by enhancing soil nutrient conversion and improving plant nutrient absorption by increasing root architecture. It has been reported that microbes control plant genes responsible for nutrient acquisition at root–soil interface (rhizosphere), but it is still not entirely apparent that what exactly is secreted and at what time the transcriptional regulation happens in the plant. Stress-induced ROS generation is handled by the activation of antioxidant enzymes, the advantageous microorganisms may also indirectly reduce the stress caused by nutritional deprivation in plants. Additionally, certain nutrients are coupled intrinsically by plants, meaning that good bacteria that might help plants better absorb one nutrient may also help plants better absorb another. For instance, adding Mg to soybean plants increased *Bradyrhizobium* nodulation and the subsequent N_2 fixation under N-sufficient and N-deficient situations since Mg is crucial for biological N_2 fixation. So, several intriguing issues include whether legumes exhibit a link between biological N_2 fixation and the Mg mobilizing microorganisms diversity and quantity in the root microbiome and whether legume plants may acquire nitrogen from microbes that promote plant magnesium acquisition.

References

Adisa IO, Pullagurala VLR, Peralta-Videa JR, Dimkpa CO, Elmer WH, Gardea-Torresdey JL, White JC (2019) Recent advances in nano-enabled fertilizers and pesticides: a critical review of mechanisms of action. Environ Sci Nano 6(7):2002–2030

Al Hasan SM, Hassan M, Saha S, Islam M, Billah M, Islam S (2016) Dietary phytate intake inhibits the bioavailability of iron and calcium in the diets of pregnant women in rural Bangladesh: a cross-sectional study. BMC Nutr 2(1):1–10

Andresen E, Peiter E, Küpper H (2018) Trace metal metabolism in plants. J Exp Bot 69(5):909–954

Atta AA, Morgan KT, Hamido SA, Kadyampakeni DM (2020) Effect of essential nutrients on roots growth and lifespan of huanglongbing affected citrus trees. Plants 9(4):483

Berendsen RL, Pieterse CM, Bakker PA (2012) The rhizosphere microbiome and plant health. Trends Plant Sci 17(8):478–486

Bhupalraj G, Patnaik MC, Khadke KM (2002) Molybdenum status in soils of Andhra Pradesh. AICRP Micro Second Nutr Soils Plants Pradesh 36:1–87

Bruns HA (2017) Soybean micronutrient content in irrigated plants grown in the Midsouth. Commun Soil Sci Plant Anal 48(7):808–817

Carrillo-Castañeda G, Muños JJ, Peralta-Videa JR, Gomez E, Tiemannb KJ, Duarte-Gardea M, Gardea-Torresdey JL (2002) Alfalfa growth promotion by bacteria grown under iron limiting conditions. Adv Environ Res 6(3):391–399

Dhaliwal SS, Naresh RK, Mandal A, Singh R, Dhaliwal MK (2019) Dynamics and transformations of micronutrients in agricultural soils as influenced by organic matter build-up: a review. Environ Sustain Indic 1:100007. https://doi.org/10.1038/s41579-019-0284-4

Etesami H, Adl SM (2020) Plant growth-promoting rhizobacteria (PGPR) and their action mechanisms in availability of nutrients to plants. In: Phyto-microbiome in stress regulation. Springer, Berlin, pp 147–203

Ferrarezi RS, Lin X, Neira ACG, Zambon FT, Hu H, Wang X, Huang JH, Fan G (2022) Substrate pH influences the nutrient absorption and rhizosphere microbiome of huanglongbing-affected grapefruit plants. Front Plant Sci 13:856937

Fitzpatrick CR, Salas-González I, Conway JM, Finkel OM, Gilbert S, Russ D, Teixeira PJPL, Dangl JL (2020) The plant microbiome: from ecology to reductionism and beyond. Ann Rev Microbiol 74:81–100

Fukushima T, Allred BE, Raymond KN (2014) Direct evidence of iron uptake by the Gram-positive siderophore-shuttle mechanism without iron reduction. ACS Chem Biol 9(9):2092–2100

Gottel NR, Castro HF, Kerley M, Yang Z, Pelletier DA, Podar M, Karpinets T, Uberbacher ED, Tuskan GA, Vilgalys R, Doktycz MJ (2011) Distinct microbial communities within the endosphere and rhizosphere of Populus deltoides roots across contrasting soil types. Appl Environ Microbiol 77(17):5934–5944

Hussain B, Li J, Ma Y, Tahir N, Ullah A (2020) Effects of Fe and Mn cations on Cd uptake by rice plant in hydroponic culture experiment. PLoS One 15(12):e0243174

Khatoon Z, Huang S, Rafique M, Fakhar A, Kamran MA, Santoyo G (2020) Unlocking the potential of plant growth-promoting rhizobacteria on soil health and the sustainability of agricultural systems. J Environ Manag 273:111118

Kramer J, Ozkaya O, Kummerli R (2020) Bacterial siderophores in community and host interactions. Nat Rev Microbiol 18:152–163

Krithika S, Balachandar D (2016) Expression of zinc transporter genes in rice as influenced by zinc-solubilizing Enterobacter cloacae strain ZSB14. Front Plant Sci 7:446

Ladha JK, Reddy CK, Padre AT, van Kessel C (2011) Role of nitrogen fertilization in sustaining organic matter in cultivated soils. J Environ Qual 40(6):1756–1766

Li B, Wang S, Zhang Y, Qiu D (2020) Acid soil improvement enhances disease tolerance in citrus infected by Candidatus Liberibacter asiaticus. Int J Mol Sci 21(10):3614

Lončarić Z, Karalić K, Popović B, Rastija D, Vukobratović M (2008) Total and plant available micronutrients in acidic and calcareous soils in Croatia. Cereal Res Commun 36:331–334. https://doi.org/10.1556/CRC.36

Lovley DR, Holmes DE, Nevin KP (2004) Dissimilatory Fe (III) and Mn (IV) reduction. Adv Microb Physiol 49(2):219–286

Mallarino AP, Kaiser DE, Ruiz DA, Laboski CAM, Camberato JJ, Vyn TJ (2017) Micronutrients for soybean production in the North Central Region. Ames, IA, USA

Marastoni L, Lucini L, Miras-Moreno B, Trevisan M, Sega D, Zamboni A, Varanini Z (2020) Changes in physiological activities and root exudation profile of two grapevine rootstocks reveal common and specific strategies for Fe acquisition. Scientific Reports 10(1), p.18839.

Mitchikpe ECS, Dossa RA, Ategbo EAD, van Raaij JM, Hulshof PJ, Kok FJ (2008) The supply of bioavailable iron and zinc may be affected by phytate in Beninese children. Journal of food composition and analysis 21(1), pp.17–25

Mohapatra KK, Singh SK, Patra A (2021) Influence of varying levels of zinc on yield and zinc biofortification in hybrid rice (Oryza sativa L.) grown in moderate zinc soil. Journal of the Indian Society of Soil Science 69(2), pp.220–223.

Mohanty S, Ghosh S, Nayak S, Das A (2017) Isolation, identification and screening of manganese solubilizing fungi from low-grade manganese ore deposits. Geomicrobiol J 34(4):309–316. https://doi.org/10.1080/01490451.2016.1189016

Monreal CM, DeRosa M, Mallubhotla SC, Bindraban PS, Dimkpa C (2016) Nanotechnologies for increasing the crop use efficiency of fertilizer-micronutrients. Biol Fertil Soils 52(3):423–437

Nguyen PH, Young MF, Tran LM, Khuong LQ, Duong TH, Nguyen HC, Truong TV, DiGirolamo AM, Martorell R, Ramakrishnan U (2021) Preconception micronutrient supplementation positively affects child intellectual functioning at 6 y of age: A randomized controlled trial in Vietnam. The American Journal of Clinical Nutrition 113(5), pp.1199–1208

Paungfoo-Lonhienne C, Rentsch D, Robatzek S, Webb RI, Sagulenko E, Näsholm T, Schmidt S, Lonhienne TG (2010) Turning the table: plants consume microbes as a source of nutrients. PLoS One 5(7):e11915

Raboy V (2003) myo-Inositol-1,2,3,4,5,6-hexakisphosphate. Phytochemistry 64(6):1033–1043

Rasheed N, Maqsood MA, Aziz T, Jabbar A (2020) Characterizing lentil germplasm for zinc biofortification and high grain output. J Soil Sci Plant Nutr 20(3):1336–1349

Santoyo G, Guzmán-Guzmán P, Parra-Cota FI, de los Santos-Villalobos S, Orozco-Mosqueda MDC, Glick BR (2021) Plant growth stimulation by microbial consortia. Agronomy 11(219):10–3390

Singh HP, Batish DR, Kaur S, Arora K, Kohli RK (2006) α-Pinene inhibits growth and induces oxidative stress in roots. Annals of Botany 98(6), pp.1261–1269

Singh MV (2001) Evaluation of current micronutrient stocks in different agro-ecological zones of India for sustainable crop production. Fertiliser News 46(2):25–42

Singh BR, Timsina YN, Lind OC, Cagno S, Janssens K (2018) Zinc and iron concentration as affected by nitrogen fertilization and their localization in wheat grain. Front Plant Sci 9:307

Singh SK, Wu X, Shao C, Zhang H (2022) Microbial enhancement of plant nutrient acquisition. Stress Biol 2(1):1–14

Siqueira Freitas D, Wurr Rodak B, Rodrigues dos Reis A, de Barros Reis F, Soares de Carvalho T, Schulze J, Carbone Carneiro MA, Guimaraes Guilherme LR (2018) Hidden nickel deficiency? Nickel fertilization via soil improves nitrogen metabolism and grain yield in soybean genotypes. Front Plant Sci 9:614

Sutradhar AK, Kaiser DE, Behnken LM (2017) Soybean response to broadcast application of boron, chlorine, manganese, and zinc. Agronomy Journal 109(3), pp.1048–1059

Tang HY, Holmes DE, Ueki T, Palacios PA, Lovley DR (2019) Iron corrosion via direct metal-microbe electron transfer. MBio 10:e00303-19. https://doi.org/10.1128/mBio.00303-19

Triveni S, Nagaraju Y (2022) Biosynthesis of silver nanoparticles from Pseudomonas fluorescens and their antifungal activity against Aspergillus niger and Fusarium udum. Annals of Applied Biology 181(2), pp.235–245

Watts-Williams SJ, Cavagnaro TR (2018) Arbuscular mycorrhizal fungi increase grain zinc concentration and modify the expression of root ZIP transporter genes in a modern barley (Hordeum vulgare) cultivar. Plant Sci 274:163–170

Williams JE, Carrothers JM, Lackey KA, Beatty NF, York MA, Brooker SL, Shafii B, Price WJ, Settles ML, McGuire MA, McGuire MK (2017) Human milk microbial community structure is relatively stable and related to variations in macronutrient and micronutrient intakes in healthy lactating women. The Journal of nutrition 147(9), pp.1739–1748

Yadav R, Ror P, Rathore P, Ramakrishna W (2020) Bacteria from native soil in combination with arbuscular mycorrhizal fungi augment wheat yield and biofortification. Plant Physiol Biochem 150:222–233

Yu S, Teng C, Bai X, Liang J, Song T, Dong L, Jin Y, Qu J (2017) Optimization of siderophore production by Bacillus sp. PZ-1 and its potential enhancement of phytoextraction of Pb from soil. J Microbiol Biotechnol 27(8):1500–1512

Zhang X, Zhang D, Sun W, Wang T (2019) The adaptive mechanism of plants to iron deficiency via iron uptake, transport, and homeostasis. Int J Mol Sci 20(10):2424

Co-shaping and Co-evolution of Microbial Biodiversity: Study for Identification of Potential Plant Growth Promoting Microbes

14

Shanu Magotra, Neelam Prabha Negi, and Hemant Kumar

Abstract

In natural and agricultural ecosystems, microbial communities play a pivotal role in shaping the health and productivity of plants in natural and agricultural ecosystems. This review article delves into the intricate relationship between plants and their associated microbiota, focusing on the co-shaping and co-evolution of microbial biodiversity. The mechanisms through which plants and microbes interact are examined, exploring the underlying principles governing the selection and maintenance of plant growth promoting microbes (PGPMs). The review also provides insights into the dynamic nature of these associations, discussing the tools and techniques employed in the identification and characterization of PGPMs. Additionally, emphasis has been provided on the significance of these beneficial microorganisms in sustainable agriculture, highlighting their potential to enhance productivity, mitigate abiotic and biotic stresses, and reduce the reliance on chemical fertilizers and pesticides, leading to a more sustainable approach. This comprehensive review serves as a valuable resource for researchers and practitioners seeking to harness the potential of PGPMs to foster plant growth and improve agricultural sustainability.

Keywords

Plants · Microbes · Plant growth promoting microbes (PGPMs) · Sustainable agriculture and microbial biodiversity

S. Magotra (✉) · N. P. Negi · H. Kumar
Department of Biotechnology, UIBT, Chandigarh University, Mohali, Punjab, India

S. Kaur et al. (eds.), *Metabolomics, Proteomics and Gene Editing Approaches in Biofertilizer Industry*, https://doi.org/10.1007/978-981-97-2910-4_14

14.1 Introduction

In nature, plants and tiny microbes work together in a complex way, showing how life on our planet can adapt and be very intricate. Microbes, these small but powerful elements of nature with diverse forms and functions, have a significant impact on how plants grow and the health of ecosystems (Adey and Loveland 2011; Singh and Sharma 2022). Among these microbial allies, bacterial species, with their diverse array of species and strains, play a crucial role in supporting plant vitality and helping them grow (Enebe and Babalola 2019). Their contributions, both direct and indirect, reverberate through the soil, forming the foundation for thriving plant communities. Recognizing the importance of these microbial associations with the plants, this chapter embarks on a compelling exploration of their symbiotic association for co-shaping and co-evolution by engaging in a dynamic exchange of influence and adaptation. Through the in-depth understanding of this microbial realm, one cannot help but be captivated by the remarkable role of plant growth promoting rhizobacteria (PGPR). These biological inoculants can be potential candidates for the promotion of plant growth and biological control of diseases (Odoh 2017; Shahrtash and Brown 2021; Kalyanasundaram et al. 2021). PGPR have also emerged as promising alternative to chemical fertilizers and pesticides (Egamberdieva and Teixeira da Silva 2015; Gupta et al. 2020; Santoyo et al. 2021). Plant growth promoting microorganisms such as bacteria, fungi, algae, and actinomycetes benefit the host plants by promoting root development for better absorption of nutrients, control of plant stress by reducing plant diseases by induction of systemic resistance, and by creating competition for nutrients and habitat (Pathak et al. 2020). The most important factor for the understanding to this co-evolutionary relationship is the ability of these microbes to manipulate the availability of essential nutrients thereby creating competition for the niche by the production of hydrolytic enzymes/antibiotic compounds and by the induction of systemic resistance (Parewa et al. 2018; Kalyanasundaram et al. 2021). PGPR not only bolster the nutritional status of their botanical partners but also tip the scales in favor of beneficial microbes, creating a fortified front against potential pathogens. Additionally, they also deploy many phytohormones such as indole acetic acid, cytokinins, gibberellins, and ethylene which have a profound influence on plant architecture, flowering, and ultimately, productivity.

Through this book chapter, we have tried to embark on the journey of co-shaping and co-evolution through the lens of microbial biodiversity, not only to understand but also to harness the profound potential that lies within the interactions between microbes and plants, a dance that has shaped the green tapestry of our planet form dawn of time.

14.2 Microbial Biodiversity and Plant Health

Microbial biodiversity is an indispensable component of healthy plant ecosystems (Liu et al. 2022). Understanding and harnessing the intricate relationships between plants and microbes can pave the way for sustainable agriculture, ensuring the vitality of our food systems while minimizing environmental impact. Soil is a thriving habitat for an immense diversity of microbes, including bacteria, fungi, nematodes, and protozoa. Beneath the soil's surface, this intricate microbial community forms symbiotic relationships with plant roots, influencing nutrient uptake, disease resistance, and overall plant well-being. Notably, mycorrhizal fungi establish mutualistic associations with plant roots, aiding in nutrient absorption, particularly phosphorus and nitrogen (Smith and Read 2010). Mycorrhizal fungi have also been reported to play a crucial role in improving soil structure, fostering aeration, and enhancing water-holding capacity, all of which contribute to overall plant health and productivity (Sharma et al. 2022). The mycorrhizal association is so profound that it is estimated nearly 90% of all plant species engage in this mutualistic relationship, underscoring its evolutionary importance.

The associated microbial communities forming the microbiome of a plant are also the key players in nutrient cycling, breaking down organic matter into forms accessible to plants (Yadav et al. 2021). Nitrogen-fixing bacteria belonging to the genus *Rhizobium* and *Bradyrhizobium* convert atmospheric nitrogen into ammonia, enriching the soil with a vital nutrient (Koshila Ravi et al. 2019). This interaction enhances plant growth and development, illustrating the mutually beneficial trade of microbial life and plant health. Additionally, microbiome diversity of a plant confers resilience against environmental stress conditions and also provides first line of defense against plant pathogens (Noman et al. 2021). Microbes (such as bacterial and fungal species) contribute to the development of robust root systems, enabling plants to withstand drought, nutrient deficiencies, and act as biocontrol agents, suppressing harmful pathogens through competition for resources, production of antimicrobial compounds, and inducing systemic resistance in plants (Phazang et al. 2020). This natural balance helps prevent disease outbreaks and reduces the need for chemical interventions, therefore enhances the overall stability and adaptability of plant ecosystems (Compant et al. 2019; Trivedi et al. 2022; Ali et al. 2023).

In the realm of sustainable agriculture, the application of PGPR has shown remarkable promise (Negi et al. 2023a, b). Ahmad et al. (2011) conducted extensive studies demonstrating the ability of these beneficial microbes to ameliorate plant growth under conditions of salt stress. This finding is of particular significance in regions where salinity poses a formidable challenge to crop production. The symbiotic relationship established between plants and PGPR proves to be a potent tool in mitigating the adverse effects of environmental stress.

14.3 Methods and Study Design

In order to decipher the intricate plant–microbe associations, it is mandatory to select the appropriate methods and study designs. The proper selection of these methodologies and study approaches will help us to understand the belowground associations between microbes and the host plants. Employing a combination of traditional and innovative methods, along with cutting-edge technologies, enhances the depth and precision of microbial identification in plant ecosystems. Integrating these techniques enables researchers to unravel the intricate relationships between microbial diversity and plant health, paving the way for more effective strategies in agriculture and environmental management. The sampling process majorly involves spatial considerations and depth profiling (Peacher and Meiners 2020). In order to capture the microbial diversity spatially, designing a sampling strategy that accounts for spatial heterogeneity in microbial communities is very important. The random or stratified sampling methods used for the sampling can help capture the diversity across the study area in a much effective way (Gamalero et al. 2022). Along with the effective sampling, depth profiling also plays a very important role for understanding the microbial diversity of a particular location. The collection of samples at various soil depths should be performed to understand how microbial communities vary in different layers, influencing plant health differently (Tokala et al. 2002). Additionally, for the collection of the samples, it is mandatory to use sterile tools and containers to prevent contamination. This is crucial for maintaining the integrity of microbial populations. In order to collect the rhizosphere samples (the region around plant roots), much attention should be paid as it harbors a significant microbial population associated with plant health (Singh et al. 2020). These techniques should involve washing of the root samples and soil core sampling for accurate representation.

After the sampling, isolation of the microbial species also has immense importance for understating the true diversity of a particular niche. There are several traditional and modern techniques, which are employed for microbial identification. Among the traditional technique, culture-based method is the most preferred one, which involved the isolation and identification of the microbes through traditional culturing techniques. This approach provides a comprehensive understanding of cultivable microorganisms but this is a very time-consuming technique (Klerks et al. 2007). For the identification of the microbes at the molecular level, use of polymerase chain reaction (PCR) is the most preferred approach. The amplification of ribosomal RNA of the unknown bacterial strains using universal primers and then sequencing the amplicons can help in the precise identification of the bacterial strain till species level (Singh et al. 2020). These approaches offer great insights into the microbial community structure and diversity, thereby helping to understand the microbiome of a plant in more appropriate manner. 16S rRNA gene sequencing is the most commonly employed approach for the identification of the bacterial species and ITS sequencing is more preferred in case of fungal strains (Knief 2014). These target specific regions help us in understanding the taxonomic position, composition of microbiome, and abundance of microbial taxa.

With the recent technological advances, next-generation sequencing (NGS) technologies have further paved the way for exploiting the microbiome in a more structured manner (Knief 2014). One of the most popular approaches for studying the microbiome of a niche through these approaches is to employ metagenomic approach. Metagenomic analysis reveals the entire microbial community's genetic content, allowing for a comprehensive assessment of diversity (Bhargava et al. 2019; Fadiji and Babalola 2020). This approach allows us to surpass the concept of great plate anomaly by directly sequencing the genetic material from the environmental sample. Figure 14.1 represents the culture dependent and culture independent approach followed in metagenomics analysis of DNA samples. In order to study the functional genes (coding regions) of microbes that are active under a given circumstance, metatranscriptomic approach is preferred over metagenomic

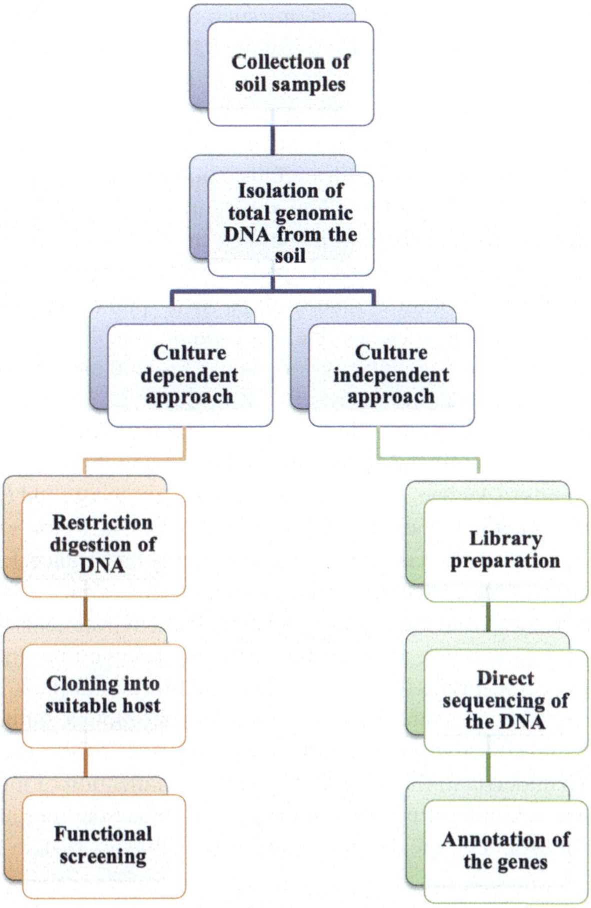

Fig. 14.1 Metagenomic analysis of DNA samples directly isolated from the soil

approach. This approach allows the researchers to analyze the total RNA pool of microbial communities to understand gene expression patterns. This helps in deciphering the functional roles of microbes in the ecosystem. Additionally, for the study of the entire set of proteins expressed by microbial communities, metaproteomics approach offers insights into microbial activity and functional dynamics. Several computational biology (bioinformatics) and statistical analysis tools are also used to integrate diverse datasets from various methods (Kumari et al. 2017). This allows for a holistic understanding of microbial communities, considering both taxonomic and functional aspects and to identify significant associations between microbial composition and plant health parameters.

14.4 Microbial Diversity in Different Ecosystems

Microbial diversity varies across ecosystems such as soil, rhizosphere, and aquatic environments, reflecting the distinct environmental conditions within each niche (Gibbons and Gilbert 2015). In soil ecosystems, a rich diversity of microorganisms exists which includes bacteria, fungi, archaea, and protozoa. Actinobacteria, known for their filamentous structure, dominate soil and contribute to the decomposition of organic matter (Prashar et al. 2014). Firmicutes, another prevalent group, play roles in nutrient cycling and soil structure. The zone surrounding the plant roots, commonly known as rhizosphere, displays a complex interplay of interactions between plants and microorganisms (Kent and Triplett 2002). Mycorrhizal fungi, such as *Glomus* species, form symbiotic relationships with plants, enhancing nutrient absorption. Nitrogen-fixing bacteria like *Rhizobium* establish associations with legumes, aiding in nitrogen availability for plants. Aquatic environments also harbor a distinct microbial panorama, where *Proteobacteria*, encompassing Alpha, Beta, Gamma, and Delta classes, dominate. *Cyanobacteria*, like *Anabaena* and *Microcystis*, contribute significantly to primary production through photosynthesis. The diversity extends to Archaea, with groups like *Crenarchaeota* and *Euryarchaeota* thriving in diverse aquatic niches (Philippot et al. 2013). Due to the varied diversity of the ecosystem, the shaping of the microbial community takes place. Extremophiles, such as *Thermus aquaticus*, excel in extreme temperatures, while *Acidobacteria* and *Nitrospirae* flourish in acidic conditions. Anaerobic environments house unique species like *Methanosarcina*, crucial for methane production. In hot springs, *Sulfolobus*, an archaeon, thrives in acidic, high-temperature conditions. *Streptomyces*, a soil-dwelling bacterium, is renowned for producing antibiotics, influencing microbial interactions and contributing to soil health (Yang et al. 2023). These mentioned examples help us understand the adaptability and specialization of microbial life in diverse ecosystems, highlighting the importance of understanding and preserving microbial diversity for ecological balance and sustainable ecosystem function.

14.5 Co-shaping and Co-evolution

Co-shaping and co-evolution in response to environmental changes refers to the dynamic relationship between plants and microbes. Plants have a very strong association with the microbes (below as well as aboveground), which results in many mutualistic benefits (Nadarajah and Abdul Rahman 2021). This interaction involves the reciprocal influence of both plants and microbes on each other's genetic, physiological, and ecological characteristics. The microbial associations with the plants can be categorized into four major categories based upon the location of this interaction. The microbes associated with the surface of the plants are known as epiphytes, microbes associated with the leaf surface are known as phyllospheric, microbes present inside the plant host tissues are known as endophytes, and microbes that inhabit the soil associated with the roots are known as rhizospheric microbes (Vishwakarma et al. 2020). All the above mentioned microbial associations greatly affect the quality and ultimately the yield of the plants, but rhizospheric microbes (microbes present in the narrow region of soil around the roots) draw special attention toward them as their association is the most dynamic that directly affects the flow of nutrients and growth of the plant (Lakshmanan et al. 2014). The rhizosphere of a plant is mainly characterized by microbes that significantly influence different growth stages of the plant and provides immunity against many stress conditions (Bandyopadhyay et al. 2022). Figure 14.2 represents various associations of plant with the rhizospheric microbes.

Due to the huge divergence in the speciation, structural complexities and interactions of the microbes with the plants, the understanding of this interactome becomes crucial. The inseparable ecological interactions between the plant host and the

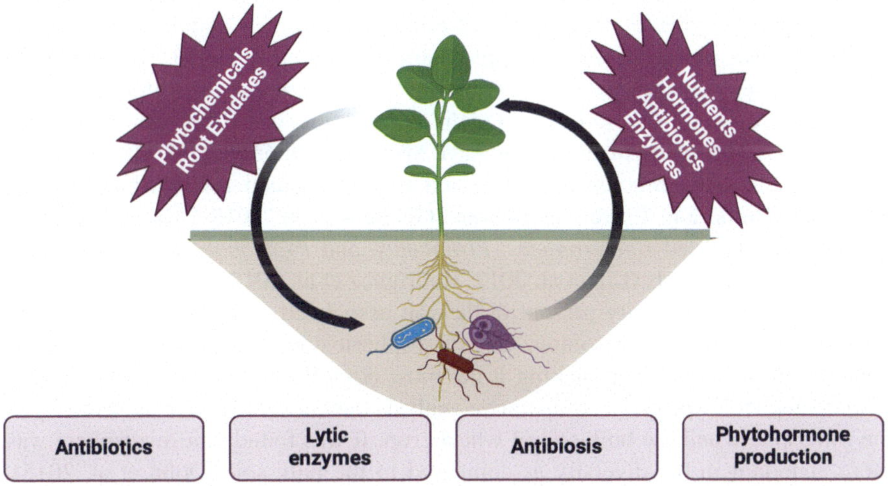

Fig. 14.2 Rhizospheric associations of plants with the microbes and their significance

microbes are of several types, including root–root, root–nematode, and root–microbes interactions, giving rise to a holobiont (Hacquard 2016). The mutualistic influence of rhizospheric microbes on the plants can be understood with techniques that help in the identification of the genome and proteome, thereby developing alternatives for more stable and sustainable agrosystems (Knapp et al. 2018).

14.6 Above and Belowground Microbiota: Associations and Interactions

Microorganisms are omnipresent and have been found to be present on the surface of the plant and also in the soil mustered in by the environment (Hardoim et al. 2015). The study of endosphere (environment present inside the host plant harboring microbes) and phyllosphere (aboveground part of the host plant harboring microbes) has resulted in the presence of diverse microbial species (Compant et al. 2019). The studies reveal the presence of specific endophytic species in various compartments (leaves, flowers, stem, and fruits) of different host plants. The specific associations help in the growth of the plant and increase immunity of the plant against various abiotic and biotic stresses (Vishwakarma et al. 2020; Negi et al. 2023a, b). Majorly the entry of microbes has been reported from the soil environment, into the roots, followed by colonization in the phyllosphere through the xylem (Compant et al. 2010, 2011; Vorholt 2012; Wallace et al. 2018). This is further affected by various environmental factors and agricultural practices. *Pseudomonas, Bacillus, Erwinia, Pantoea, Sphingomonas, Actinobacteria,* and *Methylobacterium* were found to be more dominant in the phyllosphere, whereas *Pseudomonas, Dyella, Staphylococcus, Ralstonia, Bacillus,* and *Mesorhizobium* were more dominant in the endosphere of grapevine (Compant et al. 2019; Kecskeméti et al. 2016). Steven and co-workers in 2018 reported *Enterobacteriaceae* and *Pseudomonas* as dominant taxas on the flowers of apple plant (Steven et al. 2018). Another study conducted by Wallace and co-workers in 2018 represents the dominance of *Methylobacteria* and *Sphingomonads* in the leaf cell lines of the maize crop (Wallace et al. 2018). The seeds of the plants have also been reported to be the source of microbial diversity. Studies have revealed the intertwined relationship between seed, soil, fruits, and flower microbiota (Glassner et al. 2018). Various research groups have reported *Bacteroidetes, Firmicutes,* and *Proteobacteria* as the dominant genus in the seeds (Liu et al. 2012; Rodríguez et al. 2018).

The microbial diversity present in the soil associated with the roots of a plant forms the rhizospheric microbiota. The rhizospheric area is considered to be metabolically most active area, making it the most complicated niches. Donn and co-workers in 2015 reported a comparative analysis between the microbial diversity in the rhizosphere and the bulk soil of wheat crop. It was found that rhizosphere was more abundant in the diversity as compared to the bulk soil (Donn et al. 2015). These microbial communities associated with the host plants can be transferred vertically and horizontally. The horizontal microbiome transfer means that the roots are enriched by the dominant species present in the soil such as *Bacteroidetes,*

Actinobacteria, and *Proteobacteria* (Fierer 2017), whereas the transfer of microbiota through seeds is known as the vertical transfer.

The host plant exerts rhizospheric effect on the microbes associated with the plants. In this effect, the plant releases several important compounds (such as sugars, hormones, polyphenols, organic acids, nutrients, and flavonoids) through the roots and these compounds act as chemoattractant for several microbial species (Compant et al. 2019). The exudates (phytochemicals) released by the plant roots are specific in case of specific plant hosts; therefore, these exudates facilitate the growth of specific microbes in the rhizosphere of a host and inhibit the growth of other microorganisms, thereby modifying the assembly of microbial communities in the rhizosphere. Zhalnina and co-workers in 2018 reported that the composition of exudates greatly affected the rhizospheric community of *Avena barbata* (Zhalnina et al. 2018). Similarly, Fitzpatrick and co-workers in 2018 reported differential occurrence patterns of rhizobacteria across 30 angiosperms (Fitzpatrick et al. 2018). Several microbial species, under various circumstances, through chance internalization colonize the roots and become endophytic in nature. These endophytes usually develop a symbiotic relation with the host plant. Lahrmann and co-workers in 2013 reported *Piriformospora indica* species as an important symbiotic endophyte used in agriculture (Lahrmann et al. 2013). *Piriformospora indica* increased the phosphorous uptake and was also used in tobacco plant for reducing salt stress (Trivedi et al. 2013). Also, many endophytic fungal species have also been reported. Non-pathogenic *Fusarium oxysporum* has been reported to reduce the effect of nematodes in tomato plants (Dababat and Sikora 2007). In addition to the presence of beneficial microbes, sometime pathogenic microbes are also attracted through chemotaxis. Gu and co-workers in 2017 reported the colonization of *Ralstonia solanacearum* in tomato roots by absorption of root exudates causing wilting of the plants (Gu et al. 2017; Hasegawa et al. 2019). Similarly, DeMarco and co-workers in 2020 have reported the movement of parasitic protozoan *Trypanosoma brucei* toward *Escherichia coli* colonized niches (DeMarco et al. 2020).

14.7 Plant Microbiome for Growth Enhancement and Stress Alleviation

The metabolic characteristics of a plant significantly impact the diversity of microbes in the soil, particularly in the rhizosphere. A study by Zhang and co-workers in 2014 found that cucumber roots release citric acid, attracting *Bacillus amyloliquefaciens,* while banana roots release fumaric acid, attracting *Bacillus subtilis* (Zhang et al. 2014). Moreover, the specific microbiome chosen by a plant also influences its development. This association is so precisely regulated that at times, the microbiome is referred to as the second genome (Jones et al. 2019; Magotra et al. 2021). In this mutualistic association, the microbes become involved in the production several beneficial compounds such as phytohormones, thereby regulating the plant morphology, genome, transcriptome, proteome, and metabolome (Fahad et al. 2015). These microbes have specifically been called as plant growth promoting microbes (PGPMs). PGPMs have been reported to synthesize several phytohormones for plant growth promotion and stress reduction. PGPMs release tryptophan, which is required for indole acetic acid (IAA) biosynthesis, ethylene, a stress hormone and ACC deaminase for the utilization of ACC (Haichar et al. 2012; Magotra et al. 2021).

The symbiotic co-evolution of microbes and plants can help in their better terrestrial development. Understanding the positive interactions between plants and their associated microbes can be utilized to reduce the reliance on chemical fertilizers and promote the development of a sustainable agrosystem (Werner et al. 2014; Vishwakarma et al. 2020). Various mechanisms contribute to how plant growth promoting microbes enhance plant growth, encompassing both direct and indirect mechanisms. Direct mechanisms include nutrient assimilation, the secretion of plant hormones and their signaling, production of siderophores for iron acquisition, and biological nitrogen fixation. Conversely, indirect mechanisms involve the release of volatile compounds to suppress diseases, the induction of induced systemic resistance (ISR) and systemic acquired resistance (SAR), and production of ACC deaminase to reduce ethylene levels, all employed by PGPMs.

A number of plant growth promoting microbes release several phytohormones (cytokinins, gibberellins, ACC deaminase, and auxins) and influence the plant growth. Auxin is required by the plants for root formation and growth, during stress conditions and for nodular meristem elongation. Defez and co-workers in 2017 reported the enhanced release of rubisco and osmolyte production through auxin, resulting in increased drought tolerance (Defez et al. 2017). Kim and co-workers in 2020 reported enhanced adventitious root growth by bacterial auxin resulting in less flooding (Kim et al. 2020). Cytokinin helps in plant cell division and for leaf cell development. Liu and co-workers in 2013 reported that bacterial cytokinin enhanced the relative water content, water potential of the leaves, and root exudates production during drought conditions (Liu et al. 2013). Gibberellin, another phytohormone, is helpful for the plants during germination of the seeds, elongation of the stem, production of flowers, and development of fruits. Waqas and co-workers in 2012 reported gibberellins produced by endophytic fungi, which resulted in enhanced nutrient assimilation under drought and salt stress (Waqas et al. 2012). Cohen and co-workers in 2015 reported that abscisic acid produced by the bacteria enhanced the amount of photosynthetic pigments, photoprotective pigments and reduced water loss during drought stress (Cohen et al. 2015). Wang and co-workers in 2000 reported that the bacterial ACC deaminase genes enhanced the root elongation and also increased resistance against the pathogens (Wang et al. 2000). Liu and co-workers in 2014 reported *Pseudomonas* from soybean roots which enhanced the ACC deaminase and IAA production (Liu et al. 2014). Ethylene, an important volatile phytohormone, regulates the plant growth and the process of senescence in the plants (Sukumar 2010). Thao and co-corkers in 2015 reported the association of ethylene production by the bacterial species and decreased stress conditions (Thao et al. 2015). Ethylene production has also been correlated with the shaping of the microbiome of a plant. Many researchers have created mutants (ET knockouts) and have found a shift in the bacterial diversity at the genus level; although, they could not prove that this shift is due to the ethylene levels or some other hormonal cross-talk (Doornbos et al. 2011). Another important plant growth regulation compound is jasmonic acid which also has been reported to enhance defense properties in plants. Hu and co-workers in 2018 reported enhanced jasmonic acid levels in the presence of benzoxazinoid and reduced fungal infection (Hu et al. 2018). This study opened doors for researchers in a new direction, wherein correlation between jasmonic acid and root exudates for microbial community selection and modification can be further studied.

PGPMs play a crucial role in enhancing plant growth by increasing nutrient availability. They serve as a promising alternative to chemical fertilizers and pesticides, offering safety, environmental friendly, high cost specificity, and no effect on the non-target organisms (Egamberdieva and Teixeira da Silva 2015; Gupta et al. 2020; Santoyo et al. 2021). PGPMs often referred to as biofertilizers constitute microbial concoction, which enhances the plant growth rate by making available all the required macro- and micronutrients. Plant growth promoting microorganisms such as bacteria, fungi, algae, and actinomycetes benefit the host plants by promoting root development for better absorption of nutrients, control of plant stress by reducing plant diseases by induction of systemic resistance and by creating competition for nutrients and habitat (Pathak et al. 2020). Bioinoculants, particularly from the bacterial kingdom (e.g. *Bacillus* species, *Pseudomonas* species, *Rhizobium* species) and fungal kingdom (e.g. *Mycorrhiza, Trichoderma* species and non-pathogenic *Fusarium* species) are widely preferred (Teotia et al. 2016; Parewa et al. 2018). Among plant growth promoting fungi (PGPF), mycorrhizal fungi have been reported to enhance plant growth by increasing the availability of soil nitrogen and phosphorus availability (Straker and Mitchell 1986; Marschner and Dell 1994; Naamala and Smith 2020). DaMatta and co-workers have reported 80% of phosphorus intake by the host plant *Coffea canephora* on application of arbuscular mycorrhizae to increase the storage of water in plants under drought conditions (DaMatta et al. 2002; Naamala and Smith 2020). *Trichoderma* species also enhance plant growth and inhibit soil and seed borne pathogens (Gupta et al. 2020). Overall, these biofertilizers represent a sustainable and effective approach to fostering plant health and productivity.

The plant growth promoting bacteria (PGPB) have also been successfully used as bioinoculants for better production and disease-free plants. Among PGPBs, Rhizobium, *Bradyrhizobium, Sinorhizobium,* and *Mesorhizobium* have been reported as symbionts of leguminous plants (Naamala and Smith 2020). They promote plant growth by fixation of atmospheric nitrogen, with the help of nitrogenase enzyme and in return they get the carbon fixed by the plants photosynthetically (Zhang et al. 2018). These bacterial species secrete nodulation factors (NF) upon perception of flavonoids present in the root exudates and initiate the process of nodule formation in the roots of the plants (Gage 2004). Gao and co-workers in 2012 reported *Bradyrhizobium* species in combination with *Glomus mosseae* increased the resistance and expression of defense related genes in soybean against *Cylindrocladium parasiticum* (Gao et al. 2012). Similarly, combination of *Bradyrhizobium* species and *Trichoderma* species has been reported to increase the shoot and root of cowpea plant and also provided protection against pathogenic *Rhizoctonia solani* (Paula Junior et al. 2015). *Pseudomonas* species and *Bacillus* species are the two most preferred plant growth promoting bacterial species. These bacterial species are recognized for their ability to enhance plant growth through better assimilation of essential nutrients such as phosphorus, iron, and nitrogen and production of plant hormones (Finney and Kaye 2017). In 2017, Jayashree and Jagadeesh reported the effective use of *Pseudomonas* sp. B15 and *Pseudomonas striata* to enhance plant height and increase seed biomass in brinjal, tomato, and chili (Jayashree and Jagadeesh 2017). Similarly, Molina-Romero and colleagues reported in 2017 that a consortium containing *Pseudomonas putida* KT2440 led to an increase in root–shoot dry weight, height, and diameter of maize plants (Molina-Romero et al. 2017). *Bacillus* species, when associated with plant roots, play a role in solubilizing nutrients, making them available to plants. In 2012, Alzoubi and Gaibore reported the effectiveness of *Bacillus megaterium*, which increased phosphorus availability by nearly 30% and promoted plant growth (Alzoubi and Gaibore 2012). Additionally, strains of *B. altitudinis* (BT3 and CT8), isolated from the rhizosphere of chickpea plants, were found to promote plant growth by producing siderophores, indole-3-acetic acid (IAA), ammonia, and solubilizing phosphate in pot assays (Kushwaha et al. 2021). In addition to enhancing the plant growth, the microbes associated with the plants also cause infections by the production of phytotoxic components. Several pathogenic microbes have been reported that cause diseases in many important crop plants. The diseases caused by these pathogens are characterized by several factors such as environmental conditions, population size of pathogen, and susceptibility of host. *Pseudomonas syringae* has been reported as a pathogen for tomato, green bean, and tobacco plants, *Erwinia amylovora* has been reported as a pathogen of fruit-bearing plants, *R. solanacearum* has been reported as a pathogen of banana, and *Xylella fastidiosa* has been reported as a pathogen of potato plant (Mansfield et al. 2012; Vishwakarma et al. 2020). The host plants protect themselves against the pathogens because of the presence of the associated beneficial microbes. These plant growth promoting microbes act as biological control agents by decreasing the disease occurrence in the host plants (Berg and Koskella 2018; de Vrieze et al. 2020). The biological control agents protect the host plants by the release of many lytic enzymes, production of volatile compounds, siderophores, and antibiotics that inhibit the growth and colonization of pathogenic microbes (Verma et al. 2018). The underlying mechanism behind biological control of plant diseases is through creating competition for the nutrients and niche, antagonism and release of defense response elements (Odoh 2017; Shahrtash and Brown 2021; Kalyanasundaram et al. 2021). Trichoderma species also enhance plant growth and inhibit soil and seed borne pathogens (Gupta et al. 2020). Jeyaseelan and co-workers in 2012 have reported *Trichoderma* spp. as a biocontrol agent against *Pythium aphanidermatum* pathogen of tomato (Jeyaseelan et al. 2012). Similarly, a combination of *Trichoderma harzianum* mutants and carbendazim reduced the chickpea wilt caused by *Fusarium oxysporum* f. sp. *ciceri* (Poddar et al. 2004). Application of *Trichoderma viride* (2.5 kg/ha) in the saffron cultivation soil along with subsequent treatment of corms with carbendazim (0.2%) reduced the incidence of corm rot disease of saffron (Gupta et al. 2014). Non-pathogenic *Fusarium* species have also been reported to be effective biocontrol agents. *Fusarium oxysporum* M12-4A has been reported as a biocontrol agent against *Striga hermonthica* from Nigerian savanna (Ciotola et al. 1995; Chindo et al. 2010). Similarly, Constantin and co-workers in 2020 reported the use of *Fusarium oxysporum* (Fo47) for the control of root disease caused by *Fusarium oxysporum* f. sp. *lycopersicum* in tomato by induced systemic resistance (Constantin et al. 2020). Gao and co-workers in 2012 reported *Bradyrhizobium* species in combination with *Glomus mosseae* increased the resistance and expression of defense related genes in soyabean against *Cylindrocladium parasiticum* (Gao et al. 2012). Similarly, combination of *Bradyrhizobium* species and *Trichoderma* species has been reported to increase the shoot and root of cowpea plant and also provided protection against pathogenic *Rhizoctonia solani* (Paula Junior et al. 2015).

14.8 PGPMs for Sustainable Agrosystem

Presently, for enhancing the yield of the crops and to protect them from any abiotic and biotic stresses, chemical fertilizers are the most preferred choice. The popularity of these chemical compounds is because of the easy application, bulk production, and ability to enhance the yield (Gupta et al. 2021). Although there chemical agents have been very successful for controlling the diseases, repeated applications of these chemicals have resulted in decreased nutrient concentration and fertility of the soil. These chemicals are not selective for the beneficial microflora and therefore, hamper the symbiotic relationship of beneficial microbes with the host plants. In addition to this, they also give rise to new pathotypes with enhanced disease-causing potential (Ulrich et al. 2021; Gupta et al. 2020, 2021).

Therefore, the need of the hour is to select a more viable and ecofriendly approach for enhancing the plant growth and controlling the diseases. Plant growth promoting microbes have paved a way for the establishment of sustainable agrosystems. The targeted use of beneficial microorganisms can help in the reduction of use of chemical fertilizers in agriculture (Magotra et al. 2021). These useful microbes can be directly inoculated into the soil or can be used for seed priming, and due to their ability to colonize the roots, they develop strong associations with the plants. These bioinoculants can be applied in the fields either as single formulations or as consortia of multiple microbes. Magotra and co-workers in 2021 reported the use of *Bacillus* species based bioformulation for plant growth promotion and corm rot disease reduction in *Crocus sativus* (Magotra et al. 2021). Planchamp and co-workers in 2015 reported *Pseudomonas putida* based bioformulation, which significantly reduced pathogenic fungal load and leaf necrosis in maize plants (Planchamp et al. 2015). The market for biofertilizers is primarily dominated by single inoculants but researchers have also reported several successful consortiums for growth promotion and disease control. Chiboub and co-workers in 2019 reported increased growth rate and antioxidant production in *Sulla coronaria* by using a consortium of *Rhizobium* and *Pseudomonas* species (Chiboub et al. 2020). Table 14.1 represents

Table 14.1 Bioinoculants for growth promotion and disease control in crop plants

Plant growth promoting microbe	Host plant	In vivo effect of the plant	References
Acidovorax delafieldii	Rice plant	Pot evaluation resulted in increased biomass, yield of the grains, and uptake of NPK	Cavite et al. (2021)
Bacillus subtilis MF497646 and *Pseudomonas koreensis* MG209738	Maize crop	Pot and field resulted in increased antioxidant activity and various growth parameters	Ghazy and El-Nahrawy (2021)
Bacillus amyloliquefaciens IN937a + *B. pumilus* IN937b	Tomato, pepper, and cucumber	Field evaluation resulted in increased plant growth and induction of systemic resistance in plants	Wang et al. (2021)
Bacillus mucilaginosus HQ013329	Garlic	Enhanced overall production and reduction in the hazard caused by chemical fertilizers	Mounir et al. (2020)
Pseudomonas reactans EDP28	*Zea mays*	Enhanced potassium content	Moreira et al. (2020)
Funneliformis mosseae + *Ensifer meliloti*	*Vitis vinifera*	Evaluation resulted in increased volatile compounds production	Jamily et al. (2019)
Pseudomonas stutzeri A1501	Bean plant	Enhanced dry weight and nodules	Ke et al. (2019)
Bacillus circulans	Pepper and potato plants	Pot evaluation resulted in increased total yield of pepper and tuber weight and yield of potato	Elkhatib et al. (2019)
B. subtilis, B. megaterium	*Cuminum cyminum*	Increased yield of the seeds and essential oils	Mishra et al. (2019)
Xanthomonas sp. + *Microbacterium* sp.	*Arabidopsis thaliana*	Pot evaluation resulted in increased overall growth and induction of systemic resistance in plants	Berendsen et al. (2012)
Burkholderia sp. RHT8	*Trigonella foenumgraecum*	Field evaluation resulted in increased enhanced yield and disease reduction	Kumar et al. (2018)
Fluorescent Pseudomonas sp. PF17	Sunflower plant	Enhanced overall growth	Tewari and Arora (2018)
Achromobacter xylosoxidans	*Oryza sativa*	Evaluation resulted in disease suppression	Joe et al. (2012)
Bacillus cereus BS03	Pigeon pea	Pot evaluation resulted in increased yield and induction of systemic resistance in plants	Dutta et al. (2008)

(continued)

Table 14.1 (continued)

Plant growth promoting microbe	Host plant	In vivo effect of the plant	References
Bacillus sp. HWP47	Wheat crop	Increased root dry weight (51.46%)	Patten and Glick (2002)

various bioinoculants successfully used for growth promotion and disease control in crop plants.

In India, several government bodies are providing these bioformulations to the famers under various schemes. The subsidy offered by the government on these inoculants ranges from 25% to 75%. Despite of all the favors, there is still some gap in the acceptance of these bioformulations by the farmers. Additionally, these microbial inoculants have several advantages but there are some limitations as well that are associated with them. The most important concern with the bioformulation is the shelf-life reliability and consistency. Furthermore, the microbial capacity of tolerate abiotic conditions also reduces the affectivity of the formulations. Conversely, researches have also reported pathogenicity of microbial inoculants against humans. *Burkholderia* species and *Pseudomonas* species have been reported as bioinoculants that are pathogenic to humans (Vishwakarma et al. 2020). Nonetheless, more research is required to prove the pathogenic nature of the plant growth promoting microbes.

14.9 Conclusion

The increasing population is posing a major threat in terms of food security. The pressure for enhanced yield to meet the demands of the people results in the overuse of harmful fertilizers and also giving rise to the new pathotypes. Microbial inoculants that are being used for the production of biofertilizers can be alternatives for sustainable agriculture as they have the capacity to positively impact agriculture. The mutualistic associations of these inoculants with the plants are so strong that both of these organisms help in co-shaping and co-evolution of each other.Conflict of InterestThe authors state that they have no conflicts of interest.

References

Adey WH, Loveland K (2011) Dynamic aquaria: building living ecosystems. Elsevier, Amsterdam

Ahmad F, Husain FM, Ahmad I (2011) Rhizosphere and root colonization by bacterial inoculants and their monitoring methods: a critical area in PGPR research. Microbes and microbial technology: Agricultural and environmental applications, pp.363–391

Ali S, Tyagi A, Bae H (2023) Plant microbiome: an ocean of possibilities for improving disease resistance in plants. Microorganisms 11(2):392

Alzoubi MM, Gaibore M (2012) The effect of phosphate solubilizing bacteria and organic fertilization on availability of Syrian rock phosphate and increase of triple superphosphate efficiency. World J Agric Sci 8(5):473–478

Bandyopadhyay P, Yadav BG, Kumar SG, Kumar R, Kogel KH, Kumar S (2022) Piriformospora indica and Azotobacter chroococcum consortium facilitates higher acquisition of N, P with improved carbon allocation and enhanced plant growth in Oryza sativa. J Fungi 8(5):453

Berendsen RL, Pieterse CM, Bakker PA (2012) The rhizosphere microbiome and plant health. Trends Plant Sci 17(8):478–486

Berg M, Koskella B (2018) Nutrient-and dose-dependent microbiome-mediated protection against a plant pathogen. Current Biology 28(15), pp.2487–2492

Bhargava P, Khan M, Verma A, Singh A, Singh S, Vats S, Goel R (2019) Metagenomics as a tool to explore new insights from plant-microbe interface. In: Plant microbe interface. Springer, Cham, pp 271–289

Cavite HJM, Mactal AG, Evangelista EV, Cruz JA (2021) Growth and yield response of upland rice to application of plant growth-promoting rhizobacteria. J Plant Growth Regul 40:494–508

Chiboub M, Jebara SH, Abid G, Jebara M (2020) Co-inoculation effects of Rhizobium sullae and Pseudomonas sp. on growth, antioxidant status, and expression pattern of genes associated with heavy metal tolerance and accumulation of cadmium in Sulla coronaria. J Plant Growth Regul 39:216–228

Chindo PS, Shebayan JAY, Marley PS (2010) Effect of pre-emergence herbicides on Meloidogyne spp. and Fusarium wilt of tomato in Samaru, Zaria, Nigeria. Journal of Agricultural Research (03681157), 48(4)

Ciotola M, Watson AK, Hallett SG (1995) Discovery of an isolate of Fusarium oxysporum with potential to control Striga hermonthica in Africa. Weed Research 35(4), pp.303–309

Cohen AC, Bottini R, Piccoli P (2015) Role of abscisic acid producing PGPR in sustainable agriculture. Bacterial metabolites in sustainable agroecosystem, pp.259–282.

Compant S, Clément C, Sessitsch A (2010) Plant growth-promoting bacteria in the rhizo-and endosphere of plants: their role, colonization, mechanisms involved and prospects for utilization. Soil Biology and Biochemistry 42(5), pp.669–678.

Compant S, Mitter B, Colli-Mull JG, Gangl H, Sessitsch A (2011) Endophytes of grapevine flowers, berries, and seeds: identification of cultivable bacteria, comparison with other plant parts, and visualization of niches of colonization. Microb Ecol 62:188–197

Compant S, Samad A, Faist H, Sessitsch A (2019) A review on the plant microbiome: ecology, functions, and emerging trends in microbial application. J Adv Res 19:29–37

Constantin ME, de Lamo FJ, Rep M, Takken FL (2020) From laboratory to field: applying the Fo47 biocontrol strain in potato fields. European Journal of Plant Pathology 158(3), pp.645–654.

Dababat AEFA, Sikora RA (2007) Influence of the mutualistic endophyte Fusarium oxysporum 162 on Meloidogyne incognita attraction and invasion. Nematology 9(6):771–776

DaMatta FM, Loos RA, Silva EA, Loureiro ME (2002) Limitations to photosynthesis in Coffea canephoraas a result of nitrogen and water availability. J Plant Physiol 159(9):975–981

Defez R, Andreozzi A, Dickinson M, Charlton A, Tadini L, Pesaresi P, Bianco C (2017) Improved drought stress response in alfalfa plants nodulated by an IAA over-producing Rhizobium strain. Frontiers in microbiology 8, p.2466

DeMarco SF, Saada EA, Lopez MA, Hill KL (2020) Identification of positive chemotaxis in the protozoan pathogen Trypanosoma brucei. mSphere 5(4):e00685-20

De Vrieze M, Varadarajan AR, Schneeberger K, Bailly A, Rohr RP, Ahrens CH, Weisskopf L (2020) Linking comparative genomics of nine potato-associated Pseudomonas isolates with their differing biocontrol potential against late blight. Frontiers in microbiology 11, p.857

Donn S, Kirkegaard JA, Perera G, Richardson AE, Watt M (2015) Evolution of bacterial communities in the wheat crop rhizosphere. Environ Microbiol 17(3):610–621

Doornbos RF, Geraats BP, Kuramae EE, Van Loon LC, Bakker PA (2011) Effects of jasmonic acid, ethylene, and salicylic acid signaling on the rhizosphere bacterial community of Arabidopsis thaliana. Mol Plant-Microbe Interact 24(4):395–407

Dutta S, Mishra AK, Kumar BD (2008) Induction of systemic resistance against fusarial wilt in pigeon pea through interaction of plant growth promoting rhizobacteria and rhizobia. Soil Biol Biochem 40(2):452–461

Egamberdieva D, Teixeira da Silva JA (2015) Medicinal plants and PGPR: a new frontier for phytochemicals. In: Plant-growth-promoting rhizobacteria (PGPR) and medicinal plants. Springer, Berlin, pp 287–303

Elkhatib HA, Gabr SM, Mohamed RA, Manaa A (2019) Impact of potassium fertilization rates and potassium solubilizing bacteria inoculation on the growth, yield and quality of potato (Solanum tuberosum L.). Alex Sci Exch J 40:127–142

Enebe MC, Babalola OO (2019) The impact of microbes in the orchestration of plants' resistance to biotic stress: a disease management approach. Appl Microbiol Biotechnol 103:9–25

Fadiji AE, Babalola OO (2020) Metagenomics methods for the study of plant-associated microbial communities: a review. J Microbiol Methods 170:105860

Fahad S, Hussain S, Bano A, Saud S, Hassan S, Shan D, Khan FA, Khan F, Chen Y, Wu C, Tabassum MA (2015) Potential role of phytohormones and plant growth-promoting rhizobacteria in abiotic stresses: consequences for changing environment. Environ Sci Pollut Res 22:4907–4921

Fierer N (2017) Embracing the unknown: disentangling the complexities of the soil microbiome. Nat Rev Microbiol 15(10):579–590

Finney DM, Kaye JP (2017) Functional diversity in cover crop polycultures increases multifunctionality of an agricultural system. J Appl Ecol 54(2):509–517

Fitzpatrick CR, Copeland J, Wang PW, Guttman DS, Kotanen PM, Johnson MT (2018) Assembly and ecological function of the root microbiome across angiosperm plant species. Proc Natl Acad Sci 115(6):E1157–E1165

Gage DJ (2004) Infection and invasion of roots by symbiotic, nitrogen-fixing rhizobia during nodulation of temperate legumes. Microbiol Mol Biol Rev 68(2):280–300

Gamalero E, Bona E, Glick BR (2022) Current techniques to study beneficial plant-microbe interactions. Microorganisms 10(7):1380

Gao X, Lu X, Wu M, Zhang H, Pan R, Tian J, Li S, Liao H (2012) Co-inoculation with rhizobia and AMF inhibited soybean red crown rot: from field study to plant defense-related gene expression analysis. PLoS One 7(3):e33977

Ghazy N, El-Nahrawy S (2021) Siderophore production by Bacillus subtilis MF497446 and Pseudomonas koreensis MG209738 and their efficacy in controlling Cephalosporium maydis in maize plant. Arch Microbiol 203:1195–1209

Gibbons SM, Gilbert JA (2015) Microbial diversity—exploration of natural ecosystems and microbiomes. Curr Opin Genet Dev 35:66–72

Glassner H, Zchori-Fein E, Yaron S, Sessitsch A, Sauer U, Compant S (2018) Bacterial niches inside seeds of Cucumis melo L. Plant Soil 422:101–113

Gu Y, Hou Y, Huang D, Hao Z, Wang X, Wei Z, Jousset A, Tan S, Xu D, Shen Q, Xu Y (2017) Application of biochar reduces Ralstonia solanacearum infection via effects on pathogen chemotaxis, swarming motility, and root exudate adsorption. Plant Soil 415:269–281

Gupta KJ, Mur LA, Brotman Y (2014) Trichoderma asperelloides suppresses nitric oxide generation elicited by Fusarium oxysporum in Arabidopsis roots. Molecular plant-microbe interactions 27(4), pp.307–314

Gupta V, Kumar K, Fatima K, Razdan VK, Sharma BC, Mahajan V, Rai PK, Sharma A, Gupta V, Hassan MG, Hussain R (2020) Role of biocontrol agents in management of corm rot of saffron caused by Fusarium oxysporum. Agronomy 10(9):1398

Gupta V, Sharma A, Rai PK, Gupta SK, Singh B, Sharma SK, Singh SK, Hussain R, Razdan VK, Kumar D, Paswal S (2021) Corm rot of saffron: epidemiology and management. Agronomy 11(2):339

Hacquard S (2016) Disentangling the factors shaping microbiota composition across the plant holobiont. New Phytol 209(2):454–457

Haichar FEZ, Roncato MA, Achouak W (2012) Stable isotope probing of bacterial community structure and gene expression in the rhizosphere of Arabidopsis thaliana. FEMS Microbiol Ecol 81(2):291–302

Hardoim PR, Van Overbeek LS, Berg G, Pirttilä AM, Compant S, Campisano A, Döring M, Sessitsch A (2015) The hidden world within plants: ecological and evolutionary considerations for defining functioning of microbial endophytes. Microbiol Mol Biol Rev 79(3):293–320

Hasegawa T, Kato Y, Okabe A, Itoi C, Ooshiro A, Kawaide H, Natsume M (2019) Effect of secondary metabolites of tomato (Solanum lycopersicum) on chemotaxis of Ralstonia solanacearum, pathogen of bacterial wilt disease. J Agric Food Chem 67(7):1807–1813

Hu L, Robert CA, Cadot S, Zhang XI, Ye M, Li B, Manzo D, Chervet N, Steinger T, Van Der Heijden MG, Schlaeppi K (2018) Root exudate metabolites drive plant-soil feedbacks on growth and defense by shaping the rhizosphere microbiota. Nat Commun 9(1):2738

Jamily AS, Koyama Y, Win TA, Toyota K, Chikamatsu S, Shirai T, Uesugi T, Murakami H, Ishida T, Yasuhara T (2019) Effects of inoculation with a commercial microbial inoculant Bacillus subtilis C-3102 mixture on rice and barley growth and its possible mechanism in the plant growth stimulatory effect. J Plant Protect Res 59(2):193–205

Jayashree C, Jagadeesh KS (2017) Testing the effect of the microbial consortium on growth of vegetable seedlings in a farmer's nursery. Int J Curr Microbiol Appl Sci 6(2):1636–1639

Jeyaseelan EC, Tharmila S, Niranjan K (2012) Antagonistic activity of Trichoderma spp. and Bacillus spp. against Pythium aphanidermatum isolated from tomato damping off. Archives of Applied Science Research 4(4), pp.1623–1627

Joe MM, Islam MR, Karthikeyan B, Bradeepa K, Sivakumaar PK, Sa T (2012) Resistance responses of rice to rice blast fungus after seed treatment with the endophytic Achromobacter xylosoxidans AUM54 strains. Crop Prot 42:141–148

Jones P, Garcia BJ, Furches A, Tuskan GA, Jacobson D (2019) Plant host-associated mechanisms for microbial selection. Front Plant Sci 10:862

Kalyanasundaram GT, Syed N, Subburamu K (2021) Recent developments in plant growth-promoting rhizobacteria (PGPR) for sustainable agriculture. Recent Developments in Applied Microbiology and Biochemistry pp.181–192.

Ke X, Feng S, Wang J, Lu W, Zhang W, Chen M, Lin M (2019) Effect of inoculation with nitrogen-fixing bacterium Pseudomonas stutzeri A1501 on maize plant growth and the microbiome indigenous to the rhizosphere. Syst Appl Microbiol 42(2):248–260

Kecskeméti E, Berkelmann-Löhnertz B, Reineke A (2016) Are epiphytic microbial communities in the carposphere of ripening grape clusters (Vitis vinifera L.) different between conventional, organic, and biodynamic grapes? PLoS One 11(8):e0160852

Kent AD, Triplett EW (2002) Microbial communities and their interactions in soil and rhizosphere ecosystems. Annu Rev Microbiol 56(1):211–236

Kim MJ, Shim CK, Ko BG, Kim J (2020) Effect of the microalga Chlorella fusca CHK0059 on strawberry PGPR and biological control of Fusarium wilt disease in non-pesticide hydroponic strawberry cultivation. Journal of Microbiology and Biotechnology 30(5), p.708

Klerks MM, Franz E, van Gent-Pelzer M, Zijlstra C, Van Bruggen AH (2007) Differential interaction of Salmonella enterica serovars with lettuce cultivars and plant-microbe factors influencing the colonization efficiency. ISME J 1(7):620–631

Knapp DG, Németh JB, Barry K, Hainaut M, Henrissat B, Johnson J, Kuo A, Lim JHP, Lipzen A, Nolan M, Ohm RA (2018) Comparative genomics provides insights into the lifestyle and reveals functional heterogeneity of dark septate endophytic fungi. Sci Rep 8(1):6321

Knief C (2014) Analysis of plant microbe interactions in the era of next generation sequencing technologies. Front Plant Sci 5:216

Koshila Ravi R, Anusuya S, Balachandar M, Muthukumar T (2019) Microbial interactions in soil formation and nutrient cycling. In: Mycorrhizosphere and pedogenesis. Springer, Berlin, pp 363–382

Kumar A, Singh VK, Tripathi V, Singh PP, Singh AK (2018) Plant growth-promoting rhizobacteria (PGPR): perspective in agriculture under biotic and abiotic stress. In: Crop improvement through microbial biotechnology. Elsevier, Amsterdam, pp 333–342

Kumari A, Sumer S, Jalan B, Nongbri PL, Laskar MA (2017) Impact of next-generation sequencing technology in plant–microbe interaction study. In: Microbial applications vol. 1: bioremediation and bioenergy. Springer, Berlin, pp 269–294

Kushwaha P, Srivastava R, Pandiyan K, Singh A, Chakdar H, Kashyap PL, Bhardwaj AK, Murugan K, Karthikeyan N, Bagul SY, Srivastava AK (2021) Enhancement in plant growth and zinc biofortification of chickpea (Cicer arietinum L.) by Bacillus altitudinis. J Soil Sci Plant Nutr 21:922–935

Lahrmann U, Ding Y, Banhara A, Rath M, Hajirezaei MR, Döhlemann S, von Wirén N, Parniske M, Zuccaro A (2013) Host-related metabolic cues affect colonization strategies of a root endophyte. Proc Natl Acad Sci 110(34):13965–13970

Lakshmanan V, Selvaraj G, Bais HP (2014) Functional soil microbiome: belowground solutions to an aboveground problem. Plant Physiol 166(2):689–700

Liu Y, Zuo S, Xu L, Zou Y, Song W (2012) Study on diversity of endophytic bacterial communities in seeds of hybrid maize and their parental lines. Arch Microbiol 194:1001–1012

Liu F, Xing S, Ma H, Du Z, Ma B (2013) Cytokinin-producing, plant growth-promoting rhizobacteria that confer resistance to drought stress in Platycladus orientalis container seedlings. Applied microbiology and biotechnology 97, pp.9155-9164

Liu W, Hou J, Wang Q, Ding L, Luo Y (2014) Isolation and characterization of plant growth-promoting rhizobacteria and their effects on phytoremediation of petroleum-contaminated saline-alkali soil. Chemosphere 117, pp.303–308

Liu J, Zhang J, Zhu M, Wan H, Chen Z, Yang N, Duan J, Wei Z, Hu T, Liu F (2022) Effects of plant growth promoting rhizobacteria (PGPR) strain Bacillus licheniformis with biochar amendment on potato growth and water use efficiency under reduced irrigation regime. Agronomy 12(5), p.1031

Magotra S, Bhagat N, Ambardar S, Ali T, Hurek BR, Hurek T, Verma PK, Vakhlu J (2021) Field evaluation of PGP Bacillus sp. strain D5 native to Crocus sativus, in traditional and non traditional areas, and mining of PGP genes from its genome. Sci Rep 11(1):5454

Mansfield J, Genin S, Magori S, Citovsky V, Sriariyanum M, Ronald P, Dow MAX, Verdier V, Beer SV, Machado MA, Toth IAN (2012) Top 10 plant pathogenic bacteria in molecular plant pathology. Mol Plant Pathol 13(6):614–629

Marschner H, Dell B (1994) Nutrient uptake in mycorrhizal symbiosis. Plant Soil 159:89–102

Mishra BK, Lal G, Sharma YK, Kant K, Saxena SN, Dubey PN (2019) Effect of microbial inoculants on cumin (Cuminum cyminum Linn.) growth and yield. Int J Seed Spices 9(1):53–56

Molina-Romero D, Baez A, Quintero-Hernández V, Castañeda-Lucio M, Fuentes-Ramírez LE, Bustillos-Cristales MDR, Rodríguez-Andrade O, Morales-García YE, Munive A, Muñoz-Rojas J (2017) Compatible bacterial mixture, tolerant to desiccation, improves maize plant growth. PLoS One 12(11):e0187913

Moreira H, Pereira SI, Vega A, Castro PM, Marques AP (2020) Synergistic effects of arbuscular mycorrhizal fungi and plant growth-promoting bacteria benefit maize growth under increasing soil salinity. J Environ Manag 257:109982

Mounir AM, Osman YM, Khalil OA (2020) Impact of potassium solubilizing bacteria on growth and yield of garlic. Plant Arch 20(2):8374–8388

Naamala J, Smith DL (2020) Relevance of plant growth promoting microorganisms and their derived compounds, in the face of climate change. Agronomy 10(8):1179

Nadarajah K, Abdul Rahman NSN (2021) Plant–microbe interaction: aboveground to belowground, from the good to the bad. Int J Mol Sci 22(19):10388

Negi NP, Narwal P, Sharma A (2023a) Plant–microbe interaction in alleviating drought stress. In: The chemical dialogue between plants and beneficial microorganisms. Academic Press, New York, pp 49–65

Negi NP, Prakash G, Narwal P, Panwar R, Kumar D, Chaudhry B, Rustagi A (2023b) The calcium connection: exploring the intricacies of calcium signaling in plant-microbe interactions. Front Plant Sci 14:1248648

Noman M, Ahmed T, Ijaz U, Shahid M, Azizullah, Li D, Manzoor I, Song F (2021) Plant–microbiome crosstalk: dawning from composition and assembly of microbial community to improvement of disease resilience in plants. Int J Mol Sci 22(13):6852

Odoh CK (2017) Plant growth promoting rhizobacteria (PGPR): a bioprotectant bioinoculant for sustainable agrobiology. A review. Int J Adv Res Biol Sci 4(5), pp.123–142.

Parewa HP, Meena VS, Jain LK, Choudhary A (2018) Sustainable crop production and soil health management through plant growth promoting rhizobacteria. In: Role of rhizospheric microbes in soil. Springer, Singapore, pp 299–329

Pathak D, Singh VP, Sharma J, Sheera A (2020) Plant growth promoting rhizobacteria (PGPR): a biological tool for improving plant health. Biotica Res Today 2(7):593–595
Patten CL, Glick BR (2002) Role of Pseudomonas putida indoleacetic acid in development of the host plant root system. Appl Environ Microbiol 68(8):3795–3801
Paula Junior TJ, Jesus Junior WC, Admassu B, Lehner MDS, Hau B (2015) Interactions between rhizoctonia root rot and the foliar common bean diseases anthracnose and rust. J Phytopathol 163(7–8):642–652
Peacher MD, Meiners SJ (2020) Inoculum handling alters the strength and direction of plant–microbe interactions. Ecology 101:e02994
Phazang P, Negi NP, Raina M, Kumar D (2020) Plant antimicrobial peptides: next-generation bioactive molecules for plant protection. In: Phyto-microbiome in stress regulation. Springer, Berlin, pp 281–293
Philippot L, Raaijmakers JM, Lemanceau P, Van Der Putten WH (2013) Going back to the roots: the microbial ecology of the rhizosphere. Nat Rev Microbiol 11(11):789–799
Planchamp C, Glauser G, Mauch-Mani B (2015) Root inoculation with Pseudomonas putida KT2440 induces transcriptional and metabolic changes and systemic resistance in maize plants. Front Plant Sci 5:719
Poddar RK, Singh DV, Dubey SC (2004) Integrated application of Trichoderma harzianum mutants and carbendazim to manage chickpea wilt (Fusarium oxysporum f. sp. ciceris).
Prashar P, Kapoor N, Sachdeva S (2014) Rhizosphere: its structure, bacterial diversity and significance. Rev Environ Sci Biotechnol 13:63–77
Rodríguez CE, Mitter B, Barret M, Sessitsch A, Compant S (2018) Commentary: seed bacterial inhabitants and their routes of colonization. Plant Soil 422:129–134
Santoyo G, Urtis-Flores CA, Loeza-Lara PD, Orozco-Mosqueda MDC, Glick BR (2021) Rhizosphere colonization determinants by plant growth-promoting rhizobacteria (PGPR). Biology 10(6):475
Shahrtash M, Brown SP (2021) A path forward: promoting microbial-based methods in the control of invasive plant species. Plants, 10(5), p.943.
Sharma A, Negi NP, Narwal P, Kumari P, Kumar D (2022) Arbuscular mycorrhizal fungi: a next-generation biofertilizer for sustainable agriculture. In: Beneficial microorganisms in agriculture. Springer Nature Singapore, Singapore, pp 161–176
Singh J, Sharma D (eds) (2022) Microbial resource technologies for sustainable development. Elsevier, Amsterdam
Singh RK, Singh P, Li HB, Song QQ, Guo DJ, Solanki MK, Verma KK, Malviya MK, Song XP, Lakshmanan P, Yang LT (2020) Diversity of nitrogen-fixing rhizobacteria associated with sugarcane: a comprehensive study of plant-microbe interactions for growth enhancement in Saccharum spp. BMC Plant Biol 20:1–21
Smith SE, Read DJ (2010) Mycorrhizal symbiosis. Academic press
Steven B, Huntley RB, Zeng Q (2018) The influence of flower anatomy and apple cultivar on the apple flower phytobiome. Phytobiomes 2(3):171–179
Straker CJ, Mitchell DT (1986) The activity and characterization of acid phosphatases in endomycorrhizal fungi of the Ericaceae. New Phytol 104(2):243–256
Sukumar P (2010) The role of auxin and ethylene in adventitious root formation in Arabidopsis and tomato (Doctoral dissertation, Wake Forest University).
Teotia P, Kumar V, Kumar M, Shrivastava N, Varma A (2016) Rhizosphere microbes: potassium solubilization and crop productivity—present and future aspects. In: Potassium solubilizing microorganisms for sustainable agriculture. Springer, Berlin, pp 315–325
Tewari S, Arora NK (2018) Role of salicylic acid from Pseudomonas aeruginosa PF23 EPS+ in growth promotion of sunflower in saline soils infested with phytopathogen Macrophomina phaseolina. Environ Sustain 1:49–59
Thao NP, Khan MIR, Thu NBA, Hoang XLT, Asgher M, Khan NA, Tran LSP (2015) Role of ethylene and its cross talk with other signaling molecules in plant responses to heavy metal stress. Plant Physiol 169(1):73–84

Tokala RK, Strap JL, Jung CM, Crawford DL, Salove MH, Deobald LA, Bailey JF, Morra MJ (2002) Novel plant-microbe rhizosphere interaction involving Streptomyces lydicus WYEC108 and the pea plant (Pisum sativum). Appl Environ Microbiol 68(5):2161–2171

Trivedi DK, Bhatt H, Pal RK, Tuteja R, Garg B, Johri AK, Bhavesh NS, Tuteja N (2013) Structure of RNA-interacting cyclophilin A-like protein from Piriformospora indica that provides salinity-stress tolerance in plants. Sci Rep 3(1):3001

Trivedi P, Batista BD, Bazany KE, Singh BK (2022) Plant–microbiome interactions under a changing world: responses, consequences and perspectives. New Phytol 234(6):1951–1959

Ulrich K, Kube M, Becker R, Schneck V, Ulrich A (2021) Genomic analysis of the endophytic Stenotrophomonas strain 169 reveals features related to plant-growth promotion and stress tolerance. Frontiers in microbiology, 12, p.687463.

Verma RK, Sachan M, Vishwakarma K, Upadhyay N, Mishra RK, Tripathi DK, Sharma S (2018) Role of PGPR in sustainable agriculture: molecular approach toward disease suppression and growth promotion. In: Role of rhizospheric microbes in soil: volume 2: nutrient management and crop improvement. Springer, Berlin, pp 259–290

Vishwakarma K, Kumar N, Shandilya C, Mohapatra S, Bhayana S, Varma A (2020) Revisiting plant–microbe interactions and microbial consortia application for enhancing sustainable agriculture: a review. Front Microbiol 11:560406

Vorholt JA (2012) Microbial life in the phyllosphere. Nat Rev Microbiol 10(12):828–840

Wallace JG, Kremling KA, Kovar LL, Buckler ES (2018) Quantitative genetics of the maize leaf microbiome. Phytobiomes J 2(4):208–224

Wang C, Knill E, Glick BR, Défago G (2000) Effect of transferring 1-aminocyclopropane-1-carboxylic acid (ACC) deaminase genes into Pseudomonas fluorescens strain CHA0 and its gac A derivative CHA96 on their growth-promoting and disease-suppressive capacities. Canadian Journal of Microbiology 46(10), pp.898–907

Wang H, Liu R, You MP, Barbetti MJ, Chen Y (2021) Pathogen biocontrol using plant growth-promoting bacteria (PGPR): role of bacterial diversity. Microorganisms 9(9):1988

Waqas M, Khan AL, Kamran M, Hamayun M, Kang SM, Kim YH, Lee IJ (2012) Endophytic fungi produce gibberellins and indoleacetic acid and promotes host-plant growth during stress. Molecules 17(9), pp.10754–10773

Werner GD, Cornwell WK, Sprent JI, Kattge J, Kiers ET (2014) A single evolutionary innovation drives the deep evolution of symbiotic N2-fixation in angiosperms. Nat Commun 5(1):4087

Yadav AN, Kour D, Kaur T, Devi R, Yadav A, Dikilitas M, Abdel-Azeem AM, Ahluwalia AS, Saxena AK (2021) Biodiversity, and biotechnological contribution of beneficial soil microbiomes for nutrient cycling, plant growth improvement and nutrient uptake. Biocatal Agric Biotechnol 33:102009

Yang Y, Qiu K, Xie Y, Li X, Zhang S, Liu W, Huang Y, Cui L, Wang S, Bao P (2023) Geographical, climatic, and soil factors control the altitudinal pattern of rhizosphere microbial diversity and its driving effect on root zone soil multifunctionality in mountain ecosystems. Sci Total Environ 904:166932

Zhalnina K, Louie KB, Hao Z, Mansoori N, Da Rocha UN, Shi S, Cho H, Karaoz U, Loqué D, Bowen BP, Firestone MK (2018) Dynamic root exudate chemistry and microbial substrate preferences drive patterns in rhizosphere microbial community assembly. Nat Microbiol 3(4):470–480

Zhang Q, Yu H, Tong T, Tong W, Dong L, Xu M, Wang Z (2014) Dietary supplementation of Bacillus subtilis and fructooligosaccharide enhance the growth, non-specific immunity of juvenile ovate pompano, Trachinotus ovatus and its disease resistance against Vibrio vulnificus. Fish Shellfish Immunol 38(1):7–14

Zhang L, Yuan T, Wang Y, Zhang D, Bai T, Xu S, Wang Y, Tang W, Zheng SJ (2018) Identification and evaluation of resistance to Fusarium oxysporum f. sp. cubense tropical race 4 in Musa acuminata Pahang. Euphytica 214:1–12

Lactic Acid Bacteria for Safe and Sustainable Agriculture

15

Mai N. Amer, Maissara M. K. Elmaghraby,
Aya A. M. Abdellatif, and Ibrahim M. K. Elmaghraby

Abstract

The population of the earth is currently estimated to be 8 billion, but it is predicted to increase to 8.3 billion in 2030 and 9.8 billion in 2050. The increase in population creates high demand to develop new and safe solutions that promote the yield of crops. Recently, there is an increasing interest to establish alternative bioproducts to replace chemicals and toxic pesticides which cause health risks for humans and animals and negatively affect the ecological equilibrium of the environment. Lactic acid bacteria (LAB) are significant groups of probiotic organisms isolated from different natural sources and are generally considered safe. LAB has been demonstrated as effective biofertilizers, biocontrol agents and biostimulants. This bacterial group can increase nutrient availability from compost and other organic materials. LAB are well known as biocontrol agent against a wide range of fungal and bacterial phytopathogens. As biostimulants, LAB can directly enhance plant growth or seed germination while also alleviating a variety of abiotic stressors. This review aims to underline the role of LAB in sustainable agriculture as promising alternatives to agrochemicals usage in sustaining crop productivity.

Keywords

Lactic acid bacteria · Sustainable agriculture · Natural organic acids · Biofertilizers · Biocontrol agents

M. N. Amer
Department of Chemistry of Natural and Microbial Products, Pharmaceutical and Drug Industries Research Institute, National Research Centre, Giza, Egypt

M. M. K. Elmaghraby · A. A. M. Abdellatif (✉) · I. M. K. Elmaghraby
Central Lab of Organic Agriculture, Agricultural Research Center, Giza, Egypt

S. Kaur et al. (eds.), *Metabolomics, Proteomics and Gene Editing Approaches in Biofertilizer Industry*, https://doi.org/10.1007/978-981-97-2910-4_15

15.1 Introduction

The role of microbial communities in promoting plant health has attracted more attention by recent developments in plant–microbe interaction studies (Smith et al. 2015). In order to boost plant growth effectively, soil microbiome engineering is a promising strategy for sustaining crop production in the background of climate change and increasing population. Generally, plant growth promoting microorganisms (PGPM) support plant growth by enhancing the uptake of nutrients, acting as biocontrol agents (BCAs), inducing the plant defense mechanisms against various biotic and abiotic stress and producing wide variety of active compounds that directly motivate plant growth (Elmaghraby et al. 2023). Most studies adopted PGPM applications that focused on a few classes of typical microorganisms such as *Rhizobia* spp., *Bacillus* spp., *Pseudomonas* spp., and mycorrhizal fungi (Vessey 2003).

Recently, the beneficial functions of novel PGPM species, including lactic acid bacteria (LAB), need to be further studied. LAB members are a part of soil microbiomes, but the knowledge about their practical relationships with the plant is still under study (Mazzoli et al. 2014). LAB are non-sporulating, non-motile, acid-tolerant, catalase-negative, Gram-positive cocci or rods and non-respiring but aerotolerant. They play the central role in carbohydrate fermentation by the production of lactic acid as an end product (Vinderola et al. 2019; Salvetti et al. 2018).

LAB obtains energy by substrate-level phosphorylation due to the absence of a functioning respiratory system, following two metabolic pathways for hexose fermentation, i.e., homofermentative and heterofermentative. Most LAB are part of the phylum Firmicutes, class *Bacilli* and order *Lactobacillales*. The order *Lactobacillales* includes six families, i.e., *Aerococcaceae*, *Carnobacteriaceae*, *Enterococcaceae*, *Lactobacillaceae*, *Leuconostocaceae* and *Streptococcaceae*, over 30 genera and more than 300 species, a number that continues rising as novel species are discovered (Ruiz Rodríguez et al. 2019).

LAB is used in organic farming in combination with indigenous microorganisms for compost production. The application of LAB culture enhances the decomposition of soil organic matters and increases the release of plant nutrients for absorption (Tran et al. 2019). The use of LAB as biofertilizers, biocontrol agents and biostimulants to help food production has tremendous potential.

15.2 General Uses of LAB

In fermented foods, lactic acid bacteria (LAB) are important probiotic organisms that are usually regarded as harmless. Also, LAB improve plant health, detoxify harmful pollutants and control soil organic matter and the biochemical cycle. They can be found in decaying vegetation, conventional fermented dairy products, as well as typical human gut and vaginal microbiota. Investigating LAB found in unidentified niches could result in the isolation of rare species. They are acclimated to acidic conditions and high sugar concentrations, although their classification is rather complicated. LAB strains are thought to be excellent choices for sustainable farming since they improve soil fertility and well-being (Raman et al. 2022).

As a result, they have drawn a lot of interest in sustainable agriculture. Plant growth is stimulated and shoots and root growth is boosted by LAB metabolites. LAB can be used as fertilizers to increase soil organic matter, speed up biodegradation and produce metabolites of bacteriocin and organic acid. Conversely, LAB reduces bacterial and fungal populations in the phyllosphere and rhizosphere, exhibiting an antagonistic impact against phytopathogens. The effectiveness of LAB bioremediation and the detoxification of mycotoxins and heavy metals have been suggested by numerous researches. Still, genetic engineering in LAB combined with metabolic engineering technologies creates effective cell factories specifically designed to generate useful industrial and agricultural products (Raman et al. 2022).

In fermentation systems that can be done in solid state fermentation or liquid states, LAB are primarily used in liquid state fermentation, the most investigated and industrially applied solution, which focuses on the separation of organic acids (mainly lactic acid), ethanol, bioactive peptides, etc. Thus, their isolation and further use may add value to the commodity, considering a multi-product process strategy. Examples of extracellular bacterial bioactive products which are constantly produced through growth and stay in the fermented broth are lactic acid and bacteriocins among LAB metabolites. Biomass is usually discarded at the end of fermentation; however, the curiosity in biomass as an added value component has recently gained popularity. Biomass is used to replace by-products used in green processes as a protein source. So, if bacteria used are known as a probiotic, applications can also be built in the food industry. In addition, LAB's cell wall consists of a broad multilayered sacculum formed by peptidoglycan, teichoic and Lipoteichoic acid, polysaccharides of the cell wall, and proteins of the cell surface, which are compounds of considerable potential for biotechnological applications (Vinusha et al. 2018).

Therefore, to produce more efficient and sustainable multi-product systems, the use of multiple metabolic pathways of LAB can be an advantage. Some laboratories have been distinguished by their ability for probiotics. By limiting binding sites on mucosal epithelial cells and modifying the host immune response, probiotics inhibit pathogenic bacteria and enhance the integrity of the intestinal barrier (Fusco et al. 2023). The term probiotics refers to "live microorganisms that give the host a health benefit when administered in adequate amounts" (FAO 2001).

The *Lactobacillus*, *Enterococcus*, *Streptococcus*, and *Bifidobacterium* genera compose most of the probiotic currently available (Fiocco et al. 2020). Others such as *Roseburia* spp., *Akkermansia* spp., *Propionibacterium* spp., and *Faecalibacterium* spp. show promising characteristics and are currently under evaluation (Sanders et al. 2019). There has been developing interest in the description of new LAB isolates from different origins for use as probiotics or other applications in recent years (Puebla-Barragan and Reid 2019). However, in parallel to the isolation and characterization activities, for commercial purposes, identification at the level of species (and in certain cases, at the level of strain) is required.

15.3 LAB: Plant Ecosystem

Although LAB's traditional role in the food processing and human health sectors has been extensively researched and documented, there is growing interest in their other roles, which have been validated by research and empirical evidence. These roles include acting as biofertilizers, biocontrol agents and biostimulant agents in plant production by producing organic acids, bacteriocins and other compounds. However, research on the relationship between LAB, soil and plant hosts is still limited. There is potential for greater adoption of LAB as safe and sustainable soil and plant health enhancers while also discussing the underappreciated relevance of LAB in soil–plant systems through a review of the literature and metadata (Murindangabo et al. 2023).

Fresh vegetable isolates may contain novel LAB strains with improved traits, or their metabolites may have applications in the food industry. The LAB strains' ability to adapt to vegetable and fruit habitats is hampered by the diversity of plant ecosystems, and these adaptation skills differ greatly between species and strains (Filannino et al. 2018).

That being said, fresh vegetables contain a wide variety of LAB. Among the LAB obtained from fresh fruits and vegetables are *Lactobacillus plantarum* (Borrero et al. 2018; Garzón et al. 2017; Siroli et al. 2015), currently known as *Lactiplantibacillus plantarum* comb. nov. (Zheng et al. 2020), *Leuconostoc pseudomesenteroides* (Chen et al. 2017), *Weissella confusa* (Garzón et al. 2017) and Leuconostoc mesenteroides (Liao et al. 2016).

A close association between LAB and plants suggests that LAB has the capacity to survive in the endosphere of such a wide variety of plants. Under the direct biochemical control of root exudates, the rhizosphere is known as the soil fraction. Root exudates, which contain carbohydrates, signaling agents, enzymes, and other chemicals that modify the soil climate, are used by plants to choose a large proportion of photosynthetically fixed carbon for specific microbial diversity (Bais et al. 2006). This atmosphere rich in carbohydrates would seem to be suitable for LAB, but organic acids rapidly reduce in the rhizosphere (Jones 1998; Murindangabo et al. 2023), restricting to their advantage the capacity of LAB to acidify the rhizosphere.

Although this restriction prohibits LAB in most soils from becoming a dominant bacterial species, different LAB strains have been isolated from rhizospheres (Ekundayo 2014). Microbial composition and functions in the rhizosphere, an important microbial hotspot, are among the most fascinating yet elusive topics in microbial ecology (Ling et al. 2022). Rhizosphere microbiome research has shown that plant species and age, in addition to environmental factors as temperature and humidity, are primarily affected by microbial communities in the rhizosphere (Grayston et al. 1998). For example, despite significant geographic differences between *Lactococcus lactis* subsp. *lactis* species, is a constitutive part of the mulberry microbiome and is widely detected in the rhizosphere (Chen et al. 2005).

While certain soils do not tend to contain huge amounts of LAB, they enclose a variety of LAB that generate a variety of metabolites (Chen et al. 2005; Yanagida et al. 2006). In carbon-rich soil conditions, such as those under fruit trees or animal agriculture, LAB has been reported to be the most numerous (Reyes-Escogido et al. 2010), but a rich variety of LAB has been isolated from the rhizospheres of many olive trees and desert truffles plus in desert soils (Fhoula et al. 2013). Lactobacilli can live and develop in dry conditions with water potential as low as 15 MPa (Focht and Martin 1979). There are also several LAB halotolerants, as seen by the richness of LAB in fermented brine foods. Because of their capacity to acidify their surroundings to a degree where many other microbes cannot withstand, LAB also are dominant in mixed cultures.

15.4 Applications of LAB in Agriculture

LAB, and especially *Lactobacillus* spp. as a promising agricultural input, has been appreciated by proponents of holistic and ecologically oriented agricultural systems. *Lactobacillus* is found naturally inside plants tissues, and when plant tissues become fermented and carbohydrate-rich cell material is released, LAB proliferate rapidly. The easy manner for proliferation of *lactobacilli* without the employ of laboratory facilities or microbiological knowledge has led to the extensive use of these microbes by farmers (Katz 2008).

The use of LAB in food production has a long history of avoiding microbial pathogens such as *Salmonella, E. coli*, etc. (Gomez et al. 2016). Lactic acid, a potent disinfectant, is produced efficiently by these microbes and is helpful in crop and livestock farming and could act as antimicrobial agent. LAB is capable of suppressing *Fusarium* sp. which is the causative agent *Fusarium* wilt in wide range of economic crops (Atwa et al. 2021). The antagonistic activity against a wide range of plant diseases of LAB has been reported by several studies (Yarullina et al. 2014). For example, *Lysobacter capsici* YS1215 showed obvious nematocidal and ovicidal activities against plant parasitic nematodes (Lee et al. 2014). Furthermore, the application of LAB suppressed tomato wilt caused by *Ralstonia solanacearum* (Konappa et al. 2016).

Conversely, LAB have been reported as plant growth promoting bacteria (Shrestha et al. 2014). Limanska et al. (2013) stated that *Lactobacillus plantarum* induced significant stimulation in germination of tomatoes and enhanced plant growth. The usage of microbial consortium of *L. plantarum*, *Sphaeroides*, *Rhodobacter*, and *Saccharomyces cerevisiae* improved growth of cucumber (Kang et al. 2014).

Since natural LAB strains are suitable for different substrates, their nutritional and physiological requirements can vary; acceptable strains need to be selected for different applications. Therefore, LAB from vegetable materials could be better suited for green biorefinery applications than LAB from, for example, dairy products. Several LAB strains, derived from green crops, fermented vegetables, or from herbivorous gastrointestinal tracts where plant biomass is the key source of carbon, have been extracted from plant environments. Indigenous LAB will ferment the sugars into lactic acid spontaneously, where plant material contains ample quantities of simple sugars throughout the atmosphere. Feed silages prepared from hay, legumes, and maize are the most abundant products generated by natural fermentation with LAB (Weinberg and Muck 1996; Oliveira et al. 2017; Wilkinson and Renne 2018).

15.4.1 LAB as Biofertilizers

Biofertilizers could be identified as bioproducts that contain a variety of beneficial microbes with special physiological characters to protect the plant and enhance the uptake of nutrients by host plant. LAB is one of the main microbial groups that are known for its ability to decompose wide range of organic matters in soil (Raman et al. 2022), common for manure (Partanen et al. 2010) and silage (MacDonald 1981). Successful effective microorganisms (EM) consortiums include group of yeast, LAB, photosynthetic bacteria, and actinomycetes (Higa and Kinjo 1991). LAB in mature ferments is typically the most active and dominant category of EM consortia (Ahn et al. 2014). Compost inoculated with EM enhanced its quality and promoted nutrient uptake as well as the total crop yields for wheat (Hussain et al. 1999), soybean (Javaid and Mahmood 2010), mung bean (Javaid and Bajwa 2011), rice (Javaid 2011), and cotton (Khaliq et al. 2006).

The release of organic acids is the main mechanism used by phosphate-solubilizing bacteria to dissolve inorganic phosphate in soil. Because of its ability to synthesize wide range of natural organic acids, LAB could solubilize phosphate effectively and be considered a promising alternative to agrochemicals. According to Sharma et al. (2013), phosphorus is the second most significant constituent in plant development after nitrogen. It plays a key role in all metabolic processes, such as photosynthesis and energy transfer. Usually, there is a high amount of phosphorus in the soil; however, it is unavailable. The ability of *Limosilactobacillus* sp. to solubilize phosphate and promote the growth of maize seedlings has been proved by Li et al. (2023). The inoculation of LAB strain effectively promoted the growth and chlorophyll synthesis of maize seedlings as well as increased the total phosphorus content of maize seedlings. Giassi et al. (2016) described the significant role of organic acids produced by LAB in phosphate solubilization. *Lactobacillus acidophilus* showed the ability to solubilize phosphorus and fix the atmospheric nitrogen when applied as seed treatment and as soil drench (Panetto et al. 2023).

Biological nitrogen fixation is the process by which atmospheric nitrogen (N_2) is converted into ammonia and nitrate using the nitrogenase enzyme complex. Higdon et al. (2020) reported that *Lactobacillus lactis* isolated from the mucilage microbiota of Sierra Mixe maize are diazotrophs that have the ability to fix the atmospheric nitrogen. Furthermore, Lee et al. (2015) described the role of *Enterococcus faecium* and its extract in enhancement of the nutrient uptake by the plants. The application of *E. faecium* cell-free extract greatly increased the accumulation of Mg, Ca, Na, Fe, Mn, Si, Zn and N in the oriental melon plants and accelerates its growth via gibberellin and indole-3-Acetic acid secretion.

Three strains of LAB isolated from sugarcane ferment were able to fix atmospheric nitrogen (Giassi et al. 2016). Three plant growth promoters of siderophores generated by *Lactobacillus* strains were reported by Shrestha et al. (2014).

Conversely, microbial biostimulants are well known as microbial formula that encourage plant growth by production of phytohormones that can motivate plant growth (Tsavkelova et al. 2006). Giassi et al. (2016) stated that certain strains of *Lactobacillus* are good producers of indole-3-acetic acid (IAA), which stimulate the rapid development of plants. Kang et al. (2014) studied the impact of effective microbial consortium (*Rhodobacter sphaeroides*, *Saccharomyces cerevisiae*, and *Lactobacillus plantarum*) on cucumber growth and development. Cucumber growth parameters, nutrient uptake and amino acid content were increased by application of this microbial consortium.

15.4.2 LAB as Biocontrol Agents

The damaging effects of synthetic pesticides on both environment and human health have encouraged a search for safer alternatives. Various bacteria have been identified as safe and efficient biocontrol agents (BCAs). A variety of mechanisms of actions could be achieved by BCAs (Compant et al. 2005). LAB could act as an efficient biocontrol agent by producing wide variety of antimicrobial metabolites (Trias et al. 2008) or through excluding pathogens by pre-emptively colonizing infectious plant tissues (Roselló et al. 2013) and by inducing the plant immune response (Konappa et al. 2016).

LAB produces a number of antimicrobial substances, as well as antifungal diketopiperazines, derivative of fatty acid hydroxide, 3-phenyllactate, antibacterial bacteriocins and bacteriocin bioactive compounds, and other antimicrobials substances such as organic acids, hydrogen peroxide, diacetyl and reuterin (b-OH-propionic aldehyde) pyrrolidone-5-carboxylic acid (Stoyanova et al. 2012). Organic acid production has been reported to be the main mechanism of LAB against foliar bacterial pathogen (Visser and Holzapfel 1992), but it has also been argued that LAB barrages competitor microorganisms with a variety of compounds with manifold modes of action with the diversity of antimicrobial compounds produced by LAB (Sangmanee and Hongpattarakere 2014).

In addition to direct antagonism to phytopathogens, LAB treatments can change the response of plants to pathogens, thereby enhancing their natural immunity. This result is referred to as systemic acquired resistance (SAR). Hamed et al. (2011) suggested that the improvements of morphology of tomatoes treated with various *Lactobacillus* strains as proof of a SAR response lead to resistance to fungal pathogens. In support of this theory, *L. paracasei* treated tomato seeds isolated from the tomato rhizosphere produced seedlings that were more resistant to infection with *Ralstonia solanacearum* due to degree of different expressions of defense bioactive compounds in treated seedlings (Konappa et al. 2016). *Lactobacillus* sp. recorded remarkable antifungal activity against *Rhizoctonia solani* and showed significant production of defense enzymes in basmati rice (Akhtar et al. 2023).

15.4.2.1 Antibacterial Effect of LAB

In a way like that of chlorine, the presence of LAB lowered mesophilic aerobic counts significantly. LAB strains (*Lactobacillus fermentum* and *Lactobacillus plantarum*) showed significant antibacterial capacity against *Staphylococcus aureus* (Wong et al. 2015). Antimicrobial activity of *Lactobacillus sakei Pediococcus acidilactici* and *Pediococcus pentosaceus* against pathogenic bacteria, fungi and yeasts was proved by Cizeikiene et al. (2013). The growth of *Bacillus*, *Pseudomonas*, *Listeria* and *Escherichia* was inhibited by the metabolites synthesized by the LAB strains at dissimilar degrees. Also, the many metabolites displayed antifungal activity.

Isolated from the rhizosphere of olive trees and desert truffles, *Leuconostoc mesenteroides*, *Weissella halotolerans* and *Enterococcus faecium* demonstrated high antimicrobial ability against both bacterial and fungal plant pathogens, as well as foodborne pathogens such as *Staphylococcus aureus* and *Listeria monocytogenes* (Fhoula et al. 2013). Antimicrobial metabolites against *Listeria monocytogenes*, *Bacillus cereus*, *Staphylococcus aureus*, *Streptococcus mutans,* and *Salmonella mutans* are produced from LAB strains isolated from fermented Aloreña olives (Abriouel et al. 2012). LAB strains isolated from fruit and vegetables were also tested for the capability to suppress bacterial growth in apple lesions and cut Iceberg lettuce leaves (Trias et al. 2008).

15.4.2.2 Antifungal Effect of LAB

The potential biological control activity of LAB against phytopathogenic fungi has been studied (Dalié et al. 2010). The application of LAB has been considered of great importance because they are microorganisms considered by the Food and Drug Administration (FDA, USA) as Generally recognized as Safe and with the "Qualified Presumption of Protection" status by the European Food Safety Agency (Koutsoumanis et al. 2020). The antagonistic and bio-protective potential of LAB may be related to their ability to compete for nutrients as well as their capacity to produce natural organic acids, antibiotics, bacteriocins, hydrogen peroxide and hydroxyl fatty acids (Lamont et al. 2017).

Approaches to the use of microbes to monitor the growth and reproduction of phytopathogenic fungi are beneficial because the use of pesticides and fungicides in agriculture can result in degradation of the atmosphere and should either be minimized or stopped. In addition, such microbes known as plant growth promoting bacteria such as LAB can serve as both biocontrol and regulator of plant growth (Stephane et al. 2005).

Bifidobacterium bifidum and *Lactobacillus fermentum* strains recorded significant antifungal effect against *Aspergillus parasiticus* and inhibited aflatoxin production by 88.8–99.8% (Ghazvini et al. 2016). The efficacy of LAB isolated from fruits and vegetables was proved against *Colletotrichum capsici*, a fungus that causes anthracnose as in peppers (*Capsicum annuum*) (El-Mabrok et al. 2012). Many LAB strains exhibited good inhibitory effect, while extreme inhibitory activity was displayed by their supernatants. The chitinase activity of LAB *Enterococcus faecium* has been identified and characterized as the main mechanism of action in suppression of *Fusarium solani* (Atwa et al. 2022).

15.4.2.3 Antinematode Effect of LAB

Among the most damaging crop pests, plant parasitic nematodes are responsible for significant economic losses valued at $77 billion annually worldwide (Yadav 2017). Due to their high reproductive rate and limited generation time, soil-borne diseases caused by root-knot nematodes (RKNs) can spread in very short time (Trudgill and Blok 2001). The systematic use of synthetic pesticides may be effective in controlling various genera of plant parasitic nematodes. However, the public concern about the damaging property of the continued use of chemical pesticides, especially on human health and the environment, has restricted the excessive usage of synthetic chemicals products (Aktar et al. 2009). So, there is an immediate need to find efficient and environmentally sustainable alternatives to minimize the use of chemical nematicidal products.

In the agriculture sector, LAB is currently gaining substantial interest as an alternative to chemical pesticides. Many researches also show that unlike the findings of a previous review, acetic acid could be a crucial nematicidal element for RKNs management (Choi et al. 2018).

Organic acids produced through the decomposition of organic matters in soil showed nematicidal potentiality as they can reduce the total population of nematodes (McBride et al. 2000). In addition, several studies on the nematicidal activity of microbial metabolites confirmed the nematicidal activity of fungal organic acids. Kim et al. (2016) demonstrated the role of fungal organic acids; citric acid, fumaric acid, malic acid and Kojic acid as nematicidal agents secreted by *Aspergillus oryzae*. A culture filtrate of Aspergillus *niger* F22 showed remarkable activity against *Meloidogyne incognita* as a result of its production of oxalic acid which is defined as effective nematicidal fungal product (Jang et al. 2016). Microbial acetic acid recorded ovicidal activity and regulates embryogenesis (Bansal and Bajaj 2003).

Lactic acid bacteria transform carbohydrate substrates into organic acids including propionic, formic, acetic and lactic acids, creating an unfavorable environment for propagation of phytopathogens and pests (Punia Bangar et al. 2022). Seo et al. (2019) demonstrated that *Lactobacillus brevis* showed strong nematicidal and ovicidal activities against the second-stage juveniles and eggs of *Meloidogyne incognita, M. arenaria and M. hapla.* They suggested that the production of acetic acid (11,190 μg/mL), lactic acid (7790 μg/mL), malic acid (470 μg/mL) and succinic play the central role in its antinematode potentiality. Bacterial Lactic acid which extracted from *Lysobacter capsici* was also observed to inhibit egg hatching of root-knot nematodes (Lee et al. 2014). A group of natural organic acids, i.e., acetic acid, lactic acid, malic acid and succinic acid produced by *L. brevis* WiKim0069 induced high *M. incognita* juveniles mortality and was suggested to be responsible for the nematicidal activity of the bacterial culture filtrate as reported by Seo and Kim (2014). The nematicidal activity of natural organic acids produced by *Pediococcus pentosaceus* has been proved by Ibrahim et al. (2022). The microscopic studies and the malformation pattern of the juveniles indicate that the nematicidal activity of *Pediococcus pentosaceus* may be derived mainly from their natural organic acids.

The permeation of organic acids across the membrane tends to be competitive in bipolar ion-exchange membranes, and acetic acid has even greater permeability among the carboxylic acids, followed by propionic, tartaric, lactic, oxalic and citric acids (Wódzki and Nowaczyk 2001). The exogenous acetic acid treatment was observed to destroy the cuticles, deteriorate the nuclei, and vacuolize the cytoplasm of *M. incognita* (Ntalli et al. 2016).

15.5 Conclusion

LAB have wide range of applications not only in the food industry but also has numerous agricultural uses as well. Because of its biocontrol and biostimulant properties, LAB is known as effective microbes (EM). Thus, LAB soil inculcation enriches soil biodiversity with new bioactive compounds that enhance plant health and productivity. LAB can stimulate plant development via variety of mechanisms, like acting as BCAs, boosting nutrient availability and alleviating the impacts of biotic and abiotic stresses. Similar to other types of microbial based products, LAB also has its limitations and challenges when it comes to application. There is need to more studies on LAB which focus on its biocontrol efficiency under field conditions as well as its large mass bioproduction and formulation. The integration of LAB with other management techniques in an integrated control program would be a viable way to increase efficacy against phytopathogens and pests which help in achieving sustainable food security.

References

Abriouel H, Benomar N, Cobo A, Caballero N, Fernández Fuentes MÁ, Pérez-Pulido R, Gálvez A (2012) Characterization of lactic acid bacteria from naturally-fermented manzanilla Aloreña green table olives. Food Microbiol 32(2):308–316. https://doi.org/10.1016/j.fm.2012.07.006

Ahn K, Lee K, Kim Y, Koo Y (2014) Quantitative analysis of the three main genera in effective microorganisms using qPCR. Korean J Chem Eng 31(5):849–854. https://doi.org/10.1007/s11814-013-0274-6

Akhtar M, Nosheen A, Keyani R, Yasmin H, Naz R, Mumtaz S, Hassan MN (2023) Biocontrol of Rhizoctonia solani in basmati rice by the application of Lactobacillus and Weissella spp. Sci Rep 13(1):13855. https://doi.org/10.1038/s41598-023-41058-9

Aktar W, Sengupta D, Chowdhury A (2009) Impact of pesticides use in agriculture: their benefits and hazards. Interdiscip Toxicol 2(1):1–12. https://doi.org/10.2478/v10102-009-0001-7

Atwa N, Maany D, El-Diwany A, Daba G, Elmaghraby I, Elmaghraby M, Amer M (2021) Studies on the antimicrobial activity of a locally isolated lactic acid bacterium and its application on Fusarium solani-infected plants. Egypt Pharm J 20(3):207. https://doi.org/10.4103/epj.epj_14_21

Atwa N, Wahba M, Maany D, Awad H, Abo-Alkasem M, El-Masry H, Amer M, El-Diwany A (2022) Lactic acid bacteria: economic propagation, chitinases activity, and enhancing viability by gel encapsulation. Egypt Pharm J 21(3):347. https://doi.org/10.4103/epj.epj_50_22

Bais HP, Weir TL, Perry LG, Gilroy S, Vivanco JM (2006) The role of root exudates in rhizosphere interactions with plants and other organisms. Annu Rev Plant Biol 57(1):233–266. https://doi.org/10.1146/annurev.arplant.57.032905.105159

Bansal RK, Bajaj A (2003) Effect of volatile fatty acids on embryogenesis and hatching of *Meloidogyne incognita* eggs. Nematol Mediterr 31:135–140

Borrero J, Kelly E, O'Connor PM, Kelleher P, Scully C, Cotter PD, Mahony J, Van Sinderen D (2018) Plantaricyclin a, a novel circular bacteriocin produced by Lactobacillus plantarum NI326: purification, characterization, and heterologous production. Appl Environ Microbiol 84(1):e01801-17. https://doi.org/10.1128/aem.01801-17

Chen YS, Yanagida F, Shinohara T (2005) Isolation and identification of lactic acid bacteria from soil using an enrichment procedure. Lett Appl Microbiol 40(3):195–200

Chen Y, Wu H, Kuo C, Chen Y, Ho S, Yanagida F (2017) Leucocin C-607, a novel Bacteriocin from the multiple-bacteriocin-producing Leuconostoc pseudomesenteroides 607 isolated from persimmon. Probiotics Antimicrob Proteins 10(2):148–156. https://doi.org/10.1007/s12602-017-9359-6

Choi A-R, Patra JK, Kim WJ, Kang S-S (2018) Antagonistic activities and probiotic potential of lactic acid bacteria derived from a plant-based fermented food. Front Microbiol 9:1963

Cizeikiene D, Juodeikiene G, Paskevicius A, Bartkiene E (2013) Antimicrobial activity of lactic acid bacteria against pathogenic and spoilage microorganisms isolated from food and their control in wheat bread. Food Control 31:539–545. https://doi.org/10.1016/j.foodcont.2012.12.004

Compant S, Duffy B, Nowak J, Clément C, Barka EA (2005) Use of plant growth promoting bacteria for biocontrol of plant diseases: principles, mechanisms of action, and future prospects. Appl Environ Microbiol 71(9):4951–4959

Dalié DKD, Deschamps AM, Richard-Forget F (2010) Lactic acid bacteria—potential for control of mould growth and mycotoxins: a review. Food Control 21:370–380

Ekundayo FO (2014) Isolation and identification of lactic acid bacteria from rhizosphere soils of three fruit trees, fish and ogi. Int J Curr Microbiol Appl Sci 3(3):991–998

El-Mabrok ASW, Hassan Z, Mokhtar AM, Hussain KMA, Abdul Kahar FKSB (2012) Screening of lactic acid bacteria as biocontrol against Colletotrichum capsici on Chilli Bangi. Res J Appl Sci 7:466–473. https://doi.org/10.3923/rjasci.2012.466.473

Elmaghraby MMK, Abdellatif Aya AM, Amer Mai N, Sahu Pramod K (2023) Enhancing productivity through multiple microbial inoculants. In: Kaur S, Dwibedi V, Sahu PK, Kocher GS (eds) Metabolomics, proteomes and gene editing approaches in biofertilizer industry. Springer Nature, Singapore

FAO (2001) Food and Agricultural Organization of the United Nations and World Health Organization. Health and nutritional properties of probiotics in food including powder milk with live lactic acid bacteria. World Health Organization [online], http://www.who.int/foodsafety/publications/fs_management/en/probiotics.pdf.

Fhoula I, Najjari A, Turki Y, Jaballah S, Boudabous A, Ouzari H (2013) Diversity and antimicrobial properties of lactic acid bacteria isolated from rhizosphere of olive trees and desert truffles of Tunisia. Biomed Res Int 2013:405708. https://www.hindawi.com/journals/bmri

Filannino P, Di Cagno R, Gobbetti M (2018) Metabolic and functional paths of lactic acid bacteria in plant foods: get out of the labyrinth. Curr Opin Biotechnol 49:64–72. https://doi.org/10.1016/j.copbio.2017.07.016

Fiocco D, Longo A, Arena MP, Russo P, Spano G, Capozzi V (2020) How probiotics face food stress: they get by with a little help. Crit Rev Food Sci Nutr 60(9):1552–1580

Focht DD, Martin JP (1979) Microbiological and biochemical aspects of semi-arid agricultural soils. In: Agriculture in semi-arid environments. Springer, Berlin, pp 119–147

Fusco A, Savio V, Cimini D, D'Ambrosio S, Chiaromonte A, Schiraldi C, Donnarumma G (2023) In vitro evaluation of the most active probiotic strains able to improve the intestinal barrier functions and to prevent inflammatory diseases of the gastrointestinal system. Biomedicines 11(3):865. https://doi.org/10.3390/biomedicines11030865

Garzón K, Ortega C, Tenea GN (2017) Characterization of bacteriocin-producing lactic acid bacteria isolated from native fruits of Ecuadorian Amazon. Pol J Microbiol 66(4):473–481. https://doi.org/10.5604/01.3001.0010.7037

Ghazvini R, Kouhsari E, Zibafar E, Hashemi SJ, Amini A, Niknejad F (2016) Antifungal activity and aflatoxin degradation of Bifidobacterium bifidum and Lactobacillus fermentum against toxigenic Aspergillus parasiticus. Open Microbiol J 10:197–201. https://doi.org/10.2174/1874285801610010197

Giassi V, Kiritani C, Kupper KC (2016) Bacteria as growth-promoting agents for citrus rootstocks. Microbiol Res 190:46–54. https://doi.org/10.1016/j.micres.2015.12.006

Gomez NC, Ramiro JM, Quecan BX, De Melo Franco BD (2016) Use of potential probiotic lactic acid bacteria (LAB) biofilms for the control of *Listeria monocytogenes*, *Salmonella typhimurium*, and *Escherichia coli* O157: H7 biofilms formation. Front Microbiol 7:1–15. https://doi.org/10.3389/fmicb.2016.00863

Grayston SJ, Wang S, Campbell CD, Edwards AC (1998) Selective influence of plant species on microbial diversity in the rhizosphere. Soil Biol Biochem 30(3):369–378

Hamed HA, Moustafa YA, Abdel-Aziz SM (2011) In vivo efficacy of lactic acid bacteria in biological control against Fusarium oxysporum for protection of tomato plant. Life Sci J 8(4):462–468

Higa T, Kinjo S (1991) Effect of lactic acid fermentation bacteria on plant growth and soil humus formation. In: Parr JF, Hornick SB, Whitman CE (eds) Proceedings of the first international conference on kyusei nature farming. US Department of Agriculture, Washington, DC, pp 140–147

Higdon SM, Huang BC, Bennett AB, Weimer BC (2020) Identification of nitrogen fixation genes in Lactococcus isolated from maize using population genomics and machine learning. Microorganisms 8(12):2043. https://doi.org/10.3390/microorganisms8122043

Hussain T, Javaid T, Parr JF, Jilani G, Haq MA (1999) Rice and wheat production in Pakistan with effective microorganisms. Am J Altern Agric 14(1):30–36

Ibrahim DS, Amer MN, El-Shishtawy HM, Elmaghraby MMK, Abdellatif AAM, El-Deriny MM (2022) Nematicidal activity of lactic acid bacteria against root-knot nematodes. Open Access J Microbiol Biotechnol 7(1):000216. https://doi.org/10.23880/oajmb-16000216

Jang JY, Choi YH, Shin TS, Kim TH, Shin K-S, Park HW, Kim YH, Kim H, Choi GJ, Jang KS, Cha B, Kim IS, Myung EJ, Kim J-C (2016) Biological control of *Meloidogyne incognita* by *Aspergillus niger* F22 producing oxalic acid. PLoS One 11:e0156230

Javaid A (2011) Effects of biofertilizers combined with different soil amendments on potted rice plants. Chilean J Agric Res 71(1):157–163

Javaid A, Bajwa R (2011) Field evaluation of effective microorganisms (EM) application for growth, nodulation, and nutrition of mung bean. Turk J Agric For 35(4):443–452

Javaid A, Mahmood N (2010) Growth, nodulation and yield response of soybean to biofertilizers and organic manures. Pak J Bot 42(2):863–871

Jones DL (1998) Organic acids in the rhizosphere—a critical review. Plant Soil 205(1):25–44

Kang S, Radhakrishnan R, You Y, Khan AL, Park J, Lee S, Lee I (2014) Cucumber performance is improved by inoculation with plant growth-promoting microorganisms. Acta Agric Scand Sect B Soil Plant Sci 65(1):36–44. https://doi.org/10.1080/09064710.2014.960889

Katz SE (2008) Wild fermentation: the flavor, nutrition, and craft of live-culture foods. Chelsea Green Publishing, Chelsea

Khaliq A, Abbasi MK, Hussain T (2006) Effects of integrated use of organic and inorganic nutrient sources with effective microorganisms (EM) on seed cotton yield in Pakistan. Bioresour Technol 97(8):967–972

Kim TY, Jang JY, Jeon SJ, Lee HW, Bae C-H, Yeo JH, Lee HB, Kim IS, Park HW, Kim J-C (2016) Nematicidal activity of kojic acid produced by *Aspergillus oryzae* against *Meloidogyne incognita*. J Microbiol Biotechnol 26:1383–1391

Konappa NM, Maria M, Uzma F, Krishnamurthy S, Nayaka SC, Niranjana SR, Chowdappa S (2016) Lactic acid bacteria mediated induction of defense enzymes to enhance the resistance in tomato against *Ralstonia solanacearum* causing bacterial wilt. Sci Hortic 207:183–192. https://doi.org/10.1016/j.scienta.2016.05.029

Koutsoumanis K, Allende A, Alvarez-Ordóñez A, Bolton D, Bover-Cid S, Chemaly M, Davies R, De Cesare A, Hilbert F, Lindqvist R, Nauta M, Peixe L, Ru G, Simmons M, Skandamis P, Suffredini E, Cocconcelli PS, Fernández Escámez PS, Herman L (2020) Update of the list of QPS-recommended biological agents intentionally added to food or feed as notified to EFSA 12: Suitability of taxonomic units notified to EFSA until March 2020. EFSA Journal 18(7). https://doi.org/10.2903/j.efsa.2020.6174

Lamont JR, Wilkins O, Bywater-Ekegärd M, Smith DL (2017) From yogurt to yield: potential applications of lactic acid bacteria in plant production. Soil Biol Biochem 111:1–9

Lee YS, Naning KW, Nguyen XH, Kim SB, Moon JH, Kim KY (2014) Ovicidal activity of lactic acid produced by *Lysobacter capsici* YS1215 on eggs of root-knot nematode, *Meloidogyne incognita*. J Microb Biotechnol 24:1510–1515

Lee K, Radhakrishnan R, Kang S, You Y, Joo G, Lee I, Ko J, Kim J (2015) Enterococcus faecium LKE12 cell-free extract accelerates host plant growth via gibberellin and indole-3-acetic acid secretion. J Microbiol Biotechnol 25(9):1467–1475. https://doi.org/10.4014/jmb.1502.0201

Li H, Chen S, Wang M, Shi S, Zhao W, Xiong G, Zhou J, Qu J (2023) Phosphate solubilization and plant growth properties are promoted by a lactic acid bacterium in calcareous soil. Appl Microbiol Biotechnol 108(1):24. https://doi.org/10.1007/s00253-023-12850-4

Liao X, Guo L, Ye Z, Qiu L, Gu F, Lin J (2016) Use of autochthonous lactic acid bacteria starters to ferment mango juice for promoting its probiotic roles. Prep Biochem Biotechnol 46(4):399–405. https://doi.org/10.1080/10826068.2015.1045615

Limanska N, Ivanytsia T, Basiul O, Krylova K, Biscola V, Chobert J, Ivanytsia V, Haertlé T (2013) Effect of Lactobacillus plantarum on germination and growth of tomato seedlings. Acta Physiol Plant 35(5):1587–1595. https://doi.org/10.1007/s11738-012-1200-y

Ling N, Wang T, Kuzyakov Y (2022) Rhizosphere bacteriome structure and functions. Nat Commun 13:836. https://doi.org/10.1038/s41467-022-28448-9

MacDonald P (1981) The biochemistry of silage. Wiley, New York

Mazzoli R, Bosco F, Mizrahi I, Bayer EA, Pessione E (2014) Towards lactic acid bacteria-based biorefineries. Biotechnol Adv 32:1216–1236

McBride RG, Mikkelsen RL, Barker KR (2000) The role of low molecular weight organic acids from decomposing rye in inhibiting root-knot nematode populations in soil. Appl Soil Ecol 15:243–251

Murindangabo YT, Kopecký M, Perná K, Nguyen TG, Konvalina P, Kavková M (2023) Prominent use of lactic acid bacteria in soil-plant systems. Appl Soil Ecol 189(104):955. https://doi.org/10.1016/j.apsoil.2023.104955

Ntalli N, Ratajczak M, Oplos C, Menkissoglu-Spiroudi U, Adamski Z (2016) Acetic acid, 2-undecanone, and (e)-2-decenal ultrastructural malformations on *Meloidogyne incognita*. J Nematol 48:248–260

Oliveira AS, Weinberg ZM, Ogunada IM et al (2017) Meta-analysis of effects of inoculation with homofermentative and facultative heterofermentative lactic acid bacteria on silage fermentation, aerobic stability, and the performance of dairy cows. J Dairy Sci 100:4587–4603

Panetto LD, Doria J, Santos CH, Frezarin ET, Sales LR, De Andrade LA, Rigobelo EC (2023) Lactic bacteria with plant-growth-promoting properties in potato. Microbiol Res 14(1):279–288. https://doi.org/10.3390/microbiolres14010022

Partanen P, Hultman J, Paulin L, Auvinen P, Romantschuk M (2010) Bacterial diversity at different stages of the composting process. BMC Microbiol 10(1):94

Puebla-Barragan S, Reid G (2019) Forty-five-year evolution of probiotic therapy. Microb Cell 6:184–196

Punia Bangar S, Suri S, Trif M, Ozogul F (2022) Organic acids production from lactic acid bacteria: a preservation approach. Food Biosci 46(101):615. https://doi.org/10.1016/j.fbio.2022.101615

Raman J, Kim J, Choi KR, Eun H, Yang D, Ko Y, Kim S (2022) Application of lactic acid bacteria (LAB) in sustainable agriculture: advantages and limitations. Int J Mol Sci 23(14):7784. https://doi.org/10.3390/ijms23147784

Reyes-Escogido L, Balam-Chi M, Rodríguez-Buenfil I, Valdés J, Kameyama L, Martínez-Pérez F (2010) Purification of bacterial genomic DNA in less than 20 min using chelex-100 micro-

wave: examples from strains of lactic acid bacteria isolated from soil samples. Antonie Van Leeuwenhoek 98(4):465–474
Roselló G, Bonaterra A, Francés J, Montesinos L, Badosa E, Montesinos E (2013) Biological control of fire blight of apple and pear with antagonistic Lactobacillus plantarum. Eur J Plant Pathol 137(3):621–633
Ruiz Rodríguez LG, Mohamed F, Bleckwedel J, Medina R, De Vuyst L, Hebert EM, Mozzi F (2019) Diversity and functional properties of lactic acid bacteria isolated from wild fruits and flowers present in northern Argentina. Front Microbiol 10:1091
Salvetti E, Harris HMB, Felis GE, O'Toole PW (2018) Comparative genomics of the genus Lactobacillus reveals robust phylogroups that provide the basis for reclassification. Appl Environ Microbiol 84(17):e00993-18
Sanders ME, Merenstein DJ, Reid G, Gibson GR, Rastall RA (2019) Probiotics and prebiotics in intestinal health and disease: from biology to the clinic. Nat Rev Gastroenterol Hepatol 16:605–616
Sangmanee P, Hongpattarakere T (2014) Inhibitory of multiple antifungal components produced by Lactobacillus plantarum K35 on growth, aflatoxin production and ultrastructure alterations of Aspergillus flavus and Aspergillus parasiticus. Food Control 40:224–233
Seo Y, Kim YH (2014) Control of *Meloidogyne incognita* using mixtures of organic acids. Plant Pathol J 30:450–455
Seo HJ, Park AR, Kim S, Yeon J, Yu NH, Ha S, Chang JY, Park HW, Kim J (2019) Biological control of root-knot nematodes by organic acid-producing Lactobacillus brevis WiKim0069 isolated from Kimchi. Plant Pathol J 35(6):662–673. https://doi.org/10.5423/ppj.oa.08.2019.0225
Sharma SB, Sayyed RZ, Trivedi MH, Gobi TA (2013) Phosphate solubilizing microbes: sustainable approach for managing phosphorus deficiency in agricultural soils. SpringerPlus 2(1):587. https://doi.org/10.1186/2193-1801-2-587
Shrestha A, Kim BS, Park DH (2014) Biological control of bacterial spot disease and plant growth-promoting effects of lactic acid bacteria on pepper. Biocontrol Sci Technol 24(7):763–779
Siroli L, Patrignani F, Serrazanetti DI, Tabanelli G, Montanari C, Gardini F, Lanciotti R (2015) Lactic acid bacteria and natural antimicrobials to improve the safety and shelf-life of minimally processed sliced apples and lamb's lettuce. Food Microbiol 47:74–84. https://doi.org/10.1016/j.fm.2014.11.008
Smith DL, Subramanian S, Lamont JR, Bywater-Ekegärd M (2015) Signaling in the phytomicrobiome: breadth and potential. Front Plant Sci 6:709
Stephane C, Brion DJ, Christophe N, Cessaid A (2005) Use of plant growth promoting bacteria for bio-control of plant diseases, principles, mechanism of action, and future prospects||. Appl Environ Microbiol 71(9):4951–4959
Stoyanova LG, Ustyugova EA, Netrusov AI (2012) Antibacterial metabolites of lactic acid bacteria: their diversity and properties. Appl Biochem Microbiol 48(3):229–243
Tran QN, Mimoto H, Koyama M, Nakasaki K (2019) Lactic acid bacteria modulate organic acid production during early stages of food waste composting. Sci Total Environ 687:341–347. https://doi.org/10.1016/j.scitotenv.2019.06.113
Trias R, Baneras L, Seguí EM, Romanyo EB (2008) Lactic acid bacteria from fresh fruit and vegetables as biocontrol agents of phytopathogenic bacteria and fungi. Int Microbiol 11(4):231–236
Trudgill DL, Blok VC (2001) Apomictic, polyphagous root-knot nematodes: exceptionally successful and damaging biotrophic root pathogens. Annu Rev Phytopathol 39:53–77
Tsavkelova EA, Klimova SY, Cherdyntseva TA, Netrusov AI (2006) Microbial producers of plant growth stimulators and their practical use: a review. Appl Biochem Microbiol 42(2):117–126
Vessey JK (2003) Plant growth promoting rhizobacteria as biofertilizers. Plant Soil 255(2):571–586
Vinderola G, Ouwehand AC, Salminen S, vonWright A (2019) Lactic acid bacteria: microbiological and functional aspects, 5th edn. CRC Press, Boca Raton
Vinusha KS, Deepika K, Johnson TS, Agrawal GK, Rakwal R (2018) Proteomic studies on lactic acid bacteria: a review. Biochem Biophys Rep 14:140–148
Visser R, Holzapfel WH (1992) Lactic acid bacteria in the control of plant pathogens. In: Wood BJB (ed) The lactic acid bacteria, vol 1. Springer, New York, pp 193–210

Weinberg ZG, Muck RE (1996) New trends and opportunities in the development and use of inoculants for silage. FEMS Microbiol Rev 19:53–68

Wilkinson MR, Renne M (2018) Highlights of progress in silage conservation and future perspectives. Grass Forage Sci 73:40–52

Wódzki R, Nowaczyk J (2001) Membrane transport of organics. III. Permeation of some carboxylic acids through bipolar polymer membrane. J Appl Polym Sci 80:2705–2717

Wong CB, Khoo BY, Sasidharan S, Piyawattanametha W, Kim SH, Khemthongcharoen N et al (2015) Inhibition of Staphylococcus aureus by crude and fractionated extract from lactic acid bacteria. Benefic Microbes 6:129–140. https://doi.org/10.3920/BM2014.0021

Yadav U (2017) Recent trends in nematode management practices: the Indian context. Int Res J Eng Technol 4:482–489

Yanagida F, Chen YS, Shinohara T (2006) Searching for bacteriocin-producing lactic acid bacteria in soil. J Gen Appl Microbiol 52(1):21–28

Yarullina DR, Asafova EV, Kartunova JE, Ziyatdinova GK, Ilinskaya ON (2014) Probiotics for plants: NO-producing lactobacilli protect plants from drought. Appl Biochem Microbiol 50:166–168. https://doi.org/10.1134/S0003683814020197

Zheng J, Wittouck S, Salvetti E, Franz CM, Harris HM, Mattarelli P, O'Toole PW, Pot B, Vandamme P, Walter J, Watanabe K, Wuyts S, Felis GE, Gänzle MG, Lebeer S (2020) A taxonomic note on the genus Lactobacillus: description of 23 novel genera, emended description of the genus Lactobacillus Beijerinck 1901, and union of Lactobacillaceae and Leuconostocaceae. Int J Syst Evol Microbiol 70(4):2782–2858. https://doi.org/10.1099/ijsem.0.004107

Transforming Roles of Actinobacteria in Sustainable Agriculture: From Soil Health and Plant Productivity Perspective

16

Aya A. M. Abdellatif, Doha A. S. Gebily, Maissara M. K. Elmaghraby, Pramod Kumar Sahu, Babita Thakur, and Sukhminderjit Kaur

Abstract

In Recent days the Climatic changes, including rising global temperatures and increased air pollution, has the major effect on agricultural productivity. The excessive use of artificial fertilizers and pesticides exacerbates climate stress by producing greenhouse gas emissions (GHG) and toxic soil deposits. At this crucial time, there is an urgent need to combat climate change stress via the transition to more sustainable crop production practices to ensure food security. Interest in employing beneficial microorganisms as plant or soil inoculants to improve plant health or reduce both biotic and abiotic pressures has developed in light of the negative consequences of agrochemical-related pollution. These microbes include fungus, bacteria, and Actinobacteria. Beneficial Actinobacteria are reviewed in this review along with their responsibilities in sustainable plant production. Positive effects Actinobacteria are proficient colonizers of plant tissues, encompassing both aerial and root segments. Their ability to function as antagonists against pests and plant diseases is enhanced by the fact that they are active makers of essential natural components including volatile organic compounds and antibiotics. Furthermore, even in the face of abiotic stress, a range of

A. A. M. Abdellatif (✉) · D. A. S. Gebily · M. M. K. Elmaghraby
Central Laboratory of Organic Agriculture, Agricultural Research Center, Giza, Egypt

P. K. Sahu
Agricultural Microbiology, ICAR National Bureau of Agriculturally Important Microorganisms, Mau Nath Bhanjan, Uttar Pradesh, India

B. Thakur
Department of Biotechnology, Chandigarh University, Mohali, Punjab, India

S. Kaur
Department of Biotechnology, Chandigarh University, Mohali, Chandigarh, India

S. Kaur et al. (eds.), *Metabolomics, Proteomics and Gene Editing Approaches in Biofertilizer Industry*, https://doi.org/10.1007/978-981-97-2910-4_16

commercial crops have demonstrated their ability to drive plant development through a number of mechanisms. This has led to the widespread use of Actinobacteria as biofertilizers to boost plant productivity. Actinobacteria and their natural products are important, which is why the industrial sector introduced them to create safe, inexpensive, and effective biopesticides and fertilizers. In conclusion, Actinobacteria can play a central role in agricultural sustainability, ensuring further growth in agricultural output.

Keywords
Sustainability · Actinobacteria · Plant growth-promoting rhizobacteria · Biological control · Natural compounds

16.1 Introduction

Global food insecurity is at an all-time high due to rapidly increasing world population and the lack of innovation in sustainable agricultural production techniques (Fukase and Martin 2020). The rise of phytopathogens, which seriously jeopardize crop output and quality, is another problem faced by modern agriculture (Devlet 2021). Chemical pesticides, fertilizers, and other agrochemicals can boost agricultural output, but their detrimental impacts on soil health, productivity, and environmental quality are becoming more and more well-known, and this cannot be overlooked. Because of the harmful effects of synthetic chemistry on the environment, such as soil erosion, biodiversity loss, and climate change, scientists have had to come up with creative ways to boost productivity by using beneficial microbes as viable replacements (Mitter et al. 2021). One essential agricultural strategy known as sustainability is the use of ecological principles to sustain soil production over an extended period while making use of natural resources without endangering the environment. According to Yadav et al. (2021), sustainable agriculture depends on the preservation and protection of natural resources, such as a diverse and active microbial population in the soil. When employed as microbial bio-stimulants, plant-associated microorganisms have been shown to increase agricultural output (Teixeira et al. 2022). To minimize the use of agrochemicals in an environmentally friendly way, some of them might be considered useful producers of rare secondary metabolites, biocontrol agents, and growth-promoting compounds. Microbes utilized as bioinoculants and biopesticides have enormous potential as alternatives to artificial fertilizers and pesticides that are harmful to the environment and public health (Yadav 2019).

Actinobacteria, also known as Actinomycetes, are among the most interesting groups of organisms. They belong to the biggest taxonomic groupings in the domain of bacteria. The Greek words "atkis" (a ray) and "mykes" (a fungus) are the origin of the name "Actinomycetes," as they share characteristics with both bacteria and fungi. Actinobacteria, which are mostly free-living organisms, are widely recognized for their widespread presence in soil and aquatic environments as well as their role in recycling nutrients through the breakdown of organic matter. This includes complex polymers derived from deceased plants and animals, such as chitin, starch, and lignin (Minotto et al. 2014). Actinobacteria, which include the genera *Streptomyces, Micromonospora,* and *Actinomadura,* are unique because they produce a variety of secondary metabolites such as enzymes, immunosuppressive agents, anticancer agents, and antibiotics (Raja and Prabakaran 2011). Actinobacteria are now seen to be a viable biotool for environmentally friendly farming practices. They could offer a practical and affordable substitute for manufactured insecticides and fertilizers. Numerous Actinobacteria's PGPR traits have been extensively documented. Additionally, by generating several bioactive molecules that are poisonous to plant pathogens but have no negative effects on humans or the environment, they can suppress the growth of a variety of soil-borne phytopathogens. Consequently, Boukhatem et al. (2022) state that Actinobacteria may have a direct, indirect, or both effect on the health and production of plants. Actinobacteria can significantly improve crop and soil health while also guaranteeing production in the present environment of increased demand for natural and organic agricultural methods. To ensure food security and environmental sustainability, we have evaluated the responsibilities that actinomycetes may play in revolutionizing agriculture in the ensuing decades. By highlighting the primary direct or indirect plant growth promotion qualities and the many methods of action as biological control agents, this study seeks to outline the positive and protective effects of Actinobacteria on plant growth.

16.2 Biology of Actinobacteria

With their distinctive filamentous shape and high guanine and cytosine contents in their DNA, Actinobacteria are one of the largest taxonomic families of Gram-positive bacteria. Known as aerobic filamentous Actinobacteria, a large number of Actinobacteria are capable of forming mycelium and sporulating to continue their reproduction (Barka et al. 2016).

Among Gram-positive bacteria, Actinobacteria have a distinctive morphological difference. Actinobacteria's cell structure is entirely distinct from that of fungus and is characteristic of prokaryotes. The entirety of a mycelium cell's structure is consistent with bacterial organization: the cytoplasm is home to ribosomes, genomic DNA sections, and a variety of inclusions that are most likely reserve materials like lipids, polyphosphates, or polysaccharides. Hyphae can be divided into substrate mycelium and aerial mycelium based on differences in form and function. Some Actinobacteria may form complex structures including spores, spore chains, sporangia, and sporangiospores. Important morphological traits of Actinobacteria categorization include the development and fracture modes of the substrate mycelium, the location, number, and surface features of the spore, the form of sporangia, and the presence or absence of flagella in the sporangiospore. Actinobacteria's micromorphological traits are crucial to its classification and identification. However, because of the wide range of variations, morphological characteristics cannot be solely linked to a particular Actinobacteria taxon. Actinobacteria genera may generally be found as rods, filaments, cocci, or rod-cocci (Li et al. 2016).

Based on 16S rRNA trees, a new taxonomy of the phylum Actinobacteria was released. Subclasses and suborders were raised to the levels of classes and orders, respectively, after that update; they were no longer categorized based on their taxonomic ranks (Ludwig et al. 2012). The six groups that comprise the phylum are Actinobacteria, Acidimicrobiia, Coriobacteriia, Nitriliruptoria, Rubrobacteria, and Thermoleophilia, according to Gao and Gupta (2012).

Optimal hydration conditions are necessary for germination. But when certain nutrients and triggers are present, it can happen more quickly and effectively. Before cells being able to recognize and absorb an outside energy source, spores also require sufficient amounts of internal trehalose and polyphosphates to initiate metabolic activity (Bobek et al. 2017).

An exceptionally complicated collection of bacteria that can synthesize a wide range of bioactive compounds is found in the well-known genus *Streptomyces* under the class Actinobacteria. The multicellular life cycle and massive linearized chromosomes that characterize *Streptomyces* are among their many unique traits. According to Law et al. (2019), the discovery of several major antibiotic medications has demonstrated the enormous contribution of *Streptomyces* to microbial drug development. This has undoubtedly promoted the investigation of these bacteria in various situations.

Because of their natural ability to grow in soil, *Streptomyces* are able to form substrate mycelium, which is composed of several hyphae that branch out from one another as they get bigger by extending their tips in search of food. They release a multitude of enzymes that degrade insoluble organic polymers, such as cellulose and chitin, into free sugars, which makes it easier for various ABC transporters to attach to and absorb them (Thompson et al. 2010).

Streptomyces bacteria have genes that make them resistant to other antibiotics, which helps them from taking their own lives. When these resistance genes are exposed to selection pressure, such as extended exposure to antibiotics, they can then be horizontally transmitted to other pathogenic bacteria and soil bacteria (Aminov and Mackie 2007).

Approximately two-thirds of naturally occurring antibiotics come from actinomycetes, and more than 75% of these isolates come from the *Streptomyces* species. *Streptomyces* is primarily studied since it is one of the primary suppliers of bioactive compounds (Viaene et al. 2016). Bioactive substances include those that yield extracellular enzymes and antibiotics, which are secondary metabolites; they also include agroactive and anticancer chemicals that are essential for the degradation of chitin and cellulose (Adegboye 2012).

Conversely, according to Jose and Jebakumar (2013), non-*Streptomyces actinomycetes* have invented around 30% of antibiotics (Lazzarini et al. 2000). *Thermomonosporaceae, Streptosporangiaceae, Nocardia,* and *Nocardioides* are the next most productive non-*Streptomyces Actinomycetes* after Micromonosporaceae members. Comparatively speaking to *Actinoplanes, Micromonospora*, and other non-*Streptomyces* Actinomycetes, *Streptosporangium* is less utilized at the genus level; fewer strains of this genus have been studied (Tiwari and Gupta 2012). According to Azman et al. (2015), non-*Streptomyces Actinomycetes* show variable degrees of variety, endemicity, and potential for producing new bioactive chemicals. They have been isolated from various settings. In particular, marine non-*Streptomyces Actinomycetes* with the capacity to generate bioactive chemicals have garnered significant interest in the hunt for novel molecules (Dhakal et al. 2017).

In natural ecosystems, Actinobacteria are widely dispersed and tolerant to a variety of environmental circumstances. Soil often contains communities of Actinobacteria, and by preventing the establishment of soil-borne diseases, rhizospheric *Streptomyces* can protect plant roots from them. This is mostly because of their ability to produce a wide variety of active metabolites that have antibiotic properties (Selim et al. 2021). Numerous halophilic Actinomycetes, including those from the *Streptomyces, Nocardia*, and *Micromonospora* genera, are found in soil, which is essential for plants to be able to withstand high salinity concentrations (Mohan and Rajamanickam 2018).

Aquatic settings (fresh and marine water) are rich in Actinobacteria. According to Rifaat (2003), the major organisms in the river's water were discovered to be *Streptomyces* members, while the dominating organisms in its sediments were determined to be *Micromonospora* members. Nonetheless, marine habitats are rich in these rare Actinobacteria that are capable of producing novel, powerful natural chemicals, most of which are used in the beauty and pharmaceutical industries (Dhakal et al. 2017). Actinobacteria may also be found and endure in harsh environmental settings. Thermophilic Actinobacteria are those that can grow and survive at high temperatures. According to Riquelme et al. (2015), volcanic caverns represent an unexamined extreme habitat that harbors unique and uncommon colored *Actinobacterium* strains that are capable of producing a variety of distinct secondary metabolites.

Actinobacteria, which have special biotechnological potential, are typically found in harsh environments like deserts. The *Streptomyces* and *Geodermatophilus* genera of Actinobacteria strains have been successfully isolated from the desert, and their potential to secrete novel therapeutic agents with anti-inflammatory, antiviral, antifungal, antiallergic, antibacterial, antitumor, and cytotoxic properties has been noted (Xie and Pathom-aree 2021).

16.3 Roles of Actinobacteria in Soil Health and Plant Productivity

16.3.1 Plant Nutrient Dynamics and Mineral Suppliers

Increasing agricultural output is desperately needed, as the world's population is growing at a rapid pace. The deficiency in macronutrients, particularly potassium, phosphorus, and nitrogen, in the soil may restrict this (Mekuria 2018). As a powerful ally for sustainable productivity, plant growth-promoting rhizobacteria (PGPRs), which include several Actinobacteria species, have recently been offered as possible substitutes for agrochemicals (Boukhatem et al. 2022). Like other PGPR, Actinobacteria exhibit both direct and indirect plant growth and productivity-promoting activities.

Encouraging plant growth Actinobacteria (PGPA) is a promising approach to producing microbial inoculants for use in sustainable agriculture production systems. This is due to their ability to produce essential plant growth regulators, help fix nitrogen, increase the availability of major plant nutrients in soil by solubilizing potassium and phosphorus compounds in row forms, and secrete lipases, cellulases, chitinases, and proteases.

16.3.1.1 Nitrogen (N Cycle-Mediated Functions)

Nitrogen is a major factor limiting plant development because of its significance in proteins and nucleic acids and because it makes up 2% of the dry matter in plants (Franche et al. 2008). Air nitrogen must biologically fix into ammonia for the global nitrogen cycle to operate. Nitrogen is converted to ammonia by the nitrogenase enzyme complex, which is present in a wide variety of bacteria and archaea (Canfield et al. 2010). The bacteria known as diazotrophs, which encode nitrogenase, are in charge of fixing nitrogen. The homologous nitrogenase enzyme system is classified into three primary types based on their respective metal reaction centers: iron-molybdenum, iron-vanadium, and iron-only (Newton 2007). The most prevalent kind of iron-molybdenum (MoFe) nitrogenase is encoded by the structural genes (nif HDK) and forms an octameric enzyme complex6 $(\alpha\beta\gamma 2)2$ when combined with the homodimeric nitrogenase reductase Fe-protein (Rubio and Ludden 2008).

Actinorhizal plants, which are budding, Gram-positive nitrogen-fixing organisms, are a class of plants that are related to but distinct from legume species. These plants are classified as Actinobacteria members of the genus *Frankia* (Perrine-Walker et al. 2011). The nifH gene encodes the nitrogenase reductase component, and molecular research has revealed that this gene is present in several *Frankia* species. According to Santi et al. (2013), *Frankia* exhibits a symbiotic connection whereby it exchanges fixed nitrogen for reduced carbon through the formation of root nodules. *Frankia* members have the ability to fix nitrogen, both in the free-living culture stage and inside the cells of their host plant during symbiosis, which is one of their many unique characteristics (Berry et al. 2011). *Frankia* spp. and *Frankia Casuarinae* DDNSF-02, which were isolated from *Casuarina* spp. root nodules, create a variety of significant growth regulators, including L-aspartic acid, siderophore, and IAA. Both of these isolates show remarkable abilities to generate ammonia and solubilize phosphate, leading Marappa et al. (2020) to speculate that they were plant growth enhancers.

In addition to *Frankia*, a wide range of cultivable Actinobacteria were shown to be capable of fixing nitrogen, including *Agromyces, Actinoplanes, Corynebacterium, Micromonospora, Microbispora, Streptomyces, and Propionibacteria* (Sellstedt and Richau 2013). Actinorhizal and leguminous plants' nitrogen-fixing root nodules have been shown to have a significant amount of Micromonosporae in recent years. Micromonospora species that were identified from *Medicago sativa* root nodules greatly increase the pace at which alfalfa plants nodulate and improve their capacity to absorb nitrogen. According to Martínez-Hidalgo et al. (2014), they have the potential to be rhizobia helper bacteria (RHB) agents with promising prebiotic qualities for plants.

16.3.1.2 Phosphorus (Solubilization)

The primary source of phosphorus, which is essential for plant development, is the mineralization of rock phosphates, making it a non-renewable resource. In addition to biomass accumulation and respiration chain processes, phosphorus is involved in energy transmission, signal transduction, photosynthesis, and macromolecular biosynthesis. Food production is negatively impacted by phosphorus shortages since all plant crops require adequate amounts of the mineral for growth and development (Saif et al. 2014). After being bioleached, phosphorus-containing minerals increase their chemical weathering, which raises soil fertility and yields from agriculture (Nath et al. 2018).

Numerous pathways have been studied to explore the capacity of Actinobacteria to mobilize phosphate, including the enzymatic synthesis of phosphatase and phytase (Solans et al. 2019) and the synthesis of low molecular weight organic acids that function as cation chelating agents (Kpomblekou and Tabatabai 1994). Interestingly, a strain of *Actinobacterium* may demonstrate both enzymatic dissolution of organic phosphate (Mineralization) and microbial organic acidification of inorganic phosphate (Solubilization) simultaneously (Tao et al. 2008). *Streptomyces* and *Micromonospora* comprise the bulk of Actinobacteria, with 20% or more exhibiting substantial phosphate solubilizing activity (Hamdali et al. 2008).

The synthesis of gluconic and 2-ketogluconic acid by *Streptomyces rishiriensis* 3AS4, which was isolated from wheat rhizospheres, had substantial phosphate mobilization capability. In greenhouse environments, the introduction of rock phosphate into 3AS4 led to enhanced soybean growth and decreased losses resulting from the fertilization of mineral phosphate (Vargas Hoyos et al. 2021). The *Micromonospora endolithica* isolate, 1 of 31 non-*Streptomyces* Actinobacteria isolated using *Streptomyces* phage and dry heat techniques, exhibited remarkable phosphate solubilization and generated an alkaline phosphatase and natural organic acid production when grown on liquid medium amended with powdered rock phosphate. Bean plant development was boosted throughout when *M. endolithica* was added to soil that had been treated with raw phosphate (El-Tarabily et al. 2008).

The capacity of *Streptomyces* sp. CTM396 to solubilize mineral phosphate was demonstrated in a different study, and this ability was linked to the manufacture of gluconic acid. It was discovered that supplementing with 0.05% humic acid accelerated CTM396's development and improved its capacity to dissolve Gafsa rock phosphate. An effective, sustainable, and environmentally friendly method of enhancing plant development is the application of *Streptomyces* spore suspension in conjunction with humic acids and Gafsa rock phosphate (Farhat et al. 2015).

16.3.1.3 Potassium Solubilization

Potassium is a necessary macronutrient that improves a plant's ability to withstand a variety of biotic and abiotic challenges and positively affects the metabolism of the plant (Bakhshandeh et al. 2017). Over 90% of potassium is found in soils as insoluble rock forms, and very little potassium may be made accessible (Prasad et al. 2019). Few studies have been done on microbial potassium solubilization. Still, certain PGPA isolates are active in potassium solubilization, including *Arthrobacter* sp. and *Microbacterium* sp. (Karlidag et al. 2007). Aqueous medium acidification is the outcome of microbial mechanisms of potassium solubilization, such as proton, H^+, or CO_2 excretion and the synthesis of organic acids. The release of chelating substances, including siderophores, which combine with potassium adsorbents, like calcium, iron, and aluminum, to form stable complexes, is another method (Etesami et al. 2017). For ten PGPA strains isolated from Moroccan sugar beet rhizospheric soil, varying rates of potassium solubilization capability were demonstrated. Aallam et al. (2021) found that the PGPA strains' capacity to solubilize nutrients might be linked to their generation of several organic acids, which was observed during their investigation on modified Aleksandrov agar medium. Significant solubilization rates of 3, 11, 12.75, and 17.8 mg/L were demonstrated by *Streptomyces griseorubens BC3, Streptomyces alboviridis P18, Streptomyces griseorubens BC10,* and *Nocardiopsis alba BC11* using mica. In addition, the in vivo study demonstrated a coarse root architecture and the wheat plant's maximal agronomic response to *Nocardiopsis alba BC11* and *Streptomyces griseorubens BC10* inoculation (Boubekri et al. 2022).

16.3.1.4 Iron (Siderophore Production) Nutrition

One of the most important micronutrients for plants and related microbes is iron. It serves as a crucial co-factor in several metabolic processes. In typical soil circumstances, this element has very little solubility, even though it is rather abundant in nature. Moreover, ambient oxygen quickly oxidizes iron to produce ferric oxyhydroxides, which are only sometimes soluble (Haas 2003). A crucial class of soil microorganisms for their metabolic activity is the Actinobacteria. According to Oves-Costales et al. (2009), they can produce a wide range of unique chemical classes, including siderophores, which are low molecular weight compounds that bind iron. Practically all microorganisms generate siderophores when there is an iron shortage. Siderophores are low molecular weight proteins that chelate iron and solubilize ambient iron for plant uptake (Lee et al. 2012).

A family of 500 important chemicals known as siderophores has different chemical structures (Schalk et al. 2011). However, a variety of Actinobacteria strains can create two primary categories of siderophores: hydroxamates and catechols (Wang et al. 2019). Plant development is enhanced by several helpful Actinobacteria that produce siderophores that chelate iron. According to Anwar et al. (2016), *Streptomyces nobilis* WA-3, *Streptomyces kunmingenesis* WC-3, and *Streptomyces enissocaesilis* TA-3 are very active makers of siderophores, ammonia, and HCN that promote wheat (*Triticum aestivum*) plant growth and seed germination.

Apart from its function in providing nutrients to plants, the Actinobacteria siderophore also shields plants against illnesses that are transmitted through the soil. Since the generation of harmful microbes and siderophores is dependent on iron bioavailability, Actinobacteria's iron-chelating siderophore synthesis renders iron inaccessible to the pathogen (Gopalakrishnan 2011). According to Sathya et al. (2017), siderophores have also been shown to encourage the colonization of plant roots by beneficial bacteria. Under situations of iron shortage, *Actinomadura madurae, Nocardia asteroids,* and *Streptomyces griseus* are recognized as active makers of siderophores (Kannahi and Senbagam 2014).

16.3.2 Phytohormones Production

Phytohormones, by definition, are organic molecules with low molecular weight that are synthesized in certain plant organs. These compounds can be translocated to other places where they elicit particular physiological, morphological, biochemical, and immunological responses (Sharma and Kaur 2017). Many Actinobacteria species, most notably *Streptomyces* spp., may release a range of phytohormones in the rhizospheric soil, including cytokinins, gibberellic acid, and indole-3-acetic acid. Because of this, they are crucial in controlling the growth of plants and enhancing the quantity and surface area of root tips in a variety of plants (Swarnalakshmi et al. 2016). Both tryptophan-dependent and -independent pathways can be used by Actinobacteria to produce indole-3-acetic acid (IAA), the primary auxin form. Tomato seedling development was markedly increased by *Streptomyces fradiae* NKZ-259's generation of IAA.

Carbohydrates and nitrogen sources, among other chemicals in the culture media, are essential for the cell's structural and energetic functions. Furthermore, the production of IAA by *Streptomyces fradiae* NKZ-259 may be influenced by a variety of growing circumstances, such as pH, carbon source, nitrogen source, and L-tryptophan supplementation (Myo et al. 2019). Moreover, IAA is essential for several physiological and developmental processes in plants, such as organogenesis, vascular differentiation, shoot and root growth, and fruit development (Etchells et al. 2016). The ISP-2 medium supplemented with 0.5% L-tryptophan was shown to be the most effective media for *Streptomyces* sp. VSMGT1014 to produce IAA. This strain demonstrated notable growth in this medium.

When *Streptomyces* sp. VSMGT1014 was injected into rice plants, the growth metrics were noticeably enhanced as compared to the non-inoculated plants (Harikrishnan et al. 2021). Having been isolated from marine sand and exhibiting distinctive morphological characteristics, *Agromyces mariniharenae* sp. nov. is a new *Actinobacterium* with great potential for producing IAA (Zhang et al. 2021). *Nocardiopsis* sp., an endophytic *Actinobacterium,* is known to produce large levels of IAA, which are favorable to the growth of mandarin seedlings, according to Shutsrirung et al. (2013). Cytokinins aid in the transfer of signals from roots to shoots under adverse environmental situations. They also increase the surface area of roots and encourage cell proliferation and development (Hayat et al. 2021). Gibberellic acid (GA3) is a key member of the gibberellin family of plant hormones, which acts as a growth regulator and promoter. Camara et al. (2018) state that GA3 can lead to cell elongation and division, which impacts the growth of stems, leaves, and flowers. On tomato seedlings (cv. Marmande), *Streptomyces neopeptinius* KNF2047T demonstrated impressive amounts of IAA and GA3, suggesting the possibility of promoting plant development (Goudjal et al. 2015). Another family of phytohormones that plays a significant part in plant development are called cytokinins. They regulate a variety of processes including photosynthesis, respiration activity, stem elongation, plant senescence, and seed germination (Hönig et al. 2018). Few studies stated that Actinobacteria had the potential to secrete cytokinins. Still, it has already been demonstrated that the genus *Actinoplanes*, *Frankia*, *Streptomyces*, and *Micromonospora* can synthesise zeatine, one of the most well-known cytokinins (Solans et al. 2011).

16.3.3 Aminocyclopropane-1-Carboxylate (ACC) Deaminase Activity

Under stressful situations, the ethylene phytohormone—which limits the growth of the plant generally and its root development in particular—is produced by the ACC as a precursor (Gupta and Pandey 2019). According to El-Tarabily (2008), ACC deaminase is a characteristic of *streptomycetes* and other helpful microbes that promote plant development and modulate stress. Increased soil salinity, for instance, increases stress factors such as Na^+ buildup, reactive oxygen species (ROS), and ethylene formation, which limits the development of rice and maize crops (You and Chan 2015).

ACC deaminase stimulates the hydrolysisof the amine group from ACC, resulting in two ammonia molecules and -ketobutyrate, which bacteria need as sources of nitrogen and carbon, resulting in elongation of the plant root and overall" enhancement in the plant growth (Misra et al. 2017). The ability to express ACC deaminase has been shown for *Streptomyces* sp. GMKU 336. To enhance plant development under salt stress, proline, potassium, calcium, chlorophyll, and water content in jasmine rice, *Streptomyces* sp. GMKU 336 was injected. Jaemsaeng et al. (2018) reported that it also reduced the levels of ethylene and reactive oxygen species. Further evidence that *Streptomyces venezuelae* may secrete ACC deaminase was provided by Yoolong et al. (2019).

It has been demonstrated that some halophilic Actinobacteria, such as *Micrococcus yunnanensis, Corynebacterium* variabile, and Arthrobacter nicotianae, record ACC deaminase activity. Under conditions of salt stress, these bacteria are crucial for fostering the growth of canola plants as shown in Table 16.1 (Siddikee et al. 2010). Applying *Streptomyces hydrogenans* DH16 has a good effect on the growth of pea seedlings. After colonizing the roots, *Streptomyces* hydrogenans DH16 significantly improved the lateral roots, fresh and dry weights, shoot and root lengths, and seed germination. It was also shown to have the ability to produce ACC deaminase and fix atmospheric nitrogen (Kaur and Manhas 2022).

Table 16.1 Actinobacteria in plant nutrient dynamics and growth induction

Actinobacteria species	PGP trait	Host plant	References
Micromonospora endolithica	Phosphate solubilization	Bean (*Phaseolus vulgaris* L.)	El-Tarabily et al. (2008)
Corynebacterium variabile, *Micrococcus yunnanensis*, and *Arthrobacter nicotianae*	ACC deaminase	Canola	Siddikee et al. (2010)
Actinoplanes, Frankia, Streptomyces, and *Micromonospora*	Cytokinins, gibberellins	*Ochetophilatrinervis*	Gupta and Pandey (2019)
Nocardiopsis sp.	Indole-3-acetic acid	Mandarin	Shutsrirung et al. (2013)
Micromonospora spp.	Plant probiotic	Alfalfa (*Medicago sativa*)	Martínez-Hidalgo et al. (2014)
Streptomyces neopeptinius KNF2047T	IAA, GA3	Tomato	Sharma and Kaur (2017)
Streptomyces sp.	Abscisic acid	Eggplant (*Solanum melongena*)	Rashad et al. (2015)
Streptomyces nobilis WA-3, *Streptomyces Kunmingenesis* WC-3, and *Streptomyces enissocaesilis* TA-3	Siderophore, ammonia, HCN	Wheat	Anwar et al. (2016)
Nocardiopsis yanglingensis, Streptomyces psammoticus, Streptomyces jiujiangensis, Pseudonocardia oroxyli	Phosphate solubilization, siderophores, IAA and ACC deaminase	Rice	Suksaard et al. (2017)
Streptomyces fradiae NKZ-259	Indole-3-acetic acid	Tomato	Myo et al. (2019)
Streptomyces venezuelae	ACC deaminase	Jasmine rice	Yoolong et al. (2019)
Streptomyces pseudovenezuelae	Ammonia, IAA, siderophore, phosphate solubilization	Maize	Chukwuneme et al. (2020)
Frankia sp. DDNSF-01 and *Frankia Casuarinae* DDNSF-02	Phosphate solubilization, ammonium production	–	Marappa et al. (2020)
Streptomyces sp. *A6*	HCN, siderophore	Amaranthus cruentus	Debasis et al. (2020)
Streptomyces griseorubens BC10 and *Nocardiopsis alba BC11*	Potassium solubilization	Wheat	Boubekri et al. (2022)
Streptomyces sp. strain KmiRC20A118 t	IAA production, siderophore, HCN	Oilseed rape	Cinkocki et al. (2021)
Streptomyces sp. VSMGT1014	Indole-3-acetic acid	Rice	Harikrishnan et al. (2021)

(continued)

Table 16.1 (continued)

Actinobacteria species	PGP trait	Host plant	References
Streptomyces rishiriensis 3AS4	Phosphate solubilization	Soybean	Vargas Hoyos et al. (2021)
Streptomyces hydrogenans DH16	Atmospheric nitrogen fixation, ACC deaminase, IAA	Pea (*Pisum sativum*)	Kaur and Manhas (2022)
Streptomyces spp.	Zinc solubilization	*Glycine max*	Suriyachadkun et al. (2022)
Streptomyces nanhaiensis YM4	Zinc solubilization	Rajagra (Amaranth)	Patel and Thakker (2020)
Cellulosimicrobium spp. and *Microbacterium* spp.	Zinc solubilization	Maize	Verma et al. (2022)

16.4 Actinobacteria Contributions to Plant Health and Plant Defensive System

16.4.1 Support Plant Adaptation Under Abiotic Stress Conditions

Abiotic stressors brought on by human activity and climate change have an impact on agricultural output globally, endangering food security. Microorganisms are becoming a more environmentally and commercially feasible tool for managing abiotic stress in agriculture. Actinomycetes are filamentous, Gram-positive bacteria that are frequently associated with plants as both plant endophytes and rhizosphere residents. They are being studied for possible uses in stressed environments (Grover et al. 2016). Worldwide agricultural output has drastically decreased due to salinity stress, which is one of the main abiotic factors affecting crop growth in dry and semiarid environments. The buildup of salts in soil that exceeds the critical plant limit is referred to as salinity in agriculture (Khan et al. 2019). The main reasons why plants suffer from soil salinity include osmotic, oxidative, nutrient-deficit, and ion-toxicity stresses. Salinity-induced stress results in the production of reactive oxygen species (ROS), which damage plant tissues by breaking DNA and causing base loss, double-stranded DNA breakage, and DNA–protein crosslinks (Orozco-Mosqueda et al. 2020). Netondo et al. (2004) state that respiration, protein synthesis, and lipid metabolism are impacted by the gradual accumulation of salt in plants.

Due to a reduction in chlorophyll in the leaves and their surrounding region, saline conditions significantly alter the physiology and metabolic processes of the plant, slowing down the rate of photosynthesis considerably. These biochemical and physiological alterations limit the growth of certain commercial plants and lower their overall productivity (Kumar et al. 2020). To address this issue, plants grown in saline settings can be treated with plant growth-promoting rhizobacteria (PGPR), which are microorganisms that reside in the rhizosphere of the plant and increase salt stress tolerance in crops. The Actinobacteria are a prominent group of microorganisms that have exhibited remarkable adaptability to several environmental circumstances.

Microorganisms that live in deserts have evolved a variety of survival strategies to withstand severe conditions, including a high tolerance to salt (Kothari et al. 2013). Strains of Actinobacteria from soil are resistant to a lot of salt in the medium. Three Micromonospora genus Actinobacteria strains improved the germination and growth of *Medicago sativa* seeds at a moderate salinity of 1%; at a high salinity, a little mitigating effect was noted (Alvarez-Sastre and Carro 2022). The inoculation produced a significant decrease in leaf proline concentration and an increase in plant biomass and chlorophyll content when tomato plants under salt stress were compared to those that were not infected with *Streptomyces* sp. strain PGPA39 (Palaniyandi et al. 2014).

Nocardioides sp.'s capacity to boost plant development in salt conditions and mitigate stress through the synthesis of phytohormones, which function as defense mitigators, may be the reason for the significant amount of IAA it produced (Meena et al. 2020). Tomato seedlings inoculated with the deep-sea isolate *Dermacoccus* MT2.1T exhibited greater weight and length of shoots and roots, together with a higher chlorophyll content, in comparison to non-inoculated tomato plants. Furthermore, tomato seedlings infected with *Dermacoccus* MT2.1T had lower levels of hydrogen peroxide (H_2O_2) in their leaves than control seedlings. This discovery revealed that halotrophic Dermacoccus treatment reduced oxidative stress caused by hydrogen peroxide, thereby mitigating salinity (Rangseekaew et al. 2021). Following 200 mM NaCl irrigation, the fermentation broth and *Streptomyces* sp. KLBMP5084's volatile organic compounds (VOCs) effectively suppressed *Alternaria solani*, the agent responsible for tomato early blight, and encouraged the growth of tomato seedlings, leading to a significant increase in the amount of fibrous roots (Gong et al. 2020). Significant water evaporation brought on by recent temperature rises has exacerbated drought stress by contributing to global warming. Drought stress is the main environmental element in dry and semiarid areas that affects crop quality, output, and plant development (Chatterjee and Solankey 2015). Drought may alter plant phenotypes and soil chemistry, which can have an indirect effect on root microbial populations, as well as directly altering the availability of moisture (Naylor and Coleman-Derr 2018).

According to Abdelaal et al. (2021) plants subjected to drought have evolved a range of stress-coping mechanisms, including morphological, physiological, anatomical, and biochemical responses. Plant growth-promoting microorganisms (PGPM) found in a variety of biopreparations may provide an economical and efficient way to reduce drought stress in plants, which might have an impact on food security and production (Hanaka et al. 2021). Recent research has focused on Actinobacteria that enhance plant development as a potentially useful biotool for increasing agricultural output and plant growth in drought-stressed environments (Selim et al. 2019). The methods via which *Streptomyces* sp. promotes drought tolerance in plants include osmotic pressure augmentation in plant cells, callose buildup acceleration, and cell wall lignification (Hasegawa et al. 2004).

In comparison to the control, there was a noticeable improvement in plant growth, yield, and seedling vigor when endophytic *Streptomyces coelicolor DE07, S. olivaceus DE10,* and *Streptomyces geysiriensis DE27* cultures were applied to wheat seeds. A cumulative synergistic impact that enhanced wheat plant growth in the stressed soil was produced by these endophytic Actinobacteria through the synthesis of phytohormones, traits that support plant development, and intrinsic resilience to water stress (Yandigeri et al. 2012). Proline, polysaccharides, and ectoine were among the several osmolytes that were generated by the halotrophic endophytic Actinobacteria isolate *Streptomyces albidoflavus* OsiLf-2. Rice plants that were injected with OsiLf-2 showed enhanced resistance to salt and drought by boosting the rice host's ability to adapt its osmotic pressure through changes in proline and soluble sugar levels (Niu et al. 2022).

In sugarcane plants stressed by dryness, an inoculation of *Streptomyces chartreusis* WZS021 increased the levels of phytohormones, proline, and chlorophyll. Wang et al. (2019) claim that WZS021 can improve plant development in conditions of water stress by causing sugarcane plants to become drought-tolerant through excessive root colonization and the manufacture of siderophores, indole acetic acid, and 1-aminocyclopropane-1-carboxylic deaminase.

Maize was less affected by drought stress when *Actinobacterium* isolate Ac5 produced substantial amounts of flavonoids, phytohormones, and siderophores as shown in Table 16.2. In treated maize plants in water-deficit circumstances, the Ac5 isolate enhanced growth and photosynthesis and caused a worldwide metabolic shift, decreased H_2O_2 buildup, decreased lipid peroxidation, and greater levels of molecular antioxidants have all been seen in the bacterized plants (Selim et al. 2019).

16.4.2 Enhanced Plant Defense Against Invasive Pests

Numerous biotic stressors impact plants, mostly as a result of climate change, which makes them more vulnerable to pathogen assault (Teshome et al. 2020). A pesticide-free agriculture system is becoming increasingly important because of the harmful effects that these xenobiotic compounds have on both the environment and human health. According to Olanrewaju and Babalola (2018), Actinobacteria are a class of microorganisms that are extensively dispersed in nature and have special physiological characteristics that make them ideal candidates as biological control agents for sustainable agriculture. Their amazing synthesis of antibiotics, volatile chemicals, hydrolytic enzymes, and other significant metabolites that function as antifungal, antibacterial, and antiviral agents is primarily responsible for their biological control action (Pacios-Michelena et al. 2021).

For a very long time, it has been understood that Actinobacteria improve plant resilience by inhibiting phytopathogen proliferation. A variety of biocontrol drugs have been shown to have many modes of action, such as immunomodulators, signaling molecules, enzyme synthesis, enzyme inhibitors, and/or antibiosis (Srivastava et al. 2021). In this regard, the growth and spore germination of *Fusarium oxysporum f.* sp. *Cubense* were seen to be suppressed by *Streptomyces* sp. strain H3-2,

Table 16.2 Actinobacteria-mediated abiotic stress mitigation in plants

Actinobacteria	Isolation source	Stress	Observed effect	Host crop	References
Streptomyces coelicolor DE07, *S. olivaceus* DE10, and *S. geysiriensis* DE27	Cultivated plants of arid regions	Drought	Significant auxin production resulting in intrinsic water stress tolerance	Wheat (*Triticum aestivum*)	Yandigeri et al. (2012)
Streptomyces sp. strain PGPA39	Agricultural soil	Salinity	Increase in plant biomass and chlorophyll content and a reduction in leaf proline content	Tomato (*Solanum lycopersicum*)	Palaniyandi et al. (2014)
Actinobacterium isolate Ac5	A semiarid environment	Drought	Reduced H_2O_2 accumulation and lipid peroxidation accompanied, high levels of molecular antioxidants	Maize (*Zea mays* L.)	Selim et al. (2019)
Streptomyces chartreusis WZS021	Soil	Drought	Modulating the proline and soluble sugar content, growth promotion under water stress conditions	Sugarcane, (*Saccharum officinarum*)	Wang et al. (2019)
Streptomyces sp. KLBMP5084	The root of halophyte *Limonium sinense*	Salinity	Suppressed tomato mortality due to *Alternaria solani* infection and reduced Na^+ accumulation in leaf	Tomato (*Solanum lycopersicum*)	Gong et al. (2020)
Nocardioides sp. NIMMe6	Leaves of soybean crop	Salinity	Induction of phytohormone (IAA) as defense mitigator	Wheat (*Triticum aestivum*)	Meena et al. (2020)
Dermacoccus barathri MT2.1^T	Deep-sea	Salinity	Improve in plant growth and suppress in oxidative stress caused by H_2O_2	Tomato (*Solanum lycopersicum*)	Rangseekaew et al. (2021)

(continued)

Table 16.2 (continued)

Actinobacteria	Isolation source	Stress	Observed effect	Host crop	References
Micromonospora spp.	Desert	Salinity	Increase in seed germination and salt tolerance of plant	Alfalfa (*Medicago sativa*)	Alvarez-Sastre and Carro (2022)
Streptomyces albidoflavus OsiLf-2	Rice leaves	Salinity drought	Produced abundant osmolytes, including proline, polysaccharides, and ectoine	Rice (*Oryza sativa* L.)	Niu et al. (2022)

which was isolated from a robust banana orchard. A biochemical study confirmed that strain H3-2 secreted two phenolic compounds, 2,4-bis(1,1-dimethylethyl)- and (2,2′-methylenebis)[6-(1,1-dimethylethyl)-4-methyl, which are thought to be antifungal agents (Zou et al. 2021). The microscopic study also revealed significant damage to the integrity of the cell membrane and the mycelial ultrastructure.

According to Loliam et al. (2013), *Streptomyces rubrolavendulae* S4's parasitic and enzymatic activities have both been shown to be important modes of action against *Pythium aphanidermatum.* Similar to this, a scanning electron microscopic investigation revealed that the primary mechanism of action against *Sclerotinia sclerotiorum* was the actinoparasitism potentiality of three soil strains of *Streptomyces, namely S. griseus DG5, S. rochei DG4, and S. sampsonii DG1* (Gebily Doha et al. 2021).

Microbial volatile organic chemicals (VOCs) are low molecular weight, lipophilic chemicals (often alcohols, alkanes, acids, or ketones) that are produced by microorganisms through their main and secondary metabolic biosynthetic pathways (Gotor-Vila et al. 2017).

Actinobacteria have been shown to produce distinctive volatile organic compounds (VOCs) that both promote morphological abnormalities in certain fungus and reduce the frequency and severity of several fungal plant diseases (Wang et al. 2013; Ghanem et al. 2022). The headspace volatile substances (VS) generated by *Streptomyces platensis* F-1 were found to have a noteworthy impact on mitigating the frequency and intensity of leaf blight in rice seedlings caused by *Rhizoctonia solani,* oilseed rape leaf blight caused by *Sclerotinia sclerotiorum*, and strawberry fruit rot caused by *Botrytis cinerea* (Wan et al. 2008).

According to Lyu et al. (2020), *Streptomyces yanglinensis* 3–10 produces volatile organic compounds (VOCs) such as methyl 2-methylbutyrate (M2M), 2-phenylethanol (2-PE), and β-caryophyllene (β-CA). These compounds can impede Aspergillus conidial germination and colonization on peanut kernels, making *S. yanglinensis* 3–10 a potentially effective biofumigant for biological control applications. Seven volatile organic compounds (VOCs) were found to be the main active substances released by *Streptomyces* sp. (PNM-149) and demonstrated antifungal activity against *Colletotrichum gloeosporioides*, the anthracnose causative agent that causes significant harvest losses (Gómez et al. 2021). Red pepper leaf spot (*Xanthomonas euvesicatoria*) and tomato bacterial wilt (*Ralstonia solanacearum*) were both successfully inhibited by the use of fermentation broth from *Streptomyces* sp. AN090126.

The biocontrol efficiency of this Actinobacteria strain is significantly influenced by the volatile organic compounds (VOCs) that AN090126 synthesizes, particularly dimethyl sulfide and trimethyl sulfide, according to gas chromatography–mass spectrometry studies (Le et al. 2022).

A crucial structural element of the fungal cell wall is chitin, one of the most prevalent polysaccharides in nature (Pusztahelyi 2018). Umar et al. (2021) suggest that the antagonist action against *Aspergillus niger* and *Aspergillus oryzae* may be synergistically enhanced by chitinases and other hydrolytic enzymes of *Streptomyces* isolate CT02. Following its isolation from the wheat rhizosphere, *Streptomyces* luridiscabiei U05 demonstrated potent chitinase activity as well as in vitro antifungal properties against *Penicillium verrucosum, Botrytis cinerea, Alternaria alternata, Fusarium oxysporum, F. solani,* and *Botrytis cinerea. Fusarium oxysporum f.* sp. *Lycopersici,* the causative agent of tomato wilt disease, showed minimal disease severity when chitinolytic *Streptomyces griseorubens* E44G was applied in the field, significantly increasing all tomato growth, yield, and quality parameters that were tested (Rashad et al. 2017).

A rhizospheric microbial population that functions for biocontrol induces induced systemic resistance (ISR) in plants by improving resistance to phytopathogens by inducing the creation of defense molecules (Choudhary et al. 2007). To put it another way, in tomato infected with *F. oxysporum Schlecht, f.* sp. *Lycopersici,* seedling treatments with *Streptomyces enissocaesilis* and *Streptomyces rochei* promoted antioxidant enzyme activity and systemic resistance (ISR). The results of Abbasi et al. (2019) indicate that tomato plants treated with *Streptomyces* strains exhibit increased activities of catalase (CAT) and peroxidase (POX), indicating an efficient activator of specific-induced systemic resistance (ss-ISR). In Asian nations, one of the most common rice illnesses is bacterial brilliant leaf disease, which is caused by *Xanthomonas oryzae pv. oryzicola.*

Streptomyces shenzhenensis TKSC3 and *Streptomyces* sp. SS8 consortium treatments for seed bacterization resulted in ISR mediation against *Xanthomonas oryzae pv. oryzicola.* Hata et al. (2021) discovered that the application of consortium *Streptomyces* led to the accumulation of peroxidase (POX), polyphenol oxidase, phenylalanine ammonia-lyase, and β,1-3 glucanase (GLU) in plants. These enzymes are in charge of reinforcing plant cell walls and facilitating resistant relationships. In four potato cultivars, foliar spraying crude filtrates of *Streptomyces netropsis, S. ambofaciens,* and *S. actuosus* showed a range of inhibitory effects against the tuber necrotic strain of potato virus Y (PVYNTN).

A study conducted by Nasr-Eldin et al. (2019) found that the bulk of active compounds in the crude Actinobacteria filtrates were derived from fatty acids (FAs) and their derivatives, Citroflex A, hexanoic acid, anhydride, hexadecanoic acid, ethyl ester, and diisooctyl phthalate. These compounds can function as elicitor molecules for the induction of systemic acquired resistance (SAR). It was proposed that treating tomato mosaic virus (ToMV) using the culture filtrate of *Streptomyces ovatisporus* LC597360 would be an environmentally viable method. According to Taha et al. (2021), it not only promoted plant development but also boosted the activity of defense-related enzymes such as ascorbate oxidase, catalase, peroxidase, and polyphenol oxidase.

Throughout the agricultural process, weeds are the cause of large losses in crop yield. A variety of weeds become more resistant to control when chemical pesticides are applied excessively. Certain secondary metabolites produced by Actinobacteria, including amino acids, peptides, nucleosides, lactones, and macrolides, have been developed into commercial products (Shi et al. 2019). These compounds have herbicidal properties. Herbicidal properties were examined in 16 Actinobacteria isolates that were obtained from different coastal soils. Studies by Dhanasekaran et al. (2010) showed that isolates from *Streptomyces* spp. might stop *Echinochlora crus-galli* from growing. A study conducted by Dhanasekaran et al. (2012) revealed that four isolates of Actinobacterium had noteworthy herbicidal efficacy against *Cyperus rotundus*. In particular, target-specific toxicity on the chosen weed plants is produced by *Streptomyces* strain 329's naturally occurring herbicidal metabolites, which have contact and selective action. So that, the *Streptomyces* strain-329 was suggested to provide a new molecule for an effective herbicidal product (Bo et al. 2019).

An extensive range of economically significant agricultural crops are severely harmed each year by phytonematodes, which are minute, necessary endoparasites (Ruanpanun et al. 2010). Current nematode control chemical practices are unsafe and have been further restricted in the last few years (Forghani and Hajihassani 2020). Many interesting secondary metabolites with strong nematocidal action are efficiently produced by Actinobacteria, as has been established. Hu et al. (2022) reported that *Streptomyces albogriseolus* ND41 and *S. fimicarius* D153 had a lethal impact on juvenile root knot nematode specimens by synthesizing potent nematicidal hydrosoluble macromolecule polymers that comprise several sugars and phenolic hydroxyl chemicals. A low molecular weight substance with a wide range of biological actions, fermentulin is generated by many different species.

Nonetheless, fervenulin was initially reported as a nematicidal active agent by Ruanpanun et al. (2010). A potent anthelmintic chemical, avermectin is primarily generated by *Streptomyces avermitilis* and is utilized as an efficient nematocidal and insecticidal agent (Jagannathan et al. 2021). Among the avermectins, abamectin is a biopesticide that has demonstrated strong nematostatic efficacy against two types of potato infections: *Globodera pallida* (Sasanelli et al. 2019) and *Tylenchulus semipenetrans* (Mam et al. 2019). The life cycle of Arabidopsis plant root-knot nematodes is adversely affected by both solvent extracts and batch culture supernatants of *Streptomyces monomycini* ATHUBA 220 and *S. colombiensis* ATHUBA 438 (Meidani et al. 2020).

The macrocyclic lactone Nemadectin, which has broad-spectrum action, is a highly effective antiparasitic drug generated by *Streptomyces cyaneogriseus* (Wu et al. 2020). In addition, it has been shown that the biological nematicide emamectin benzoate, which is derived from the fermentation product of *Streptomyces avermitilis* in the rhizosphere, effectively suppresses the growth of *M. incognita* in tomato plants when applied in a field setting (Cheng et al. 2015). It has been reported that, in greenhouse settings, *Streptomyces galilaeus* strain KPS-C004, isolated from strongly contaminated nematode soil, may be used as a biocontrol agent against chilli infected with *Meloidogyne incognita.* According to Nimnoi et al. (2017), *Streptomyces galilaeus* strain KPS-C004 was able to endure and grow in the competitive soil environment while maintaining its metabolic activity over the culture period.

The incidence of illness induced by root-knot nematodes on chillies was inhibited by both the spore suspension and culture filtrate of *Streptomyces* sp. KPS-E004 isolated from earthworms. Accordingly, it was proposed that this particular Actinobacteria isolate was both an efficient generator of nematocidal compounds and an active parasite of plant nematodes (Ruanpanun and Chamswarng 2015). Plant parasitic nematodes' eggshells are mostly made of chitin, which is the main target of several biological control techniques (Fanelli et al. 2005). The synthesis of chitinase by *Streptomyces rubiginosus* SP24 has been effectively optimized and purified. When pure chitinase was used to coat Vigna radiata seeds, it demonstrated potent ovicidal and nematocidal properties that contributed to the suppression of root-knot nematode infection and the promotion of plant development (Jha and Modi 2018).

Naturally occurring products and culture masses derived from Actinobacteria have been proposed and developed as safe substitutes for chemical insecticides, given their numerous drawbacks, which include environmental risks and pest resistance (Meena and Mishra 2020; El-khawaga and Megahed 2012). Known to cause significant harm to over 150 commercially significant vegetable crops, *Helicoverpa falciparum* and *Spodoptera litura* are regarded as the most economically significant insect pests globally. The marine actinomycete *Streptomyces* sp. AP-123 was used to extract, identify, and manufacture a new polyketide metabolite with antifeedant action against *Helicoverpa armigera* and *Spodoptera litura* as an insecticidal agent as shown in Table 16.3 (Arasu et al. 2013).

As a promising option for the synthesis of bioinsecticides, the crude extract of *Streptomyces albidoflavus* strain G30 shows notable insecticidal potentiality against green peach aphid (Chen et al. 2018). Kim et al. (2020) have shown that *Streptomyces* sp. AN120537 produces antimycins, which are insect juvenile hormone antagonists. At lower concentrations, the naturally occurring beneficial chemicals produced by *Streptomyces hydrogenans* in the fermentation broth dramatically reduced the rate of reproduction of *Spodoptera litura*, whereas at higher concentrations they had a deadly effect on the larvae (Kaur et al. 2014).

Table 16.3 Biocontrol activity of Actinobacteria against different plant pests

Actinobacteria species/strain	Mode of action	Active compound	Plant	Target pathogens	References
Streptomyces sp. CMU-MH021	Antibiosis	Fervenulin	–	Meloidogyne incognita	Ruanpanun et al. (2010)
Streptomyces sp. *AP-123*	Antibiosis	Polyketide metabolite	–	*Helicoverpa armigera* and *Spodoptera litura*	Arasu et al. (2013)
Streptomyces avermitilis	Antibiosis	Emamectin benzoate	Tomato (*Solanum lycopersicum* L.)	*Meloidogyne incognita*	Cheng et al. (2015)
Streptomyces antibioticus strain M7	Antibiosis	Actinomycins	Tomato	*Meloidogyne incognita*	Kaur et al. (2014)
Streptomyces galilaeus strain KPS-C004	Competition/antagonism	Spores	Chili	*Meloidogyne incognita*	Nimnoi et al. (2017)
Streptomyces rubiginosus SP24	Lytic enzymes	Chitinase	Mung bean Vigna radiate	Root-knot nematodes	Jha and Modi (2018)
Streptomyces sp.	Antibiosis	Spectinabilin	Pinus densiflora tree	Bursaphelenchus xylophilus	Liu et al. (2019)
Streptomyces avermitilis	Antibiosis	Abamectin	Orange tree	Tylenchulus semipenetrans	Mam et al. (2019)
Streptomyces netropsis DSM 40093 (SCF7), *S. ambofaciens* (SCF11), and *S. actuosus* (SCF20)	Induction of systematic resistance by production of elicitor molecules	–	Potato	Tuber necrotic strain of potato virus Y (PVYNTN)	Nasr-Eldin et al. (2019)
Streptomyces yanglinensis 3–10	Antibiosis	Methyl 2-methylbutyrate (M2M)	Peanut kernels	*Aspergillus flavus* and *Aspergillus parasiticus*	Lyu et al. (2020)
Streptomyces sp. (PNM-149)	Antibiosis	Methyl anthranilate and 2-methyl isoborneol	–	*Colletotrichum gloeosporioides*	Gómez et al. (2021)
Streptomyces isolate CT02	Lytic enzymes	Chitinase	–	*Aspergillus niger* and *Aspergillus oryzae*	Umar et al. (2021)

Actinobacteria species/strain	Mode of action	Active compound	Plant	Target pathogens	References
Streptomyces sp. strain H3-2	Antibiosis	Phenol,2,4-bis(1,1-dimethylethyl)- and (phenol,2,2′-methylenebis)[6-(1,1-dimethylethyl)-4-methyl-	Banana	*Fusarium oxysporum* f. sp. Cubense	Zou et al. (2021)
Streptomyces shenzhenensis TKSC3 and *Streptomyces* sp. SS8	Induction of systematic resistance	–	Rice	*Xanthomonas oryzae pv. oryzicola*	Hata et al. (2021)
Streptomyces ovatisporus LC597360	Induction of systematic resistance	–	Tomato	Tomato mosaic virus (ToMV)	Taha et al. (2021)
Streptomyces sp. AN090126	Antibiosis	Dimethyl sulfide and trimethyl sulfide	Tomato and pepper	*Ralstonia solanacearum* and *Xanthomonas euvesicatoria*	Le et al. (2022)

16.5 Biotechnological Products from Actinobacteria: Current and Beyond

Primary and secondary metabolites produced in large quantities by Actinobacteria have intriguing qualities including antibacterial, antiviral, and anticancer effects. As such, they have important uses in agriculture, biotechnology, and medicine (Barka et al. 2016). Severe environmental circumstances encourage Actinobacteria to evolve new secondary metabolic pathways, which raises the prospect of finding new biological uses for bioactive substances (Qi et al. 2019). It has been estimated that there are around 23,000 microbial bioactive secondary metabolites. More than 10,000 of these substances are produced by Actinobacteria, which make up over 45% of all known bioactive microbial metabolites (Kalyani et al. 2019).

16.5.1 Antimicrobial Compounds

Actinobacteria are a great subject for future research and development because they provide an endless supply of antibiotics. The majority of currently recognized antibiotics were initially identified from the *Streptomyces* genus. All significant drug families now utilized in clinics, such as aminoglycosides, glycopeptides, macrolides, β-lactams, and tetracyclines, are among the compounds generated (Mast and Stegmann 2019). A family of compounds known as antibiotics works to specifically stop the development of bacteria while sparing eukaryotic species. The chemical structure of the antibiotic determines its action. Actinobacteria generate many antibiotic families, such as glycopeptides, β-lactams, tetracyclines, macrolides, and aminoglycosides, from a chemical perspective (De Simeis and Serra 2021).

A significant class of natural compounds known as glycopeptide antibiotics are primarily generated by a broad range of filamentous Actinomycetes and target the cell walls of prokaryotes. They consist chemically of nonribosomal peptides that have been glycosylated (Yim et al. 2013). By attaching to the acyl-D-alanyl-D-alanine (D-Ala-D-Ala) terminus of the developing peptidoglycan on the cytoplasmic membrane's outer surface, they were utilized to treat infections that were potentially fatal and brought on by *Staphylococcus aureus, Enterococcus species, and Clostridium difficile* (Binda et al. 2014). Essential glycopeptides, vancomycin and teicoplanin, were effectively generated by *Streptomyces toyocaensis* NRRL 15009 and show remarkable efficacy against Gram-positive bacterial pathogens, such as those belonging to the genera *Staphylococcus, Streptococcus, and Enterococcus* (Pootoolal et al. 2002).

According to Fernández-Martínez et al. (2014), chloramphenicol (CHL) is a bactericidal agent that is generated by *Streptomyces venezuelae*. It functions as an inhibitor of ribosomal peptidyl transferase activity in a variety of Gram-positive and Gram-negative bacteria, including many strains that are resistant to drugs. Acidophilic Actinobacteria isolates from the genera *Streptomyces* and Kitasatospora are rich sources of novel bioactive compounds that are effective against *Helicobacter pylori*, the causative agent of severe diseases like gastric and colorectal cancer.

These compounds include carbamic acid, maltol, 2.4-di-tert-butylphenol, methyl dimendone, prolyl-leucyl, and oleamide (Eftekharivash et al. 2021). Frigocyclinone is a vital antibiotic including an aminodeoxysugar osamine and an angucyclinone core connected by an uncommon C-glycosidic bond.

According to Mo et al. (2019), *Streptomyces griseus* NTK 97 and marine sponge-associated *Streptomyces* sp. M7_15 efficiently generated it, and even at low concentrations, it had a strong antibacterial action against *Bacillus subtilis* and *Staphylococcus aureus*. According to Vicente et al. (2015), frigocyclinone has the potential to be employed in the development of anticancer medications since it can kill rhabdomyosarcoma cancer cells with an EC50 of 5.2 M after 48 h of treatment. An abundant supply of new bacteriostatic and anticancer drugs can be found in marine actinomycetes. According to Ko et al. (2014), lajollamycin (1) is a nitro-tetraene spiro-beta-lactone-gamma-lactam antibiotic that is generated by a marine strain of *Streptomyces nodosus* (NPS007994).

The body of research suggests that many Actinobacteria isolates are tightly linked to the manufacture of antifungal drugs. *Kutzneria* sp. strain TSII, which promotes plant development, was isolated from mangrove silt soil and demonstrated strong activity in the production of 2,4-bis(1,1-dimethylethyl) phenol, an antifungal agent (Devi et al. 2021).

16.5.2 Antitumor Compounds

Reducing the detrimental effects of chemotherapeutic medicines and increasing the antitumor agents' specificity is the aim of producing antitumor agents from microorganisms so that human usage is safe (Naggar et al. 2017). Numerous naturally occurring compounds with potent anticancer effects are produced by Actinobacteria. These consist of nonribosomal peptides, macrolides, isoprenoides, indolocarbazoles, enediynes, and anthracyclines. Their antitumor action can be achieved by blocking key signal transduction enzymes such as proteases, mitochondrial permeabilization, inhibiting topoisomerase I or II, causing DNA cleavage, blocking cellular metabolism, or, in certain situations, by preventing tumor-induced angiogenesis (Olano et al. 2009).

Isolated from a saline soil, *Streptomyces* sp. KML-2 produces chromomycin SA and 1-(1H-indol-3-yl)-propane-1,2,3-triol, which have anticancer activities against MCF-7 and Hela cancer cell lines (Aftab et al. 2015). This production is extremely stable. Three marine isolates, identified as *Streptomyces* sp. strain AMS11, *Nocardiopsis halotolerans* strain AMS10b, and *Nocardiopsis halotolerans* strain AMS10a, exhibited encouraging results when cultured against two cancer cell lines: HEPG2 (liver cancer) and HCT116 (colon cancer) (Naggar et al. 2017). According to Davies-Bolorunduro et al. (2019), a metabolic extract of *Streptomyces bingchenggensis* ULS14 showed cytotoxicity on the following cell lines: HL-60 (human acute promyelocytic leukemia), HeLa (cervical carcinoma), AGS (human gastric), MCF-7 (breast adenocarcinoma), and K562 (human acute myeloid leukemia).

Straitjacket al. (2004) reported that a novel compound called (6*R*,10*S*)-10-methyl-6-dodecanolide, a member of the caprolactones, was extracted from marine *Streptomyces* sp. B6007, which was isolated from the mangrove sediment. This compound has demonstrated activity against cell lines of gastric adenocarcinoma (HM02), hepatocarcinoma (HepG2), and breast cancer (MCF-7).

16.5.3 Enzymes

Actinobacteria are widely known for their capacity to generate distinct enzymes that may find use in industry, agriculture, and medicine. With the help of their hydrolytic enzymes, certain Actinobacteria operate on renewable biomass and biopolymers including cellulose, xylan, and chitin to produce biofuels and biochemicals.

The main enzymes involved in handling wastes formed from keratin are called keratinases, and they have a wide range of biotechnological uses, including animal feed (Holkar et al. 2018). Actinobacteria representatives of the Nocardiopsis and *Streptomyces* taxa (Al-jelawi 2017; González et al. 2020) have been identified as active keratinase producers. Enzymes known as lipases catalyze the hydrolysis of triacylglycerol at the oil–water interface, converting it into glycerol and free fatty acids. The food, cosmetics, and detergent sectors all use lipases extensively. The maximal lipase production capability of *Streptomyces* sp. Al-Dhabi-49 was demonstrated under submerged fermentation growth conditions with glucose supplementation as the only carbon source (Al-Dhabi et al. 2020). Proteases are important in the dairy, leather, and pharmaceutical sectors because they catalyze proteolytic processes.

Moreover, it has been demonstrated that the protease enzyme generated by *Streptomyces* spp. have antiviral and antibacterial properties (Ananya et al. 2021). Purified and identified as a functional metabolite of halophilic *Actinopolyspora* sp. VITSDK2, thermostable serine protease was discovered (Suthindhiran et al. 2013). Microbial chitinases are involved in parasitism and the control of various phytopathogens on a trophic level. Thus, according to Rathore and Gupta (2015), chitinolytic bacteria are attractive options for use as biopesticides. Maltose was shown to be an effective carbon source for *Streptomyces* albus, which in turn produced chitinase with antifungal properties against *Magnaporthe oryzae, Rhizoctonia solani,* and *Botrytis cinerea* (Ekundayo et al. 2022).

16.5.4 Probiotics

Probiotics are by definition live microbial supplements that help their hosts develop, survive, and maintain better health (Thakur et al. 2023a, b). It has been demonstrated that feeding shrimp and fish with enhanced sorghum can shield them from infections and promote the growth of aquatic life (Tan et al. 2016). According to Das et al. (2008), probiotic Actinobacteria enhance feed digestion and feed utilization while also producing antagonistic chemicals like bacteriocin and competing for nutrients and attachment sites. Potential sources of probiotics are marine Actinobacteria due to their ability to combat pathogenic *Vibrio* spp. Many *Streptomyces* and Propionibacteria strains can produce B vitamins and may be used as probiotics to boost the immune system and growth performance of chickens, hence increasing productivity in the poultry sector (Shah and Dwivedi 2022; Thakur et al. 2023a, b).

Exoenzymatic and anti-vibrio properties were demonstrated by three Actinobacteria strains from the *Streptomyces* species that were isolated from the Caspian Sea. According to Norouzi et al. (2018), these strains are thus deserving of being presented as possible probiotics for aquatic culture. Freshwater fish of great commercial significance is Nile tilapia. *Streptomyces* strains LNW002 and YNW004 are useful as probiotics in aquaculture, effectively combating bacterial infections in Nile tilapia to eliminate synthetic antibiotic residue and enhance animal nutrition, immunity, and productivity (Jirayut et al. 2020).

16.5.5 Pigments

Any material that can absorb light might be referred to as microbial pigment. Many pigments from different microorganisms have been identified and separated; some of these pigments have anticancer, antioxidant, and antibacterial properties. Microbial pigments are superior to plant pigments in several ways, including the quick growth rate of microorganisms, an economical manufacturing method, and the ease of use of the necessary production procedures. Many of the microbial pigments have strong antibiotic qualities and can be either endopigments or exopigments (Fernandes et al. 2021). Actinobacteria, particularly *Streptomyces* species, are biotechnologically significant microorganisms that generate a wide range of industrial pigments, such as melanoid pigments, or dark-brown melanin, which find widespread use in pharmacology, cosmetics, medicine, and the agricultural sector (Manivasagan et al. 2013).

It has been observed that *Streptomyces glaucescens* NEAE-H has the ability to generate a significant quantity of extracellular melanin on protease-peptone extract iron agar. An in vitro experiment was used to demonstrate the anticancer properties of isolated melanin pigment against a skin cancer cell line (El-Naggar and El-Ewasy 2017). With many uses in agriculture and medicine, prodigiosin is a red pigment that is one of the most significant commercial pigments. At its lowest inhibitory concentration, the powerful bioactive metabolite undecylprodigiosin red pigment—produced by *Streptomyces* sp. JAR6—was efficacious against both *Staphylococcus aureus* and *Escherichia coli* (Abraham and Chauhan 2017).

A significant class of natural compounds known as glycopeptide antibiotics are primarily generated by a broad range of filamentous Actinomycetes and target the cell walls of prokaryotes. They consist chemically of nonribosomal peptides that have been glycosylated (Yim et al. 2013). By attaching to the acyl-D-alanyl-D-alanine (D-Ala-D-Ala) terminus of the developing peptidoglycan on the cytoplasmic membrane's outer surface, they were utilized to treat infections that were potentially fatal and brought on by *Staphylococcus aureus, Enterococcus species,* and *Clostridium difficile* (Binda et al. 2014). Essential glycopeptides, vancomycin and teicoplanin, were effectively generated by *Streptomyces toyocaensis* NRRL 15009 and show remarkable efficacy against Gram-positive bacterial pathogens, such as those belonging to the genera *Staphylococcus, Streptococcus, and Enterococcus* (Pootoolal et al. 2002).

In many Gram-positive and Gram-negative bacteria, including many multidrug-resistant strains, chloramphenicol (CHL), a bactericidal chemical generated by *Streptomyces venezuelae,* inhibits ribosomal peptidyl transferase activity (Fernández-Martínez et al. 2014). A variety of novel bioactive compounds, including carbamic acid, maltol, 2.4-di-tert-butylphenol, methyl dimendone, prolyl-leucyl, and oleamide, have been found to be present in acidophilic Actinobacteria isolates from the *Streptomyces* and Kitasatospora genera. These isolates have also been shown to be effective against *Helicobacter pylori*, which is known to be the causative agent of severe diseases, including gastric and colorectal cancer (221). Angucyclinone cores joined to aminodeoxysugaro samines by an uncommon C-glycosidic bond constitute the essential antibiotic frigocyclinone.

According to Mo et al. (2019), *Streptomyces griseus* NTK 97 and marine sponge-associated *Streptomyces* sp. M7_15 efficiently generate it, and even at low concentrations, it had a strong antibacterial action against *Bacillus subtilis* and *Staphylococcus aureus*. Frigocyclinone has the ability to kill rhabdomyosarcoma cancer cells with an EC50 of 5.2 μM after 48 h of treatment, indicating that it may be utilized in the development of anticancer medications (Vicente et al. 2015). An abundant supply of new bacteriostatic and anticancer drugs can be found in marine actinomycetes. According to Ko et al. (2014), lajollamycin (1) is a nitro-tetraene spiro-beta-lactone-gamma-lactam antibiotic that is generated by a marine strain of *Streptomyces nodosus* (NPS007994).

The body of research suggests that many Actinobacteria isolates are tightly linked to the manufacture of antifungal drugs. *Kutzneria* sp. strain TSII, which promotes plant development, was isolated from mangrove silt soil and demonstrated strong activity in the production of 2,4-bis(1,1-dimethylethyl) phenol, an antifungal agent as shown in Table 16.4 (Devi et al. 2021).

16.6 Conclusions

Looking at the vast potential of Actinobacteria in addressing modern-day challenges in crop production and protection. It would be rather more fruitful in the utilization of Actinobacteria antagonism against the deadliest pathogens. Actinobacteria have a treasure trove of physiological potential that could be utilized in fighting against different kinds of stresses. Soil nutrient dynamics, nutrient fortification in plants, abiotic stress, biotic stress, etc. could be less devastating with the Actinobacteria application. The biotechnological interventions in utilizing

Table 16.4 Biotechnological applications of actinobacteria

Biotechnological natural product	Application	Producer actinobacteria	References
Vancomycin and teicoplanin	Antibacterial	*Streptomyces toyocaensis* NRRL15009	Balagurunathan et al. (2009)
Chandrananimycin	Antibiotic, anticancer	*Actinomadura* sp. isolate M048	Maskey et al. (2003)
(6*R*,10*S*)-10-methyl-6-dodecanolide	Anticancer	*Streptomyces* sp. B6007	Stritzke et al. (2004)
Hydroxy-1-norresistomycin	Anticancer	*Streptomyces chinensis*	Gorajana et al. (2005)
1,6-Phenazinediol	Anticancer	*Actinomadura* sp. M048	Lombó et al. (2006)
Vitamin B12	Essential in the formation of red blood cells	*Streptomyces Fulvissimus*	Atta (2007)
Tartrolon D	Anticancer	*Streptomyces* sp. MDG-04-17-069	Pérez et al. (2009)
Access rhodomycin	Antimicrobial	*Streptomyces purpurascens*	Holkar et al. (2013)
Serine protease	Leather and fabric processing	*Actinopolyspora* sp. VITSDK2	Suthindhiran et al. (2013)
Chloramphenicol	Antibacterial	*Streptomyces venezuelae* strain ATCC 10712	Fernández-Martínez et al. (2014)
Novobiocin	Antibacterial	*Streptomyces niveus*	Flinspach et al. (2014)
Asparaginase, chitinase	Antinematode	*Streptomyces* sp.	Rashad et al. (2015)
Undecylprodigiosin	Anticancer, antibacterial	*Streptomyces* strain JAR6	Abraham and Chauhan (2017)
Keratinases	Degradation of keratin containing wastes	*Nocardiopsis* sp. 28ROR	González et al. (2020)
Melanin	Anticancer, antioxidant	*Streptomyces glaucescens* NEAE-H	El-Naggar et al. (2017)
Bleomycin	Anticancer	*Streptomyces verticillus*	Kong et al. (2018)
Staurosporine and kigamicin	Anticancer	*Streptomyces bingchenggensis* ULS14	Davies-Bolorunduro et al. (2019)
Frigocyclinone	Antibacterial, anticancer	*Streptomyces griseus* NTK 97	Mo et al. (2019)
Cellulase	Production of fermentable sugars from lignocellulose	*Streptomyces coelicoflavus* NBRC 15399	Waheeb et al. (2021)

(continued)

Table 16.4 (continued)

Biotechnological natural product	Application	Producer actinobacteria	References
Tetradecanoic acid and pentadecanoic acid	Antifungal	*Streptomyces* sp. SCA3-4	Qi et al. (2019)
Lipase	Detergent industry	*Streptomyces* sp. Al-Dhabi-49	Al-Dhabi et al. (2020)
Granaticins	Antibacterial, anticancer	*Streptomyces vietnamensis* GIMV4.0001	Deng et al. (2021)
Phenol, 2,4-bis(1,1-dimethylethyl)	Antifungal	*Kutzneria* sp. strain TSII	Devi et al. (2021)
Methyl dimendone, prolyl-leucyl, and oleamide	Antibacterial	*Streptomyces spinoverrucosus* and *Streptomyces cirratus*	Eftekharivash et al. (2021)
Chitinase	Antifungal	*Streptomyces albus*	Ekundayo et al. (2022)
N-hexadecanoic acid	Anti-inflammatory	*Streptomyces* sp. 1S1	Mothana et al. (2022)

Actinobacteria potential could play a pivotal role in transforming agriculture with sustainability.

References

Aallam Y, Maliki BE, Dhiba D, Lemriss S, Souiri A, Haddioui A, Tarkka M, Hamdali H (2021) Multiple potential plant growth promotion activities of endemic *Streptomyces* spp. from Moroccan sugar beet fields with their inhibitory activities against *Fusarium* spp. Microorganisms 9:1429. https://doi.org/10.3390/microorganisms9071429

Abbasi S, Safaie N, Sadeghi A, Shamsbakhsh M (2019) *Streptomyces* strains induce resistance to *Fusarium oxysporum* f. sp. *lycopersici* race 3 in tomato through different molecular mechanisms. Front Microbiol 10:1505. https://doi.org/10.3389/fmicb.2019.01505

Abdelaal K, AlKahtani M, Attia K, Hafez Y, Király L, Künstler A (2021) The role of plant growth-promoting bacteria in alleviating the adverse effects of drought on plants. Biology 10:520. https://doi.org/10.3390/biology10060520

Abraham J, Chauhan R (2017) Profiling of red pigment produced by *Streptomyces* sp. JAR6 and its bioactivity. 3 Biotech 8:22. https://doi.org/10.1007/s13205-017-1044-7

Adegboye MF (2012) Taxonomy and ecology of antibiotic producing actinomycetes. Afr J Agric Res 7:2255–2261. https://doi.org/10.5897/ajarx11.071

Aftab U, Zechel DL, Sajid I (2015) Antitumor compounds from *Streptomyces* sp. KML-2, isolated from Khewra salt mines, Pakistan. Biol Res 48:58. https://doi.org/10.1186/s40659-015-0046-3

Al-Dhabi NA, Esmail GA, Ghilan AM, Arasu MV (2020) Isolation and screening of *Streptomyces* sp. Al-Dhabi-49 from the environment of Saudi Arabia with concomitant production of lipase and protease in submerged fermentation. Saudi J Biol Sci 27:474–479. https://doi.org/10.1016/j.sjbs.2019.11.011

Al-jelawi RO (2017) Production of keratinases from *Nocardiopsis* sp. 28ROR as a novel Iraqi strain. Asian J Pharm Clin Res 10:160. https://doi.org/10.22159/ajpcr.2017.v10i4.16427

Alvarez-Sastre C, Carro L (2022) Desert actinobacteria strains increase salt stress resilience in crops. Environ Sci Proc 16(1):17. https://doi.org/10.3390/environsciproc2022016017

Aminov RI, Mackie RI (2007) Evolution and ecology of antibiotic resistance genes. FEMS Microbiol Lett 271:147–161. https://doi.org/10.1111/j.1574-6968.2007.00757.x

Ananya R, Sivaperumal P, Roy A, Lakshmi T (2021) Production of protease enzyme and antimicrobial activities from marine actinobacterium of *Streptomyces* sp. J Pharm Res Int 33:302–311. https://doi.org/10.9734/jpri/2021/v33i63a35634

Anwar S, Ali B, Sajid I (2016) Screening of rhizospheric actinomycetes for various in vitro and in vivo plant growth promoting (PGP) traits and for agroactive compounds. Front Microbiol 7:1334. https://doi.org/10.3389/fmicb.2016.01334

Arasu MV, Al-Dhabi NA, Saritha V, Duraipandiyan V, Muthukumar C, Kim S (2013) Antifeedant, larvicidal and growth inhibitory bioactivities of novel polyketide metabolite isolated from *Streptomyces* sp. AP-123 against *Helicoverpa armigera* and *Spodoptera litura*. BMC Microbiol 13:105. https://doi.org/10.1186/1471-2180-13-105

Atta HM (2007) Production of Vitamin B 12 by *Streptomyces Fulvissimus*. Egypt J Basic Appl Sci 23:166–184. https://doi.org/10.4314/ejbs2.v23i1.40302

Azman A, Othman I, Velu SS, Chan K, Lee L (2015) Mangrove rare actinobacteria: taxonomy, natural compound, and discovery of bioactivity. Front Microbiol 6:856. https://doi.org/10.3389/fmicb.2015.00856

Bakhshandeh E, Pirdashti H, Lendeh KS (2017) Phosphate and potassium-solubilizing bacteria effect on the growth of rice. Ecol Eng 103:164–169. https://doi.org/10.1016/j.ecoleng.2017.03.008

Balagurunathan R, Selvameenal L, Radhakrishnan M (2009) Antibiotic pigment from desert soil actinomycetes; biological activity, purification and chemical screening. Indian J Pharm Sci 71:499. https://doi.org/10.4103/0250-474x.58174

Barka EA, Vatsa P, Sanchez L, Gaveau-Vaillant N, Jacquard C, Klenk H, Clément C, Ouhdouch Y, Van Wezel GP (2016) Taxonomy, physiology, and natural products of actinobacteria. Microbiol Mol Biol Rev 80:1–43. https://doi.org/10.1128/mmbr.00019-15

Berry AM, Mendoza-Herrera A, Guo Y, Hayashi J, Persson T, Barabote R, Demchenko K, Zhang S, Pawlowski K (2011) New perspectives on nodule nitrogen assimilation in actinorhizal symbioses. Funct Plant Biol 38:645. https://doi.org/10.1071/fp11095

Binda E, Marinelli F, Marcone G (2014) Old and new glycopeptide antibiotics: action and resistance. Antibiotics 3:572–594. https://doi.org/10.3390/antibiotics3040572

Bo AB, Kim JD, Kim YS, Sin HT, Kim HJ, Khaitov B, Ko YK, Park KW, Choi JS (2019) Isolation, identification and characterization of *Streptomyces* metabolites as a potential bioherbicide. PLoS One 14:e0222933. https://doi.org/10.1371/journal.pone.0222933

Bobek J, Šmídová K, Čihák MA (2017) A waking review: old and novel insights into the spore germination in *Streptomyces*. Front Microbiol 8:2205. https://doi.org/10.3389/fmicb.2017.02205

Boubekri K, Soumare A, Mardad I, Lyamlouli K, Ouhdouch Y, Hafidi M, Kouisni L (2022) Multifunctional role of actinobacteria in agricultural production sustainability: a review. Microbiol Res 261:127059. https://doi.org/10.1016/j.micres.2022.127059

Boukhatem ZF, Merabet C, Tsaki H (2022) Plant growth promoting actinobacteria, the most promising candidates as bioinoculants. Front Agron 4:849911. https://doi.org/10.3389/fagro.2022.849911

Camara MC, Vandenberghe LP, Rodrigues C, De Oliveira J, Faulds C, Bertrand E, Soccol CR (2018) Current advances in gibberellic acid (GA3) production, patented technologies and potential applications. Planta 248:1049–1062. https://doi.org/10.1007/s00425-018-2959-x

Canfield DE, Glazer AN, Falkowski PG (2010) The evolution and future of earth's nitrogen cycle. Science 330:192–196. https://doi.org/10.1126/science.1186120

Chatterjee A, Solankey S (2015) Functional physiology in drought tolerance of vegetable crops: an approach to mitigate climate change impact. In: Choudhary ML, Patel VB, Siddiqui MW, Mahdl SS, Verma RB (eds) Climate dynamics in horticultural science. Apple Academic Press, New York, pp 169–192. https://doi.org/10.1201/b18035-15

Chen Y, Shafi J, Li M, Fu D, Ji M (2018) Insecticidal activity of endophytic actinomycetes isolated from *Azadirachta indica* against *Myzus persicae*. Arch Biol Sci 70:349–357. https://doi.org/10.2298/abs170729052c

Cheng X, Liu X, Wang H, Ji X, Wang K, Wei M, Qiao K (2015) Effect of emamectin benzoate on root-knot nematodes and tomato yield. PLoS One 10:e0141235. https://doi.org/10.1371/journal.pone.0141235

Choudhary DK, Prakash A, Johri BN (2007) Induced systemic resistance (ISR) in plants: mechanism of action. Indian J Microbiol 47:289–297. https://doi.org/10.1007/s12088-007-0054-2

Chukwuneme CF, Babalola OO, Kutu FR, Ojuederie OB (2020) Characterization of actinomycetes isolates for plant growth promoting traits and their effects on drought tolerance in maize. J Plant Interact 15:93–105. https://doi.org/10.1080/17429145.2020.1752833

Cinkocki R, Lipková N, Javoreková S, Petrová J, Maková J, Medo J, Ducsay L (2021) The impact of growth-promoting *Streptomycetes* isolated from rhizosphere and bulk soil on oilseed rape (*Brassica napus* L.) growth parameters. Sustainability 13:5704. https://doi.org/10.3390/su13105704

Das S, Ward LR, Burke C (2008) Prospects of using marine actinobacteria as probiotics in aquaculture. Appl Microbiol Biotechnol 81:419–429. https://doi.org/10.1007/s00253-008-1731-8

Davies-Bolorunduro OF, Adeleye IA, Akinleye MO, Wang PG (2019) Anticancer potential of metabolic compounds from marine actinomycetes isolated from Lagos Lagoon sediment. J Pharm Anal 9:201–208. https://doi.org/10.1016/j.jpha.2019.03.004

De Simeis D, Serra S (2021) Actinomycetes: a never-ending source of bioactive compounds—an overview on antibiotics production. Antibiotics 10:483. https://doi.org/10.3390/antibiotics10050483

Debasis M, Ganeshamurthy AN, Komal S, Radha TK, Rupa TR (2020) Plant growth promotion and biocontrol activity of some typical harsh environment rhizo-microbes and their effects on *Amaranthus cruentus* plants. Res J Biotechnol 15:35–45

Deng M, Li Y, Luo X, Zheng X, Chen Y, Zhang Y, Zhang W, Zhou H, Zhu H (2021) Discovery of mycothiogranaticins from *Streptomyces vietnamensis* GIMV4.0001 and the regulatory effect of mycothiol on the granaticin biosynthesis. Front Chem 9:802279. https://doi.org/10.3389/fchem.2021.802279

Devi TS, Vijay K, Vidhyavathi RM, Kumar P, Govarthanan M, Kavitha T (2021) Antifungal activity and molecular docking of phenol, 2,4-bis(1,1-dimethylethyl) produced by plant growth-promoting mangrove actinobacteria; *Kutzneria* sp. strain TSII. Arch Microbiol 203:4051–4064. https://doi.org/10.21203/rs.3.rs-291369/v1

Devlet A (2021) Modern agriculture and challenges. Front Life Sci Relat Technol 2:21–29. https://doi.org/10.51753/flsrt.856349

Dhakal D, Pokhrel AR, Shrestha B, Sohng JK (2017) Marine rare actinobacteria: isolation, characterization, and strategies for harnessing bioactive compounds. Front Microbiol 8:1106. https://doi.org/10.3389/fmicb.2017.01106

Dhanasekaran D, Thajuddin N, Panneerselvam A (2010) Herbicidal agents from actinomycetes against selected crop plants and weeds. Nat Prod Res 24:521–529. https://doi.org/10.1080/14786410802299281

Dhanasekaran D, Ambika K, Thajuddin N, Panneerselvam A (2012) Allelopathic effect of actinobacteria isolates against selected weeds. Arch Phytopathol Plant Prot 45:505–521. https://doi.org/10.1080/03235408.2011.587988

Eftekharivash L, Hamedi J, Zarrini G, Bakhtiari R (2021) Acidophilic and acid tolerant actinobacteria as new sources of antimicrobial agents against *Helicobacter pylori*. Arch Razi Inst 76:261–272. https://doi.org/10.22092/ari.2019.128039.1401

Ekundayo FO, Folorunsho AE, Ibisanmi TA, Olabanji OB (2022) Antifungal activity of chitinase produced by *Streptomyces* species isolated from grassland soils in FUTA area, Akure. Bull Natl Res Cent 46:95. https://doi.org/10.1186/s42269-022-00782-4

El-khawaga MA, Megahed MM (2012) Antibacterial and insecticidal activity of actinomycetes isolated from sandy soil of (Cairo-Egypt). Egypt Acad J Biol Sci 4:53–67. https://doi.org/10.21608/eajbsg.2012.16661

El-Naggar NE, El-Ewasy SM (2017) Bioproduction, characterization, anticancer and antioxidant activities of extracellular melanin pigment produced by newly isolated microbial cell factories *Streptomyces glaucescens* NEAE-H. Sci Rep 7:42129. https://doi.org/10.1038/srep42129

El-Tarabily KA (2008) Promotion of tomato (*Lycopersicon esculentum* mill.) plant growth by rhizosphere competent 1-aminocyclopropane-1-carboxylic acid deaminase-producing *streptomycete* actinomycetes. Plant Soil 308:161–174. https://doi.org/10.1007/s11104-008-9616-2

El-Tarabily KA, Nassar AH, Sivasithamparam K (2008) Promotion of growth of bean (*Phaseolus vulgaris* L.) in a calcareous soil by a phosphate-solubilizing, rhizosphere-competent isolate of *Micromonospora endolithica*. Appl Soil Ecol 39:161–171. https://doi.org/10.1016/j.apsoil.2007.12.005

Etchells JP, Smit ME, Gaudinier A, Williams CJ, Brady SM (2016) A brief history of the TDIF-PXY signalling module: balancing meristem identity and differentiation during vascular development. New Phytol 209:474–484. https://doi.org/10.1111/nph.13642

Etesami H, Emami S, Alikhani HA (2017) Potassium solubilizing bacteria (KSB) mechanisms, promotion of plant growth, and future prospects—a review. J Plant Nutr Soil Sci 17:897–911. https://doi.org/10.4067/s0718-95162017000400005

Fanelli E, Di Vito M, Jones JT, De Giorgi C (2005) Analysis of chitin synthase function in a plant parasitic nematode, *Meloidogyne artiellia*, using RNAi. Gene 349:87–95. https://doi.org/10.1016/j.gene.2004.11.045

Farhat MB, Boukhris I, Chouayekh H (2015) Mineral phosphate solubilization by *Streptomyces* sp. CTM396 involves the excretion of gluconic acid and is stimulated by humic acids. FEMS Microbiol Lett 362:fnv008. https://doi.org/10.1093/femsle/fnv008

Fernandes CJ, Doddavarapu B, Harry A, Dilip SP, Ravi L (2021) Isolation and identification of pigment producing actinomycete *Saccharomonospora azurea* SJCJABS01. Biomed Pharmacol J 14:2261–2269. https://doi.org/10.13005/bpj/2326

Fernández-Martínez LT, Borsetto C, Gomez-Escribano JP, Bibb MJ, Al-Bassam MM, Chandra G, Bibb MJ (2014) New insights into chloramphenicol biosynthesis in *Streptomyces venezuelae* ATCC 10712. Antimicrob Agents Chemother 58:7441–7450. https://doi.org/10.1128/aac.04272-14

Flinspach K, Rückert C, Kalinowski J, Heide L, Apel AK (2014) Draft genome sequence of *Streptomyces niveus* NCIMB 11891, producer of the aminocoumarin antibiotic novobiocin. Genome Announc 2(1):e01146-13. https://doi.org/10.1128/genomea.01146-13

Forghani F, Hajihassani A (2020) Recent advances in the development of environmentally benign treatments to control root-knot nematodes. Front Plant Sci 11:1125. https://doi.org/10.3389/fpls.2020.01125

Franche C, Lindström K, Elmerich C (2008) Nitrogen-fixing bacteria associated with leguminous and non-leguminous plants. Plant Soil 321:35–59. https://doi.org/10.1007/s11104-008-9833-8

Fukase E, Martin W (2020) Economic growth, convergence, and world food demand and supply. World Dev 132:104954. https://doi.org/10.1016/j.worlddev.2020.104954

Gao B, Gupta RS (2012) Phylogenetic framework and molecular signatures for the main clades of the phylum actinobacteria. Microbiol Mol Biol Rev 76:66–112. https://doi.org/10.1128/mmbr.05011-11

Gebily Doha AS, Ghanem GAM, Ragab Mona M, Ali Ayat M, Soliman NK, Abd El-Moity TH (2021) Characterization and potential antifungal activities of three *Streptomyces* spp. as biocontrol agents against *Sclerotinia sclerotiorum* (Lib.) de Bary infecting green bean. Egypt J Biol Pest Control 31:33. https://doi.org/10.1186/s41938-021-00373-x

Ghanem GAM, GebilyDoha AS, Ragab MM, Ali AM, Soliman NK, Abd El-Moity TH (2022) Efficacy of antifungal substances of three *Streptomyces* spp. against different plant pathogenic fungi. Egypt J Biol Pest Control 32:112. https://doi.org/10.1186/s41938-022-00612-9

Gómez ÁG, Ramos FA, Sinuco DC (2021) Screening of volatile organic compounds from actinobacteria for the control of phytopathogen *Colletotrichum gloeosporioides*. Biocontrol Sci Technol 31:1067–1079. https://doi.org/10.1080/09583157.2021.1918635

Gong Y, Chen L, Pan S, Li X, Xu M, Zhang C, Xing K, Qin S (2020) Antifungal potential evaluation and alleviation of salt stress in tomato seedlings by a halotolerant plant growth-promoting actinomycete *Streptomyces* sp. KLBMP5084. Rhizosphere 16:100262. https://doi.org/10.1016/j.rhisph.2020.100262

González V, Vargas-Straube MJ, Beys-da-Silva WO, Santi L, Valencia P, Beltrametti F, Cámara B (2020) Enzyme bioprospection of marine-derived actinobacteria from the Chilean coast and new insight in the mechanism of keratin degradation in *Streptomyces* sp. G11C. Mar Drugs 18:537. https://doi.org/10.3390/md18110537

Gopalakrishnan S (2011) Biocontrol of charcoal-rot of sorghum by actinomycetes isolated from herbal vermicompost. Afr J Biotechnol 10:18142–18152. https://doi.org/10.5897/ajb11.2710

Gorajana A, Kurada BV, Peela S, Jangam P, Vinjamuri S, Poluri E, Zeeck A (2005) 1-Hydroxy-1-norresistomycin, a new cytotoxic compound from a marine actinomycete, *Streptomyces chibaensis* AUBN1/7. J Antibiot 58:526–529. https://doi.org/10.1038/ja.2005.72

Gotor-Vila A, Teixidó N, Di Francesco A, Usall J, Ugolini L, Torres R, Mari M (2017) Antifungal effect of volatile organic compounds produced by *Bacillus amyloliquefaciens* CPA-8 against fruit pathogen decays of cherry. Food Microbiol 64:219–225. https://doi.org/10.1016/j.fm.2017.01.006

Goudjal Y, Zamoum M, Meklat A, Sabaou N, Mathieu F, Zitouni A (2015) Plant-growth-promoting potential of endosymbiotic actinobacteria isolated from sand truffles (*Terfezia leonis* Tul.) of the Algerian Sahara. Ann Microbiol 66:91–100. https://doi.org/10.1007/s13213-015-1085-2

Grover M, Bodhankar S, Maheswari M, Srinivasarao C (2016) Actinomycetes as mitigators of climate change and abiotic stress. In: Subramaniam G, Arumugam S, Rajendran V (eds) Plant growth promoting actinobacteria. Springer, Singapore, pp 203–212. https://doi.org/10.1007/978-981-10-0707-1_13

Gupta S, Pandey S (2019) ACC deaminase producing bacteria with multifarious plant growth promoting traits alleviates salinity stress in French bean (*Phaseolus vulgaris*) plants. Front Microbiol 10:1506. https://doi.org/10.3389/fmicb.2019.01506

Haas H (2003) Molecular genetics of fungal siderophore biosynthesis and uptake: the role of siderophores in iron uptake and storage. Appl Microbiol Biotechnol 62:316–330. https://doi.org/10.1007/s00253-003-1335-2

Hamdali H, Bouizgarne B, Hafidi M, Lebrihi A, Virolle MJ, Ouhdouch Y (2008) Screening for rock phosphate solubilizing actinomycetes from Moroccan phosphate mines. Appl Soil Ecol 38:12–19. https://doi.org/10.1016/j.apsoil.2007.08.007

Hanaka A, Ozimek E, Reszczyńska E, Jaroszuk-Ściseł J, Stolarz M (2021) Plant tolerance to drought stress in the presence of supporting bacteria and fungi: an efficient strategy in horticulture. Horticulturae 7:390. https://doi.org/10.3390/horticulturae7100390

Harikrishnan H, Shanmugaiah V, Balasubramanian N (2021) Screening and optimization of indole-3-acetic acid production by bacterial strain isolated from rice rhizosphere and its effects on plant growth. Environ Exp Bot 19:141–149. https://doi.org/10.22364/eeb.19.14

Hasegawa S, Meguro A, Nishimura T, Kunoh H (2004) Drought tolerance of tissue-cultured seedlings of mountain Laurel (*Kalmia latifolia* L.) induced by an endophytic actinomycete: enhancement of osmotic pressure in leaf cells. Actinomycetologica 18:43–47. https://doi.org/10.3209/saj.18_43

Hata EM, Yusof MT, Zulperi D (2021) Induction of systemic resistance against bacterial leaf streak disease and growth promotion in rice plant by *Streptomyces shenzhenesis* TKSC3 and *Streptomyces* sp. SS8. Plant Pathol J 37:173–181. https://doi.org/10.5423/ppj.oa.05.2020.0083.2021

Hayat S, Ashraf A, Aslam B, Asif R, Muzammil S, Asif Zahoor M, Waseem M, Riaz Malik I, Khurshid M, Afzal M, Saqalein M, Hussnain Siddique M, Muzammil A, Sabir S (2021) Actinobacteria: potential candidate as plant growth promoters. In: Plant stress physiology. IntechOpen, London. https://doi.org/10.5772/intechopen.93272

Holkar S, Begde D, Nashikkar N, Kadam T, Upadhyay A (2013) Rhodomycin analogues from *Streptomyces purpurascens*: isolation, characterization and biological activities. Springerplus 2:93. https://doi.org/10.1186/2193-1801-2-93

Holkar CR, Jain SS, Jadhav AJ, Pinjari DV (2018) Valorization of keratin based waste. Process Saf Environ Prot 115:85–98. https://doi.org/10.1016/j.psep.2017.08.045

Hönig M, Plíhalová L, Husičková A, Nisler J, Doležal K (2018) Role of cytokinins in senescence, antioxidant defence and photosynthesis. Int J Mol Sci 19:4045. https://doi.org/10.3390/ijms19124045

Hu Q, Yang M, Bo T, Li Y, Wu C, Mo M, Liu Y (2022) Soluble macromolecules from two *Streptomyces* strains with potent nematicidal activity against *Meloidogyne incognita*. Rhizosphere 22:100529. https://doi.org/10.21203/rs.3.rs-199348/v1

Jaemsaeng R, Jantasuriyarat C, Thamchaipenet A (2018) Molecular interaction of 1-aminocyclopropane-1-carboxylate deaminase (ACCD)-producing endophytic *Streptomyces* sp. GMKU 336 towards salt-stress resistance of Oryza sativa L. CV. KDML105. Sci Rep 8:1950. https://doi.org/10.1038/s41598-018-19799-9

Jagannathan SV, Manemann EM, Rowe SE, Callender MC, Soto W (2021) Marine actinomycetes, new sources of biotechnological products. Mar Drugs 19:365. https://doi.org/10.3390/md19070365

Jha S, Modi HA (2018) Statistical optimization of chitinase production by *Streptomyces rubiginosus* SP24 and efficacy of purified chitinase to control root-knot nematode infection in *Vignaradiata* under controlled conditions. Chem Biol Technol Agric 5:20. https://doi.org/10.1186/s40538-018-0133-0

Jirayut E, Varissara C, Wacharaporn N, Sutassa T, Archiraya P, Suppakit T, Pongsan C, Bungonsiri I (2020) Isolation and characterization of aerobic actinomycetes with probiotic properties in Nile tilapia. J Appl Pharm Sci 10:40–49. https://doi.org/10.7324/japs.2020.10905

Jose PA, Jebakumar SR (2013) Non-streptomycete actinomycetes nourish the current microbial antibiotic drug discovery. Front Microbiol 4:240. https://doi.org/10.3389/fmicb.2013.00240

Kalyani BS, Krishna P, Sreenivasulu K (2019) Screening and identification of novel isolate *Streptomyces* sp., NLKPB45 from Nellore costal region for its biomedical applications. Saudi J Biol Sci 26:1655–1660. https://doi.org/10.1016/j.sjbs.2018.08.027

Kannahi M, Senbagam N (2014) Studies on siderophore production by microbial isolates obtained from rhizosphere soil and its antibacterial activity. J Chem Pharm Res 6:1142–1145

Karlidag H, Esitken A, Turan M, Sahin F (2007) Effects of root inoculation of plant growth promoting rhizobacteria (PGPR) on yield, growth and nutrient element contents of leaves of. Apple Sci Hortic 114:16–20. https://doi.org/10.1016/j.scienta.2007.04.013

Kaur T, Manhas RK (2022) Evaluation of ACC deaminase and indole acetic acid production by *Streptomyces hydrogenans* DH16 and its effect on plant growth promotion. Biocatal Agric Biotechnol 42:102321. https://doi.org/10.1016/j.bcab.2022.102321

Kaur T, Vasudev A, Sohal SK, Manhas RK (2014) Insecticidal and growth inhibitory potential of *Streptomyces hydrogenans* DH16 on major pest of India *Spodoptera litura* (Fab.) (*Lepidoptera: Noctuidae*). BMC Microbiol 14:227. https://doi.org/10.1186/s12866-014-0227-1

Khan F, Ahmed KBM, Shariq M, Siddiqui MA (2019) Potentiality of plant growth-promoting rhizobacteria in easing of soil salinity and environmental sustainability. In: Akhtar M (ed) Salt stress, microbes, and plant interactions: causes and solution. Springer, Singapore, pp 21–58. https://doi.org/10.1007/978-981-13-8801-9-2

Kim JH, Choi JY, Park DH, Park D, Park MG, Kim SY, Ju YJ, Kim JY, Wang M, Kim C, Je YH (2020) Isolation and characterization of the insect growth regulatory substances from actinomycetes. Comp Biochem Physiol C Toxicol Pharmacol 228:108651. https://doi.org/10.1016/j.cbpc.2019.108651

Ko K, Lee S, Kim S, Kim E, Oh K, Shin J, Oh D (2014) Lajollamycins, nitro group-bearing spiro-β-lactone-γ-lactams obtained from a marine-derived *Streptomyces* sp. J Nat Prod 77:2099–2104. https://doi.org/10.1021/np500500t

Kong J, Yi L, Xiong Y, Huang Y, Yang D, Yan X, Shen B, Duan Y, Zhu X (2018) The discovery and development of microbial bleomycin analogues. Appl Microbiol Biotechnol 102:6791–6798. https://doi.org/10.1007/s00253-018-9129-8

Kothari VV, Kothari RK, Kothari CR, Bhatt VD, Nathani NM, Koringa PG, Joshi CG, Vyas BR (2013) Genome sequence of salt-tolerant *Bacillus safensis* strain VK, isolated from saline desert area of Gujarat, India. Genome Announc 1:e00671-13. https://doi.org/10.1128/genomea.00671-13

Kpomblekou K, Tabatabai M (1994) Effect of organic acids on release of phosphorus from phosphate rocks. Soil Sci 158:442–453. https://doi.org/10.1097/00010694-199415860-00006

Kumar A, Singh S, Gaurav AK, Srivastava S, Verma JP (2020) Plant growth-promoting bacteria: biological tools for the mitigation of salinity stress in plants. Front Microbiol 11:1216. https://doi.org/10.3389/fmicb.2020.01216

Law JW, Pusparajah P, Ab Mutalib N, Wong SH, Goh B, Lee L (2019) A review on mangrove actinobacteria diversity: the roles of *Streptomyces* and novel species discovery. Prog Microb Mol Biol 2:a0000024. https://doi.org/10.36877/pmmb.a0000024

Lazzarini A, Cavaletti L, Toppo G, Marinelli F (2000) Rare genera of actinomycetes as potential producers of new antibiotics. Antonie van Leeuwenhoek. 78:399–405. https://doi.org/10.1023/a:1010287600557

Le KD, Yu NH, Park AR, Park D, Kim C, Kim J (2022) *Streptomyces* sp. AN090126 as a biocontrol agent against bacterial and fungal plant diseases. Microorganisms 10:791. https://doi.org/10.3390/microorganisms10040791

Lee J, Postmaster A, Soon HP, Keast D, Carson KC (2012) Siderophore production by actinomycetes isolates from two soil sites in Western Australia. BioMetals 25:285–296. https://doi.org/10.1007/s10534-011-9503-9

Li Q, Chen X, Jiang Y, Jiang C (2016) Morphological identification of actinobacteria. In: Dhanasekaran D, Jiang Y (eds) Actinobacteria—basics and biotechnological applications. IntechOpen, London. https://doi.org/10.5772/61461

Liu M, Hwang B, Jin C, Li W, Park D, Seo S, Kim C (2019) Screening, isolation and evaluation of a nematicidal compound from actinomycetes against the pine wood nematode, *Bursaphelenchus xylophilus*. Pest Manag Sci 75:1585–1593. https://doi.org/10.1002/ps.5272

Loliam B, Morinaga T, Chaiyanan S (2013) Biocontrol of *Pythium aphanidermatum* by the cellulolytic actinomycetes *Streptomyces rubrolavendulae* S4. Sci Asia 39:584. https://doi.org/10.2306/scienceasia1513-1874.2013.39.584

Lombó F, Velasco A, Castro A, De la Calle F, Braña AF, Sánchez-Puelles JM, Méndez C, Salas JA (2006) Deciphering the biosynthesis pathway of the antitumor thiocoraline from a marine actinomycete and its expression in two *Streptomyces* species. Chembiochem 7:366–376. https://doi.org/10.1002/cbic.200500325

Ludwig W, Euzéby J, Schumann P, Buss HJ, Trujillo ME, Kämpfer P, Whiteman WB (2012) Road map of the phylum actinobacteria. In: Goodfellow M, Kämpfer P, Busse HJ, Trujillo ME, Suzuki KI, Ludwig W, Whitman WB (eds) Bergey's manual of systematic bacteriology. Springer-Verlag, New York, pp 1–28

Lyu A, Yang L, Wu M, Zhang J, Li G (2020) High efficacy of the volatile organic compounds of *Streptomyces yanglinensis* 3-10 in suppression of *Aspergillus* contamination on peanut kernels. Front Microbiol 11:142. https://doi.org/10.3389/fmicb.2020.00142

Mam E, Sandy EH, Sherin FA (2019) Nematicidal effect of abamectin, boron, chitosan, hydrogen peroxide and Bacillus thuringiensis against citrus nematode on Valencia orange trees. J Plant Sci Phytopathol 3:111–117. https://doi.org/10.29328/journal.jpsp.1001041

Manivasagan P, Venkatesan J, Sivakumar K, Kim S (2013) Actinobacteria melanins: current status and perspective for the future. World J Microbiol Biotechnol 29:1737–1750. https://doi.org/10.1007/s11274-013-1352-y

Marappa N, Ramachandran L, Dharumadurai D, Nooruddin T (2020) Plant growth-promoting active metabolites from *Frankia* spp. of actinorhizal casuarina spp. Appl Biochem Biotechnol 191:74–91. https://doi.org/10.1007/s12010-020-03243-8

Martínez-Hidalgo P, Galindo-Villardón P, Trujillo ME, Igual JM, Martínez-Molina E (2014) *Micromonospora* from nitrogen fixing nodules of alfalfa (*Medicago sativa* L.). A new promising plant probiotic bacteria. Sci Rep 4:6389. https://doi.org/10.1038/srep06389

Maskey RP, Li FC, Qin S, Fiebig HH, Laatsch H (2003) Chandrananimycins A-C: production of novel anticancer antibiotics from a marine *Actinomadura* sp. isolate M048 by variation of medium composition and growth conditions. J Antibiot 56:622–629. https://doi.org/10.7164/antibiotics.56.622

Mast Y, Stegmann E (2019) Actinomycetes: the antibiotics producers. Antibiotics 8:105. https://doi.org/10.3390/antibiotics8030105

Meena RK, Mishra P (2020) Bio-pesticides for agriculture and environment sustainability. In: Kumar S, Meena RS, Jhariya MK (eds) Resources use efficiency in agriculture. Springer, Singapore, pp 85–107. https://doi.org/10.1007/978-981-15-6953-13

Meena KK, Bitla UM, Sorty AM, Singh DP, Gupta VK, Wakchaure GC, Kumar S (2020) Mitigation of salinity stress in wheat seedlings due to the application of phytohormone-rich culture filtrate extract of methylotrophic *Actinobacterium Nocardioides* sp. NIMMe6. Front Microbiol 11:2091. https://doi.org/10.3389/fmicb.2020.02091

Meidani C, Savvidis A, Lampropoulou E, Sagia A, Katsifas E, Monokrousos N, Hatzinikolaou DG, Karagouni AD, Giannoutsou E, Adamakis IS, Ntalli NG (2020) The nematicidal potential of bioactive *Streptomyces* strains isolated from Greek rhizosphere soils tested on arabidopsis plants of varying susceptibility to *Meloidogyne* spp. Plants 9:699. https://doi.org/10.3390/plants9060699

Mekuria W (2018) The link between agricultural production and population dynamics in Ethiopia: a review. Adv Plants Agric Res 8:348–353. https://doi.org/10.15406/apar.2018.08.00336

Minotto E, Milagre LP, Oliveira MT, Van Der Sand ST (2014) Enzyme characterization of endophytic actinobacteria isolated from tomato plants. J Adv Sci Res 5:16–23

Misra S, Dixit VK, Khan MH, Kumar Mishra S, Dviwedi G, Yadav S, Lehri A, Singh Chauhan P (2017) Exploitation of agro-climatic environment for selection of 1-aminocyclopropane-1-carboxylic acid (ACC) deaminase producing salt tolerant indigenous plant growth promoting rhizobacteria. Microbiol Res 205:25–34. https://doi.org/10.1016/j.micres.2017.08.007

Mitter EK, Tosi M, Obregón D, Dunfield KE, Germida JJ (2021) Rethinking crop nutrition in times of modern microbiology: innovative biofertilizer technologies. Front Sustain Food Syst 5:29. https://doi.org/10.3389/fsufs.2021.606815

Mo J, Ye J, Chen H, Hou B, Wu H, Zhang H (2019) Cloning and identification of the frigocyclinone biosynthetic gene cluster from *Streptomyces griseus* strain NTK 97. Biosci Biotechnol Biochem 83:2082–2089. https://doi.org/10.1080/09168451.2019.1638755

Mohan KD, Rajamanickam U (2018) Biodiversity of actinomycetes and secondary metabolites. Inn Orig Inter J Sci 5:21–27

Mothana AA, Al-Shamahy HA, Mothana RA, Khaled JM, Al-Rehaily AJ, Al-Mahdi AY, Lindequist U (2022) *Streptomyces* sp. 1S1 isolated from southern coast of the Red Sea as a renewable natural resource of several bioactive compounds. Saudi Pharm J 30:162–171. https://doi.org/10.1016/j.jsps.2021.12.012

Myo EM, Ge B, Ma J, Cui H, Liu B, Shi L, Jiang M, Zhang K (2019) Indole-3-acetic acid production by *Streptomyces fradiae* NKZ-259 and its formulation to enhance plant growth. BMC Microbiol 19:155. https://doi.org/10.1186/s12866-019-1528-1

Naggar MM, El Assar SA, Ali Shata AM (2017) Production of antitumor agents from novel marine actinomycetes strains isolated from Alexandria, Egypt. J Antimicrob Agents 3:145. https://doi.org/10.4172/2472-1212.1000145

Nasr-Eldin M, Messiha N, Othman B, Megahed A, Elhalag K (2019) Induction of potato systemic resistance against the potato virus Y (PVYNTN), using crude filtrates of *Streptomyces* spp. under greenhouse conditions. Egypt J Biol Pest Control 29:62. https://doi.org/10.1186/s41938-019-0165-1

Nath M, Bhatt D, Bhatt MD, Prasad R, Tuteja N (2018) Microbe-mediated enhancement of nitrogen and phosphorus content for crop improvement. In: Prasad R, Sarvajeet S, Tuteja N (eds) Crop improvement through microbial biotechnology. Elsevier, Amsterdam, pp 293–304. https://doi.org/10.1016/b978-0-444-63987-5.00014-1

Naylor D, Coleman-Derr D (2018) Drought stress and root-associated bacterial communities. Front Plant Sci 8:2223. https://doi.org/10.3389/fpls.2017.02223

Netondo GW, Onyango JC, Beck E (2004) Sorghum and salinity. Crop Sci 44:806–811. https://doi.org/10.2135/cropsci2004.8060

Newton WE (2007) Physiology, biochemistry, and molecular biology of nitrogen fixation. In: Biology of the nitrogen cycle. Elsevier, Amsterdam, pp 109–129. https://doi.org/10.1016/b978-044452857-5.50009-6

Nimnoi P, Pongsilp N, Ruanpanun P (2017) Monitoring the efficiency of *Streptomyces galilaeus* strain KPS-C004 against root knot disease and the promotion of plant growth in the plant-parasitic nematode infested soils. Biol Control 114:158–166. https://doi.org/10.1016/j.biocontrol.2017.08.016

Niu S, Gao Y, Zi H, Liu Y, Liu X, Xiong X, Yao Q, Qin Z, Chen N, Guo L, Yang Y, Qin P, Lin J, Zhu Y (2022) The osmolyte-producing endophyte *Streptomyces albidoflavus* osilf-2 induces drought and salt tolerance in rice via a multi-level mechanism. Crop J 10:375–386. https://doi.org/10.1016/j.cj.2021.06.008

Norouzi H, Danesh A, Mohseni M, Rabbani Khorasgani M (2018) Marine actinomycetes with probiotic potential and bioactivity against multidrug-resistant bacteria. Int J Mol Cell 7:44–52. https://doi.org/10.22088/IJMCM.BUMS.7.1.44

Olano C, Méndez C, Salas J (2009) Antitumor compounds from marine actinomycetes. Mar Drugs 7:210–248. https://doi.org/10.3390/md7020210

Olanrewaju OS, Babalola OO (2018) *Streptomyces*: implications and interactions in plant growth promotion. Appl Microbiol Biotechnol 103:1179–1188. https://doi.org/10.1007/s00253-018-09577-y

Orozco-Mosqueda MD, Glick BR, Santoyo G (2020) ACC deaminase in plant growth-promoting bacteria (PGPB): an efficient mechanism to counter salt stress in crops. Microbiol Res 235:126439. https://doi.org/10.1016/j.micres.2020.126439

Oves-Costales D, Kadi N, Challis GL (2009) The long-overlooked enzymology of a non-ribosomal peptide synthetase in dependent pathway for virulence-conferring siderophore biosynthesis. Chem Commun 21:6530–6541. https://doi.org/10.1039/b913092f

Pacios-Michelena S, Aguilar González CN, Alvarez-Perez OB, Rodriguez-Herrera R, Chávez-González M, Arredondo Valdés R, Ascacio Valdés JA, Govea Salas M, Ilyina A (2021) Application of *Streptomyces* antimicrobial compounds for the control of phytopathogens. Front Sustain Food Syst 5:696518. https://doi.org/10.3389/fsufs.2021.696518

Patel K, Thakker N, J. (2020) Deliberating plant growth promoting and mineral-weathering proficiency of streptomyces nanhaiensis strain ym4 for nutritional benefit of millet crop (Pennisetum glaucum). Journal of microbiology, biotechnology and food sciences 9(4):721–726. https://doi.org/10.15414/jmbfs.2020.9.4.721-726

Palaniyandi S, Damodharan K, Yang S, Suh J (2014) *Streptomyces* sp. strain PGPA39 alleviates salt stress and promotes growth of 'Micro Tom' tomato plants. J Appl Microbiol 117:766–773. https://doi.org/10.1111/jam.12563

Pérez M, Crespo C, Schleissner C, Rodríguez P, Zúñiga P, Reyes F, Tartrolon D (2009) A cytotoxic macrodiolide from the marine-derived actinomycete *Streptomyces* sp. MDG-04-17-069. J Nat Prod 72:2192–2194. https://doi.org/10.1021/np9006603

Perrine-Walker F, Gherbi H, Imanishi L, Hocher V, Ghodhbane-Gtari F, Lavenus J, Benabdoun FM, Nambiar-Veetil M, Svistoonoff S, Laplaze L (2011) Symbiotic signaling in actinorhizal symbioses. Curr Protein Pept Sci 12:156–164. https://doi.org/10.2174/138920311795684896

Pootoolal J, Thomas MG, Marshall CG, Neu JM, Hubbard BK, Walsh CT, Wright GD (2002) Assembling the glycopeptide antibiotic scaffold: the biosynthesis of from *Streptomyces toyocaensis* NRRL15009. Proc Natl Acad Sci 99:8962–8967. https://doi.org/10.1073/pnas.102285099

Prasad M, Srinivasan R, Chaudhary M, Jat LK (2019) In: Singh AK, Kumar A, Singh PK (eds) Plant growth promoting rhizobacteria (PGPR) for sustainable agriculture: perspectives and challenges. Woodhead Publishing, Sawston, pp 129–157. https://doi.org/10.1016/B978-0-12-815879-1.00007-0

Pusztahelyi T (2018) Chitin and chitin-related compounds in plant–fungal interactions. Mycology 9:189–201. https://doi.org/10.1080/21501203.2018.1473299

Qi D, Zou L, Zhou D, Chen Y, Gao Z, Feng R, Zhang M, Li K, Xie J, Wang W (2019) Taxonomy and broad-spectrum antifungal activity of *Streptomyces* sp. SCA3-4 isolated from rhizosphere soil of *opuntia stricta*. Front Microbiol 10:1390. https://doi.org/10.3389/fmicb.2019.01390

Raja A, Prabakaran P (2011) Actinomycetes and drug—an overview. Am J Drug Discov Dev 1:75–84. https://doi.org/10.3923/ajdd.2011.75.84

Rangseekaew P, Barros-Rodríguez A, Pathom-aree W, Manzanera M (2021) Deep-sea actinobacteria mitigate salinity stress in tomato seedlings and their biosafety testing. Plants 10:1687. https://doi.org/10.3390/plants10081687

Rashad FM, Fathy HM, El-Zayat AS, Elghonaimy AM (2015) Isolation and characterization of multifunctional *Streptomyces* species with antimicrobial, nematicidal and phytohormone activities from marine environments in Egypt. Microbiol Res 175:34–47. https://doi.org/10.1016/j.micres.2015.03.002

Rashad YM, Al-Askar AA, Ghoneem KM, Saber WI, Hafez EE (2017) Chitinolytic *Streptomyces griseorubens* E44G enhances the biocontrol efficacy against *Fusarium* wilt disease of tomato. *Phytoparasitica* 45:227–237. https://doi.org/10.1007/s12600-017-0580-3

Rathore AS, Gupta RD (2015) Chitinases from bacteria to human: properties, applications, and future perspectives. Enzyme Res 2015:791907. https://doi.org/10.1155/2015/791907

Rifaat HM (2003) The biodiversity of actinomycetes in the River Nile exhibiting antifungal activity. J Mediterran Ecol 4:3–4

Riquelme C, Marshall Hathaway JJ, Enes Dapkevicius MD, Miller AZ, Kooser A, Northup DE, Jurado V, Fernandez O, Saiz-Jimenez C, Cheeptham N (2015) Actinobacteria diversity in volcanic caves and associated geomicrobiological interactions. Front Microbiol 6:1342. https://doi.org/10.3389/fmicb.2015.01342

Ruanpanun P, Chamswarng C (2015) Potential of actinomycetes isolated from earthworm castings in controlling root-knot nematode *Meloidogyne incognita*. J Gen Plant Pathol 1:43–50. https://doi.org/10.1007/s10327-015-0637-4

Ruanpanun P, Laatsch H, Tangchitsomkid N, Lumyong S (2010) Nematicidal activity of fervenulin isolated from a nematicidal actinomycete, *Streptomyces* sp. CMU-MH021, on *Meloidogyne incognita*. World J Microbiol Biotechnol 27:1373–1380. https://doi.org/10.1007/s11274-010-0588-z

Rubio LM, Ludden PW (2008) Biosynthesis of the iron-molybdenum cofactor of nitrogenase. Annu Rev Microbiol 62:93–111. https://doi.org/10.1146/annurev.micro.62.081307.162737

Saif S, Khan MS, Zaidi A, Ahmad E (2014) Role of phosphate-solubilizing actinomycetes in plant growth promotion: current perspective. In: Khan M, Zaidi A, Musarrat J (eds) Phosphate solubilizing microorganisms. Springer, Cham. https://doi.org/10.1007/978-3-319-08216-5-6

Santi C, Bogusz D, Franche C (2013) Biological nitrogen fixation in non-legume plants. Ann Bot 111:743–767. https://doi.org/10.1093/aob/mct048

Sasanelli N, Toderas I, Veronico P, Iurcu-Straistaru E, Rusu S, Melillo MT, Caboni P (2019) Abamectin efficacy on the potato cyst nematode *Globodera pallida*. Plants 9:12. https://doi.org/10.3390/plants9010012

Sathya A, Vijayabharathi R, Gopalakrishnan S (2017) Plant growth-promoting actinobacteria: a new strategy for enhancing sustainable production and protection of grain legumes. 3 Biotech 7:102. https://doi.org/10.1007/s13205-017-0736-3

Schalk IJ, Hannauer M, Braud A (2011) New roles for bacterial siderophores in metal transport and tolerance. Environ Microbiol 13:2844–2854. https://doi.org/10.1111/j.1462-2920.2011.02556.x

Selim S, Hassan YM, Saleh AM, Habeeb TH, AbdElgawad H (2019) Actinobacterium isolated from a semi-arid environment improves the drought tolerance in maize (*Zea mays* L.). Plant Physiol Biochem 142:15–21. https://doi.org/10.1016/j.plaphy.2019.06.029

Selim MSM, Abdelhamid SA, Mohamed SS (2021) Secondary metabolites and biodiversity of actinomycetes. J Genet Eng Biotechnol 19:72. https://doi.org/10.1186/s43141-021-00156-9

Sellstedt A, Richau KH (2013) Aspects of nitrogen-fixing actinobacteria, in particular free-living and symbiotic *Frankia*. FEMS Microbiol Lett 342:179–186. https://doi.org/10.1111/1574-6968.12116

Shah F, Dwivedi M (2022) Vitamin B12 producing actinobacteria as probiotics for poultry production. In: Dharumadurai D (ed) Methods in actinobacteriology, Springer protocols handbooks. Humana, New York, pp 571–588. https://doi.org/10.1007/978-1-0716-1728-1-85

Sharma S, Kaur M (2017) Plant hormones synthesized by microorganisms and their role in biofertilizer—a review article. Int J Adv Res 5:1753–1766. https://doi.org/10.21474/ijar01/6144

Shi L, Wu Z, Zhang Y, Zhang Z, Fang W, Wang Y, Wan Z, Wang K, Ke S (2019) Herbicidal secondary metabolites from actinomycetes: structure diversity, modes of action, and their roles in the development of herbicides. J Agric Food Chem 68:17–32. https://doi.org/10.1021/acs.jafc.9b06126

Shutsrirung A, Chromkaew Y, Pathom-Aree W, Choonluchanon S, Boonkerd N (2013) Diversity of endophytic actinomycetes in Mandarin grown in northern Thailand, their phytohormone production potential and plant growth promoting activity. Soil Sci Plant Nutr 59:322–330. https://doi.org/10.1080/00380768.2013.776935

Siddikee MA, Chauhan PS, Anandham R, Han G, Sa T (2010) Isolation, characterization, and use for plant growth promotion under salt stress, of ACC deaminase-producing halotolerant bacteria derived from coastal soil. J Microbiol Biotechnol 20:1577–1584. https://doi.org/10.4014/jmb.1007.07011

Solans M, Vobis G, Cassán F, Luna V, Wall LG (2011) Production of phytohormones by root-associated saprophytic actinomycetes isolated from the actinorhizal plant Ochetophila trinervis. World J Microbiol Biotechnol 27:2195–2202. https://doi.org/10.1007/s11274-011-0685-7

Solans M, Messuti MI, Reiner G, Boenel M, Vobis G, Wall LG (2019) Exploring the response of actinobacteria to the presence of phosphorus salts sources: metabolic and co-metabolic processes. J Basic Microbiol 59:487–495. https://doi.org/10.1002/jobm.201800508

Srivastava DA, Harris R, Breuer G, Levy M (2021) Secretion-based modes of action of biocontrol agents with a focus on *Pseudozyma aphidis*. Plants 10:210. https://doi.org/10.3390/plants10020210

Stritzke K, Schulz S, Laatsch H, Helmke E, Beil W (2004) Novel caprolactones from a marine *Streptomycete*. J Nat Prod 67:395–401. https://doi.org/10.1021/np030321z

Suksaard P, Pathom-aree W, Duangmal K (2017) Diversity and plant growth promoting activities of actinomycetes from mangroves. Chiang Mai J Sci 44:1210–1223

Suriyachadkun C, Chunhachart O, Srithaworn M, Tangchitcharoenkhul R, Tangjitjareonkun J (2022) Zinc-solubilizing Streptomyces spp. as Bioinoculants for promoting the growth of soybean (Glycine max (L.) Merrill). Journal of Microbiology and Biotechnology 32(11):1435–1446. https://doi.org/10.4014/jmb.2206.06058

Suthindhiran K, Jayasri MA, Dipali D, Prasar A (2013) Screening and characterization of protease producing actinomycetes from marine saltern. J Basic Microbiol 54:1098–1109. https://doi.org/10.1002/jobm.201300563

Swarnalakshmi K, Senthilkumar M, Ramakrishnan B (2016) Endophytic actinobacteria: nitrogen fixation, phytohormone production, and antibiosis. In: Subramaniam G, Arumugam S, Rajendran V (eds) Plant growth promoting actinobacteria. Springer, Singapore, pp 123–145. https://doi.org/10.1007/978-981-10-0707-1-8

Taha M, Ghaly M, Atwa H, Askoura M (2021) Evaluation of the effectiveness of soil *Streptomyces* isolates for induction of plant resistance against tomato mosaic virus (ToMV). Curr Microbiol 78:3032–3043. https://doi.org/10.1007/s00284-021-02567-w

Tan LT, Chan K, Lee L, Goh B (2016) *Streptomyces* bacteria as potential probiotics in aquaculture. Front Microbiol 7:79. https://doi.org/10.3389/fmicb.2016.00079

Tao G, Tian S, Cai M, Xie G (2008) Phosphate-solubilizing and -mineralizing abilities of bacteria isolated from soils. Pedosphere 18:515–523. https://doi.org/10.1016/s1002-0160(08)60042-9

Teixeira LP, Cristina BA, Nogueira CF (2022) Plant growth promotion and biocontrol by endophytic and rhizospheric microorganisms from the tropics: a review and perspectives. Front Sustain Food Syst 6:796113. https://doi.org/10.3389/fsufs.2022.796113

Teshome DT, Zharare GE, Naidoo S (2020) The threat of the combined effect of biotic and abiotic stress factors in forestry under a changing climate. Front Plant Sci 11:601009. https://doi.org/10.3389/fpls.2020.601009

Thakur B, Kaur S, Tripathi M, Upadhyay SK (2023a) Exploring the potential of lactic acid bacteria and its molecular mechanism of action in the development of biosurfactants: current finding and future outlook. In: Biotechnology and genetic engineering reviews. Taylor & Francis, Milton Park, pp 1–32

Thakur B, Kaur S, Rani N, Kaur R, Upadhyay SK, Tripathi M (2023b) Exploring microbial contributions to nutraceutical production: from natural to designed foods. In: Molecular biotechnology. Springer, Berlin, pp 1–16

Thompson BJ, Widdick DA, Hicks MG, Chandra G, Sutcliffe IC, Palmer T, Hutchings MI (2010) Investigating lipoprotein biogenesis and function in the model Gram-positive bacterium *Streptomyces coelicolor*. Mol Microbiol 77(4):943–957. https://doi.org/10.1111/j.1365-2958.2010.07261.x

Tiwari K, Gupta RK (2012) Diversity and isolation of rare actinomycetes: an overview. Crit Rev Microbiol 39:256–294. https://doi.org/10.3109/1040841x.2012.709819

Umar AA, Hussaini AB, Yahayya J, Sani I, Aminu H (2021) Chitinolytic and antagonistic activity of *Streptomyces* isolated from Fadama soil against phytopathogenic fungi. Trop Life Sci Res 32:25–38. https://doi.org/10.21315/tlsr2021.32.3.2

Vargas Hoyos HA, Chiaramonte JB, Barbosa-Casteliani AG, Fernandez Morais J, Perez-Jaramillo JE, Nobre Santos S, Nascimento Queiroz SC, Soares Melo I (2021) An actinobacterium strain from soil of Cerrado promotes phosphorus solubilization and plant growth in soybean plants. Front Bioeng Biotechnol 9:579906. https://doi.org/10.3389/fbioe.2021.579906

Verma S, Kumar M, Kumar A, Das S, Chakdar H, Varma A, Saxena AK (2022) Diversity of bacterial endophytes of maize (Zea mays) and their functional potential for micronutrient Biofortification. Current Microbiology 79(1). https://doi.org/10.1007/s00284-021-02702-7

Viaene T, Langendries S, Beirinckx S, Maes M, Goormachtig S (2016) *Streptomycesas* a plant's best friend. FEMS Microbiol Ecol 92:119. https://doi.org/10.1093/femsec/fiw119

Vicente J, Stewart A, Van Wagoner R, Elliott E, Bourdelais A, Wright J (2015) Monacyclinones, new angucyclinone metabolites isolated from *Streptomyces* sp. M7_15 associated with the Puerto Rican sponge *Scopalina ruetzleri*. Mar Drugs 13:4682–4700. https://doi.org/10.3390/md13084682

Waheeb MS, Elkhatib W, Yassien M, Hassouna N (2021) Production of cellulase by soil isolated *Streptomyces* sp. Arch Pharm Sci Ain Shams Univ 5:225–233. https://doi.org/10.21608/aps.2021.92280.1067

Wan M, Li G, Zhang J, Jiang D, Huang H (2008) Effect of volatile substances of *Streptomyces platensis* F-1 on control of plant fungal diseases. Biol Control 46:552–559. https://doi.org/10.1016/j.biocontrol.2008.05.015

Wang Z, Wang C, Li F, Li Z, Chen M, Wang Y, Qiao X, Zhang H (2013) Fumigant activity of volatiles from *Streptomyces alboflavus* TD-1 against *Fusarium moniliforme* Sheldon. J Microbiol 51:477–483. https://doi.org/10.1007/s12275-013-2586-y

Wang Z, Solanki MK, Yu Z, Yang L, An Q, Dong D, Li Y (2019) Draft genome analysis offers insights into the mechanism by which *Streptomyces chartreusis* WZS021 increases drought tolerance in sugarcane. Front Microbiol 9:3262. https://doi.org/10.3389/fmicb.2018.03262

Wu Y, Yang S, Zhang Z, Chen S (2020) Improvement of nemadectin production by overexpressing the regulatory gene nemR and nemadectin biosynthetic gene cluster in *Streptomyces cyaneogriseus*. Pharm Fronts 2:e151–e159. https://doi.org/10.1055/s-0040-1722746

Xie F, Pathom-aree W (2021) Actinobacteria from desert: diversity and biotechnological applications. Front Microbiol 12:765531. https://doi.org/10.3389/fmicb.2021.765531

Yadav N (2019) Actinobacteria for sustainable agriculture. J Appl Biotechnol Bioeng 6:38–41. https://doi.org/10.15406/jabb.2019.06.00172

Yadav AN, Kaur T, Kour D, Devi R, Guleria G, Negi R, Yadav A, Ahluwalia AS (2021) Functional annotation and biotechnological applications of soil microbiomes: current research and future challenges. In: Yadav AN (ed) Soil microbiomes for sustainable agriculture. Springer, Cham, pp 605–634

Yandigeri MS, Meena KK, Singh D, Malviya N, Singh DP, Solanki MK, Yadav AK, Arora DK (2012) Drought-tolerant endophytic actinobacteria promote growth of wheat (*Triticum aestivum*) under water stress conditions. Plant Growth Regul 68:411–420. https://doi.org/10.1007/s10725-012-9730-2

Yim G, Thaker MN, Koteva K, Wright G (2013) Glycopeptide antibiotic biosynthesis. J Antibiot 67:31–41. https://doi.org/10.1038/ja.2013.117

Yoolong S, Kruasuwan W, Thanh Phạm HT, Jaemsaeng R, Jantasuriyarat C, Thamchaipenet A (2019) Modulation of salt tolerance in Thai Jasmine rice (*Oryza sativa* L. CV. KDML105) by *Streptomyces venezuelae* ATCC 10712 expressing ACC deaminase. Sci Rep 9:1275. https://doi.org/10.1038/s41598-018-37987-5

You J, Chan Z (2015) ROS regulation during abiotic stress responses in crop plants. Front Plant Sci 6:1092. https://doi.org/10.3389/fpls.2015.01092

Zhang Y, Zhao J, Hu J, Han C, Yu B, Wu Y, Xiao J, Wang X, Xiang W (2021) Agromyces marini-harenae sp. nov., a novel indole-acetic acid producing actinobacterium isolated from marine sand. Int J Syst Evol 71. https://doi.org/10.1099/ijsem.0.005026

Zou N, Zhou D, Chen Y, Lin P, Chen Y, Wang W, Xie J, Wang M (2021) A novel antifungal actinomycete *Streptomyces* sp. strain H3-2 effectively controls banana *Fusarium* wilt. Front Microbiol 12:706647. https://doi.org/10.3389/fmicb.2021.706647

Commercialization and Market Perspectives of Biofertilizers Through Advanced Approaches

17

Deepak Kumar, Ritik Kumar, Sakshi, Deepshikha, and Anil Kumar

Abstract

The growing interest in lowering the dependency on agrochemicals, increasing consumer preferences toward more nutritious and organic food, coupled with the pressing need for sustainable agricultural practices, has necessitated the deployment of biofertilizers as an alternative to chemical fertilizers. Biofertilizers are eco-friendly products consisting of living organisms that improve soil health, enhance nutritional uptake to plants as well as protect them from phytopathogens. The market for biofertilizers is segmented based on product type, crop type, microorganism type, mode of application, and the region where they are applied. According to estimates, the global biofertilizer market was valued at USD 2.8 billion in 2022 and is anticipated to reach USD 5.2 billion by 2028.The key factors driving the biofertilizer market are increased awareness and adoption of sustainable agriculture practices supported by government initiatives and policies, however, challenged by high production cost, technicality, and skill need. Nevertheless, the industry is in nascent stage but holds a promising future in the years to come.

Keywords

Biofertilizer · Solid formulation · Liquid formulation

D. Kumar (✉) · R. Kumar · Sakshi · Deepshikha
Department of Biotechnology, Chandigarh University Gharuan, Mohali, India
e-mail: deepak.e12689@cumail.in

A. Kumar
School of Chemical and Metallurgical Engineering, University of the Witwatersrand, Johannesburg, South Africa

S. Kaur et al. (eds.), *Metabolomics, Proteomics and Gene Editing Approaches in Biofertilizer Industry*, https://doi.org/10.1007/978-981-97-2910-4_17

17.1 Introduction

Pollution has grown to be a significant environmental issue over time. Numerous industrial, agricultural, and human activities produce pollution (Sakshi et al. 2023). In developing nations like India, agriculture is a major source of income for more than 60% of the population and is the backbone of the national economy. However, there is a pressing need to boost agricultural output using sustainable methods due to the world's expanding population. Many techniques have been modified in order to increase soil fertility, agricultural productivity, and sustainability (Sagar et al. 2021; Patel et al. 2023). When talking about the current and future state of the world, the terms "food security" and "food insecurity" are used frequently. The availability and accessibility of adequate nutrition that satisfies dietary needs and supports an active and healthy lifestyle is referred to as food security. People experience food scarcity when they lack sufficient access to wholesome food, which prevents them from consuming enough to lead a healthy and active lifestyle (Stewart and Roberts 2012). The reduction of soil nutrients as a result of ongoing cultivation and the application of synthetic or chemical fertilizers to maintain agricultural productivity are two factors that may be to blame for food insecurity. As a result, the soil has become less fertile, and the majority of the food produced is unsafe to eat due to chemical residues (Daniel et al. 2022).

The majority of modern farming methods employ chemical-based fertilizers, which have previously been connected to a number of harmful effects like greenhouse gas emissions and eutrophication of water resources. Eco-friendly and practical substitutes for chemical fertilizers are provided by biofertilizers (Aloo et al. 2021). Biofertilizers are defined as inoculants containing living microbiological species that improve soil fertility by contributing nitrogen through biological fixation or by breaking down and dissolving intricate organic compounds. They enhance plant development and proliferation by synthesizing substances that promote the growth of plants. They promote several soil microbiological procedures that increase the amount of nutrients that are accessible to plants in a form they can easily absorb (Okur 2018). Unlike chemical fertilizers, biofertilizers are a distinctive, affordable, and environmentally beneficial option that enhances the health of the soil while at the same time increasing yield in an eco-friendly way (Bisen et al. 2015). It can be clearly demonstrated in both controlled and field settings that numerous agronomically significant crops have shown the favorable effects of rhizobacteria that promote plant growth (Aloo et al. 2021).

Broadly, two categories of commercialized biofertilizers are there: one of them is solid biofertilizers, and the other is liquid biofertilizers, depending on the state of the carrier material used in the formulations. It is preferable to use either solid or liquid biofertilizers with multipurpose qualities (Patel et al. 2023). A good carrier material, appropriate packaging, the generation of the inoculum, the addition of cell protectants like starch, glycerol, and lactose, and the best distribution methods are all included in the formulation of a liquid- or solid-based bio-inoculant. Recent advancements in formulation include biofilm-based biofertilizers, nano-biofertilizers, pelletized biofertilizers, and the immobilization and co-mobilization

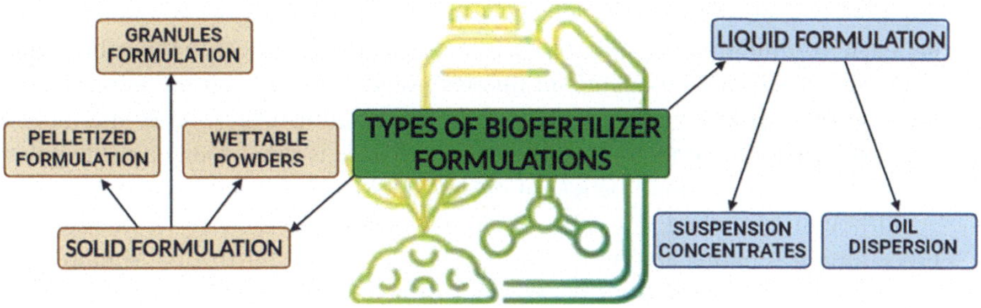

Fig. 17.1 Biofertilizer formulations types and subtypes

of microbial bio-inoculants, which are used as commercialized biofertilizers (Rani et al. 2023). Biofertilizers are being developed and utilized globally to boost crop yields and preserve soil health because of growing awareness of the role they play in preserving soil fertility. Since organic farming requires the use of biofertilizers, the global biofertilizer industry is expanding rapidly (Kumawat et al. 2021).

17.2 Types of Biofertilizer Formulations

In present time, various types of biofertilizers present in the market depend on their application and their usage type. Because of commercialization of biofertilizer, different formulations of biofertilizers exist in the market. Different types of biofertilizers are shown in Fig. 17.1.

17.2.1 Solid Formulation

The integration of an appropriate quantity of inoculum with a solid carrier characterizes the composition known as the solid formulation of biofertilizer. The carrier, being an inert substance, has the ability to increase phosphate contents to plants and degrade the various organic pollutants (Warren 2009). In granular or powder form, solid formulations encompass microbial cultures and include carrier molecules of either organic or inorganic composition. The immobilization of plant growth promoting rhizobacteria (PGPR) serves to mitigate the impact of adverse conditions through either the formation of a biofilm or the adherence of PGPR to a support, or alternatively, by encapsulating it within alginate beads (Rabin et al. 2015).

17.2.1.1 Types of Solid Formulation

(a) ***Pelletized Biofertilizer Formulation*:** One type of carrier-based biofertilizer is pelletized formulations. Pellet biological fertilizer is a solid, spherical, compressed particle of biofertilizer composed of a compacted mixture of carrier material and microorganisms. The idea behind the biofertilizer pellet formulation is to force the formulation until it takes the shape of pellets (Pathirana and

Yapa 2020). Starting with the selection and isolation of target microbial strains, the basic formulation of pellet biofertilizer is nearly identical to other formulations. The microbial inoculants in this formulation need a carrier substance to give the biofertilizer a solid state and optimal circumstances. After the inoculants and carrier material are combined, the mixture is run through a pellet press machine to produce small pellets of biofertilizer (Sarbani and Yahaya 2022).

Although spreading livestock manure is a low-productivity technological procedure, it remains among the most valuable natural fertilizers. Manure fertilization also results in considerable nutritional losses. With the primary goal of turning high-moisture biological matter, such as meat and manure, combined to create field-spreadable pellets, the manufacturing of pelletized biological fertilizers has consequently gained popularity (Möller 2015). Usually, the pellets were created with a 4 or 6 mm diameter so that spreaders for organic or mineral fertilizer can easily disperse them. Biofertilizer pellets get moist in the soil, break down, and release nutrients (Staugaitis et al. 2016; Šarauskis et al. 2021).

(b) ***Granules Formulation of Biofertilizer*****:** The process in which biofertilizer is formulated into small grain like particles is known as granular formulation. The formulation plays a role in upholding the homogeneity of the biofertilizer constituents within every granular particle (Bashan et al. 2014). After the creation of a powdered inoculum derived from proficient strains within the microbiome, the optimal carrier material undergoes meticulous processing and is conjoined with the bacterial inoculum through the employment of a mixer, thereby ensuring the attainment of homogeneity (Herrmann and Lesueur 2013). The resulting homogeneous mixture is subsequently introduced into a rotary drum for granulation, followed by the air-drying of the particles in preparation for storage. The granular formulation incorporating phosphate-solubilizing *Burkholderia*, in conjunction with clay, rock phosphate, and rice bran as carrier materials, exhibited heightened bacterial viability throughout the storage duration (Anandham et al. 2007).

(c) ***Wettable Powders*****:** This type of formulation is the oldest type of formulation. These desiccated formulations elicit substantial interest due to their facile miscibility with water and seamless integration into a liquid carrier, typically water, immediately prior to application (Brar et al. 2006). Water-dispersible powders (WPs) exhibit prolonged shelf life, surpassing 18 months (about 1 and a half years) when moisture content is effectively controlled, a characteristic contributing to their sustained prominence in the market, and this protracted shelf life is also correlated with their robust market presence (Mishra and Arora 2016). Constituent materials in these formulations encompass agricultural resources and industrial by-products, such as mixtures like cereal husk–gravel and wood shaving-silica-cane syrup blend. Compared to the use of chemicals, this compound exhibited efficacy in the postharvest disease's biological management (Brar et al. 2006).

17.2.1.1.1 Advantages of Solid Formulation

The effectiveness of *Trichoderma harzianum* in pellet formulation within a wheat carrier is evident, as it can enhance the growth and productivity of both mustard and tomato crops. Under conditions of soil stress, such as cold fields, wet fields, high acid stress, and conditions requiring elevated moisture, granular formulations are employed (Saif et al. 2021). Solid formulation of biofertilizers plays an important role in enhancing the provision of phosphorus to plants, bolstering resistance against soil-borne plant pathogens, and fostering the biological degradation of organic pollutants. The solid formulation known as peat exhibits several advantages for its high water-holding capacity and extensive surface area, which contribute to the viability of inoculants, providing notable benefits. The granular formulation, which contains *Trichoderma virens* as active ingredients, is helpful in the eradication of soil-borne diseases caused by *Pythium*, *Rhizoctonia*, and *Fusarium* (Mishra and Arora 2016).

17.2.1.1.2 Disadvantages of Solid Formulation

Solid biofertilizers suffer from significant drawbacks, including a brief shelf life and subpar quality due to the organisms' limited survival in unfavorable conditions, elevated contamination risks, and erratic field performance (Velineni and Brahmaprakash 2011). Additionally, the production of these biofertilizers incurs high costs, attributed to the substantial expenses, processes of extracting, dehydrating, grinding, sifting, and adjusting the acidity level of the substrate (Ansari et al. 2015) The higher density of granular formulation which is type of solid formulation makes transporting them to the fields challenging (Saif et al. 2021).

17.2.2 Liquid Formulation

The viable microorganisms in liquid form are known as liquid biofertilizer (Sarbani and Yahaya 2022). This liquid inoculant fosters the development of specific microorganism strains, excluding any solid carrier material. Following the screening and identification of proficient microbes intended for biofertilizer application, the cells undergo cultivation within a nutrient-rich broth medium (Santos et al. 2019). After the successful culturing and extensive propagation of cells, liquid inoculants are poised for application onto seeds or host plants (de Souza et al. 2015). Flowable or aqueous suspensions are other names for liquid formulation. Liquid inoculants are composed of a blend or amalgamation of microorganism cultivation and diverse fluids, such as H_2O, fat, and polymers. This mixture is integrated with culture to enhance secureness, adherence, and diffusion capacity. Liquid inoculants currently dominate the market due to their ease of preparation and cost-effectiveness compared to solid formulations (Dey 2021). Following the cultivation and harvesting of bacterial cultures, the liquid medium is fortified with protective agents. These agents include natural polymers like corn sugar gum, Irish moss, and gel as well as artificial polymers such as polyvidone (Bernabeu et al. 2018). To prolong the viability of microbial cultures during preservation, various substances such as glycerin and simple carbohydrates are also introduced into the liquid medium (Valetti et al. 2016).

17.2.2.1 Types of Liquid Formulation

(a) ***Suspension Concentrates (SC)***: The formulation of suspension concentrates (SCs) entails introducing solid constituents characterized by restricted water dissolvability while demonstrating sufficient resistance to hydrolysis. Prior to application, SCs undergo dilution in water. The enhancement of retention and dissolution attributes may be achieved through the incorporation of surface-active agents and diverse additives (Tadros 2013). Typically, farmers exhibit a preference for suspension concentrates over wettable powders due to their non-dusty nature and ease of measurement and pouring into the spray tank (Mishra and Arora 2016).

(b) ***Oil Dispersion (ODs)***: Oil dispersion refers to a durable suspension of an active component in a solvent or oil that is not soluble in water, ensuring stability. Over the past decade, ODs have gained increasing significance. A recently developed formulation, based on soybean oil, revealed that an oil dispersion containing *Trichoderma asperellum* exhibited notable ability for the management of disease, namely cacao black pod, demonstrating an extended life span of conidia compared to water-based solution. It is imperative to adopt precautionary measures in handling formulations containing fungi in oil dispersion, as prolonged storage may lead to the settling or dense compaction of the active ingredient (conidia) at the bottom of the container (Mishra and Arora 2016). Oil-infused formulations have demonstrated superior efficacy when utilized in foliar spray applications, particularly in bolstering the activity of entomopathogens and due to its diminished evaporation rate, oil persists for an extended duration, affording prolonged contact (Luz and Batagin 2005).These formulations can be administered as emulsions, either as oil dispersed in water or, in certain instances, as invert emulsions with water dispersed in (Batta 2007).

17.2.2.1.1 Advantages of Liquid Formulation

Liquid biofertilizers exhibit greater appeal in comparison to solid inoculants due to their extended shelf life spanning 1.5–2 years, absence of contamination, lack of reliance on adhesive substances, and compatibility with contemporary agricultural machinery (Mahanty et al. 2017). Liquid inoculants, characterized by elevated microorganism concentrations, facilitate the achievement of equivalent effects at reduced dosages in contrast to solid inoculants (Mącik et al. 2020). The utilization of liquid biofertilizers can reduce reliance on chemical fertilizers. Additionally, due to their lower dosages relative to solid biofertilizers, a reduced quantity is required, thereby facilitating the utilization of smaller storage spaces (Riaz et al. 2020). An effective formulation of biofertilizer is observed in liquid form, particularly suitable for the cultivation of numerous crops, especially those grown hydroponically where the soil is not employed (Sarbani and Yahaya 2022).

17.2.2.1.2 Disadvantages of Liquid Formulation

The elevated production costs primarily restrict utilization to developed countries, thereby reducing its application in other regions. For extended storage, certain conditions, such as cool temperatures, are necessary for liquid biofertilizers and as indicated in reports, liquid biofertilizers demonstrate heightened susceptibility to environmental stresses (John et al. 2011). Bacteria with low survival rates, such as *Azospirillum* sp., receive no assistance in creating a protective environment, rendering them essentially ineffective. While liquid biofertilizers possess the capability for prolonged storage, microorganisms within them may encounter challenges such as nutrient depletion, hypoxia, and environmental stresses. These factors contribute to a significant reduction in the microbial population. Hence, specific storage conditions, such as maintaining cool temperatures, become imperative.

17.3 Potential Biofertilizers in Agriculture

The biofertilizers which are produced in laboratory contain living or dormant cells of various organisms. Plants are administered with these biofertilizers to enhance their growth. Unlike man-made fertilizers, these fertilizers feature microorganisms that do not directly provide nutrients but facilitate the absorption of available nutrients in the rhizosphere (Itelima et al. 2018). Microbial biofertilizers play a vital role in sustaining ideal soil richness and enhancing its composition by promoting soil particle aggregation. They also aid in improving plant–water relations, offering protection against drought, reducing the susceptibility of plants to certain terrestrial diseases comprising conditions resulting from fungi that generate mycotoxins and minimizing the incidence of insect pests. Regarded as an appealing and sustainable biotechnological option, biofertilizers present a viable alternative to boost crop yield, enhance fertility of soil, and encourage the development of plants while reducing both production expenses and environmental impact associated with the use of chemical fertilizers. Utilizing potential biofertilizers is crucial for enhancing soil efficiency, sustainability, and environmental conservation, mitigating the negative effects of agricultural practices, improving food quality, solubilizing key nutrients for plant accessibility, promoting the formation of root hairs to enhance water uptake, and generating phytohormones to boost soil fertility and facilitate plant growth and development without harmful side effects (Nosheen et al. 2021). Various microorganisms, such as nitrogen-fixing soil bacteria (e.g., *Azotobacter*, *Rhizobium*), nitrogen-fixing cyanobacteria (e.g., *Anabaena*), phosphate-solubilizing bacteria (e.g., *Pseudomonas*), and arbuscular mycorrhizal fungi, are frequently employed as biofertilizers. In potato cultivation, the impact of inoculating phosphate-solubilizing bacteria, namely *Pseudomonas putida*, *Microbacterium laevaniformans*, and *Pantoea agglomerans* on production and development of potatoes was assessed. The combination of *Pantoea agglomerans* or *Microbacterium laevaniformans* with *Pseudomonas putida* significantly enhanced biomass and promoted tuber growth. The observed increase in yield may be attributed to the enhanced phosphorus (P) supply from the bacteria to the growing plants (Malboobi et al. 2009). Similar to many agricultural plants, the application of biofertilizers that serve as growth enhancers affects tomatoes production. Through the effective inoculation of seedlings with *Burkholderia tropica*, roots were colonized, and this colonization extended to aerial tissues (Bernabeu et al. 2015). Consequently, this successful colonization resulted in an upsurge in tomato production. Forest species also make use of biofertilizers. In the greenhouse, there is potential for the development of *Pinus radiata* seedlings using the biofertilizer with sample that consists of symbiotic relation between *Suillus luteus* fungi and roots of *Rhizopogon luteolus*, along with *Trichoderma harzianum*. The inclusion of *Coriolopsis rigida* and *Rhizopogon luteolus* in the inoculant resulted in *Pinus radiata* plants displaying improved quality parameters. This provides alternative for incorporation into the production of *Pinus* spp. (Chávez et al. 2014). The production of maize, wheat, barley, beans, and orange increases with usage of biofertilizer inoculated with bacteria *Azospirillum brasilense* and *Rhizobium etli* and fungus *Glomus intraradices*.

17.4 World Biofertilizers Market

By 2023, the biofertilizer market has grown at a yearly compound growth rate of over 14%. The worldwide biofertilizer market in 2016 was projected to be worth 1106.4 million USD. It is anticipated to have grown at a 14.2% annual rate to USD 3124.5 million by the end of 2024. In contrast, the sector is still in the beginning stages but has a bright future ahead of it (Joshi and Gauraha 2022). Europe has the greatest commercial potential for biostimulants and biofertilizers overall, with North America, Asia-Pacific, and Latin American nations following next (Peter et al. 2020).

17.5 Asian Market

The third-biggest market for biofertilizers is Asia-Pacific, due to rising demand from local markets including Taiwan, China, and India. Asia-Pacific was anticipated to have tenfold growth between 2013 and 2020 because of expanding agricultural operations and consumer preferences for organic food. But in contrast to other areas, soil treatment—rather than seed treatment—is the main use of biological fertilizers in the Asia-Pacific region. Local farmers in the majority of Asian countries often mix different organic compounds into their own customary biofertilizer formulations, which are then fermented or composted to enhance their activities and functions (Badgley et al. 2007).

Asia is heavily impacted by the government's policies that back sustainable and ecologically friendly farming practices. Approximately US $1.5 billion has been spent on the development of biopesticides and biofertilizers (Sekar et al. 2016). Approximately 1,000,000 ha are currently under organic production, reflecting the country's growing use of organic agricultural practices (Sekar et al. 2016). Technologies for biofertilizers are being developed, tested, and adopted at different stages in Asia (Atieno et al. 2020).

Organic farming now occupies more than 1 lakh hectares in developing nations like India and are backed by about 100 biofertilizer manufacturers nationwide

(Sekar et al. 2016). However, India's overall biofertilizer production is significantly lower than what is needed, indicating that a rise in biofertilizer production is required. Over 300 factories may be found throughout China, where microbial inoculants have been cultivated over 167 million hectares. Farmer's income increased significantly by over USD 27 million between 1987 and 2006 because of the use of *Rhizobium* bio-inoculants, specifically in Taiwan and Japan (Machado and Serralheiro 2017). Some of the most used biofertilizers in Asia are listed in Table 17.1.

17.6 African Market

Significant nutrient loss is occurring in African soils as a result of the numerous ecological harms brought on by the overuse of synthetic fertilizers, that is becoming more and more unpredictable and often irreversible (Savci 2012; García-Fraile et al. 2015). Furthermore, there has been a suggestion of a change in the use of agrochemical inputs, or at the very least, a reduction in them, due to the desire for high-quality food supplies and green agriculture (Adeleke et al. 2019).

The first recorded commercial use of biofertilizer was in 1952, according to published research from South Africa (Strijdom 1998). In Southern Africa, South Africa, Zambia, Malawi, and Zimbabwe are the countries with the greatest number of inoculant manufacturing and use (Raimi et al. 2021). Among the nations in East Africa, Kenya has the most advanced systems for producing and applying biofertilizers. Nonetheless, successful inoculant production and application have been documented in Tanzania, Rwanda, Mozambique, and Uganda, according to published studies (Mukhongo et al. 2016).

The establishment of agro markets, research and agricultural institutions, and adequate and efficient extension programs could all help to increase Africa's production and adoption of biofertilizers. To attain heightened commercial yield and optimal biofertilizer application among African farmers, the African government must consciously intervene through the creation of biofertilizer policies, funding, and effective strategies for the market and methods for quality control. More specifically, significant biofertilizer uptake and commercialization in Africa will result

Table 17.1 Most commonly used biofertilizers companies in the world

Market	Most commonly used biofertilizers companies
Asia-Pacific	Agrinos AS; National Fertilizers Limited; Kiwa Bio-Tech; BioAg Pty Ltd; Australian Bio Fert Pty Ltd., etc.
Africa	Rizobacter Argentina S.A.; Novozymes A/S; Symborg; International Panaacea Limited; MBFI, etc.
United States	Kula Bio Inc.; Novozymes; Rizobacter; Sustane Natural Fertilizer Inc.; Symborg Inc., etc.
Europe	Agronutrition; Biolchim SPA; Koppert Biological Systems Inc.; Lallemand Inc.; Symborg, Inc., etc.

from public and the private sector's participation in technical support and sustainable investment (Raimi et al. 2021).

Africa still produces relatively little biofertilizer and uses very little of it, despite the fact that many industrialized nations have recognized the significance of biofertilizers in environmentally friendly farming. Therefore, farmers in Africa should be encouraged to use biofertilizer, particularly in order to tackle issues related to soil fertility and food insecurity (Raimi et al. 2021). Some of the most used biofertilizers in Africa are listed in Table 17.1.

17.7 American Market

Latin America and the Caribbean saw a 73.59% growth in synthetic fertilizer consumption from 2005 to 2018, rising from 15.2 to 26.3 million tons. Utilizing inorganic fertilizers has the potential to raise production costs, harm water supplies, and pollute the environment. It indicates a significant challenge for production systems reliant on small-scale family farming, which have encountered challenging circumstances in recent years (Carlos 2005; Martínez Reyes et al. 2018). The creation of a biofertilizer decreased production costs by raising yield of crop and cutting the usage of chemical fertilizers by about 50%. This showed that using biofertilizers instead of synthetic fertilizers is practical, creating the possibility of a production system that is both ecologically beneficial and sustainable.

The highest market proportion of the global biofertilizer market is held by the biofertilizer market in North America (Balasubramanian and Karthickumar 2017). Between 2020 and 2025, the biofertilizer market in Latin America is probably going to expand at a 10% annual rate. After Brazil, most prominent and fastest-growing market is Argentina. The primary driving forces behind the markets in these two nations are the expansion of organic farms, the ease of the registration process, and the government policies. Some of the most used biofertilizers in America are listed in Table 17.1.

17.8 European

Agriculture in Europe regularly imports chemical fertilizers (Tur-Cardona et al. 2018); however, the sole and excessive use of chemical fertilizers is creating certain problems. Certain types of rock phosphate can accumulate cadmium, which can be detrimental to people's health and water sources (Marini et al. 2020). The goal set by the European Union is to use bio-based fertilizers (BBFs) as a substitute for up to 30% of chemical-based fertilizers. In order to facilitate the implementation of biofertilizer, the government creates the strategy, policies, and laws. One example of this is the Regulation (EU) 2019/1009 of the European Parliament and Council on fertilizing goods. The goal of BBFs is to reduce the European Union's (EU) dependency on chemical fertilizers that are imported (Kurniawati et al. 2023).

Because they are circular and resource-efficient, organic-based fertilizers are among the new fertilizing products that the new law is targeting. In response to the energy crisis and within the framework of the EU Green Deal, EU has developed new regulations pertaining to fertilizing goods that will make it easier to produce bio-based fertilizers (Kurniawati et al. 2023). Manure is already utilized by farmers as a component of bio-based fertilizers. Majority of farmers acknowledged applying manure to their crops, and almost 50% said they had experienced micronutrients and macronutrients related problems (Tur-Cardona et al. 2018). For BBFs to be recognized in society, their manufacture must therefore meet comparable requirements to those of mineral fertilizers. It will also help with the cost and the fast-release nutrients in addition to concentrated items (Kurniawati et al. 2023).

By 2030, 25% of land for agriculture is ought to be used for organic farming, according to the EU's farm-to-fork policy. Although nutrient recovery and recycling are the main goals of the circular, biological based products economy, the application of bio-based fertilizers to sustainable agriculture is now being assessed (Huygens et al. 2019). Some of the most used biofertilizers in Europe are listed in Table 17.1.

17.9 Advancement in Biofertilizer

Due to their promising impacts on crop productivity and yield, modern fertilizer formulations including chemical, synthetic, and insecticides have been developed. Modern fertilizers, however, have the potential to have long-term negative consequences for the environment and reduce the availability of natural resources and soil bacteria for agricultural use (Anandham et al. 2015). For the biofertilizer industry, maintaining efficiency and maintaining the amount of microorganisms in the product have proven to be a major challenge (Schoebitz et al. 2013; Reetha et al. 2014). To protect the environment and ecology, numerous researchers have developed a variety of biofertilizers over the past few decades that have wide and advanced effects on crop development (Sarbani and Yahaya 2022). The primary obstacles to the development of biofertilizers are the product's relatively short lifespan and low microbial density. Therefore, the development of sustainable biofertilizers that can win over the public with their effectiveness is crucial (Stella et al. 2019).

Recent advancements in the formulation of microbial biological inoculations include immobilization and co-mobilization, biofilm-based biological fertilizers, and encapsulation of microbial bio-inoculants (Rani et al. 2023). Recent advancement in biofertilizers is presented in Fig. 17.2.

17.9.1 Use of Biofilms as Biofertilizers (BFBFs)

The area where the activity of microorganisms is strongest, surrounding the roots, is referred to as the rhizosphere. Rich in nutrients and attractive to microorganisms, roots release a number of metabolites in the rhizosphere (Meena et al. 2018). These

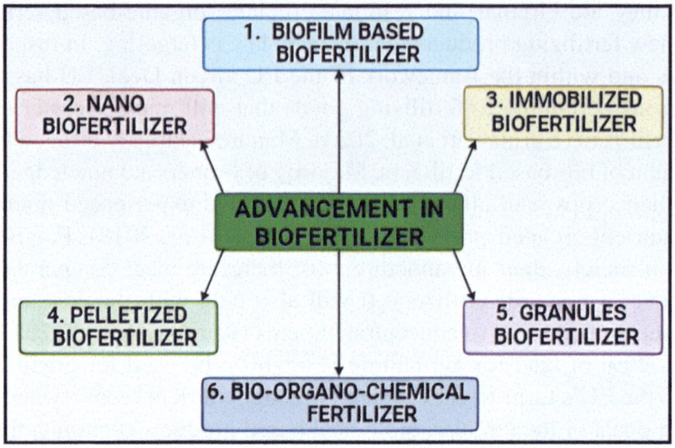

Fig. 17.2 Recent advancements in biofertilizers

compounds produce microbial biofilms, which are necessary for the health of the soil and plant growth. In harsh conditions, it is believed that microbial biofilm in the root system of the plant acts as a survival mechanism. In addition to giving nutrition to plants, microbes can also help plants fight against diseases and pests (Dewi et al. 2023). Beneficial microbial species can also be employed to create biofilms in vitro, which can then be utilized as biofilm biofertilizers also known as BFBFs. When used, the BFBFs can restore agriculture-related ecosystems that have been damaged by overuse of chemical treatments by providing sub-network elements of the soil–plant–microbe bulk network characteristics. Because of this, the bulk network of damaged ecosystems can be restored by substituting sub-networks for better interactions (Premarathna et al. 2021).

According to recent research, biofilm biofertilizers are the next generation of bioweapons, outperforming all other traditional biofertilizers in terms of performance in overcoming biotic and abiotic obstacles and enhancing crop yields (Sudadi et al. 2018).

17.9.2 Nano-biofertilizers

Utilizing nano-biofertilizers, which are made to get beyond the drawbacks of both traditional fertilizers and biofertilizer formulations (BFFs), is a novel strategy for promoting crop growth. Nutrients coated or encapsulated in nano-materials particle sizes usually ranging from 1 to 100 nm are included in nano-fertilizers. One notable feature of nano-biofertilizers is their ability to distribute nutrients gradually and in a controlled manner (Patel et al. 2023). These microscopic particles, which are obtained from readily accessible organic and inorganic resources, have remarkable properties that make them quite potent even at lower concentrations micronutrients and macronutrient immobilized nanoparticles. This emphasizes how much more effective they are than conventional fertilizers (Carvajal-Muñoz and Carmona-Garcia 2012).

Nano-biofertilizers are first adsorbed onto the root surface of plants when they are applied to the rhizosphere. Here, when the positively charged surface interacts with the negatively charged surface of root, absorption and accumulation occur (Zhou et al. 2011). Nanoparticles have to go through a number of root barriers before they can be lifted and transported by the xylem to the plant (Lv et al. 2019). Dispersion and plant build-up of nano-fertilizers, which eventually support the plants' growth and general development, are significantly impacted by the way in which they are administered (Patel et al. 2023).

17.9.3 Immobilized Biofertilizers

Immobilized biofertilizer is acknowledged as the most advanced and modern biofertilizer formulation. Microorganism cells are enclosed and linked to insoluble and inert substances in this formulation. The creation of biofertilizer encapsulation increased a cell's tolerance to environmental stresses and deteriorating soil conditions (Vassilev et al. 2015). Therefore, they are more stable with respect to pH and temperature and more resilient to changes in the environment (Pour et al. 2019).

Apart from that, immobilization helps release microbes or enzyme into the soil gradually and steadily. A macro or microform may be included in the encapsulation, depending on the purpose for which it is used. This formulation involves ejecting the mixture dropwise into calcium chloride to produce stiff capsules from a prepared culture of biofertilizer combined with an encapsulating matrix composition (alginate). The microbe or the enzyme it produces during fermentation can both be immobilized using this technique (Sarbani and Yahaya 2022). Comparing other formulations of biofertilizer, alginate-based nanocapsules provide superior functionality and benefit to the host plant (Ivanova et al. 2006). At the intended application site in the rhizosphere, an abundance of microbiological cells must be maintained by the gradual and continual release of immobilized microbes. Additionally, it can be challenging to extend shelf life, but immobilization is a useful technique, particularly for gram-negative microbes that do not produce spores (Pereg and McMillan 2015).

Nevertheless, encapsulation and microencapsulation technology in biofertilizer is still in early stages and is only being tested in lab settings (Sarbani and Yahaya 2022).

17.9.4 Bio-organo-Chemical Fertilizer

In order to address environmental and economic concerns, agricultural production systems have placed a strong emphasis on fertilizer use efficiency (FUE). However, there has not been much progress in fertilizer technology over the past few decades; therefore, the method now in use to increase FUE is somewhat ineffective. A new method adopting a novel approach in improving soil capacity can be developed by utilizing the strongest elements from chemical, biological, and organic sources to produce bio-organo-chemical fertilizer (Goenadi et al. 2018).

The innovative strategy of fertilizer (chemical) enhanced with organic and biological agents can be used to create bio-organo-chemical fertilizer. These mechanisms cause an epigenetic transformation in the organisms, which leads to the expression of particular traits and produces noteworthy performance across a wide array of applications. In order to increase FUE and improve soil capacity, this is a very essential approach. For plants to produce food as efficiently as possible and to maximize soil fertility, bio-organo fertilizer is an essential part of integrated management of soil fertility. The basic concept of nutrient management is using the correct source in the right amount, time, and at the appropriate place. This approach may be applied to all cropping systems worldwide to maximize profitability, productivity, as well as environmental preservation (Reetz 2016).

17.10 Conclusion

Biofertilizers, with their utilization of microbial synergy, hold the potential to transform agriculture through the improvement of soil fertility, encouragement of sustainable crop development, and reduction of environmental consequences. Demonstrated in research on the cultivation of potatoes and tomatoes, the incorporation of various microorganisms, including nitrogen-fixing bacteria and phosphate-solubilizing bacteria, highlights their ability to substantially increase crop yields and enhance nutrient accessibility for plants. The prospective assessment of biofertilizers underscores the transformative capacity inherent in solid formulations, such as pelletized and granular formulations, providing effective nutrient delivery alongside environmental advantages. Advancements in the future of biofertilizers are on the horizon, encompassing transformative developments like biofilm-based formulations, nano-biofertilizers featuring controlled nutrient release, and innovative immobilized formulations. These strides are targeted at overcoming challenges such as limited lifespan and low microbial density, heralding the era of sustainable and efficient biofertilizers. A comprehensive strategy to improve soil capacity, fertilizer efficiency, and overall environmental sustainability in agriculture is evident in the evolution of pelletized and granular formulations, coupled with bio-organo-chemical fertilizers.

References

Adeleke RA, Nunthkumar B, Roopnarain A, Obi L (2019) Applications of plant–microbe interactions in agro-ecosystems. In: Microbiome in plant health and disease. Springer Singapore, Singapore, pp 1–34. https://doi.org/10.1007/978-981-13-8495-0_1

Aloo BN, Makumba BA, Mbega ER (2021) Status of biofertilizer research, commercialization, and practical applications: a global perspective. In: Biofertilizers. Elsevier, Amsterdam, pp 191–208. https://doi.org/10.1016/B978-0-12-821667-5.00017-8

Anandham R, Choi KH, Gandhi PI, Yim WJ, Park SJ, Kim KA, Madhaiyan M, Sa TM (2007) Evaluation of shelf life and rock phosphate solubilization of Burkholderia sp. in nutrient-amended clay, rice bran and rock phosphate-based granular formulation. World J Microbiol Biotechnol 23:1121–1129. https://doi.org/10.1007/s11274-006-9342-y

Anandham R, Premalatha N, Jee HJ, Weon HY, Kwon SW, Krishnamoorthy R, Gandhi PI, Kim YK, Gopal NO (2015) Cultivable bacterial diversity and early plant growth promotion by the traditional organic formulations prepared using organic waste materials. Int J Recycl Org Waste Agric 4(4):279–289. https://doi.org/10.1007/s40093-015-0107-1

Ansari MF, Tipre DR, Dave SR (2015) Efficiency evaluation of commercial liquid biofertilizers for growth of Cicer arietinum (chickpea) in pot and field study. Biocatal Agric Biotechnol 4(1):17–24. https://doi.org/10.1016/j.bcab.2014.09.010

Atieno M, Herrmann L, Nguyen HT, Phan HT, Nguyen NK, Srean P, Than MM, Zhiyong R, Tittabutr P, Shutsrirung A, Bräu L, Lesueur D (2020) Assessment of biofertilizer use for sustainable agriculture in the Great Mekong Region. J Environ Manag 275:111300. https://doi.org/10.1016/j.jenvman.2020.111300

Badgley C, Moghtader J, Quintero E, Zakem E, Chappell MJ, Avilés-Vázquez K, Samulon A, Perfecto I (2007) Organic agriculture and the global food supply. Renew Agric Food Syst 22(2):86–108. https://doi.org/10.1017/S1742170507001640

Balasubramanian P, Karthickumar P (2017) Biofertilizers and biopesticides: a holistic approach for sustainable agriculture. In: Sustainable utilization of natural resources. CRC Press, Boca Raton, pp 255–284. https://doi.org/10.1201/9781315153292-9

Bashan Y, de Bashan LE, Prabhu SR, Hernandez J-P (2014) Advances in plant growth-promoting bacterial inoculant technology: formulations and practical perspectives (1998–2013). Plant Soil 378(1–2):1–33. https://doi.org/10.1007/s11104-013-1956-x

Batta YA (2007) Biocontrol of almond bark beetle (Scolytus amygdali Geurin-Meneville, Coleoptera: Scolytidae) using Beauveria bassiana (Bals.) Vuill. (Deuteromycotina: Hyphomycetes). J Appl Microbiol 103(5):1406–1414. https://doi.org/10.1111/j.1365-2672.2007.03369.x

Bernabeu P, Pistorio M, Torres-Tejerizo G, Estrada-de los Santos P, Galar M, Boiardi J, Luna M (2015) Colonization and plant growth-promotion of tomato by Burkholderia tropica. Sci Hortic 191:113–120. https://doi.org/10.1016/j.scienta.2015.05.014

Bernabeu PR, García SS, López AC, Vio SA, Carrasco N, Boiardi JL, Luna MF (2018) Assessment of bacterial inoculant formulated with Paraburkholderia tropica to enhance wheat productivity. World J Microbiol Biotechnol 34:81. https://doi.org/10.1007/s11274-018-2461-4

Bisen K, Keswani C, Mishra S, Saxena A, Rakshit A, Singh HB (2015) Unrealized potential of seed biopriming for versatile agriculture. In: Nutrient use efficiency: from basics to advances. Springer India, New Delhi, pp 193–206. https://doi.org/10.1007/978-81-322-2169-2_13

Brar SK, Verma M, Tyagi RD, Valéro JR (2006) Recent advances in downstream processing and formulations of Bacillus thuringiensis based biopesticides. Process Biochem 41(2):323–342. https://doi.org/10.1016/j.procbio.2005.07.015

Carlos CL (2005) Fertirrigación: cultivoshortícolas, frutales y ornamentales

Carvajal-Muñoz JS, Carmona-Garcia C (2012) Benefits and limitations of biofertilization in agricultural practices. Livest Res Rural Dev 24(3):1–8

Chávez D, Pereira Cancino G, Machuca Á (2014) Estimulación del crecimiento en plántulas de Pinus radiata utilizando hongos ectomicorrícicos y saprobios como biofertilizantes. Bosque (Valdivia) 35:11–12. https://doi.org/10.4067/S0717-92002014000100006

Daniel AI, Fadaka AO, Gokul A, Bakare OO, Aina O, Fisher S, Burt AF, Mavumengwana V, Keyster M, Klein A (2022) Biofertilizer: the future of food security and food safety. Microorganisms 10(6):1220. https://doi.org/10.3390/microorganisms10061220

de Souza R, Ambrosini A, Passaglia LMP (2015) Plant growth promoting bacteria as inoculants in agricultural soils. Genet Mol Biol 38:401–419. https://doi.org/10.1590/S1415-475738420150053

Dewi WS, Amalina DD, Romadhon MR (2023) Microbial biofilm for soil health, plant growth, and productivity under multi stress. A review. IOP Conf Ser Earth Environ Sci 1162(1):012008. https://doi.org/10.1088/1755-1315/1162/1/012008

Dey A (2021) Liquid biofertilizers and their applications: an overview. In: Environmental and agricultural microbiology. Wiley, New York, pp 275–292. https://doi.org/10.1002/9781119525899

García-Fraile P, Menéndez E, Rivas R (2015) Role of bacterial biofertilizers in agriculture and forestry. AIMS Bioeng 2(3):183–205. https://doi.org/10.3934/bioeng.2015.3.183

Goenadi DH, Mustafa AB, Santi LP (2018) Bio-organo-chemical fertilizers: a new prospecting technology for improving fertilizer use efficiency (FUE). IOP Conf Ser Earth Environ Sci 183:012011. https://doi.org/10.1088/1755-1315/183/1/012011

Herrmann L, Lesueur D (2013) Challenges of formulation and quality of biofertilizers for successful inoculation. Appl Microbiol Biotechnol 97(20):8859–8873. https://doi.org/10.1007/s00253-013-5228-8

Huygens D, Saveyn H, Tonini D, Eder P, Delgado Sancho L (2019) Technical proposals for selected new fertilising materials under the fertilising products regulation (Regulation (EU) 2019/1009)

Itelima JU, Bang WJ, Onyimba IA, Sila MD, Egbere OJ (2018) Bio-fertilizers as key player in enhancing soil fertility and crop productivity: a review. https://doi.org/10.26765/DRJAFS.2018.4815

Ivanova E, Teunou E, Poncelet D (2006) Alginate based macrocapsules as inoculant carriers for the production of nitrogen fixing bio fertilizers. Chem Ind Chem Eng Q 12(1):31–39. https://doi.org/10.2298/CICEQ0601031I

John RP, Tyagi RD, Brar SK, Surampalli RY, Prevost D (2011) Bio-encapsulation of microbial cells for targeted agricultural delivery. Crit Rev Biotechnol 31:211–226. https://doi.org/10.3109/07388551.2010.513327

Joshi SK, Gauraha AK (2022) Global biofertilizer market: emerging trends and opportunities. In: Trends of applied microbiology for sustainable economy. Elsevier, Amsterdam, pp 689–697. https://doi.org/10.1016/B978-0-323-91595-3.00024-0

Kumawat KC, Keshani, Nagpal S, Sharma P (2021) Present scenario of bio-fertilizer production and marketing around the globe. In: Biofertilizers. Elsevier, Amsterdam, pp 389–413. https://doi.org/10.1016/B978-0-12-821667-5.00028-2

Kurniawati A, Stankovics P, Hilmi YS, Toth G, Smol M, Toth Z (2023) Understanding the future of bio-based fertilisers: the EU's policy and implementation. Sustain Chem Clim Action 3:100033. https://doi.org/10.1016/j.scca.2023.100033

Luz C, Batagin I (2005) Potential of oil-based formulations of Beauveria bassiana to control Triatoma infestans. Mycopathologia 160(1):51–62. https://doi.org/10.1007/s11046-005-0210-3

Lv J, Christie P, Zhang S (2019) Uptake, translocation, and transformation of metal-based nanoparticles in plants: recent advances and methodological challenges. Environ Sci Nano 6(1):41–59. https://doi.org/10.1039/C8EN00645H

Machado R, Serralheiro R (2017) Soil salinity: effect on vegetable crop growth. Management practices to prevent and mitigate soil salinization. Horticulturae 3(2):30. https://doi.org/10.3390/horticulturae3020030

Mącik M, Gryta A, Frąc M (2020) Biofertilizers in agriculture: an overview on concepts, strategies and effects on soil microorganisms. Adv Agron 162:31–87. https://doi.org/10.1016/bs.agron.2020.02.001

Mahanty T, Bhattacharjee S, Goswami M, Bhattacharyya P, Das B, Ghosh A, Tribedi P (2017) Biofertilizers: a potential approach for sustainable agriculture development. Environ Sci Pollut Res Int 24(4):3315–3335. https://doi.org/10.1007/s11356-016-8104-0

Malboobi MA, Behbahani M, Madani H, Owlia P, Deljou A, Yakhchali B, Moradi M, Hassanabadi H (2009) Performance evaluation of potent phosphate solubilizing bacteria in potato rhizosphere. World J Microbiol Biotechnol 25:1479–1484. https://doi.org/10.1007/s11274-009-0038-y

Marini M, Caro D, Thomsen M (2020) The new fertilizer regulation: a starting point for cadmium control in European arable soils? Sci Total Environ 745:140876. https://doi.org/10.1016/j.scitotenv.2020.140876

Martínez Reyes L, Aguilar Jiménez CE, Carcaño-Montiel MG, Galdámez Galdámez J, Morales Cabrera JA, Martínez Aguilar FB, Llaven Martínez J, Gómez Padilla E (2018) Biofertilización y fertilizaciónquímica en maíz (Zea mays l.) en Villaflores, Chiapas, México. Siembra 5(1):26–37. https://doi.org/10.29166/siembra.v5i1.1425

Meena RS, Vijayakumar V, Yadav GS, Mitran T (2018) Response and interaction of Bradyrhizobium japonicum and arbuscular mycorrhizal fungi in the soybean rhizosphere. Plant Growth Regul 84(2):207–223. https://doi.org/10.1007/s10725-017-0334-8

Mishra J, Arora NK (2016) Bioformulations for plant growth promotion and combating phytopathogens: a sustainable approach. In: Bioformulations: for sustainable agriculture. Springer, Berlin, pp 3–33. https://doi.org/10.1007/978-81-322-2779-3_1

Möller K (2015) Assessment of alternative phosphorus fertilizers for organic farming: meat and bone meal

Mukhongo RW, Tumuhairwe JB, Ebanyat P, AbdelGadir AH, Thuita M, Masso C (2016) Production and use of arbuscular mycorrhizal fungi inoculum in Sub-Saharan Africa: challenges and ways of improving. Int J Soil Sci 11(3):108–122. https://doi.org/10.3923/ijss.2016.108.122

Okur N (2018) A review: bio-fertilizers—power of beneficial microorganisms in soils. Biomed J Sci Tech Res 4(4):4028–4029. https://doi.org/10.26717/BJSTR.2018.04.0001076

Patel C, Singh J, Karunakaran A, Ramakrishna W (2023) Evolution of nano-biofertilizer as a green technology for agriculture. Agriculture 13(10):1865. https://doi.org/10.3390/agriculture13101865

Pathirana BKW, Yapa PN (2020) Evaluation of different carrier substances for the development of an effective pelleted biofertilizer for rice (Oryza sativa L.) using co-inoculated bacteria and arbuscular mycorrhizal fungi. Asian J Biotechnol Bioresour Technol 6:1–10. https://doi.org/10.9734/ajb2t/2020/v6i130070

Pereg L, McMillan M (2015) Scoping the potential uses of beneficial microorganisms for increasing productivity in cotton cropping systems. Soil Biol Biochem 80:349–358. https://doi.org/10.1016/j.soilbio.2014.10.020

Peter AJ, Amalraj ELD, Talluri VR (2020) Commercial aspects of biofertilizers and biostimulants development utilizing rhizosphere microbes: global and Indian scenario. In: Rhizosphere microbes. Springer, Berlin, pp 655–682. https://doi.org/10.1007/978-981-15-9154-9_27

Pour MM, Saberi-Riseh R, Mohammadinejad R, Hosseini A (2019) Investigating the formulation of alginate- gelatin encapsulated Pseudomonas fluorescens (VUPF5 and T17-4 strains) for controlling Fusarium solani on potato. Int J Biol Macromol 133:603–613. https://doi.org/10.1016/j.ijbiomac.2019.04.071

Premarathna M, Seneviratne G, Ketipearachchi KG, Pathirana A, Karunaratne RKC, Balasooriya WK, Fonseka K (2021) Biofilm biofertilizer can reinstate network interactions for improved rice production. Ceylon J Sci 50(3):235. https://doi.org/10.4038/cjs.v50i3.7904

Rabin N, Zheng Y, Opoku-Temeng C, Du Y, Bonsu E, Sintim HO (2015) Biofilm formation mechanisms and targets for developing antibiofilm agents. Future Med Chem 7(4):493–512. https://doi.org/10.4155/fmc.15.6

Raimi A, Roopnarain A, Adeleke R (2021) Biofertilizer production in Africa: current status, factors impeding adoption and strategies for success. Sci Afr 11:e00694. https://doi.org/10.1016/j.sciaf.2021.e00694

Rani P, Rajput S, Thakur B, Kaur S (2023) Immobilization and co-mobilization: an unexploited biotechnological tool for enhancing efficiency of biofertilizers. In: Metabolomics, proteomes and gene editing approaches in biofertilizer industry. Springer Nature, Singapore, pp 219–236. https://doi.org/10.1007/978-981-99-3561-1_13

Reetha D, Kumaresan G, John Milton D (2014) Studies to improve the shelf life of Azospirillum lipoferum immobilized in alginate beads. Int J Sci Res 5:1871–1878

Reetz HF (2016) Fertilizers and their efficient use. International Fertilizer Industry Association, Paris

Riaz U, Mehdi SM, Iqbal S, Khalid HI, Qadir AA, Anum W et al (2020) Bio-fertilizers: eco-friendly approach for plant and soil environment. In: Bioremediation and biotechnology: sustainable approaches to pollution degradation. Springer, Berlin, pp 189–213. https://doi.org/10.1007/978-3-030-35691-0_9

Sagar A, Rathore P, Ramteke PW, Ramakrishna W, Reddy MS, Pecoraro L (2021) Plant growth promoting rhizobacteria, arbuscular mycorrhizal fungi and their synergistic interactions to counteract the negative effects of saline soil on agriculture: key macromolecules and mechanisms. Microorganisms 9(7):1491. https://doi.org/10.3390/microorganisms9071491

Saif S, Abid Z, Ashiq MF, Altaf M, Ashraf RS (2021) Biofertilizer formulations. In: Biofertilizers. Wiley, New York, pp 211–256. https://doi.org/10.1002/9781119724995.ch7

Sakshi RK, Kaur H, Kumar D, Chauhan S (2023) Microbial endophytes: a novel approach for emerging pollutants. In: George N, Dwibedi V, Rath SK, Chauhan PS (eds) Management and mitigation of emerging pollutants. Springer, Cham. https://doi.org/10.1007/978-3-031-41005-5_9

Santos MS, Nogueira MA, Hungria M (2019) Microbial inoculants: reviewing the past, discussing the present and previewing an outstanding future for the use of beneficial bacteria in agriculture. AMB Exp 9:205. https://doi.org/10.1186/s13568-019-0932-0

Šarauskis E, Naujokienė V, Lekavičienė K, Kriaučiūnienė Z, Jotautienė E, Jasinskas A, Zinkevičienė R (2021) Application of granular and non-granular organic fertilizers in terms of energy, environmental and economic efficiency. Sustainability 13(17):9740. https://doi.org/10.3390/su13179740

Sarbani NMM, Yahaya N (2022) Advanced development of bio-fertilizer formulations using microorganisms as inoculant for sustainable agriculture and environment—a review. Malays J Sci Health Technol 8(1):92–101. https://doi.org/10.33102/mjosht.v8i1.228

Savci S (2012) An agricultural pollutant: chemical fertilizer. Int J Environ Sci Dev 3:73–80. https://doi.org/10.7763/IJESD.2012.V3.191

Schoebitz M, López MD, Roldán A (2013) Bioencapsulation of microbial inoculants for better soil–plant fertilization. A review. Agron Sustain Dev 33(4):751–765. https://doi.org/10.1007/s13593-013-0142-0

Sekar J, Raj R, Prabavathy VR (2016) Microbial consortial products for sustainable agriculture: commercialization and regulatory issues in India. In: Agriculturally important microorganisms. Springer Singapore, Singapore, pp 107–132. https://doi.org/10.1007/978-981-10-2576-1_7

Staugaitis G, Masevičienė A, Mažeika R, Arbačiauskas J (2016) Granuliuotosorganinėstrąšos

Stella M, Theeba M, Illani ZI (2019) Organic fertilizer amended with immobilized bacterial cells for extended shelf-life. Biocatal Agric Biotechnol 20:101248. https://doi.org/10.1016/j.bcab.2019.101248

Stewart WM, Roberts TL (2012) Food security and the role of fertilizer in supporting it. Procedia Eng 46:76–82. https://doi.org/10.1016/j.proeng.2012.09.448

Strijdom BW (1998) South African studies on biological nitrogen-fixing systems and the exploitation of the nodule bacterium-legume symbiosis. South Afr J Sci 94:11–23

Sudadi, Suryono, Triharyanto E (2018) The application of biofilm biofertilizer-based organic fertilizer to increase available soil nutrients and spinach yield on dry land (a study case in lithosol soil type). IOP Conf Ser Earth Environ Sci 200:012006. https://doi.org/10.1088/1755-1315/200/1/012006

Tadros TF (2013) Encyclopedia of colloid and interface science. Springer, Berlin. https://doi.org/10.1007/978-3-642-20665-8

Tur-Cardona J, Bonnichsen O, Speelman S, Verspecht A, Carpentier L, Debruyne L, Marchand F, Jacobsen BH, Buysse J (2018) Farmers' reasons to accept bio-based fertilizers: a choice experiment in seven different European countries. J Clean Prod 197:406–416. https://doi.org/10.1016/j.jclepro.2018.06.172

Valetti L, Angelini JG, Taurian T, Ibáñez FJ, Muñoz VL, Anzuay MS, Ludueña LM, Fabra A (2016) Development and field evaluation of liquid inoculants with native bradyrhizobial strains for peanut production. Afr Crop Sci J 24:1–13. https://doi.org/10.4314/acsj.v24i1.1

Vassilev N, Vassileva M, Lopez A, Martos V, Reyes A, Maksimovic I, Eichler-Löbermann B, Malusà E (2015) Unexploited potential of some biotechnological techniques for biofertilizer production and formulation. Appl Microbiol Biotechnol 99(12):4983–4996. https://doi.org/10.1007/s00253-015-6656-4

Velineni S, Brahmaprakash GP (2011) Survival and phosphate solubilizing ability of Bacillus megaterium in liquid inoculants under high temperature and desiccation stress. J Agric Sci Tech 13:795–802. 20.1001.1.16807073.2011.13.5.6.7

Warren JR (2009) Socioeconomic status and health across the life course: a test of the social causation and health selection hypotheses. Social forces 87(4):2125–2153. https://doi.org/10.1353/sof.0.0219

Zhou D, Jin S, Li L, Wang Y, Weng N (2011) Quantifying the adsorption and uptake of CuO nanoparticles by wheat root based on chemical extractions. J Environ Sci 23(11):1852–1857. https://doi.org/10.1016/S1001-0742(10)60646-8

Intellectual Property Rights in Biofertilizer Industry

18

Priyanka Verma and Sakshi Manhas

Abstract

IPR in biofertilizer is of great significance in today's rapidly evolving agricultural demands. To improve soil fertility and plant development while minimizing environmental consequences, this developing industry significantly depends on breakthroughs in microbial strains, manufacturing techniques, formulations, and applications. Moreover, biofertilizers have gained immense popularity due to environmental friendly practices over sustainable chemical derivative fertilizers but face a unique challenge in terms of intellectual property protection due to biological nature. Patents, trademarks, trade secrets, copyrights provide inventors with exclusive rights by safe guarding their innovations of novel microbial stains, processes, formulations, and applications. It helps companies to commercialize the biofertilizer product without the fear of imitation and can establish distinctive image, logo, name. Hence it can maintain the confidentiality of the product and facilitates the sharing of knowledge that contributes to socio-economic growth and advancement in biofertilizer industry. One of the main drawback of IPR is it provides patent for limited time period, thus reducing effectiveness of maintaining competitive advantage in sector as natural processes and its development are time consuming. Overall, IPR provides legal mechanism to safeguard the inventions, branding, confidentiality information, and creativity of work associated with biofertilizers.

Keywords

Biofertilizer · Intellectual property rights · Agriculture · Sustainable

P. Verma (✉) · S. Manhas
Department of Forensic Science, Chandigarh University, Mohali, Punjab, India
e-mail: priyanka.pharma@cumail.in

S. Kaur et al. (eds.), *Metabolomics, Proteomics and Gene Editing Approaches in Biofertilizer Industry*, https://doi.org/10.1007/978-981-97-2910-4_18

18.1 Introduction

18.1.1 Introduction to Intellectual Property Rights (IPR)

To thrive in the era of globalization's strong rivalry in technology and trade, it is crucial to maintain a leading position in invention and creativity. Particularly in disciplines like software engineering, missile technology, lunar and extraterrestrial expeditions, and numerous other technological domains, India has become known for its intellectual strength. Conversely, India lags behind in the production of registered patents, industrial designs, trademarks, and other IPR assets. India ranks an unsettling 29th out of 30 countries globally in the IP index, according to a new analysis from the US Chamber of Commerce. The nation as a whole and policymakers are concerned about this predicament (http://articles.economictimes.indiatimes.com/2015-02-04/news/58795926_1_ip-environment-gipc-intellectual- property-index). Any society's progress is directly impacted by IPR and its governing framework (Jajpura 2015).

Intellectual property rights (IPR) have suffered from a lack of understanding, which has resulted in the death of new ideas, an increase in the risk of infringement, economic setbacks, and a fall in the era of intellectual advancement in the country. Therefore, there is a pressing need to increase awareness of IPR in order to encourage domestic innovation and breakthroughs in the fields of science and technology (Samaddar and Chaudhary 2008; Sinha et al. 2009). This manuscript intends to highlight the various facets of intellectual property rights in regard to their importance and role within the biofertilizer industry.

18.1.2 Intellectual Property Rights and Their Classification

The idea of intellectual property is strongly related to the creative and innovative abilities of the human mind. The act of inventing or producing something new necessitates large amounts of human labor, time, energy, knowledge, resources, and other factors. The underlying idea that inspires an innovation or piece of art is an intangible resource owned by the person who put out the effort to make it. As a result, the creator or innovator is given legal rights—or exclusive rights—to benefit financially from their invention or work in line with the law (Narayanan 2010; Sharma 2014). Intellectual property rights (IPR) are geographical rights that allow the owner to sell, purchase, or license their intellectual property (IP) just like physical property (Cuts International Jaipur 2007). Each kind of intellectual property rights (IPR) bestows special rights upon its originator or inventor, enabling them to protect and receive economic benefits, even though it is essential to formally register IPR with a legal body in a tangible or perceptible manner to partake of their advantages. In turn, this encourages skill advancement and social advancement (http://www.wipo.int/edocs/pubdocs/en/intproperty/450/wipo_pu b_450.pdf; http://www.wipo.int/wipo_magazine/en/2009/01/article_0003.html; http://www.wipo.

int/edocs/pubdocs/en/intproperty/489/wipo_pub_489.pdf; www.commerce.nic.in/trade/international_trade_ip_trips1.asp).

The types of ideas and creations that come from the human mind and their numerous uses are used to categorize intellectual property rights. Patents, trademarks, industrial designs, semiconductor integrated circuit layout designs, geographical indications of source, and copyright and related rights (covering literary and artistic works, musical compositions, visual and performing arts, photography, motion pictures, computer programs, and broadcast content) are some of these categories (www.copyright.gov.in/Documents/handbook.html; www.legalservicesindia.com/articles/cop.html; WIPO 2004; Nair 2010).

18.1.3 Importance of IPR in Biofertilizer Industry

The application of biofertilizers has been increasing, because of the benefits, such as higher yields and the capacity to handle problems that crop diversification presents, like disease outbreaks and unfavorable weather patterns. Major corporations are essential to the development and dissemination of modern facilities, biotechnological innovations, as Lyson has shown in his study (Lyson 2002). They also profit from genetically modified goods by using intellectual property rights. As a result, the options that less developed countries have with regard to intellectual property (IP) in agricultural biotechnology are linked to the use and procurement of both public and private technologies, the creation and preservation of domestic institutional innovations, and the investigation of options for technology transfer (Wijk et al. 1993).

Only some of the underdeveloped countries have practical intellectual property (IP) strategies to protect these discoveries. The most viable options include using trade secrets, making agreements for the transfer of these technologies, gaining access to technical expertise, securing intellectual property rights across international borders, and developing technology transfer frameworks that support a particular country's technological goals (Barragán-Ocaña and del Carmen del-Valle-Rivera 2016). Furthermore, it is yet unclear if social institutions will effectively adopt and use biotechnology to meet societal requirements and improve overall social well-being, despite the technology's significant technological ability to solve a wide range of issues (Phillips 2002). Intellectual property laws must be put in place in order to provide domestic inventions made in the agriculture industry legal protection. This necessity is especially clear when it comes to developing technological capacities, as demonstrated by the example of biotechnological goods, and in the context of biofertilizers, which is the subject of this discussion (Chavas 2008).

By registering intellectual property rights (IPR), one may ensure the protection of intellectual properties generated both domestically and abroad. In conformity with the Patents Act and the Patents Regulation, Intellectual Property Corporation of Malaysia (MyIPO) oversees and defends IPRs in Malaysia. Thirty of the forty intellectual property rights (IPRs) covering biopesticides and ten IPRs devoted to biofertilizers were submitted by MyIPO. For biopesticides, four intellectual property rights (IPRs) were issued, but only three for biofertilizers. Roughly 20–47% of the IPRs for biopesticides and biofertilizers were submitted by universities and public research institutes (PRIs), respectively. The remaining IPRs in both categories were owned by private businesses. The greater number of IPRs filed for biopesticides can be attributed to the comparatively greater quantity of research done on them as opposed to biofertilizers. Before being released into the market, imported biopesticide products had their IPRs protected by MyIPO.

Significant breakthroughs were made in the fields of biopesticides and biofertilizers, many of which remain maintained as industry trade secrets rather than being officially registered with MyIPO. As such, it is anticipated that the true number of inventions created for both technologies will be higher than the number reported in this study (Dardak and Adham 2014).

18.2 Overview of Biofertilizer Industry

18.2.1 Biofertilizers

The term "biofertilizers" describes fertilizers derived from microbiological sources that improve plant growth, productivity, and resistance to both biotic and abiotic stresses. The food demands of a growing global population are mostly met by traditional agriculture. But this has also resulted in a greater dependence on chemical pesticides and fertilizers (Santos et al. 2012). Chemical fertilizers are synthetic materials with precise ratios of potassium, phosphorus, and nitrogen that are created in an industrial setting. Their usage pollutes the environment, causing air pollution and eutrophication of aquatic bodies (Youssef and Eissa 2014). Recent efforts have focused toward the production of "nutrient-rich, high-quality food" with an emphasis on biosafety and sustainability in response to these difficulties (Raja 2013; Bhardwaj et al. 2014).

Additional advantages of biofertilizers include longer shelf life without endangering the environment (Sahoo et al. 2014). Through processes like nitrogen fixation, phosphate and potassium solubilization, mineralization, the release of compounds that regulate plant growth, the synthesis of antibiotics, and the biodegradation of organic matter within the soil, they maintain the quality of the soil by promoting the availability of various micro- and macronutrients (Sinha et al. 2014).

18.2.2 Types of Biofertilizers

There are several uses for biofertilizers as shown in Fig. 18.1. Different varieties of biofertilizers can be distinguished according to their nature, mode of action, and accessibility. The several kinds of biofertilizers available for increased crop productivity are shown in Fig. 18.2.

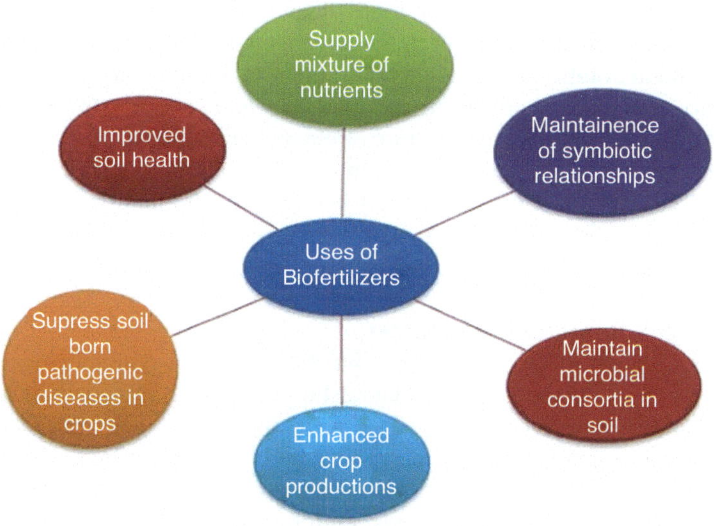

Fig. 18.1 Different uses of biofertilizers (Kaur and Purewal 2019)

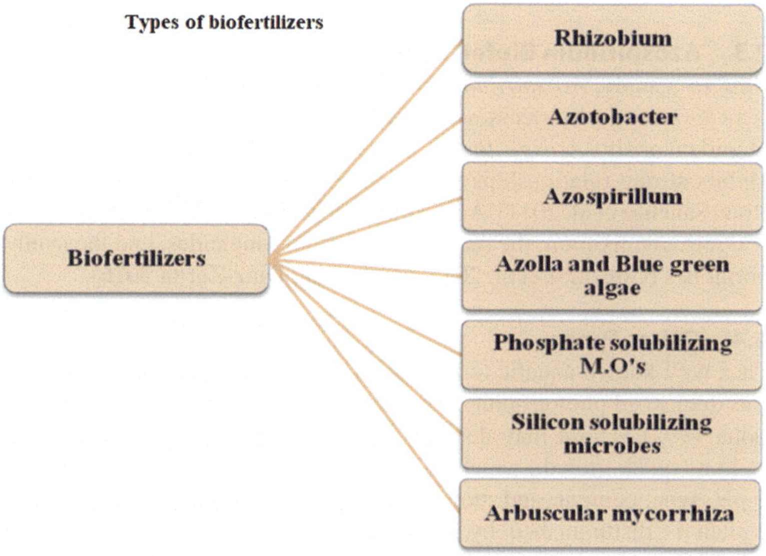

Fig. 18.2 Different types of biofertilizers (Kaur and Purewal 2019)

18.2.2.1 Rhizobium Biofertilizer

Rhizobium is an endosymbiont that fixes nitrogen and is a member of the *Rhizobiaceae* family. It infects plant roots and causes specialized root nodules to grow (Gouda et al. 2018). Researchers reported a greater incidence of isolates BHU

B13-398 and BHU M strains in the mung bean crop. It is well known that these strains promote plant development in the rhizosphere. Their actions lead to longer roots and shoots, higher plant heights, and more biomass in dry weight (Kumari et al. 2018). Few researchers also found that *Rhizobium* controls the expression of genes linked to MT and the manufacture of phytochelatins in *Medicago sativa*. The purpose of this rule is to shield crops from undue copper (Cu) stress. According to their research, *Rhizobium* inoculation reduced the growth inhibition brought on by too much Cu, which eventually raised the nitrogen contents in *Medicago sativa* seedlings (Chen et al. 2018).

18.2.2.2 Azotobacter Biofertilizer

Azotobacter, a gram-positive diazotroph, functions as a non-symbiotic microbe that offers several benefits to crops. Its relationship to plants in growth enhances their general health and increases yield. *Azotobacter* belongs to the family *Azotobacteraceae* and is distinguished by its aerobic nature (Sethi and Adhikary 2012). In order to maximize crop yields, a number of scientific studies support the use of *Azotobacter* in agricultural areas. Different studies have demonstrated that the application of *Azotobacter* and related strains increases plant dry matter and the generation of secondary metabolites (Paul et al. 2002; Nagananda et al. 2010; Damir et al. 2011).

18.2.2.3 Azospirillum Biofertilizer

According to a study, *Azospirillum* is a kind of biofertilizer that is essential for a number of biochemical processes required in the creation of food (Llorente et al. 2016). Fundamentally, *Azospirillum* is an important *Rhodospirillales* order member that exhibits strong relationships with grasses, especially monocots like rice and corn (Ruiz-Sanchez et al. 2011). According to few studies, this connection is directly related to nitrogen fixation, the secretion of certain fungicides, and the synthesis of phytohormones (Gonzalez et al. 2015; Cassan and Diaz-Zorita 2016).

18.2.2.4 Other Types

Azolla is a well-known aquatic fern belonging to the *Salviniaceae* family. It is also known as duckweed phototrophic fern and comes in seven different species (Roger and Ladha 1992). *Azolla* may develop significant biomass in as little as 10 days because to its quick growth, which is impacted by a number of parameters, including the pH, type, content, and moisture levels of the soil. The small, free-floating *Azolla* plant is distinguished by its unique scaly leaves and floating roots. Its acknowledged symbiotic association with *Anabaena azollae*, which aids in nitrogen fixation, is one of its noteworthy characteristics (Emrooz et al. 2018).

Cyanobacteria, or blue-green algae (BGA), are filamentous microorganisms with specialized cell structures called heterocysts, or micronodules. Heterocysts are essential to the process of nitrogen fixation. According to studies, these bacteria work in symbiotic relationships with ferns, fungal strains, and flowering plants to aid in nitrogen fixation (Soma et al. 2018; Sarker et al. 2018; Islam and Shamsuddoha 2018). Phosphorus has a special and important function among macronutrients since it regulates signal transduction, protein synthesis, respiration, and nitrogen fixation in plants (Khan et al. 2010; Pande et al. 2017). Phosphorus is insoluble in the soil when it is first found, which prevents plants from using it. It has to be transformed from its complicated, bound condition to a free, usable form in order to be available for regular consumption (1996) (Corona et al. 1996). Studies have shown that changes in the composition of soil can result from natural events such as the weathering of silica and rocks containing silica derivatives. Particular kinds of microbial communities play key roles in the decomposition, modification, and control of silicon and its derivatives (Kang et al. 2017).

18.3 Understanding IPR in Biofertilizer Sector

18.3.1 Patent

Patents provide exclusive rights to inventors, allowing them to prevent others from making, using, or selling their invention for a specified period (usually 20 years). Due to the need of growing global concern about pollution, greenhouse gas accumulation, and increased need for plant-based foods, the demand for biofertilizers and biocontrol agents is expected to grow (Vasanthi et al., 2018; Pirttilä et al. 2021). Thus, the patents can be obtained for novel microbial strains, novel biofertilizer formulations, or innovative production methods. Patents encourage investment in research and development and promote innovation by ensuring that inventors can benefit from their creations. The geographical analysis of an Asian countries related to biofertilizer innovations shows China (642) to be top most in innovation and granted patent, followed by USA (192) and India (140), as it holds top-most position in terms of dominance of patent families. The World Intellectual Property Organization (WIPO) database and the respective patent offices of different countries can be referred to for specific patents related to biofertilizers.

In a study in the biofertilizer sector, 2047 patent families were created between 2007 and 2017. These patent families included subjects such as the invention, manufacturing, formulation, use of microorganisms, and innovative techniques. One thousand two hundred eighty-eight patents were subjected to a thorough comparison study. The chronology study showed that interest in the field of biofertilizers had been growing steadily, with 2 significant development phases: from 2007 to 2010 (24% compound annual growth rate percentage; CAGR) and from 2011 to 2016 (31% CAGR). A CAGR of 39.53% for published applications and 54.82% for granted applications was seen between 2007 and 2017, during this time 247 patents were issued and 1047 applications were published. Ten patents were awarded by the World Intellectual Property Organization (WIPO) and the European Patent Office (EPO); out of the 247 granted patents, 237 of which were from 27 different nations. After China with 86 patents, the USA (42), Australia (10), Russia (32), Korea (18), and Ukraine (6) came next. A number of other nations each possessed one patent, while Japan, India, and Canada each owned four patents (Srivastava and Adholeya 2019).

18.3.2 Trademarks

Trademarks are used to protect distinctive signs or symbols associated with a particular biofertilizer brand (Gupta et al. 2016), ensuring that consumers can easily identify and differentiate products in the market. Trademarks can safeguard the reputation and branding efforts of biofertilizer companies, preventing others from using similar marks that may lead to confusion (Caradonia et al. 2019). Details about registered trademarks in different countries can be accessed through the respective trademark offices (Deorsola et al. 2017).

18.3.3 Copyrights

While copyrights primarily apply to original artistic or literary works, they may also extend to certain forms of biofertilizer-related content, such as research articles, instructional materials, or software programs (Reichman 1989). Copyright protection grants the creator exclusive rights to reproduce, distribute, display, or modify their work (Ginsburg 1997). To cross-reference copyrighted content related to biofertilizers, academic databases, research journals, and relevant publications can be consulted.

18.3.4 Trade Secrets

Trade secrets protects confidential or proprietary information that provides a competitive advantage to biofertilizer companies. This can include manufacturing processes, formulation details, quality control methods, or customer databases. Unlike patents, trade secrets can be perpetual as long as the information remains confidential. Legal frameworks, such as non-disclosure agreements (NDAs) (https://gowlingwlg.com/en/insights-resources/articles/2017/protecting-trade-secrets-using-non-disclosure-agre/), can help protect trade secrets in practice.

It is worth mentioning that IPR laws and regulations may vary from country to country.

18.4 Conclusion

Due to the growing global concern because of greenhouse effect and pollution by inorganic residuals of fertilizers, involving effects of synthetic plant protection compounds on human and animal health lead the need of microbiological tools, namely biofertilizers and biocontrol agents has to be superseded in global market. When it comes to biofertilizers, which are organic substances that enhance soil fertility and nutrient availability, IPR plays a vital role in the biofertilizer industry by providing legal protection and incentives for innovation, fostering collaborations, facilitating technology transfer, and attracting investments. By securing IPR for their biofertilizer formulations and technologies, individuals and organizations can safeguard their intellectual property and contribute to the sustainable development of agriculture. It is worth noting that IPR is a complex and evolving field, and obtaining and protecting IPR for biofertilizers may involve various legal mechanisms, such as patents, trademarks, copyrights, or trade secrets. The specific requirements and procedures for obtaining IPR may vary across different jurisdictions, so it is crucial for researchers and companies to consult with legal professionals specializing in intellectual property.

References

Barragán-Ocaña A, del Carmen del-Valle-Rivera M (2016) Rural development and environmental protection through the use of biofertilizers in agriculture: an alternative for underdeveloped countries? Technol Soc 46:90–99

Bhardwaj D, Ansari MW, Sahoo RK, Tuteja N (2014) Biofertilizers function as key player in sustainable agriculture by improving soil fertility, plant tolerance and crop productivity. Microb Cell Fact 13:66

Caradonia F, Battaglia V, Righi L, Pascali G, La Torre A (2019) Plant biostimulant regulatory framework: prospects in Europe and current situation at international level. J Plant Growth Regul 38:438–448

Cassan F, Diaz-Zorita M (2016) Azospirillum sp. in current agriculture: from the laboratory to the field. Soil Biol Biochem 103:117–130

Chavas J-P (2008) On the economics of agricultural production. Aust J Agric Resour Econ 52(4):365–380

Chen J, Liu YQ, Yan XW, Wei GH, Zhang JH, Fang LC (2018) Rhizobium inoculation enhances copper tolerance by affecting copper uptake and regulating the ascorbate-glutathione cycle and phytochelatin biosynthesis-related gene expression in Medicago sativa seedlings. Ecotoxicol Environ Saf 162:312–323

Corona MEP, Klundert IVD, Verhoeven JTA (1996) Availability of organic and inorganic phosphorus compounds as phosphorus sources for carex species. New Phytol 133:225–231

Cuts International Jaipur (2007) Intellectual property rights, biodiversity and traditional knowledge. Monographs on globalisation and Indian-myths and realities, vol 13, pp 20–22

Damir O, Mladen PI, Bozidar S, Sran N (2011) Cultivation of the bacterium Azotobacter chroococcum for preparation of biofertilizers. Afr J Biotechnol 10:3104–3111

Dardak RA, Adham KA (2014) Transferring agricultural technology from government research institution to private firms in Malaysia. Procedia Soc Behav Sci 115:346–360

Deorsola AB, Leal MCMR, Cavalcante MD, Schmidt IJ, Braga EJ (2017) Intellectual property and trademark legal framework in BRICS countries: a comparative study. World Patent Inf 49:1–9

Emrooz HBM, Maleki M, Rahmani A (2018) Azolla-derived hierarchical nanoporous carbons: from environmental concerns to industrial opportunities. J Taiwan Inst Chem Eng 91:281–290

ET Bureau: India ranked second last in Intellectual Property Index. http://articles.economictimes.indiatimes.com/2015-02-04/news/58795926_1_ip-environment-gipc-intellectual- property-index. Accessed 4 Feb 2015

Ginsburg JC (1997) Authors and users in copyright. J Copyright Soc USA 45:1–19

Gonzalez AJ, Larraburu EE, Llorente BE (2015) Azospirillum brasilense increased salt tolerance of jojoba during in vitro rooting. Ind Crop Prod 76:41–48

Gouda S, Kerry RG, Das G, Paramithiotis S, Shin HS, Patra JK (2018) Revitalization of plant growth promoting rhizobacteria for sustainable development in agriculture. Microbiol Res 206:131–140

Gupta V, Sengupta M, Prakash J, Tripathy BC (2016) Intellectual property rights. In: Basic and applied aspects of biotechnology. Springer, Berlin, pp 487–501

http://www.wipo.int/wipo_magazine/en/2009/01/article_0003.html. Accessed 4 Dec 2014

Islam MM, Shamsuddoha MD (2018) Coastal and marine conservation strategy for Bangladesh in the context of achieving blue growth and sustainable development goals (SDGs). Environ Sci Pollut 87:45–54

Jajpura L (2015) In: Mor S (ed) Microfinance and microentrepreneurship: a paradigm shift for societal development, 1st edn. Vista International Publication House, Delhi, pp 263–271

Kang S, Waqas M, Shahzad R, You Y, Asaf S, Khan MA, Lee K, Joo G, Kim S, Lee I (2017) Isolation and characterization of a novel silicate-solubilizing bacterial strain Burkholderia eburnea CS4-2 that promotes growth of japonica rice (Oryza sativa L. cv. Dongjin). Soil Sci Plant Nutr 63(3):233–241

Kaur P, Purewal SS (2019) Biofertilizers and their role in sustainable agriculture. In: Soil biology. Springer, Berlin, pp 285–300

Khan MS, Zaidi A, Ahemad M, Oves M, Wani PA (2010) Plant growth promotion by phosphate solubilizing fungi current perspective. Arch Agron Soil Sci 56:73–98

Kumari P, Meena M, Gupta P, Dubey MK, Nath G, Upadhyay RS (2018) Plant growth promoting rhizobacteria and their biopriming for growth promotion in mung bean (Vigna radiata (L.) R. Wilczek). Biocatal Agric Biotechnol 16:163–171

Llorente BE, Alasia MA, Larraburu EE (2016) Biofertilization with Azospirillum brasilense improves in vitro culture of Handroanthus ochraceus, a forestry, ornamental and medicinal plant. New Biotechnol 33:32–40

Lyson TA (2002) Advanced agricultural biotechnologies and sustainable agriculture. Trends Biotechnol 20(5):193–196

Nagananda GS, Das A, Bhattacharya S, Kalpana T (2010) In vitro studies on effect of biofertilizers (Azotobacter and Rhizobium) on seed germination and development of Trigonella foenumgraecum L. using a novel glass marble containing liquid medium. Int J Bot 6:394–403

Nair MD (2010) TRIPS, WTO and IPR—world patents. J Intellect Prop Rights 15:151–153

Narayanan S (2010) Intellectual property rights economy vs. science and technology. Int J Intellect Prop Rights 1(1):6–10

Pande A, Pandey P, Mehra S, Singh M, Kaushik S (2017) Phenotypic and genotypic characterization of phosphate solubilizing bacteria and their efficiency on the growth of maize. J Genet Eng Biotechnol 15:379–391

Paul S, Verma OP, Rathi MS, Tyagi SP (2002) Effect of Azotobacter inoculation on seed germination and yield of onion (Allium cepa). Ann Agric Res 23:297–299

Phillips PWB (2002) Biotechnology in the global agri-food system. Trends Biotechnol 20(9):376–381

Pirttilä AM, Mohammad Parast Tabas H, Baruah N, Koskimäki JJ (2021) Biofertilizers and biocontrol agents for agriculture: how to identify and develop new potent microbial strains and traits. Microorganisms 9(4):817

Raja N (2013) Biopesticides and biofertilizers: ecofriendly sources for sustainable agriculture. J Biofertil Biopest 4:e112

Reichman JH (1989) Computer programs as applied scientific know-how: implications of copyright protection for commercialized university research. Vanderbilt Law Rev 42:639

Roger PA, Ladha JK (1992) Biological N_2 fixation in wetland rice fields: estimation and contribution to nitrogen balance. Plant Soil 141:41–55

Ruiz-Sanchez M, Armada E, Munoz Y, Salamone IE, Aroca R, Ruiz-Lozano JM, Azcon R (2011) Azospirillum and arbuscular mycorrhizal colonization enhance rice growth and physiological traits under well-watered and drought conditions. J Plant Physiol 168:1031–1037

Sahoo RK, Ansari MW, Pradhan M, Dangar TK, Mohanty S, Tuteja N (2014) Phenotypic and molecular characterization of efficient native Azospirillum strains from rice fields for crop improvement. Protoplasma 251(4):943–953

Samaddar SG, Chaudhary BD (2008) Practical insights into intellectual property strategy for technical institute. J Intellect Prop Rights 13:590–600

Santos VB, Araujo SF, Leite LF, Nunes LA, Melo JW (2012) Soil microbial biomass and organic matter fractions during transition from conventional to organic farming systems. Geoderma 170:227–231

Sarker S, Bhuyan MAH, Rahman MM, Islam MA, Hossain MS, Basak SC, Islam MM (2018) From science to action: exploring the potentials of blue economy for enhancing economic sustainability in Bangladesh. Ocean Coast Manag 157:180–192

Sethi SK, Adhikary SP (2012) Azotobacter: a plant growth promoting rhizobacteria used as biofertilizer. Dyn Biochem Process Biotechnol Mol Biol 6:68–74

Sharma DK (2014) Intellectual property and the need to protect it. Indian J Sci Res 9(2014):84–87

Sinha B, Joshi H, Ghosh PK (2009) Challenges in creation and management of knowledge capital in technical educational institutions. J Intellect Prop Rights 14(2009):340–345

Sinha RK, Valani D, Chauhan K, Agarwal S (2014) Embarking on a second green revolution for sustainable agriculture by vermiculture biotechnology using earthworms: reviving the dreams of Sir Charles Darwin. Int J Agric Saf Health Saf 1:50–64

Soma K, Burg SWK, Hoefnagel EWJ, Stuiver M, Heide M (2018) Social innovation—a future pathway for blue growth. Mar Policy 87:363–370

Srivastava S, Adholeya A (2019) Intellectual property: data mapping in the area of biofertilizers. In: Singh H, Keswani C, Singh S (eds) Intellectual property issues in microbiology. Springer, Singapore, pp 193–221

Vasanthi N, Saleena LM, Raj SA (2018) Silica solubilization potential of certain bacterial species in the presence of different silicate minerals. Silicon 10:267–275

Wijk JV, Cohen JI, Komen J (1993) Intellectual property rights for agricultural biotechnology. In: Options and implications for developing countries, research report 3. IBS-ISNAR, The Hague

WIPO (2004) IPO Intellectual property handbook, WIPO Publication no. 489 (E). ISBN 978-92-805-1291-5, second edition reprinted 2008

WIPO manual: what is intellectual property? http://www.wipo.int/edocs/pubdocs/en/intproperty/450/wipo_pu b_450.pdf

www.commerce.nic.in/trade/international_trade_ip_trips1.asp. Accessed 23 Dec 2014

www.copyright.gov.in/Documents/handbook.html. Accessed 30 Dec 2014

www.legalservicesindia.com/articles/cop.html. Accessed 30 Dec 2014

Youssef MMA, Eissa MFM (2014) Biofertilizers and their role in management of plant parasitic nematodes. A review. E3 J Biotechnol Pharm Res 5:1–6

The manufacturer's authorised representative in the EU is Springer Nature Customer Service Centre GmbH, Europaplatz 3, 69115 Heidelberg, Germany. If you have any concerns regarding our products, please contact ProductSafety@springernature.com

Printed and bound by CPI Group (UK) Ltd, Croydon, CR0 4YY
01/10/2026
02225862-0010